NUMERICAL MODELING IN APPLIED PHYSICS AND ASTROPHYSICS

OTHER JONES AND BARTLETT TITLES OF INTEREST

Introduction to Cosmology

Jayant V. Narlikar
Tata Institute of Fundamental Research, Bombay, India

The Comet Book

Robert D. Chapman and John C. Brandt
NASA, Goddard Flight Center, Maryland

Astrophysics
I: **Stars**
II: **Interstellar Matter and Galaxies**

Richard L. Bowers
Los Alamos National Laboratory

Terry Deeming
Digicon Geophysical Corporation

Numerical Astrophysics

Edited by Joan Centrella
University of Texas

James LeBlanc
Lawrence Livermore Laboratory

Richard L. Bowers
Los Alamos National Laboratory

NUMERICAL MODELING IN APPLIED PHYSICS AND ASTROPHYSICS

Richard L. Bowers
Los Alamos National Laboratory

James R. Wilson
Lawrence Livermore National Laboratory

JONES AND BARTLETT PUBLISHERS
BOSTON

Editorial, Sales, and Customer Service Offices

Jones and Bartlett Publishers
20 Park Plaza
Boston, MA 02116

Library of Congress Cataloging-in-Publication Data

Bowers, Richard L.,
Numerical modeling in applied physics and astrophysics
Richard L. Bowers, James R. Wilson.
p. cm.
ISBN 0-86720-123-1
1. Astrophysics—Data processing. 2. Supercomputers. I. Wilson, James R. (James Ricker) II. Title.
QB461.B65 1991
523.01—dc20 90-4479
CIP

Printed in the United States of America
95 94 93 92 91 10 9 8 7 6 5 4 3 2 1

Contents

Preface

Numerical Modeling in Applied Physics and Astrophysics is intended to be a self-contained introductory presentation of selected, practical computational methods which have been extensively and successfully applied to a wide range of astrophysical problems. It may be used as a primary text for a course on computational methods, or as a source for self-study by students in applied physics and astrophysics. Although the approach has been developed for students and researchers engaged in modelling astrophysical phenomena who have access to scientific computing facilities, it should also be of use to theorists interested primarily in the results of numerical modeling. No attempt has been made to be mathematically rigorous. Instead, physical intuition and physical arguments are emphasized wherever possible. We have chosen to present a relatively simple yet flexible approach to the solution of complex problems which has been found by experience to be easily modified to include new physics, or different problem geometries.

A complete survey of computational methods which have been or could be employed in astrophysical research might include finite difference methods, finite element methods, free Lagrange type methods and spectral methods. A complete discussion of all of these topics, even at the introductory level, would result in more material than could be covered in a one year introductory course. Consequently, we have chosen to limit our discussions to a single method, finite differences, as applied to a wide range of processes. The topics which have been selected for discussion here represent only one of many approaches that have been used to model astrophysical phenomena. Indeed, a detailed discussion of all of the many methods which have been developed would fill many volumes. Our purpose here is to present at a coherent introductory level one method which can be used to describe a wide range of astrophysical phenomena. A thorough understanding of this material should enable the student to pursue newer areas of computational research in the literature, or to develop an independent approach to computational physics and astrophysics. To assist the reader in this direction we include a number of references to other computational methods. In general, these methods represent quite different approaches which have either been successfully used in astrophysics, or which demonstrate a fundamentally different approach.

We shall discuss numerical (finite difference) methods for equations that are common to astrophysics, but many of the difference equations and methods are of more general applicability to many areas of physics as well.

Experience has shown that an experimental approach to numerical modeling is often efficient and successful. In particular, large scale scientific computers may be thought of as experimental facilities; computational programs represent controlled numerical experiments the results of which must be compared with physical observations and with the results of theoretical analyses. As with any experimental program, large scale computing requires that adequate computer capability and computer time be available. However, this does not always mean that a computation program should be made as large as the machines in use will bear; on the contrary it is often advisable to carry out preliminary work on a modest or even small scale until the computational program and the physical approximations upon which it is based have been thoroughly explored. Large scale (and therefore time consuming) computations may then be undertaken. Furthermore, it is often more useful to test approximations on simple problems with known solutions by a method of trial and error guided by experience, than to try first to prove formally that the methods are, for example, stable or of a given accuracy.

The text is organized around a series of individual topics, each of which is developed in such a manner that it may be incorporated with any of the other topics in a workable manner. Thus, a numerical program which includes all of the physics capabilities covered here may be developed in one or two spatial dimensions.

The subject matter and level of presentation have been developed for upper division undergraduate or first year graduate students in astronomy and the physical sciences who have completed course work in lower division physics, chemistry or astronomy, including partial differential equations at the level of sophomore electromagnetic theory. Background material in stellar and nonstellar astrophysics presented at a comparable level may be found in, for example, Bowers and Deeming (1984). A series of references are included which complement or extend the topics discussed in the text; they are not intended to be exhaustive. A detailed discussion of numerical methods in mathematics may be found in *Numerical Recipies* (Press, et. al., 1986).

Notation

The use of index notation to denote independent variables can become quite complicated, particularly in two dimensional problems. In order to simplify notation we shall frequently omit certain indicies which are common to all quantities in an equation. In such cases a phrase such as "common subscript k, j omitted" followed or preceded by an expression of the form $A^{n+1} - A^{n}$ will be understood to stand for $A^{n+1}{}_{k,j} - A^{n}{}_{k,j}$. A similar convention will also be used for superscripts, or for other indicies.

Acknowledgements

A number of colleagues have contributed to the evolution of this text, either through discussions or from their reading of developing versions of the manuscript. We wish, in particular, to thank Dimitri Mihalas for a critical reading of the text. We are also indebted to Willy Benz, who used material from the chapters on hydrodynamics at Harvard Univeristy, and who offered numerous comments and suggestions from the trenches. Sean Clancy, using material from Chapter 5 to develop an Eulerian hydrodyanmic code, contributed the figures in Chapter 5 demonstrating mass advection alogrithms. Our discussion of implicit methods has benefited from numerous discussions with Tom Oliphant. We also thank Stan Woosley for discussions about thermonuclear reactions, and for contributing unpublished material on minimal nuclear networks. Finally, we wish to thank Paul Whalen for sharing his insights into, and experience in general with numerical methods.

In recognition of his many contributions to numerical physics, and for his support and encouragement over the years, we dedicate this text to James LeBlanc.

1

ASTROPHYSICS ON A GRID

The physics of many astronomical objects can be described by coupled (generally nonlinear) systems of partial differential equations, supplemented by constituitive relations describing the composition and properties of the matter out of which the objects are made. Solutions to systems of equations of this type in terms of known functions can usually be constructed only by making restrictive approximations and by invoking simplifying assumptions about the underlying physics of the material. Solutions of this type can be extremely useful when, for example, they reveal how large-scale structures depend on general trends in equations of state, energy generation rates or boundary conditions, or as a means of showing, for example, how observable parameters, such as radius and luminosity vary with the mass of the system. In many cases the simplifying assumptions needed to construct these solutions may be of such severity that the problem ultimately solved may have little in common with the original problem that was posed. The use of numerical methods, such as those to be developed in subsequent chapters, offers a way of solving complicated systems of coupled partial differential equations and constituitive relations.

There are several obvious advantages to solutions which can be expressed in terms of known functions. If the equations contain arbitrary parameters (corresponding, for example, to initial conditions) their values may be changed at will in the solution to obtain answers to an entire family of problems. And various limits may often be taken to determine the relative importance of each aspect of the underlying physics models. Furthermore the solution may (in principle) be substituted back into the partial differential equations to verify that it is, in fact, a correct solution.

It may be thought that when we make recourse to numerical methods we automatically lose the ability to perform all of the in principle simple "analytic" procedures such as those mentioned above, but this is not always true. For example, once a numerical program has been developed it can be used repeatedly to solve problems with varying initial conditions, different geometries or with changes in values of parameters determining the underlying physics models. Furthermore, it is often possible to make major changes in aspects of the physics or the problem being solved with only limited additional development of an existing program. Therefore, a properly developed program can in fact be used to solve a family of problems. Limiting procedures may also be carried out by, for example, modifying coefficients which couple terms in the equations (in this way entire physics processes, such as radiation transport or magnetic effects, may be turned off), or by modifying parameters entering into constituitive relations such as an equation of state, opacity or electrical resistivity.

Unfortunately, there is no obvious process when using numerical methods that is equivalent to direct substitution of the solution into the equations to see if the answer is correct. This is, perhaps, the major drawback to numerical astrophysics. Nevertheless, as we shall see in the next section the significance of this difficulty can often be greatly reduced in practice.

Computational methods should not be envisioned as a substitute for traditional analysis but should whenever possible be used in conjunction with those methods to gain a better understanding of the behavior and properties of complicated systems. For example, the results of a detailed computer calculation may lead to analytic approximations for physical variables which may then be used to develop simplified models describing the basic behavior of the system (see, for example, Stein 1966). An analysis of these simple models may also suggest where improvements in the underlying physics models of the computational program may be needed, or even where errors and serious omission may have occured.

There are a number of methods which have been used extensively to solve complex equations in physics and astrophysics. These include finite difference methods, finite element methods, free

Lagrange type methods and spectral analysis. Each approach has unique advantages and disadvantages. Probably the most extensively used (though not the newest) approach in areas where dynamics is important is the method of finite differences. Since it would be impossible to cover adequately all methods in an introductory text, we have chosen to limit our discussions to the application of finite difference methods to a large range of situations.

The primary purpose of this chapter is to present an overview of some basic aspects of a numerical approach to large scale computational modeling of astrophysical phenomena. The first section contains an overview of the structure of computational programs. The second section is intended to give the reader a glimpse of the machinery of finite difference approximations to partial differential equations. The concluding section is a brief review of the properties of linear second order partial differential equations. Readers familiar with these topics may wish to skip to Chapter 2.

1.1 CODE ARCHITECTURE

The following chapters will present a selection of numerical algorithms which have been developed for a variety of problems in astrophysics, and which have been used to successfully investigate phenomena ranging from stellar evolution to cosmology. The topics to be considered include hydrodynamics, radiation transport, magnetic fields (magnetohydrodynamics), various types of energy production, capture and emission processes, and gravitational fields in the Newtonian regime. Relativistic gravitation and hydrodynamics will also be discussed in the final chapter. Each topic (physics unit) will be discussed individually, but in such a way that various or all of these phenomena (with the exception of relativistic gravitation and hydrodynamics) may be combined in a single approach.

A considerable amount of time and effort can be spent developing a computational program to solve complicated problems of this type. For this reason, it is advantageous to develop a systematic approach to the organization and, in fact, philosophy of large scale computer program development. As the organization of the subsequent chapters suggests, the equations describing a well defined aspect of physics (such as hydrodynamics or radiation transport) can often be treated as a single unit. It is then possible to develop in a more or less independent fashion approximations to the equations corresponding to each physics unit (it is generally necessary, however, to keep in mind any other physical units which may be coupled with the one under consideration). Figure 1.1 shows schematically one approach to a computational program that is extremely general, and allows relatively straightforward modifications of an existing program to include, for example, new physics capabilities. In Figure 1.1 each box may represent a single subroutine or a collection of subroutines. The blocks, called drivers, usually contain a series of subroutine calls and logic to determine which (based in part on input information) lower level subroutine is to be called and the calling order. The *SETUP* block represents information necessary to generate a specific problem. This may include zoning information and the initial conditions, as well as parameters which control how the computational program is to be executed by the computer. The latter might include the times at which output is to be calculated and how long this particular problem

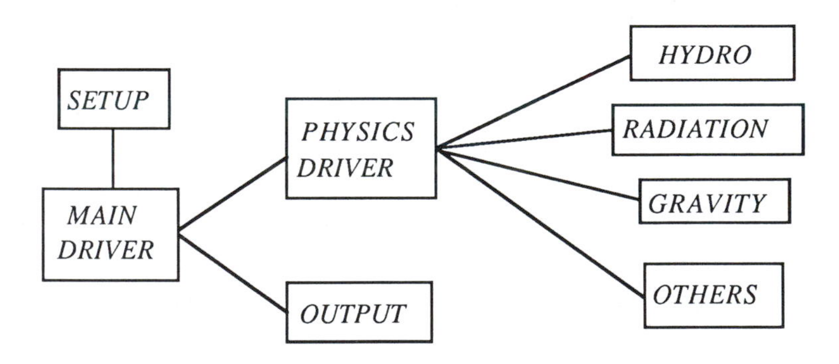

Figure 1.1 Schematic layout of a typical computational program. Labeled boxes on the right represent independent physics units.

is to run. The *MAIN DRIVER* block is a subroutine or collection of subroutines that actually control the program during the course of a calculation, calling the *PHYSICS DRIVER* and the *OUTPUT* block in the appropriate order. These portions of a code, once written, may undergo relatively little change. The *PHYSICS DRIVER* calls each physics unit in its proper temporal order. Calls to individual physics blocks may be based on the value of an input parameter, thereby allowing units to be skipped for certain calculations simply by means of an input parameter. The completion of a single sequence of physics units (hydrodynamics, radiation, gravity and others in Figure 1.1) represents a full computational cycle. This may be accomplished by either solving all of the physics sub blocks by iteration or by operator splitting (which will be discussed in Chapter 2). The *OUTPUT* block supplies information about the current state of the system being calculated, such as values of the dependent variables on a zone by zone basis and in the form of graphics. Each one of the blocks on the far right represents individual physics units. These portions of the program contain most if not all of the equations representing the basic physics, including the constituitive relations.

There are many advantages to this type of structured approach. For example, a set of relatively simple test problems involving a single physics unit in a complicated program to which the answer is known can be solved to test a particular numerical algorithm, or to verify that a specific algorithm has been properly implemented. Test problems can also be chosen to exercise specific aspects of a given physics unit (for example, test problems involving advection in the absence of fluid acceleration, or shock wave formation and propagation can be run to check aspects of a hydrodynamics unit). If the hydrodynamics portion of a computer program is developed in a general form and can successfully calculate a series of test problems by simply modifying the input, it is reasonable to expect that this portion of the program will yield reasonable results when applied to similar though more complicated problems. Exercises of this type when applied to each physics unit are in some ways equivalent to substituting the answer back into the equations as a way of checking solutions in terms of known functions. Another advantage of this approach is associated with the frequency with which various physics units require modification. For many problems there are basic physics units (such as the hydrodynamics) which, once developed, generally do not need to be changed during the course of many calculations. Others (such as the equation of state, opacity, electrical resistivity or reaction blocks) may require daily modifications as the results of calculations reveal a sensitivity to these aspects of the physics, or as improvements in the underlying physics supplies better approximations for the constitutive relations. A modular structure, such as the one shown in Figure 1.1, allows extensive modification to some blocks without affecting the reliability of the rest of the computational program.

Hydrodynamics represents a typical example of a physics unit. The equations of inviscid fluid hydrodynamics consist of three dynamical equations describing the conservation of momentum, mass and energy, appropriate boundary conditions and initial conditions, and equations of state. The dynamical equations in three spatial dimensions may be written in the following (Lagrangian) form:

Momentum

$$\rho \frac{d\mathbf{v}}{dt} = -\nabla P \tag{1.1}$$

Mass continuity

$$\frac{d\rho}{dt} = -\rho \nabla \cdot \mathbf{v} \tag{1.2}$$

Internal energy

$$\frac{d\varepsilon}{dt} = \frac{P}{\rho^2} \frac{d\rho}{dt} \tag{1.3}$$

where, instead of an equation for the total fluid energy (kinetic plus internal), we have given an equation describing the change in internal energy (1.3). Equations (1.1)-(1.3) represent a coupled set of nonlinear partial differential equations which must be solved for the dependent variables ρ, ε, P and $\mathbf{v}$ which denote the mass density, specific internal energy, pressure and fluid velocity respectively. Equations (1.1)-(1.3) represent five equations containing six unknowns. An equation of state, which represents the sixth equation, is necessary to complete the formulation. In general the equation of state will be a complicated function (or a tabular array) representing P in terms of ρ and ε and the material composition.

The latter may be determined from the solution of additional equations, such as the Saha equation for composition, or equations describing reactions among the particles in the material, which will usually depend on the variables ρ and ε. A typical physics unit might consist of one or more subroutines containing finite difference equations approximating (1.1)-(1.3), and a subroutine which constructs the equation of state.

1.2 FINITE DIFFERENCES

The method of finite differences is a relatively simple and highly successful approach to the numerical solution of coupled partial differential equations in physics. The approach to finite difference methods which will be emphasized in this text is based in part on the application of physical intuition to the equations of theoretical physics. Many of the partial differential equations encountered in theoretical physics may be derived by thinking in terms of finite elements of matter. Take, for example, a star such as the sun which, to good approximation, may be considered to be spherically symmetric. Then the star may be described as a one-dimensional system in the sense that all variations occur in the radial direction. For a system of this type the fluid lying within a conical section extending between $r - \Delta r/2$ and $r+\Delta r/2$ whose apex angle is $\Delta\theta$ may conveniently be called a fluid element (see Figure 1.2). The net force acting on this fluid element is given by the difference between the gravitational force exerted on it by the star, and the force due to pressure differences acting on its surfaces:

$$\Delta V \rho \left(\begin{matrix} \textit{acceleration of} \\ \textit{mass element at } r \end{matrix} \right) = F_{Pressure} + F_{gravity} \tag{1.4}$$

where ΔV is the volume of the fluid element, and its average mass density is denoted by ρ. Denote the total mass inside the sphere of radius $r - \Delta r/2$ by $m(r - \Delta r/2)$. Then the gravitational force acting on this fluid element is given by the product of the gravitational acceleration at r and the mass of the fluid element, and may be approximated by

$$F_{gravity} = -\frac{m(r - \Delta r/2)G}{r^2}\rho(r)\,\Delta V. \tag{1.5}$$

Next, we consider the pressure acting on the surface of the fluid element. Denote the pressure on the outer surface of the element by $P(r+\Delta r/2)$ and that on the inner surface of the element by $P(r-\Delta r/2)$. Finally, the pressure acting on the conical sides of the fluid element

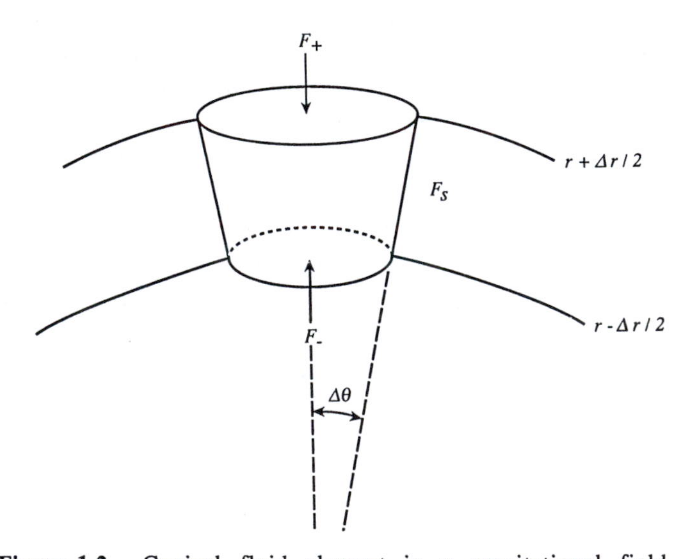

Figure 1.2 Conical fluid element in a gravitational field.

can be approximated by $P(r)$. The net force exerted by $P(r)$ over this part of the surface has an upward radial component given by

$$F_{S,\,rad} = 2\pi r\,\Delta r\;\Delta\theta\,P(r)\sin\Delta\theta$$

$$\approx 2\pi r \, P(r) \, \Delta r \, \Delta\theta^2$$

where in the last step $\Delta\theta << 1$ has been assumed. The nonradial components clearly cancel by symmetry Combining this force with the net force acting on the upper and lower surfaces yields the total pressure force acting on the fluid element:

$$F_{Pressure} = P(r + \Delta r/2) \, \pi \, (r + \Delta r/2)^2 \Delta\theta^2 - P(r - \Delta r/2) \, \pi \, (r - \Delta r/2)^2 \Delta\theta^2$$

$$+ 2\pi r \, \Delta r \, P(r) \, \Delta\theta^2 .$$

Expanding the radius in powers of Δr and retaining only first order terms,

$$F_{Pressure} = \pi \, \Delta\theta^2 r^2 \left[P(r - \Delta r/2) - P(r + \Delta r/2) \right]$$

$$\approx -\frac{\partial P}{\partial r} \Delta V . \tag{1.6}$$

In the second line of (1.6) the pressure has been expanded in a Taylor series about the point r, and terms of second and higher order in Δr have been dropped.

Problem 1.1

Supply the intermediate details of the derivation of (1.6) for the net pressure force acting on a conical fluid element in a spherical symmetric mass distribution.

If we now take the limit $\Delta r \to 0$, and assume that the star is in hydrostatic equilibrium (a very good approximation for a star like the sun), we obtain the partial differential equation

$$\frac{\partial P}{\partial r} = -\frac{m(r)\,G\,\rho}{r^2}\,.$$

(1.7a)

The approach used above should be familiar to the reader, since it is frequently encountered when constructing partial differential equations in mechanics, electromagnetic theory, fluid dynamics and thermodynamics.

It is also useful in constructing finite difference representations for the physical processes described by the partial differential equation (1.7a). Suppose that we equate (1.5) and (1.6), which is equivalent to assuming that the acceleration of the fluid element is zero, integrate over all angles for fixed radius, and denote variables at the point r by subscript k, and those at $r+\Delta r/2$ by $k+1/2$. Then it is easily shown that

$$(P_{k+1/2} - P_{k-1/2})\,4\pi r^2 = -\frac{m_{k-1/2}\,G\,\rho_k}{r_k^2}\,4\pi r^2\,\Delta r_k\,.$$

Cancelling common factors and dividing by the finite interval Δr_k, we obtain the finite difference equation

$$\frac{P_{k+1/2} - P_{k-1/2}}{\Delta r_k} = -\frac{m_{k-1/2}\,G\,\rho_k}{r_k^2}$$

(1.7b)

which is a possible finite difference representation of the partial differential equation (1.7a). It should be noted that in (1.7b) the pressure is associated with the surface of the fluid element (at $k+1/2$) while the density is located at the center of the fluid element. Both of these results seems eminently reasonable. However, the density and pressure are intrinsic properties of the fluid that are related by an equation of state, such as $P = P(\rho)$. This implies that from ρ_k we can obtain P_k. In order to solve (1.7b), however, we must find $P_{k+1/2}$ and $P_{k-1/2}$. We shall see that issues of this type are by no means uncommon when formulating finite difference approximations to partial differential equations.

Problem 1.2

Discuss the centering of the dependent variables appearing in the finite difference equation (1.7b). How could the results above be extended to obtain a finite difference representation of (1.4) when the acceleration is not small relative to the difference between the pressure and gravitational accelerations?

As a second example, consider the flow of radiation in a region of a spherically symmetric star, such as the sun, which is assumed to be in radiative equilibrium. The luminosity at a radius r can be defined as the net outward flux of radiation energy times the surface area through which it flows:

$$\begin{pmatrix} \text{luminosity} \\ \text{at radius } r \end{pmatrix} = 4\pi r^2 \left[\begin{pmatrix} \text{outward} \\ \text{flux} \end{pmatrix} - \begin{pmatrix} \text{inward} \\ \text{flux} \end{pmatrix} \right]. \tag{1.8}$$

The quantity in square brackets above is the net flux crossing the surface at r, and in the diffusion regime is essentially given by Ficks' law. In the interior of most stars the radiation mean free path λ_R is small compared with the length scale associated with temperature variations, and radiative energy transport occurs primarily by diffusion. Consider a spherical surface of radius r in a stellar interior as shown schematically in Figure 1.3. The net outward flux of radiative energy can be found from the radiative transfer equation (see, for example, Mihalas and Mihalas 1984). For simplicity, however, consider instead the following heuristic argument. In the radiative diffusion regime photons whose mean free path is λ_R and which cross the surface at r in either the outward or the inward direction can be shown to originate approximately within a distance $l = (2/3)\lambda_R$ of the surface. The photons emitted at any point behind the surface are emitted isotropically, and of those emitted within a distance l behind the surface at r, approximately half will be emitted

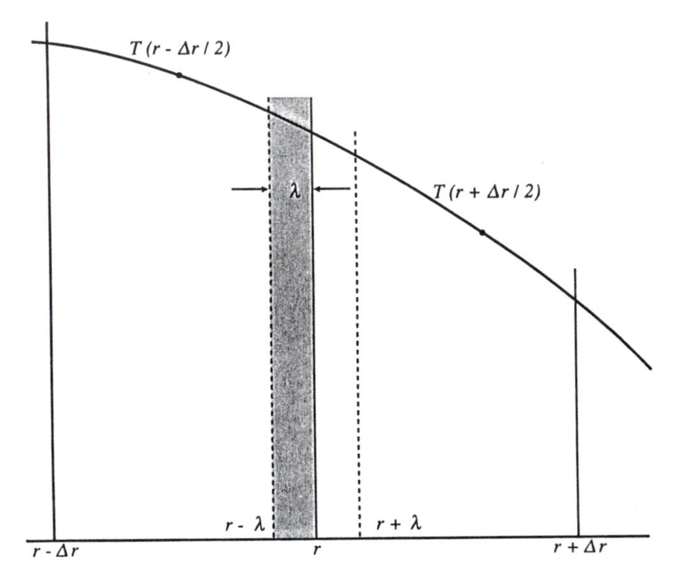

Figure 1.3 Temperature versus radius in a spherical star shown schematically.

in the outward direction, and the average energy density of these photons can be taken to be approximately $u(r-l/2)$. Consequently, the radiation energy flux originating within a distance l behind the surface which crosses it in the outward direction can be approximated by

$$\begin{pmatrix} \text{outward radiation} \\ \text{energy flux at } r \end{pmatrix} \equiv \frac{1}{4} u(r - l/2)\, c\, .$$

Here $u = aT^4$ is the radiation energy density, a is the radiation constant and c is the speed of light. The temperature in a star like the sun decreases outward as shown schematically in Figure 1.3. Using the results above, the outward flux defined above may be approximated by $F_+ = (1/4)acT(r - \lambda_R/3)^4$. The temperature at $r \pm \lambda_R/3$ may be approximated by linear interpolation using the temperature at $r \pm \Delta r/2$

$$T(r \pm \lambda_R/3) = T(r) \pm \frac{2\lambda_R}{3}\frac{\Delta T}{\Delta r}. \tag{1.9}$$

It is readily shown that to lowest order in Δr (where the subscript on the mean free path has been omitted for simplicity),

$$T(r) \equiv \frac{T(r-\lambda/3) + T(r+\lambda/3)}{2} \tag{1.10}$$

and

$$\Delta T(r) \equiv T(r - \Delta r/2) - T(r + \Delta r/2). \tag{1.11}$$

Using (1.9) in the definition of the outward flux, it follows that

$$F_+ = \frac{c}{4}u(r-\lambda/3) \approx \frac{ac}{4}[T(r) - \frac{2\lambda}{3}\frac{\Delta T}{\Delta r}]^4$$

$$\approx \frac{ac}{4}[T(r)^4 - \frac{8}{3}\lambda T(r)^3 \frac{\Delta T}{\Delta r}]. \tag{1.12}$$

A similar expression is obtained for the inward flux of photons which originates within a distance l ahead of the surface at r, and is denoted by F_-. It is easily shown to be obtained by reversing the signs in (1.12). Substituting the results for the inward and outward flux into (1.8) leads to the following expression for the steady state luminosity:

$$L(r) = 4\pi r^2 (F_+ - F_-) = -16\pi r^2 a \frac{\lambda c}{3} T^3 \frac{\Delta T}{\Delta r} . \tag{1.13}$$

Taking the limit $\Delta r \rightarrow 0$ and rewriting the result give the equation of radiative equilibrium in differential form:

$$\frac{\partial T}{\partial r} = - \frac{3}{16\pi r^2 ac\, \lambda} \frac{L}{T^3} . \tag{1.14a}$$

A finite difference representation of the partial differential equation (1.14a) may be obtained from (1.13) and the definitions (1.10) and (1.11). It is readily shown that

$$\frac{T_{k+1/2} - T_{k-1/2}}{\Delta r_k} = - \frac{3}{16\pi ac\, r_k^2 \lambda_k} \frac{2^3 L_k}{(T_{k+1/2} + T_{k-1/2})^3} \tag{1.14b}$$

where quantities centered at r and $r+\Delta r/2$ are denoted by k and $k+1/2$ respectively. It should be noted that the derivation leading to (1.14b) has supplied a definition (1.10) of the temperature at a point between two fluid elements (see Problem 1.3).

Problem 1.3

Verify that linear interpolation between the temperatures defined at $r + \Delta r/2$ and $r - \Delta r/2$ (see Figure 1.2) gives the result (1.9). How does this compare with the average appearing in (1.10)?

The methods above are not always sufficient to specify completely all definitions of the dependent variables entering into the finite difference equations. For example, the radiation mean free path λ_R entering into (1.14b) is centered at k, but it is in general a function of the density and temperature, which are centered at $k+1/2$. Arriving at suitable definitions of quantities such as λ_k which are not naturally centered in the finite difference equations is in general nontrivial, as will be seen in subsequent chapters.

An approach to numerical methods suggested by the two simple examples above is based on the concept of fluid elements of finite volume. Fluid properties and thermodynamic properties can be assigned to these elements, and fluxes of energy into and out of the fluid element can be calculated. A natural extention of this approach when fluid motion occurs is to attribute a velocity to the boundary surfaces of the element, and to use these velocities to describe dynamically the changes in position and volume of the element. The question of how to center dependent physical variables in terms of finite volume fluid elements in a consistent and mathematically acceptable way will enter prominently in each of the subsequent chapters. For these reasons, further discussion of finite difference methods will be postponed until Chapter 2, where a more detailed discussion is presented which establishes the framework for the subsequent chapters.

1.3 Structure of Second Order Partial Differential Equations

Many of the equations that arise in physics are second order linear partial differential equations. In this section we will discuss the important properties of these types of equations for two dimensional systems. A great deal of insight into the nature of the solutions to these types of partial differential equations (and the finite difference approximations to them) can also be obtained in this way. In this section we shall briefly review their classifications and properties. A more complete discussion of these topics from a physical viewpoint can be found, for example, in Morse and Feshbach (1953), or in Mathews and Walker (1964), while a rigorous mathematical discussion can be found in, for example, Garabedian (1964). Consider the partial differential equation for a function $F(x,y)$ of two independent variables

$$a\frac{\partial^2 F}{\partial x^2} + 2b\frac{\partial^2 F}{\partial x \partial y} + c\frac{\partial^2 F}{\partial y^2} = f\left(\frac{\partial F}{\partial x}, \frac{\partial F}{\partial y}, F, x, y\right) \tag{1.15}$$

where the real coefficients a, b and c may be functions of the independent variables x and y (one of which may represent time). It can be shown (subject to suitable restrictions on the three coefficients) that a coordinate transformation can be found from the variables (x,y) to the new set $[\xi(x,y), \eta(x,y)]$ under which (1.15) reduces to one of the three following canonical forms. In the first case

$$\frac{\partial^2 F}{\partial \xi^2} - \frac{\partial^2 F}{\partial \eta^2} = f_1\left(\frac{\partial F}{\partial \xi}, \frac{\partial F}{\partial \eta}, F, \eta, \xi\right) \tag{1.16a}$$

when

$$b^2 > ac\,; \tag{1.16b}$$

a partial differential equation of this type is called hyperbolic. The partial differential equation takes the form

$$\frac{\partial^2 F}{\partial \xi^2} = f_3\left(\frac{\partial F}{\partial \xi}, \frac{\partial F}{\partial \eta}, F, \eta, \xi\right) \tag{1.17a}$$

when

$$b^2 = ac\,; \tag{1.17b}$$

the partial differential equation is called parabolic. And finally,

$$\frac{\partial^2 F}{\partial \xi^2} + \frac{\partial^2 F}{\partial \eta^2} = f_2\left(\frac{\partial F}{\partial \xi}, \frac{\partial F}{\partial \eta}, F, \eta, \xi\right) \tag{1.18a}$$

when

$$b^2 < ac\,; \tag{1.18b}$$

an equation of this type is called elliptic. It can be shown that the conditions (1.16b), (1.17b) and (1.18b) satisfied by the coefficients remain invariant under coordinate transformations, and thus are intrinsic properties of the partial differential equation. Finally, the condition that (1.15) may be transformed into one of the forms above may be expressed as

$$c(\xi,\eta)\, d\xi^2 - 2b(\xi,\eta)\, d\xi\, d\eta + a(\xi,\eta)\, d\eta^2 \neq 0\,. \tag{1.19}$$

When (1.19) is satisfied, it can be shown that the values of $F(\xi,\eta)$ and its normal derivative along a curve C are, in combination with the original partial differential equation, sufficient to determine the value of $F(\xi,\eta)$ at neighboring points, and thus that a solution $F(\xi,\eta)$ can be constructed in a finite domain (see Problem 1.4). In general two families of curves, which are called characteristic curves of the partial differential equation, are defined by the solutions

$$c(\xi,\eta)\, d\xi^2 - 2b(\xi,\eta)\, d\xi\, d\eta + a(\xi,\eta)\, d\eta^2 = 0\,. \tag{1.20a}$$

This ordinary differential equation may be rewritten to exhibit the two independent equations:

$$a\frac{d\eta}{d\xi} = b \pm \sqrt{b^2 - ac} \tag{1.20b}$$

Hyperbolic Equations

For hyperbolic equations the constraint on the coefficients (1.16b) guarantees that there will be two real solutions to (1.20), and thus two real families of characteristics exist (see Problem 1.5). When one of the independent variables is chosen to represent time, initial conditions and discontinuities in the initial conditions can be shown to propagate along these characteristics (they may propagate in either the positive or negative temporal direction). Furthermore, wave motion is typically represented by hyperbolic equations. In fact, a typical example of a hyperbolic equation is the wave equation which, in one spatial dimension, takes the familiar form

$$\frac{\partial^2 F}{\partial x^2} - \frac{1}{c^2}\frac{\partial^2 F}{\partial t^2} = 0 , \tag{1.21}$$

where $F(x,t)$ is the wave amplitude, c is the wave velocity, and we have set $\xi = x$ and $\eta = ct$. Hyperbolic equations describe, for example, the propagation of initial data specified as the value of the function and its first derivative normal to an open boundary curve (Cauchy data) as long as the boundary curve is nowhere tangent to a characteristic curve (see Problem 1.5). If some portion of the boundary curve along which initial conditions are to be set is tangent to a characteristic then a suitable set of boundary conditions consists of a combination of Cauchy conditions and Neumann or Dirichlet conditions.

Parabolic Equations

For parabolic equations (1.17b) indicates that there is only one characteristic equation and consequently only one family of (real) characteristic curves. The heat equation is a typical example of a parabolic partial differential equation and is obtained from (1.17a) if $f_2 = 0$, $\xi = x$ and $\eta = t$:

$$\frac{\partial^2 F}{\partial x^2} = K \frac{\partial F}{\partial t}. \tag{1.22}$$

Here $F(x,t)$ represents the temperature, and K is the ratio of the heat capacity to the thermal conductivity. In the case of parabolic equations the initial data diffuses along a single family of characteristics in such a manner as to smooth out irregularities.

Elliptic Equations

Finally, when (1.18b) is satisfied (1.15) can be reduced to elliptic form. Laplace's equation, which is a typical example of an elliptic partial differential equation, results by setting $\xi = x$ and $\eta = y$ in (1.18a) with $f_3 = 0$:

$$\frac{\partial^2 F}{\partial x^2} + \frac{\partial^2 F}{\partial y^2} = 0. \tag{1.23}$$

Other well known examples of elliptic equations are Helmholtz's equation (in which case the right side in (1.23) is proportional to F itself) and Poisson's equation for the gravitational potential (in which case the right hand side of (1.23) is proportional to the mass density). Because of (1.18b) the characteristics for elliptic partial differential equations are complex. Furthermore, it can be shown that there is a close relationship between solutions of elliptic equations in two dimensions and the behavior of analytic functions. Typically, the initial data (either Neumann or Dirichlet or a linear combination of the two) must be specified along a closed boundary surrounding the region of interest (this may include conditions at spatial infinity). The initial data does not propagate along the characteristics of the partial differential equation (which are complex). Interior to this boundary the solution can have neither a maximum nor a minimum since once $F(x,y)$ establishes a tendency either to increase or decrease, it must continue to do so until it reaches a singularity. As one consequence wave motion is not possible in the solution of elliptic equations.

The discussion above touches on some of the main features of the three basic types of partial differential equations which will be encountered in subsequent chapters. Table 1.1 gives a short summary of these properties.

Table 1.1
Features of linear, second order partial differential equations of the form (1.15).

	Hyperbolic	Parabolic	Elliptic
Relationship between coefficients of the second derivatives	$b^2 > ac$	$b^2 = ac$	$b^2 < ac$
Nature of characteristic curves	two real	one real	two complex
Boundary conditions	Cauchy	Neumann or Dirichlet	Neumann or Dirichlet
Boundary	open	open	closed
Typical properties of the solution	wave motion	diffusion; irreversibility	no wave motion possible

Problem 1.4

It is well known that the value of a function and its first derivative at a nonsingular point are sufficient to uniquely determine the solution to an ordinary second order linear differential equation. This problem considers the conditions under which this theorem can be extended to partial differential equations of the form (1.15). Suppose that the initial data $F(s)$ and $N(s) = \mathbf{n} \cdot \nabla F$ are known along a curve C parameterized by $x = x(s)$ and $y = y(s)$ (see Figure 1.4).

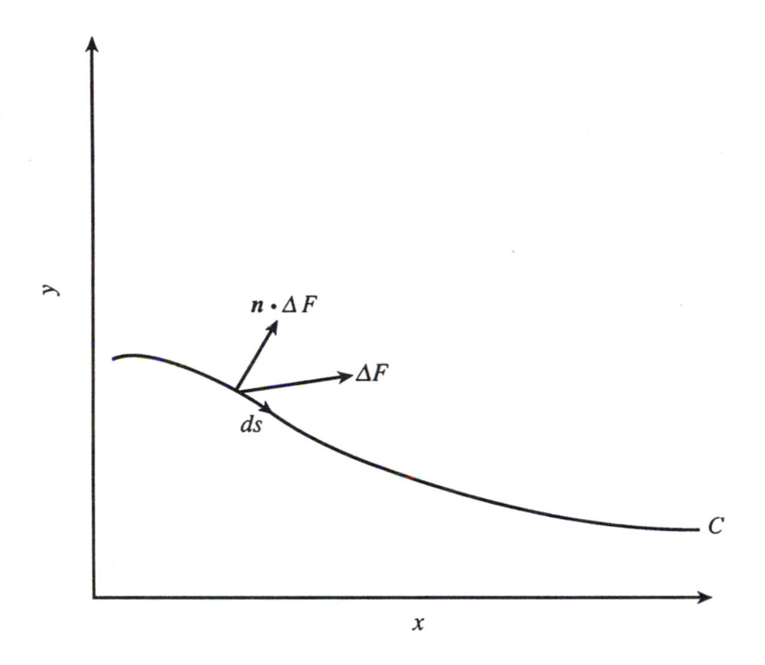

Figure 1.4 Initial data along an arbitrary curve.

The value $F(x+\Delta x, y+\Delta y)$ can be found uniquely in terms of the initial (Cauchy) data if all of the partial derivatives of $F(x,y)$ occurring in a Taylor series expansion about the curve C can be found. Thus write

$$F(x+\Delta x, y+\Delta y) = F(s) + F_x \Delta x + F_y \Delta y + \frac{1}{2}\left(F_{xx} \Delta x^2 + 2F_{xy} \Delta x \Delta y + F_{yy} \Delta y^2 \right) + \ldots \tag{1.24}$$

where the dots denote third and higher order derivatives. Noting that

$$\frac{dF}{ds} = F_x \frac{dx}{ds} + F_y \frac{dy}{ds}$$

use the initial data to solve for F_x and F_y. Show that the equations can always be solved for these first derivatives. Next find two equations for the second order derivatives F_{xx}, F_{yy} and F_{xy} using the first derivatives found above. Show that these two equations and the original partial differential equation (1.15) can be solved for the second derivatives as long as (1.19) is satisfied. Finally explain why knowledge of the first and second partial derivatives of F are sufficient to determine the function at nearby points.

Problem 1.5

Find the characteristic curves ξ and η for the wave equation (1.21) assuming that the wave velocity c is a constant. Transform the wave equation to characteristic coordinates, and show that its solution is

$$F(x,t) = f(\xi - \eta) + g(\xi + \eta) \tag{1.25}$$

where f and g are arbitrary functions. Finally suppose that Cauchy initial data are given along the open curve $t=0,\ x_o < x < x_1$:

$$F(x, 0) = \psi(x)$$

$$\frac{1}{c}\left(\frac{\partial F}{\partial t}\right)_{t=0} = \phi(x) \tag{1.26}$$

Find the general form of the solution $F(x,t)$ in terms of the initial data $\psi(x)$ and $\phi(x)$. Notice that the solution to the wave equation (as is common of hyperbolic equations) can be extrapolated backward in time.

Problem 1.6

Plot a series of characteristic curves in the x,y plane for the wave equation discussed in Problem 1.5, showing the domain of the initial conditions. Pick an arbitrary value of the spatial coordinate between

x_0 and x_1 and discuss the domain of initial conditions that determines its value. What is the solution for times $t > (x_1 - x_0)/2c$?

Problem 1.7

The temperature along a rod of length L is given by the solution to the diffusion equation (1.22) subject to suitable boundary conditions. Suppose that

$$F(x,0) = f(x) \qquad 0 \le x \le L$$

$$F(0,t) = 0\,, \quad F(L,t) = 0.$$

Use separation of variables to show that the solution for $t \ge 0$ is given by

$$F(x,t) = \sum_{n=1}^{\infty} A_n \, e^{-n^2\pi^2 t/L^2} \sin\frac{n\pi x}{L}$$

where

$$A_n = \frac{2}{L}\int_0^L f(x) \sin\frac{n\pi x}{L}\, dx \qquad n = 1, 2, \ldots$$

Is the solution valid for times $t < 0$?

Problem 1.8

Show that the hydrodynamic equations in one spatial dimension are equivalent to a second order hyperbolic equation.

References

P. R. Garabedian, *Partial Differential Equations* (New York: Wiley, 1964).

J. Mathews and R. L. Walker, *Mathematical Methods of Physics* (New York: Benjamin, 1964).

D. Mihalas and B. W. Mihalas, *Foundations of Radiation Hydrodynamics* (New York: Oxford University Press, 1984).

P. M. Morse and H. Feshbach, *Methods of Theoretical Physics,* Vol. II (New York: McGraw-Hill, 1953).

A. Sommerfeld, *Partial Differential Equations in Physics* (New York: Academic Press, 1949).

R. F. Stein, "Stellar Evolution: A Survey with Analytical Models," in *Stellar Evolution* edited by R. F. Stein and A. G. W. Cameron (New York: Plenum Press, 1966).

2

OVERVIEW OF FINITE DIFFERENCE METHODS

Very few systems of partial differential equations describing astrophysical phenomena can be solved by analytic means, and those systems which can be solved analytically often result from simplified assumptions about the underlying physics. The goal of finite difference methods is to replace the system of partial differential equations by a system of finite difference equations which are defined on a dependent coordinate grid, such as a space-time grid. A space-time grid represents a framework in which the continuum properties of matter may be represented in discrete form. It is often convenient, then, to physically associate a discrete unit of matter with a mass or volume element as in fluid dynamics. Dynamics then consists of the interaction of this element with its neighbors, and its resulting evolution in space and time.

This chapter contains definitions, and establishes a formalism and the mathematical machinery needed for further discussions of finite difference approximations. Two critical aspects of finite difference methods, accuracy and stability, will also be considered. The order of approximation of the finite difference schemes raises difficult questions which are still under active investigation. Numerical stability of finite difference schemes will also be discussed with emphasis on the physical basis of the concept whenever possible. Finally, a practical method for solving a complicated system of coupled finite difference equations known as operator splitting will be discussed. The results developed in this chapter will be used extensively in the remaining chapters.

2.1 FINITE DIFFERENCE APPROXIMATIONS

There are many ways of approximating partial differential equations which are suitable for use on scientific computers. The use of finite difference representations will be emphasized in the following chapters. A finite difference representation of a system of time dependent equations consists of a space-time grid upon which are defined finite difference approximations to the functions and their derivatives which constitute the partial differential equations under study. The formulation of these concepts will now be considered.

Computational Grid

The concept of finite difference approximations to time evolving systems can be illustrated by considering a problem in one spatial dimension in Cartesian coordinates (x,t). The space-time continuum upon which a set of partial differential equations is defined will be replaced by a finite coordinate grid, or computational grid, whose discrete elements may be denoted by discrete spatial coordinates at each instant of time. One type of computational grid, shown for example in Figure 2.1a, consists of a set of fixed coordinate positions x_k, which may or may not be uniformly spaced, that are defined at the discrete time t^n. The temporal variable will change in a discrete fashion such that $t^{n+1} = t^n + \Delta t^{n+1/2}$, where $\Delta t^{n+1/2}$ need not necessarily be constant. The dependent variables are defined on this space-time grid, and the evolution equations take them from one spatial level to the next. In the example above subscript k represents the spatial index and superscript n represents the temporal index. Since the spatial coordinates are fixed, they do not carry a temporal index. Another type of grid consists of spatial coordinates each of which may change in time. At time t^n they may be denoted by $x^n{}_k$ (see Figure 2.1b). The spatial coordinates at the next time interval t^{n+1} = $t^n + \Delta t^{\,n+1/2}$ will be denoted by $x^{n+1}{}_k$. The difference between the two spatial coordinates $x^n{}_{k+1}$ and $x^n{}_k$ will be denoted by $\Delta x^n{}_{k+1/2}$ and the time difference $t^{n+1} - t^n$ will be denoted by $\Delta t^{n+1/2}$. Here k lies between one and K (the total number of spatial zones), and n lies between zero and N (the total number of computational time steps). Unless otherwise stated, subscripts will be used to denote spatial indices, and superscripts will be used to denote the temporal index.

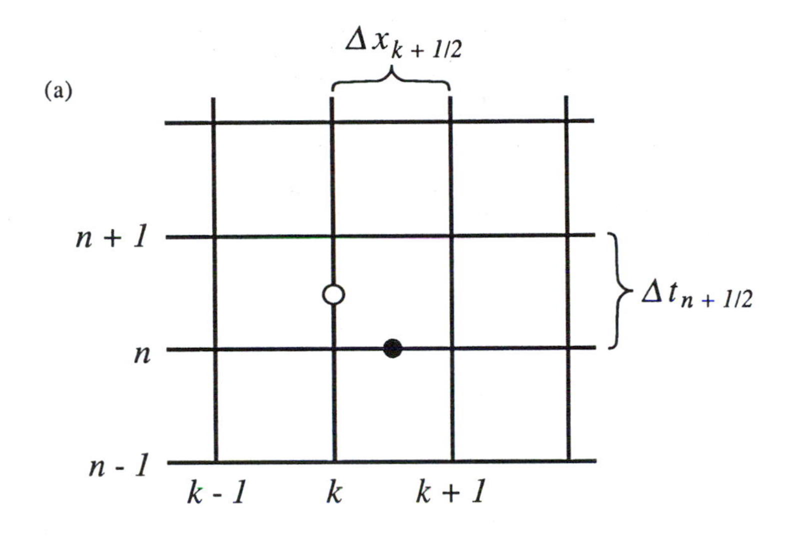

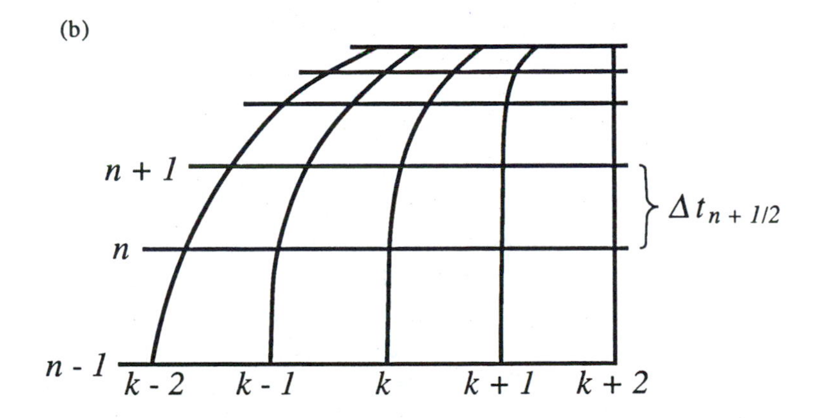

Figure 2.1. Space-time grids: (a) Uniform space-time grid with $\Delta x^n{}_{k+1/2} = \Delta x$ and $\Delta t^{n+1/2} = \Delta t$ where Δx and Δt are constants. (b) An initially uniform spatial grid which becomes nonuniform at later times. Note that the time step is not a constant.

In two spatial dimensions, separate subscript will be used to label each coordinate direction. In general, the spatial coordinates will depend on time. The number of zones in the spatial dimension K may also change in time during the course of some applications (this occurs, for example, when dynamic rezoning is employed). The spatial coordinates at time t^n are given simply by

$$x^n_k = \sum_{l=1}^{K-1} \Delta x^n_{l+1/2} \qquad k = 2,\, 3,\, \ldots\, K \tag{2.1}$$

where

$$t^n = \sum_{l=0}^{n-1} \Delta t^{l+1/2}. \tag{2.2}$$

A portion of a typical space-time grid is shown in Figure 2.1. Note that the coordinates $x^n{}_k$ are defined at the zone edges, while the increments $\Delta x^n{}_{k+1/2}$ are zone centered. The finite boundaries of the spatial grid are $x^n{}_1$ and $x^n{}_K$. In general $\Delta x^n{}_{k+1/2}$ and $\Delta t^{n+1/2}$ will change with time. Finally the spatial grid need not be uniform; that is, $\Delta x^n{}_{k+1/2}$ need not equal $\Delta x^n{}_{k-1/2}$ (see Figure 2.1b). Whenever possible uniform zoning should be used, since it generally yields better results. However many systems of astrophysical interest involve variables (such as density and pressure) which range over many orders of magnitude during the course of a calculation. In problems of this kind uniform zoning becomes impractical and some form of variable zoning, grid motion or rezoning may be necessary. In addition to the space-time grid defined in terms of the coordinates $x^n{}_k$, it is often necessary to define a staggered grid whose coordinates lie between $x^n{}_{k+1}$ and $x^n{}_k$. These coordinates will be defined by $x^n{}_{k+1/2}$. It should be noted that the subscript $1/2$ does not necessarily imply that the staggered coordinates lie halfway between $x^n{}_{k+1}$ and $x^n{}_k$. The spatial increments defined in terms of these coordinates are $\Delta x^n{}_k \equiv x^n{}_{k+1/2} - x^n{}_{k-1/2}$, and represent a second space-time grid. The necessity of two staggered spatial grids is associated with the centering arising from the approximation of functions and their derivatives on the grid itself.

Functions

In the interests of notational convenience the following convention will be used: the phrase "common subscript k,j omitted" followed by an expression of the form $A^{n+1} - A^n$ will be understood to stand for $A^{n+1}{}_{k,j} - A^n{}_{k,j}$. A similar convention will also be used for superscripts, or for other combinations of indices.

Finite difference approximations to physical variables may be associated with zone centers (in some instances zone interiors), zone edges or with zone corners, depending on the spatial dimensionality, the type of variable and the nature of the scheme used to approximate the partial differential equations. For example, in one spatial dimension a dependent variable may be defined on the grid as illustrated by the following examples. If the variable $f(x,t)$ is associated with a zone center at time t^n, then

$$f(x,t) = f(x^n_{k+1/2}\ ,\ t^n) \rightarrow f^n_{k+1/2} \tag{2.3}$$

as shown by the filled circle in Figure 2.1a. A function representing another type of dependent variable, $g(x,t)$, associated for example with a zone-edge at time $t^{n+1/2}$ will be represented by

$$g(x,t) = g(x^{n+1/2}_k,\ t^{n+1/2}) \rightarrow g^{n+1/2}_k , \tag{2.4}$$

as shown by the open circle in Figure 2.1a. Finite difference notation for functions of two or more spatial dimensions may be defined in a similar manner.

In the differential limit all variables will be centered at the same point (or are moved to the same point when compared) so the distinction between (2.3) and (2.4), for example, does not arise. Because the limit $\Delta x \rightarrow 0, \Delta t \rightarrow 0$ is never actually taken in finite difference representations the distinction can be extremely important.

Partial Derivatives

Finite difference approximations to the partial derivatives of $f(x,t)$ can be constructed from Taylor series expansions. For fixed t, expanding about x_o gives

$$f(x+\Delta x) = f(x_o) + \left(\frac{\partial f}{\partial x}\right)_{x_o}\Delta x + \frac{1}{2}\left(\frac{\partial^2 f}{\partial x^2}\right)_{x_o}\Delta x^2 + O(\Delta x^3) \tag{2.5}$$

where $O(\Delta x^3)$ represents terms of order Δx^3 and higher. Suppose that $f(x,t)$ corresponds to an edge centered variable on a spatial grid: $f(x_o) \rightarrow f_k$ and $x_o \rightarrow x_k$. For simplicity, we omit the time index since it plays no role in the analysis below. Setting $\Delta x = \Delta x_{k+1/2}$ and rewriting (2.5),

$$\left(\frac{\partial f}{\partial x}\right)_k = \frac{f_{k+1} - f_k}{\Delta x_{k+1/2}} + O(\Delta x) \tag{2.6}$$

which represents a first order (in space) forward differenced approximation to $(\partial f/\partial x)_t$ at the grid point k. A first order backward differenced approximation can be obtained by setting $\Delta x = -\Delta x_{k-1/2}$ in (2.5). Finally a centered difference can be found by subtracting (2.5) with $\Delta x = \Delta x_{k+1/2}$ from (2.5) with $\Delta x = -\Delta x_{k-1/2}$ to obtain

$$\left(\frac{\partial f}{\partial x}\right)_k = \frac{f_{k+1} - f_{k-1}}{2\,\Delta x_k} - \frac{1}{2}\left(\frac{\partial^2 f}{\partial x^2}\right)_k \frac{(\Delta x_{k+1/2}^2 - \Delta x_{k-1/2}^2)}{\Delta x_k} + O(\Delta x^2), \tag{2.7}$$

where

$$\Delta x_k = \frac{1}{2}(\Delta x_{k+1/2} + \Delta x_{k-1/2}). \tag{2.8}$$

For uniform zoning, (2.7) is second order accurate in space.

Problem 2.1

Show that an edge centered first order approximation to $(\partial f/\partial x)$ when $f(x)$ is zone centered is given by

$$\left(\frac{\partial f}{\partial x}\right)_k = \frac{f_{k+1/2} - f_{k-1/2}}{\Delta x_k}, \tag{2.9}$$

with Δx_k given by (2.8). What order of accuracy results if uniform zoning is assumed?

Finally, consider an approximation to $(\partial f/\partial t)_x$ which is centered at $t^{n+1/2}$ when $f(t)$ is centered at t^n. The spatial index will be omitted for simplicity in the following discussion. Expanding $f(t+\Delta t)$ in a Taylor series about t we find

$$f^{n+1} = f^{n+1/2} + \left(\frac{\partial f}{\partial t}\right)^{n+1/2} \frac{\Delta t^{n+1/2}}{2} + \frac{1}{2}\left(\frac{\partial^2 f}{\partial t^2}\right)^{n+1/2} \left(\frac{\Delta t^{n+1/2}}{2}\right)^2 + O(\Delta t^3)$$

$$f^n = f^{n+1/2} - \left(\frac{\partial f}{\partial t}\right)^{n+1/2} \frac{\Delta t^{n+1/2}}{2} + \frac{1}{2}\left(\frac{\partial^2 f}{\partial t^2}\right)^{n+1/2} \left(\frac{\Delta t^{n+1/2}}{2}\right) + O(\Delta t^3). \tag{2.10}$$

Subtracting the second equation above from the first one gives

$$\left(\frac{\partial f}{\partial t}\right)_{k+1/2} = \frac{f^{n+1} - f^n}{\Delta t^{n+1/2}} + O(\Delta t^2), \tag{2.11}$$

which is second order accurate, even for variable $\Delta t^{n+1/2}$. A similar analysis for the spatial derivative of an edge centered variable f_k gives

$$\left(\frac{\partial f}{\partial x}\right)_{k+1/2} = \frac{f_{k+1} - f_k}{\Delta x_{k+1/2}} + O(\Delta x^2). \tag{2.12}$$

In this case the temporal superscript has been omitted for convenience.

The examples above demonstrate several different ways in which variables and their derivatives can be centered on a computational grid. We shall return frequently to the issue of centering in later chapters, where physical arguments will often be used to determine the type of centering required.

Problem 2.2

Use the Taylor series expansion (2.5) to derive a finite difference approximation for $(\partial^2 f/\partial x^2)_t$ centered at $k+1/2$ assuming that $f(x,t)$ is centered at $k+1/2$. What order accuracy is the result?

The discussion above considered finite difference approximations to partial derivatives in Cartesian coordinates of a single spatial dimension. These results may be extended to higher spatial dimensions in a straightforward manner. We now consider finite difference approximations in curvilinear coordinates (orthonormality will, however, be assumed), and suppose that the partial differential equations have been written in vector or tensor form. Expressions must then be found for quantities such as $\nabla \cdot \boldsymbol{F}$, $\nabla \times \boldsymbol{F}$ and $\nabla \Phi$ where $\boldsymbol{F}$ is a vector function, and Φ is a scalar function.

The gradient of a scalar function $\Phi(x,t)$ is defined in terms of the total derivative along a path s whose tangent vector is ds by

$$du = \nabla\Phi \cdot ds \ .$$

For the systems considered here the components of $\nabla\Phi$ will generally be coordinate derivatives $(\partial\Phi/\partial\xi_i)$ where ξ_i represent the coordinates.

Finite difference approximations to the divergence of $\boldsymbol{F}$ can be found by the following method. Consider a spherically symmetric system described on a one dimensional grid with zone boundaries r_k, zone widths $\Delta r_{k+1/2} \equiv r_{k+1} - r_k$ and zone volumes

$$\Delta V_{k+1/2} = \frac{4\pi}{3}\left(r_{k+1}^3 - r_k^3\right). \tag{2.13}$$

The assumption of spherical symmetry means that all partial derivatives with respect to θ and ϕ vanish identically. Since all quantities depend only on the spatial variable r (and possibly on the time) the system is often called one dimensional. Further suppose that $\boldsymbol{F}(r,t)$ is to be represented by the zone-edge centered quantity F_k (the time variable is to be held constant, and is thus omitted for simplicity), and that $\nabla \cdot \boldsymbol{F}$ is constant across a zone. Then

$$(\nabla \cdot \boldsymbol{F})_{k+1/2} = \frac{1}{\Delta V_{k+1/2}} \int \nabla \cdot \boldsymbol{F} dV = \frac{1}{\Delta V_{k+1/2}} \int_S \boldsymbol{F} \cdot ds \tag{2.14}$$

where the integrals are over the volume and surface, respectively, of zone $k+1/2$. Evaluating the right hand side of (2.14) over the surface of $\Delta V_{k+1/2}$, and substituting into (2.14) gives

$$(\nabla \cdot \boldsymbol{F})_{k+1/2} = \frac{F_{k+1}A_{k+1} - F_k A_k}{\Delta V_{k+1/2}}, \tag{2.15}$$

where the area of a zone boundary is denoted by A_k [note that (2.15) holds for any one dimensional, orthonormal coordinate system]. For spherical coordinates

$$A_k = 4\pi r_k^2 . \tag{2.16}$$

In the differential limit $\Delta r \to 0$, (2.15) reduces to the usual expression $\nabla \cdot F = (1/r^2)(\partial r^2 F/\partial r)$. Although (2.15) could be rewritten in terms of Δr, the use of zone areas and volumes has advantages, as will be seen in later chapters.

Problem 2.3

Rewrite (2.15) in terms of $\Delta r_{k+1/2}$ for spherical coordinates, being careful to define all quantities on the grid. Is the definition of $r_{k+1/2}$ unique?

Problem 2.4

Construct a finite difference approximation to $\nabla \cdot \boldsymbol{F}$ which is zone centered on a two dimensional cylindrical grid (r,z). Assume that the components of $\boldsymbol{F}$ are zone-edge centered, and that they are independent of θ (two dimensionality of $\boldsymbol{F}$).

Problem 2.5

Use the definition of the curl of $\boldsymbol{F}$,

$$\nabla \times \boldsymbol{F} = - \frac{1}{\Delta V} \int_S \boldsymbol{F} \times ds \ , \quad \Delta V \to 0 \tag{2.17}$$

to construct a zone centered finite difference approximation to $\nabla \times \boldsymbol{F}$ on the grid defined in Problem 2.4.

Partial Differential Equations

The examples above show ways of constructing finite difference approximations to partial derivatives. Now we use these methods to construct finite difference representations of partial differential equations, and comment on the limiting relationship between finite difference equations and the partial differential equations that they are supposed to approximate. In the following examples we will emphasize the centering of variables in the finite difference equations. The issue of centering is fundamental to the development of finite difference approximations to partial differential equations. Unfortunately many systems of partial differential equations do not lend themselves to naturally centered finite difference equations. Nevertheless it is always desirable to strive for natural centering. This is important because centering affects both the accuracy and stability of the equations. Finally it must be remembered that the relationship between partial differential equations and finite difference equations is not unique.

While there is no general formal approach that can be used to construct finite difference representations to partial differential equations, three principles can be identified (largely on the basis of experience and hindsight) which may be used as a guide. These are that the finite difference representation should:

I. Reduce to the correct partial differential equation in the limit $\Delta x \to 0$ and $\Delta t \to 0$;

II. Satisfy acceptable stability criteria and be reasonably accurate;

III. Reflect the intrinsic properties of the partial differential equations.

Examples of each of these principles will be considered in the discussions below. It should be noted that few finite difference representations completely satisfy I-III above, and that some of the conditions are clearly problem (and even machine) dependent. Nevertheless they can serve as useful guides to developing or understanding specific examples.

As an example, suppose that we wish to construct a finite difference representation of the partial differential equation

$$\left(\frac{\partial u}{\partial t}\right)_x + u\left(\frac{\partial u}{\partial x}\right)_t = 0, \tag{2.18}$$

where $u = u(x,t)$. If u is zone-edge centered, then the partial derivative

$(\partial u/\partial t)_x$ can be approximated by the forward time difference (2.11), and the spatial derivative $(\partial u/\partial x)_t$ could be represented by (2.12). In order to complete the approximation we must decide where to center u, the coefficient of the spatial derivative. One choice would be $u \to u^n{}_k$; then one finite difference equation corresponding to (2.18) would be

$$\frac{u_k^{n+1} - u_k^n}{\Delta t^{n+1/2}} + u_k^n \frac{(u_{k+1}^n - u_k^n)}{\Delta x_{k+1/2}} = 0. \tag{2.19}$$

But the gradient in (2.18) can equally well be written as $\partial(u^2/2)/\partial x$, in which case we could try as a finite difference approximation to (2.18)

$$\frac{u_k^{n+1} - u_k^n}{\Delta t^{n+1/2}} + \left(\frac{u_{k+1}^n + u_k^n}{2}\right)\frac{u_{k+1}^n - u_k^n}{\Delta x_{k+1/2}} = 0. \tag{2.20}$$

In this case, the form of the partial differential equation suggests another way of centering the coefficient of $(\partial u/\partial x)_t$ in (2.18). In the differential limit $\Delta x \to 0$ and $\Delta t \to 0$ both (2.19) and (2.20) reduce to the same partial differential equation (2.18). However each finite difference approximation has different properties (as we shall see later neither is particularly good). As is evident from this example, there are many (in fact an infinite number of) finite difference approximations which can be made to a given partial differential equation.

We might expect any finite difference representation of a partial differential equation to reduce to the correct partial differential equation in the differential limit. A simple example shows that this is not always the case. Consider the diffusion equation

$$\frac{\partial \theta}{\partial t} = D\frac{\partial^2 \theta}{\partial x^2} \tag{2.21}$$

where θ can be the temperature, or the radiation energy density, and D is a (constant) diffusion coefficient. A possible finite difference approximation to (2.21) assuming uniform zoning and a constant Δt is

$$\frac{\theta^{n+1}_{k+1/2} - \theta^{n-1}_{k+1/2}}{2\Delta t} = D\frac{\theta^{n}_{k+3/2} - 2\theta^{n}_{k+1/2} + \theta^{n}_{k-1/2}}{\Delta x^2} . \tag{2.22}$$

This approximation reduces to (2.21) in the differential limit, but can be shown to be numerically unstable. Another finite difference approximation which is stable results by replacing $\theta^{n}{}_{k+1/2}$ in the right hand term of (2.22) by the time average $(1/2)(\theta^{n+1}{}_{k+1/2}+\theta^{n-1}{}_{k+1/2})$. Expanding $\theta^{n\pm1}$ in a Taylor series [see for example (2.10)], substituting into (2.22) and taking the limit $\Delta x \to 0$ and $\Delta t \to 0$ yields the result

$$\frac{\partial \theta}{\partial t} = D\,\frac{\partial^2 \theta}{\partial x^2} - D\,\frac{\partial^2 \theta}{\partial t^2}\left(\frac{\Delta t}{\Delta x}\right)^2 , \tag{2.23}$$

which reduces to (2.21) only if the ratio $(\Delta t/\Delta x) \to 0$ as $\Delta x \to 0$ and $\Delta t \to 0$. It should be noted that the original partial differential equation is parabolic, while (2.23) is hyperbolic. As this example demonstrates, the construction of a finite difference approximation may lead to a result which has a different character from that of the original differential equation.

Problem 2.6

Show that (2.19) and (2.20) each reduce to (2.18) in the limit $\Delta x \to 0$, $\Delta t \to 0$ independent of the order in which the limit is taken.

Problem 2.7

Find a finite difference approximation for the following partial differential equations, where ρ is the density, T is the temperature, the diffusion coefficient is $D(T)$, $K(T)$ is the coupling coefficient and P

is the pressure (all zone centered). The velocity u is zone-edge centered. The quantity A (assumed for simplicity to be a constant) can be thought of as an emission term.

$$\frac{\partial \rho}{\partial t} + \frac{\partial (\rho u)}{\partial x} = 0 \tag{2.24}$$

$$\frac{\partial T}{\partial t} = \frac{\partial}{\partial x}\left[D(T)\frac{\partial T}{\partial x}\right] \tag{2.25}$$

$$\frac{\partial T}{\partial t} = K(T)\,(A - T^4) \tag{2.26}$$

$$\rho\,\frac{\partial u}{\partial t} = -\,\frac{\partial P}{\partial x}\,. \tag{2.27}$$

Assume uniform zoning, and carefully denote the centering of all quantities in the grid. All variables are functions of x and t.

2.2 STABILITY AND ACCURACY

The concepts of stability and accuracy of time evolving finite difference equations will now be introduced, and several simple examples will be developed to illustrate them. For simplicity the discussion will focus on time evolving systems in one spatial dimension with independent variables (x,t), but the extension to higher spatial dimensions can usually be made without difficulty.

The accuracy of a finite difference equation is often obtained as follows. First all quantities in it are expanded in a Taylor series about a convenient space-time grid point. Then, a linearized form of the partial differential equation is used to eliminate as many terms in the series as possible, leaving quantities which are proportional to powers of Δx and Δt. In most analyses it is further assumed that the spatial grid is uniform, and that a single (constant) time step is used. Under this special set of circumstances it is relatively easy to establish a definition of the order of accuracy of the method. Unfortunately, these definitions generally break down whenever the equations are nonlinear, whenever the spatial grid is nonuniform, or whenever more than one time step occurs which changes during the evolution. Nevertheless, one can use this simplified approach as a gauge of the method's accuracy, although it should be borne in mind that such estimates can often be misleading. Other definitions of accuracy can be proposed; however, their relative merit in complex problems is often difficult to judge. In the final analysis a method may be judged to be of greater accuracy if it gives better results when applied to a problem to which the solution is known.

Finite difference approximations can be found which lead under some or all circumstances to apparently nonsensical results. In these methods errors of any kind (even those associated with machine roundoff) propagate and are amplified to such an extent that they dominate the true solution. These methods are termed unstable. Approximations which are unstable for all choices of Δt and Δx are termed unconditionally unstable and must be abandoned. Schemes which are stable for restricted values of Δt are termed conditionally stable. Finally, schemes which remain stable for any value of the time step Δt are called unconditionally stable. A simple physical example of the concept of stablility may be used to illustrate these points (Roach 1972). Assume that a steady state solution exists to a partial differential equation, and suppose that this state is subject to a small perturbation. If the initial perturbation dies out, then the steady state represents a stable solution. An acceptable finite difference approximation should also possess this property. For example, consider the partial differential equation with $u>0$

$$\frac{\partial \psi}{\partial t} + u\frac{\partial \psi}{\partial x} = 0, \tag{2.28}$$

and the finite difference approximation to it,

$$\frac{\psi_k^{n+1} - \psi_k^n}{\Delta t} + u\,\frac{\psi_{k+1}^n - \psi_{k-1}^n}{2\,\Delta x} = 0. \tag{2.29}$$

(The assumption of constant Δt and Δx results in no loss of generality.) We further assume that a steady state solution denoted by

$$\hat{\psi}^n_k$$

exists at time t^n: then by assumption (2.29) becomes

$$\hat{\psi}^n_{k+1} - \hat{\psi}^n_{k-1} \equiv 0.$$

Now impose on the steady state solution a perturbation ε_k defined such that $|\varepsilon_k| < |\varepsilon_{k+1}|$, with $\varepsilon_k > 0$ for even k, and $\varepsilon_k < 0$ for odd k (see Figure 2.2a).

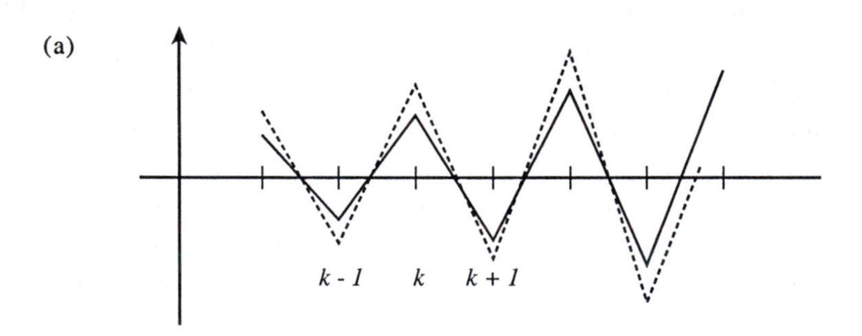

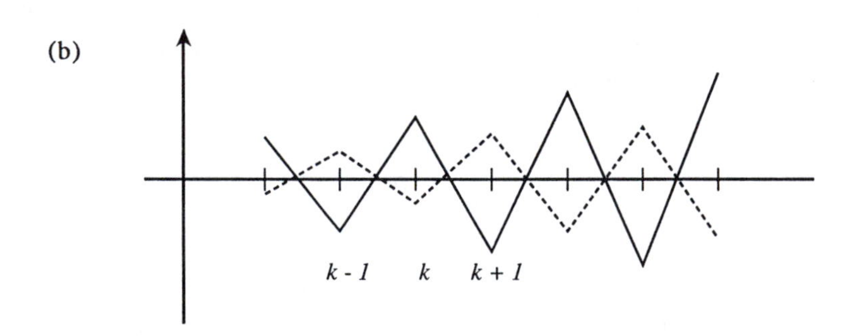

Figure 2.2. Numerical instabilities. (a) The initial perturbation $\psi^n_k - \psi^n_k$ at time t^n (solid line). The dashed curve shows $\psi^{n+1}_k - \psi^n_k$ at the next time t^{n+1}. (b) Initial perturbation (solid) $\theta^n_k - \theta^n_k$ and the value one time step later (dashed) using (2.30) subject to (2.31).

Taking

$$\psi_k^n = \hat{\psi}_k^n + \varepsilon_k,$$

and substituting into the finite difference equation we find

$$\psi_k^{n+1} = \hat{\psi}_k^n + \varepsilon_k - \sigma(\varepsilon_{k+1} - \varepsilon_{k-1}).$$

For even values of k the perturbation is positive, and $|\varepsilon_{k+1}| > |\varepsilon_{k-1}|$ so the change in ψ may be written in the form

$$\psi_k^{n+1} - \hat{\psi}_k^n = \varepsilon_k + \sigma(|\varepsilon_{k+1}| - |\varepsilon_{k-1}|).$$

The right side is positive for all values of σ. For odd values of k the perturbation is negative, and the change in ψ may be written as

$$\psi_k^{n+1} - \hat{\psi}_k^n = |\varepsilon_k| - \sigma(\varepsilon_{k+1} - \varepsilon_{k-1}).$$

Now the right side is negative for all values of σ. Thus, the perturbation increases in magnitude at each point and exceeds the steady state choice for any value of $\sigma = u\Delta t/2\Delta x$. It also follows that

$$|\psi_{k\pm 1}^{n+1}| > |\psi_{k\pm 1}^n|.$$

at each point. Repeated application (though more complicated because more zones are coupled with each increase in t) shows that the amplitude at each point continues to increase in magnitude without limit. Thus a perturbation, no matter how small, will be propagated forward in time with increasing amplitude. In a short period of time by (2.29) the effects of the perturbation will dominate the steady state solution. The method (2.29) is therefore unconditionally unstable.

The difference between an unstable method and a stable one (and vice versa) can often result from making a small change in the structure of a finite difference equation. For example, simply replacing $\psi^n{}_{k-1}$ and $2\Delta x$ by $\psi^n{}_k$ and Δx on the right side of (2.29) gives the conditionally stable representation for $u > 0$

$$\frac{\psi_k^{n+1} - \psi_k^n}{\Delta t} + u \frac{\psi_k^n - \psi_{k-1}^n}{\Delta x} = 0.$$

As another example, consider the diffusion equation (2.21) and the finite difference approximation to it

$$\frac{\theta_k^{n+1} - \theta_k^n}{\Delta t} = D \frac{(\theta_{k+1}^n - 2\theta_k^n + \theta_{k-1}^n)}{\Delta x^2}. \tag{2.30}$$

This should be compared with (2.22). Assuming a steady state solution, denoted by

$$\hat{\theta}_k^n$$

and imposing the perturbation of Figure 2.2, it follows that

$$\theta_k^{n+1} = \hat{\theta}_k^n + \varepsilon_k + (D\Delta t/\Delta x^2)(\varepsilon_{k+1} + \varepsilon_{k-1} - 2\varepsilon_k).$$

The last term in parentheses is opposite in sign to the first two terms, so the deviation from steady state is now reversed if $D(\Delta t/\Delta x^2)$ is too large. We also find that

$$\theta_{k\pm 1}^{n+1} - \hat{\theta}_{k\pm 1}^n$$

has reversed sign. If the time step is too large, then

$$|\theta_k^{n+1}| > |\theta_k^n|$$

and the method is unstable. However, if Δt is small (see Problem 2.8) then

$$|\theta_k^{n+1}| < |\theta_k^n|$$

and the method is conditionally stable (see Figure 2.2).

Problem 2.8

Assume that $|\varepsilon_k| = \varepsilon$ in Figure 2.2, and show that (2.30) is stable if

$$\frac{2D\Delta t}{\Delta x^2} \leq 1 \, . \tag{2.31}$$

Also derive expressions for $\theta^{n\pm 1}{}_{k\pm 1}$ and show that they are decreasing in magnitude.

In any numerical calculation perturbations may arise from grid boundary effects, irregularities in zoning, or from machine roundoff. In a scheme such as (2.29) these perturbations will always grow until eventually (often only in a few tens of cycles) they dominate the true solution. In schemes such as (2.30), the perturbations will be damped (as are perturbations of the differential equations) if suitable controls on the time step are observed, and a reasonable solution may be constructed.

Linearized Stability Analysis

Although the method of perturbations discussed above gives physical insight into the nature of some numerical instabilities, it is usually impractical when applied to a system of coupled finite difference equations. A more general approach, due to von Neumann, can be applied to systems of linear finite difference equations (for a detailed discussion see Richtmyer and Morton 1967). To show how this method works, consider a linear finite difference equation and boundary conditions for $F^n{}_k$, whose solution may be constructed from a linear combination of modes of wavelength λ and wave vector $2\pi/\lambda$:

$$F_k^n = \sum_{\lambda} A(\lambda)\xi(\lambda)^n e^{ik\theta(\lambda)} \, , \tag{2.32}$$

$$\theta(\lambda) = 2\pi\Delta x/\lambda \ . \tag{2.33}$$

The temporal growth of mode λ is governed by the magnitude of $\xi(\lambda)$. A stable solution can be shown to be one for which each mode is bounded in time; that is, $|\xi(\lambda)| \leq 1$ for all modes for all times. If $|\xi(\lambda)| > 1$ for any mode, the solution will be unstable. When (2.32) is substituted into the finite difference equation, a coupled set of algebraic equations results which connects the value of F at (n,k) to (usually) a few nearby spatial and temporal grid points. Then another way of expressing the stability constraint is to rewrite the finite difference equation in the form

$$F_k^{n+1} = \sum_i G_{ki} F_i^n \tag{2.34}$$

where G_{ki} is a matrix whose coefficients are all defined at time t^n, and which connect the spatial point k to its neighbors (nearest neighbors is common, but not necessary). Stability requires that the spectral radius of the eigenvalues of G_{ki} be less than unity. This implies that the determinant of the matrix G be less than unity: $det(G) \leq 1$.

A rigorous analysis of stability can only be carried out for relatively simple systems of finite difference equations, and should include the effects of boundary conditions. Unfortunately, most astrophysical problems involve nonlinear effects which cannot be handled by the method above. Nevertheless, stability analysis of linearized equations offers insight and guidance when working with more nearly realistic systems of equations. The von Neumann analysis leads to necessary, but not sufficient conditions for stability. In practice the best way to assess the stability of a specific finite difference method is to try it on a problem of interest. As noted earlier, perturbations will arise naturally in any computational grid and will expose unstable finite difference approximations if they exist. Fortunately, most instabilities show up immediately and dramatically.

To illustrate the concepts above, we consider a simple model which can be solved analytically, and for which finite difference approximations can be developed and solved with the aid of a hand calculator. The model is expressed in terms of an ordinary differential equation, but the results obtained from its analysis yield valuable insights into the behavior of finite difference approximations to partial differential equations as well. In particular, questions of centering and the relation of centering to stability and accuracy will be developed. Suppose that the temperature T of a material is found to relax from its initial value T_o to an equilibrium value T_{eq} on a time scale $\tau(T)$ according to the law

$$\frac{dT}{dt} = - \frac{T - T_{eq}}{\tau(T)} . \tag{2.35}$$

The relaxation time $\tau(T)$ may depend on the temperature; (2.35) is then quasi-linear in T. Although (2.35) is an ordinary differential equation, most of the methods discussed below and the insight and conclusions drawn from them apply to more complicated systems of partial differential equations. The simple model (2.35) has a number of interesting applications, including the thermalization of neutrinos, the thermal coupling of electrons and ions in an ionized plasma, and (if T is replaced by $E*$, the specific energy of stars in a globular cluster) the evolution of stellar distributions in globular clusters.

One of the major problems which arises in constructing finite difference approximations to a differential equation is centering of the dependent variables. Consider (2.35) integrated over the time interval t to $t+\Delta t$:

$$\frac{1}{\Delta t}\int_t^{t+\Delta t} \frac{dT}{dt'}dt' = \frac{T^{n+1} - T^n}{\Delta t} = \frac{1}{\Delta t}\int_t^{t+\Delta t} \frac{T(t') - T_{eq}}{\tau[T(t')]} dt'$$

$$T^{n+1} = T(t+\Delta t), \quad T^n = T(t).$$

The right side cannot be as easily reduced as can the left side, since T and τ will in general vary with time over the finite interval Δt. It is particularly important in the discussions to follow not to assume that $\Delta t \rightarrow 0$. Three ways of approximating the right side of this equation are considered below. In the first $T(t')$ is replaced by T^n, giving

$$\frac{T^n - T_{eq}}{\tau(T^n)}.$$

Alternately, $T(t')=T^{n+1}$ at the end of the time interval could be used, in which case the right side is

$$\frac{T^{n+1} - T_{eq}}{\tau(T^{n+1})}.$$

Finally, and more naturally, $T(t')$ could be approximated by an average value, in which case the integral would be

$$\frac{T^{n+1/2} - T_{eq}}{\tau(T^{n+1/2})},$$

where

$$T^{n+1/2} = \frac{1}{2}(T^{n+1} + T^n).$$

The first example above corresponds to backward time differencing, the second to forward time differencing and the third to time centering. Each approximation has its own advantages and disadvantages, as is seen in the following examples. Clearly more sophisticated averages may be employed, although we will not consider them here.

As our first example, assume that $T_{eq} = 0$ and $\tau(T) \equiv \tau_o$ is a constant. Then (2.35) becomes the simple linear differential equation

$$\frac{dT}{dt} = -\frac{T}{\tau_o} \qquad (2.36)$$

the solution to which, subject to the boundary condition $T(0) = T_o$, is given by

$$T(t) = T_o e^{-t/\tau_o}. \tag{2.37}$$

We now consider several finite difference approximations to (2.36), investigate their stability, and compare the resulting solutions to the exact solution (2.37).

A simple finite difference approximation to (2.36) is given by the forward time difference

$$\frac{T^{n+1} - T^n}{\Delta t} = -\frac{T^n}{\tau_o} \tag{2.38}$$

which when solved for T^{n+1} gives

$$T^{n+1} = T^n(1 - \Delta t/\tau_o). \tag{2.39}$$

This is an example of an explicit method since new values of T appear only on the left hand side of (2.38). Since (2.36) is assumed to contain no spatial dependence we apply (2.34) with the matrix G (in this example a scalar) restricted to satisfy $-1 \leq G \leq 1$ for stability

$$-1 \leq 1 - \sigma \leq 1,$$

where $\sigma \equiv \Delta t/\tau_o$. The right hand inequality simply implies that

$$\Delta t \geq \tau_o.$$

The left hand inequality leads to the requirement that

$$\Delta t \leq 2\tau_o.$$

While this constraint leads to stable solutions in the sense that the amplitude does not grow in time, it can be unacceptable because it overshoots the correct solution if the time step is too large (see Figure 2.3 for $\Delta t / \tau_o = 1.25$). It should be noted that overshoot is not an intrinsic property of the solution to the differential equation. Thus stability constraints are not always sufficient to guarantee that the solution is physically acceptable. If, for example, $T(t)$ represents the temperature, then clearly $T(t) \geq 0$ for all times. If we apply this constraint on physical grounds, then the matrix G is restricted to satisfy $0 \leq G \leq 1$. We then find the time step constraint

$$\Delta t \leq \tau_o. \tag{2.40}$$

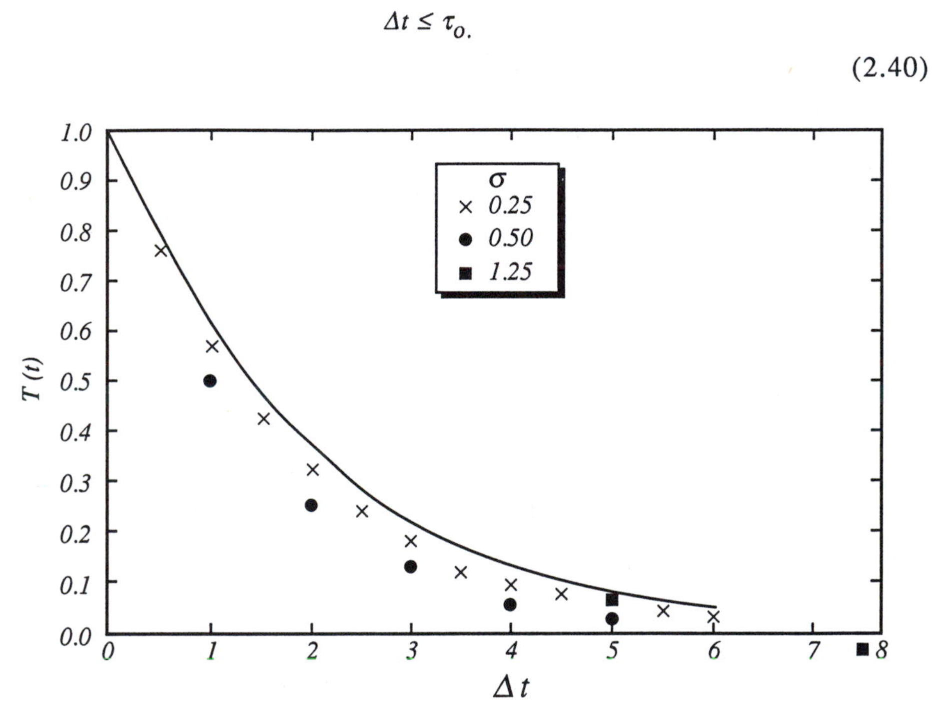

Figure 2.3a. Explicit solutions to (2.36) assuming $T(0) = 1$, and $\tau_o = 2$ for several choices of the time step. The exact solution is shown by the solid curve. The values of $\sigma = \Delta t/\tau_o$ are shown above each curve. Instability (manifested by oscillations of opposite sign in T) occur for $\sigma > 1$. All stable explicit solutions underestimate $T(t)$.

Thus the finite difference approximation (2.38) is conditionally stable, with the time step constraint (2.40). Figure 2.3a shows the exact solution (2.37), and several solutions obtained from (2.39) with different values of $\Delta t/\tau_o$. The stability constraint (2.32) has the simple physical interpretation that the computational time steps should not exceed the relaxation time τ_o, which sets the rate for the equilibration process. As shown in Figure 2.3a, the smaller the ratio $\Delta t/\tau_o$, the better the agreement between (2.39) and the exact solution. Finally, note that the solution for $\Delta t/\tau_o = 1.25$ is in fact oscillatory.

Next consider the forward time difference in which T^n is replaced by T^{n+1} on the right hand side of (2.38):

$$\frac{T^{n+1} - T^n}{\Delta t} = -\frac{T^{n+1}}{\tau_o}. \tag{2.41}$$

This represents one aspect of a fully implicit finite difference equation, because all values of T on the right hand side are evaluated

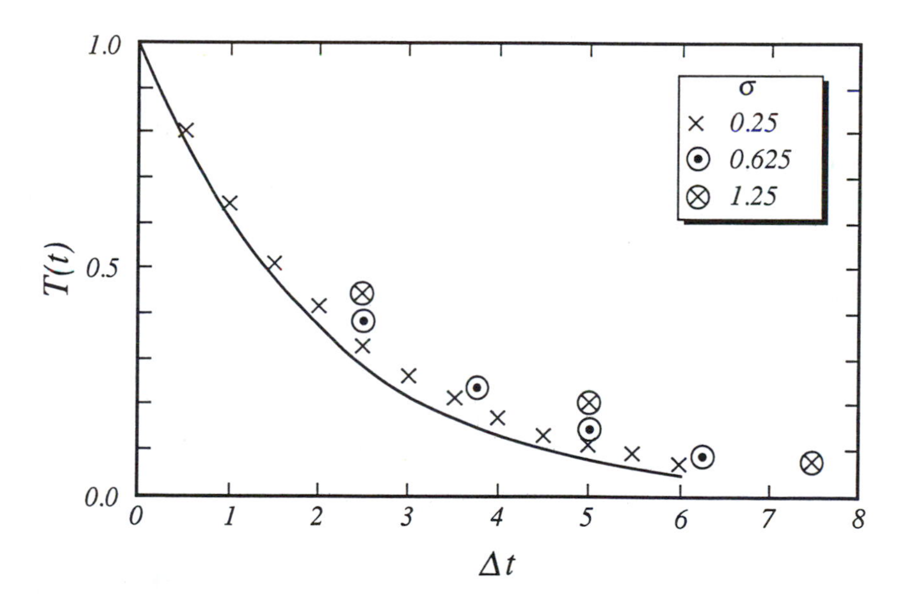

Figure 2.3b. Fully implicit solutions to (2.36) assuming $T(0) = 1$, and $\tau_o = 0$ for several choices of the time step. The exact solution is shown by the solid curve. All solutions overestimate $T(t)$. Note that the larger Δt is, the less accurate the result. As $\Delta t \to 0$, the implicit scheme approaches the explicit one.

at the new time. The solution to (2.41) is

$$T^{n+1} = \frac{T^n}{1+\Delta t/\tau_o} . \tag{2.42}$$

Applying (2.34) to (2.42) leads to the result $0 \leq G < 1$ for any (positive) value of $\Delta t/\tau_o$, which means that the finite difference approximation (2.42) is unconditionally stable. Figure 2.3b shows solutions (2.42) for several values of the time step. An advantage of the fully implicit scheme is that it places no time step limitations on the calculation. The disadvantage is a loss (sometimes substantial) in accuracy when the time step becomes large. If the exact shape of $T(t)$ as determined by (2.36) is important in a calculation, then the explicit scheme may be the best choice and the expense of running with a relatively small time step must be incurred. On the other hand, if the major importance of the relaxation process is to reduce T to T_{eq} on about the right time scale, then the implicit scheme may be adequate. Thus, the selection of a specific finite difference representation will often be guided by the physics of the problem under investigation. Note that the implicit scheme can give reasonable accuracy simply by reducing the time step (even though this is not needed for stability), as shown in Figure 2.3b.

The accuracy of the finite difference approximation (2.39) and (2.41) clearly depend on the magnitude of the time step. The order of accuracy of the methods can be determined by expanding each finite difference approximation in a Taylor series in Δt about the value t^n. Thus (2.39) becomes

$$\frac{1}{\Delta t}\left[T^n + \left(\frac{dT}{dt}\right)^n \Delta t + \frac{1}{2}\left(\frac{d^2T}{dt^2}\right)^n \Delta t^2 + O(\Delta t^3) - T^n\right] = -\frac{T^n}{\tau_o}$$

which reduces to

$$\left(\frac{dT}{dt} + \frac{T}{\tau_o}\right)^n = -\frac{1}{2}\left(\frac{d^2T}{dt^2}\right)\Delta t + O(\Delta t^2).$$

But the left hand quantity in parentheses vanishes according to (2.36), and we are left with terms of first order and higher in Δt. A similar analysis of (2.41) shows that it too is first order accurate.

Problem 2.9

Show that the time centered scheme given by

$$\frac{T^{n+1} - T^n}{\Delta t} = -\frac{1}{2}\frac{T^{n+1} + T^n}{\tau_o} \tag{2.43}$$

is a second order accurate finite difference approximation to (2.36), and that it is unconditionally stable. Compare the solution to (2.43) with the exact solution and the explicit solution for $\Delta t/\tau_o = 0.25$, and $\Delta t/\tau_o = 1.25$. Discuss the behavior of the solution for large time steps.

We now consider the more general equation (2.35) and assume that the relaxation time is given by

$$\tau(T) = K^{-1}\, T^{3/2} \tag{2.44}$$

which is characteristic of collisional relaxation of electrons on ions. Since $\tau(T)$ depends on T, the partial differential equation (2.35) is quasi-linear. For initial conditions assume that $T = 1$, $T_{eq} = 0.25$ and that $K = 0.2$. Figure 2.4a shows the exact solution, and several solutions obtained from the explicit finite difference approximation

$$\frac{T^{n+1} - T^n}{\Delta t} = -K\,\frac{(T^n - T_{eq})}{(T^n)^{3/2}}\,. \tag{2.45}$$

Note that the right hand side is evaluated at T^n. For $K\Delta t < 0.1$ the solution is reasonable, but for values larger than 0.2 oscillations occur whose amplitude increases with increasing $K\Delta t$.

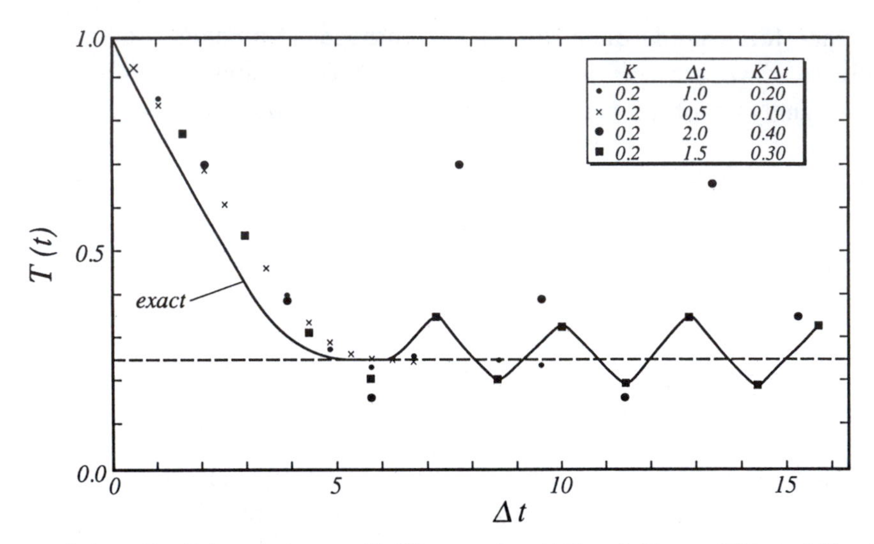

Figure 2.4a. Explicit solution to (2.45) assuming $T(0) = 1$, $T_{eq} = 0.25$ and $K = 0.2$. The exact solution is shown by the solid curve. The stability analysis of (2.38) suggests that a time step satisfying $K\Delta t < T^{3/2} \leq 1$ might be reasonable here. In fact, solutions with $K\Delta t > 1$ become unstable near $T = T_{eq}$. Note that $K\Delta t = 1$ also shows small amplitude oscillations near steady state.

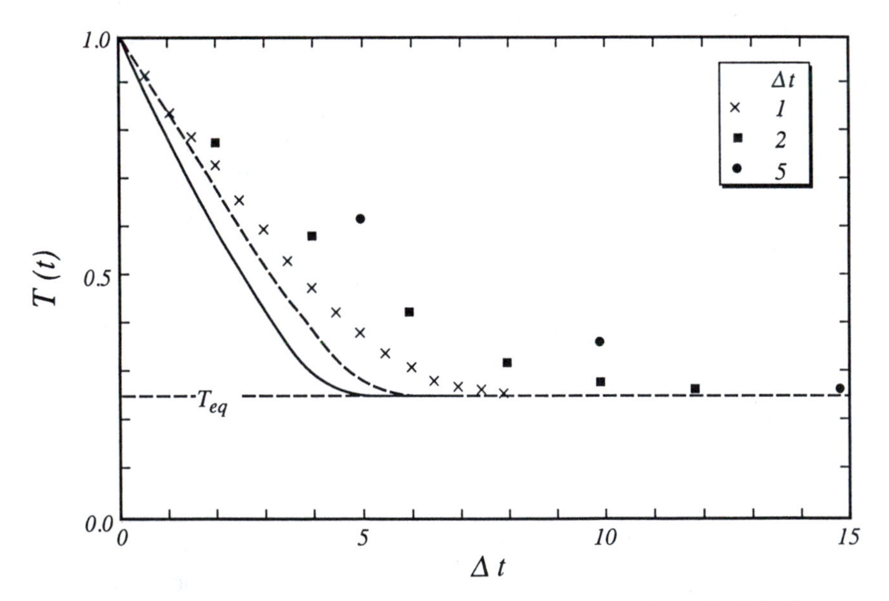

Figure 2.4b. Implicit solution to the linearized equation (2.46) for several values of the initial time step. The exact solution is shown by the solid curve, and the explicit solution with $K\Delta t = 1$ from Figure 2.4a is shown by the dashed curve. $K = 0.2$ for all cases.

Replacing T^n in the numerator of the right hand side of (2.45) with T^{n+1} gives an implicit finite difference approximation whose solution is

$$T^{n+1} = \frac{T^n + \alpha^n T_{eq}}{1 + \alpha^n} \tag{2.46}$$

where

$$\alpha^n = K\Delta t/(T^n)^{3/2}$$

which is stable for all Δt (see Figure 2.4). Note that the method converges to the equilibrium temperature for all time steps, but that the accuracy and the rate of convergence to equilibrium are poor for large time steps.

The finite difference approximations to (2.35) considered thus far have used the current values of the temperature, T^n, in the relaxation time $\tau(T)$. Both explicit and implicit schemes shown in Figure 2.3 and Figure 2.4 overestimate the temperature at any instant. This is not surprising since decreasing T implies, through (2.44) a reduced relaxation time, whereas the use of a fixed initial T for each cycle in (2.44) gives too large a relaxation time. One way to obtain improved accuracy is to use, for example, a time scheme obtained from (2.45) by replacing T^n on the right by $(T^{i-1}+T^n)/2$, and iterating on the temperature appearing in $\tau(T)$. One way to accomplish this is to take for each iteration during a single time step

$$T^i = \frac{T^n(1 - \sigma^{i-1}/2) + \sigma^{i-1}T_{eq}}{1 + \sigma^{i-1}/2}$$

where

$$\sigma^{i-1} = \frac{K\,\Delta t}{[\frac{1}{2}(T^{i-1} + T^n)]^{3/2}}. \tag{2.47}$$

We may take $T^{i-1} = T^n$ for the first iteration $i = 1$, in which case

$$\sigma^0 = \frac{K\, \Delta t}{(T^n)^{3/2}} .$$

Then T^1 is constructed from σ^0 and (2.47). This process is repeated so that the new value T^i found from T^n and T^{i-1} is used for T^{i-1} in the next iteration, and the process is continued until the solution converges in the sense that the difference $T^i - T^{i-1}$ becomes sufficiently small (if the time step is too large, the iterations will diverge, and the solution is unstable). We then set $T^{n+1} = T^i$, the final iterated value, and move to the next cycle obtaining T^{n+2} from T^{n+1} in a similar manner. The results of the iterative time centered scheme (2.47) shown in Figure 2.5 agree well with the analytic solution. We note that the iterative scheme is considerably more accurate than an implicit scheme modeled after the second order scheme (2.43); this suggests that including the effects of nonlinearity due to variable coefficients can be important in some cases.

Problem 2.10

Use the explicit scheme (2.45) and the implicit scheme (2.46) to solve (2.35) with initial conditions $T_o=0.10$, $T_{eq}=0.50$, and $K=0.2$. Compare the results with the exact solution.

In the examples considered above the time step has been assumed constant. Examination of the explicit solutions to (2.45) shown in Figure 2.4 indicate that large time steps produce reasonably accurate solutions for $T>T_{eq}$ even though instabilities result when $T \approx T_{eq}$. In fact, an acceptably accurate solution can be obtained by using an initially large time step which decreases as $T \rightarrow T_{eq}$. The stability analysis of the linear equation (2.36) indicates that Δt should be less than τ_o. This suggests as a tentative stabilty criterion

$$\Delta t^{n+1/2} = C\, \tau(T^n) = C\, \frac{(T^n)^{3/2}}{K}, \quad C<1.$$

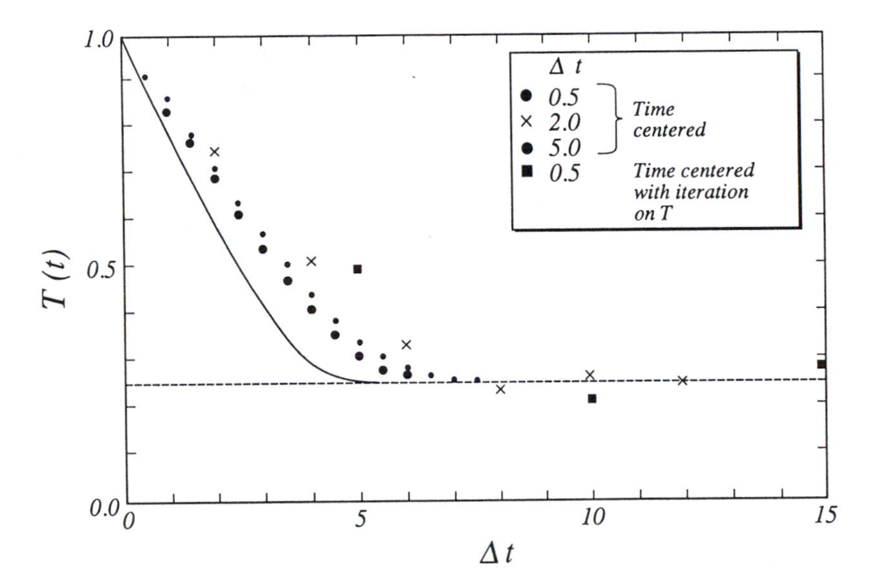

Figure 2.5. Time centered solution to (2.47) using the old temperature in $\tau(T)$ for three values of the time step. Also shown is a time centered solution including iterations on the temperature (circled dots). The iterative solution changes in the third figure only for the time steps equal to *0.5*, *1.0* and *2.0*. In all cases $T(0) = 1$, $T_{eq} = 0.25$ and $k = 0.2$.

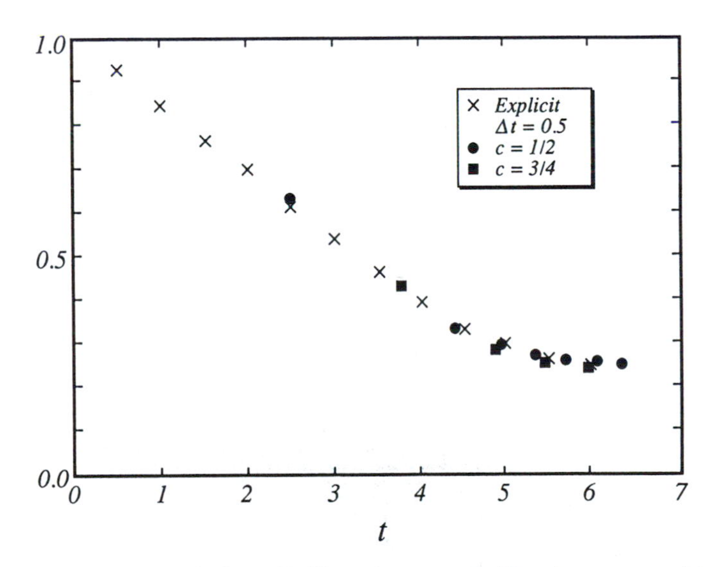

Figure 2.6. Explicit solution (2.45) using a variable time step with $C = 0.5$ and $C = 0.75$, and the explicit solution (x) with $\Delta t = 0.5$. The solution with variable time step converges rapidly to the steady state solution.

For the example shown in Figure 2.4 ($K=0.2$ and $T_o=1.0$) $\tau(1)=5$ and $\tau(0.25)=0.63$. Figure 2.6 shows the explicit solution (2.45) using the variable time step above with $C=0.5$.

Our discussion of stability will be concluded with an example involving spatial gradients as well as time derivatives. Consider a fluid moving in one dimension with spacially uniform velocity, u_o, whose density is $\rho(x,t)$. The rate of change of the density is given by

$$\frac{\partial \rho}{\partial t} + \frac{\partial (\rho u_o)}{\partial x} = 0 \tag{2.48}$$

with the initial condition

$$\rho(x,0) = f(x). \tag{2.49}$$

A solution of (2.48) which satisfies the initial condition (2.49) is

$$\rho(x,t) = f(x-u_o t). \tag{2.50}$$

Equation (2.48) is the mass continuity equation of hydrodynamics for a fluid moving with constant velocity. We will return to a more general form later. A possible finite difference approximation to (2.48) is the explicit scheme

$$\frac{\rho^{n+1}_{k+1/2} - \rho^{n}_{k+1/2}}{\Delta t^{n+1/2}} = -u_o \frac{\rho^{n}_{k+1/2} - \rho^{n}_{k-1/2}}{\Delta x_{k+1/2}}. \tag{2.51}$$

This should be compared with the unstable scheme (2.29). Since (2.51) is linear, the discussion leading to (2.32) applies and we find for each mode λ that (2.51) reduces to

$$(\xi_\lambda - 1) = -\sigma (1 - e^{i\theta_\lambda}) \tag{2.52}$$

where it is customary to define

$$\sigma \equiv \frac{u_o \Delta t^{n+1/2}}{\Delta x_{k+1/2}},$$

or

$$\xi_\lambda = (1 - \sigma) + \sigma e^{-i\theta}. \tag{2.53}$$

The stability constraint $|\xi_\lambda|^2 \leq 1$ is then given by

$$|\xi_\lambda|^2 = 1 - 2\sigma + 2\sigma^2 - 2\sigma(1 - \sigma)\cos\theta_\lambda \leq 1.$$

First consider the case $u_o<0$; it follows that $|\xi_\lambda|^2$ takes on its maximum value for $\cos\theta_\lambda=1$, and that $|\xi_\lambda|^2 > 1$ for all $\Delta t^{n+1/2}$. Thus the solution to (2.43) is unstable if $u_o < 0$. For $u_o > 0$ and $\sigma \leq 1$, the solution is stable. Finally if $u_o > 0$ and $\sigma > 1$ the solution is again unstable. Thus we find that (2.51) is acceptable only as long as $0 \leq \sigma \leq 1$:

$$u_o \Delta t^{n+1/2} \leq \Delta x_{k+1/2}. \tag{2.54}$$

The physical interpretation of (2.54) is that fluid cannot cross more than one zone in one time step. In particular $\Delta t^{n+1/2}$ must be chosen as the smallest value of $\Delta x_{k+1/2}/u_o$ when nonuniform zoning is employed.

Von Neumann's stability analysis may be extended to coupled systems of linear partial differential equations (for example, linearization of the inviscid fluid equations of hydrodynamics results in three coupled equations describing the rate of change of momentum, energy and density). For a system of N coupled partial differential equations, construct a vector V^n the elements of which are the N dependent variables at time t^n. Then if a matrix G can be constructed from the finite difference equations such that

$$V^{n+1} = \mathbf{G} V^n, \tag{2.55}$$

the system will be stable if for each eigenvalue λ_i of **G**

$$|\lambda_i| \leq 1 + O(\Delta t) \,. \tag{2.56}$$

The additional term $O(\Delta t)$ allows for possible real exponential growth in a solution.

Problem 2.11

Show that the forward time difference and centered space difference equation

$$\frac{\rho^{n+1}_{k+1/2} - \rho^{n}_{k+1/2}}{\Delta t} + u_o \frac{\rho^{n}_{k+3/2} - \rho^{n}_{k-1/2}}{2\Delta x} = 0$$

is unconditionally unstable. Assume for simplicity that Δt and Δx are constant, and use the von Neumann method.

The finite difference approximation (2.51) expresses the requirement that the time rate of change of the zone mass $\Delta m_{k+1/2} = \rho_{k+1/2} \Delta V_{k+1/2}$ is equal to the flux of mass into the zone, which guarantees that the total mass in the grid will be constant.

Problem 2.12

Find a finite difference approximation to (2.48) that is stable when $u_o < 0$. What is the stability constraint?

Problem 2.13

Show that (2.48) is only first order accurate.

Another interesting property of (2.51) can be seen by solving for $\rho^{n+1}{}_{k+1/2}$:

$$\rho^{n+1}_{k+1/2} = \rho^{n}_{k+1/2}\,(1-\sigma) + \sigma\rho^{n}_{k-1/2} \tag{2.57}$$

where σ is defined below (2.52). In the special case of uniform zoning and $\sigma=1$, (2.57) reduces to the exact solution

$$\rho^{n+1}_{k+1/2} = \rho^{n}_{k-1/2} \;;$$

that is, the mass simply moves ahead one zone per time step. In practice, however, σ will be less than one (particularly if nonlinearities are present, or if nonuniform zoning is used), and the transport of matter from zone to zone will be accompanied by a spreading, or diffusion, of the initial mass profile (2.49) Figure 2.7 shows this effect on an initial profile after four time steps. The grid is assumed to be uniform, and $\sigma=1/2$. This is an example of dispersion or damping error, which results in the decay of the amplitude of Fourier components, and occurs whenever $|\xi_\lambda| < 1$.

Two other types of error associated with finite difference equations are phase error and aliasing error. Phase error occurs when Fourier components propagate with different speeds than do components of the partial differential equations (Roach 1972, page 75). Aliasing errors are associated with the minimum wavelength ($\lambda \approx 2\Delta x$) of the computational grid. For example, in a finite difference model of fluid motion, nonlinearities couple different modes in such a way as to transfer energy from long wavelength modes to shorter wavelength modes. The process continues until the minimum wavelength $\lambda \approx 2\Delta x$ is reached. It has been found that if the difference scheme conserves energy, the energy which builds up in the λ_{min} mode eventually couples back into the longest wavelength modes of the problem. In real fluids, energy also cascades from long wavelength to shorter wavelengths, but it is eventually dissipated by molecular processes into thermal energy. Fortunately any type of damping in a numerical scheme (whether numerical or from physical effects such as viscosity) will usually accomplish the same thing, thereby removing alaising errors.

An experimental approach to stability of a nonlinear system which has been found to work well in practice will be adopted here. First, linearize the equations and determine the linear stability

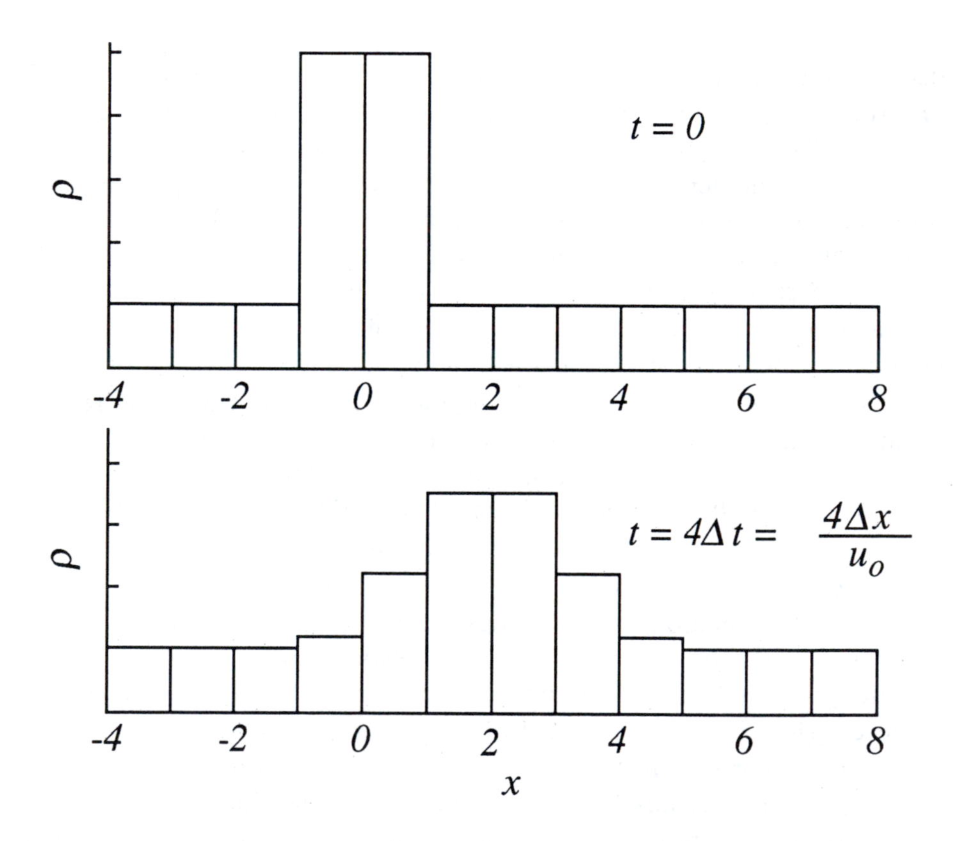

Figure 2.7. Eulerian diffusion on a uniform grid. The initial density maximum is $\rho_0 = 5.0$ and is centered at $x = 0$. After four time steps (b) the density peak has moved forward a distance $2\Delta x$, its maximum value is $\rho = 3.5$, and the mass originally confined to two zones has diffused so as to cover six zones.

constraints. If the linearized equations are unstable, the method must be modified or abandoned. If the linearized equations are stable, the analysis gives the maximum time step Δt_L. Usually a suitable time step for the nonlinear equations can be taken to be $\Delta t_I = c_I \Delta t_L$, with $c_I < 1$ (initially $c_I \approx 1/2$ is a reasonable choice). The finite difference equations are then used to solve test problems subject to the time step constraint $\Delta t \le \Delta t_I$. If a representative set of simple test problems can be computed with reasonable accuracy, then the method is in all likelihood stable. Fortunately most instabilities will evolve very rapidly and are easily recognized, as we shall see in later chapters.

Problem 2.14

Apply von Neumann's stability analysis to the diffusion equation (2.30) assuming D is a constant and verify the heuristic result (2.31).

2.3 Operator Splitting

The methods developed above can be used to express most of the partial differential equations encountered in physics in finite difference form. However we are still faced with the task of solving the finite difference equations, a process which in general involves solving N coupled nonlinear equation for N unknowns. One approach to a solution employs matrix inversion of the complete set of equations, augmented possibly by successive iterations to take nonlinear effects into consideration. A simpler approach is called operator splitting and is illustrated below.

Consider a typical partial differential equation for the variable F which can be written in the symbolic form

$$\frac{\partial F}{\partial t} = A + B + C + \ldots, \tag{2.58}$$

where the terms A, B, C, ... are in general functions of F, of other dependent variables, and of the independent variables x,t. They may also contain spatial derivatives of various orders. We shall attempt to replace (2.58) by successive operations which may be written symbolically as

$$\frac{\partial F}{\partial t} = A$$

$$\frac{\partial F}{\partial t} = B$$

$$\frac{\partial F}{\partial t} = C$$

$$\vdots$$

$$(2.59)$$

These equations should be thought of as representing FORTRAN processes in which the values which appear on the right side of the equal sign are evaluated and the results inserted into the variables on the left hand side. Thus (2.59) indicates that F is advanced in time in several steps, the first due to all processes denoted by A, the second by all processes B, and so forth until all the processes in (2.58) have been included. The dependent variables (in this case F) are changed due to each process, and the new value is used in the following process wherever it appears. Operator splitting represents an approximation whose accuracy, as might be expected, depends in part on the size of the computational time step. A straightforward generalization of this scheme can often be used for a system of N coupled partial differential equations where F in the example above is replaced by F_i, $(1 \le i \le N)$, and the processes $\{A, B, C, \ldots\}$ are replaced by the set $\{A_i, B_i, C_i, \ldots\}$.

The concept of operator splitting can be illustrated by a simple example involving an ordinary differential equation describing the rate of change of a variable T resulting from two processes, A and B, which may depend on T :

$$\frac{dT}{dt} = A(T)T + B(T)T. \tag{2.60}$$

The results obtained from the analysis of this equation yield insight into the properties of finite difference approximations to partial differential equations. The exact solution of this equation may be written in the integral form

$$T = T_o \, exp\left[\int_{t_o}^{t+\Delta t} A(T)dt' + \int_{t_o}^{t+\Delta t} B(T)dt'\right.$$

$$T_o = T(t_o) \tag{2.61}$$

For simplicity $A(T)$ and $B(T)$ will be assumed to be essentially constant over the time interval Δt. We now construct a finite difference representation of (2.60) in which the process A and B are operator split, and compare the results with a method which does not employ operator splitting (nonsplit method). These two methods can also be compared with the exact solution. First, consider the nonsplit explicit finite difference representation of (2.60)

$$\frac{T^{n+1} - T^n}{\Delta t} = (A + B)T, \tag{2.62}$$

whose solution is

$$T^{n+1} = T^n \, (1 + A\Delta t + B\Delta t). \tag{2.63}$$

The change in T after one time step Δt can be compared with the exact solution, and is

$$T^{n+1} = T^n \, [1 + (A+B)\Delta t + \tfrac{1}{2}(A+B)^2 \Delta t^2 + \ldots] \tag{2.64}$$

where terms of third and higher order have been dropped.

Next consider the operator split explicit scheme for (2.60) in which T is first changed due to the process A,

$$\frac{T' - T^n}{\Delta t} = AT^n \, , \tag{2.65a}$$

and then as a result of the process B,

$$\frac{T^{n+1} - T'}{\Delta t} = BT' . \tag{2.65b}$$

Eliminating T' from (2.65) leads to the solution (note that this corresponds to using the intermediate value of T' obtained from the first step as input to the second step)

$$\begin{aligned} T^{n+1} &= T'(1+B\Delta t) = T^n(1+A\Delta t)(1+B\Delta t) \\ &= T^n[1 + (A+B)\Delta t + AB\Delta t^2]. \end{aligned} \tag{2.66}$$

Two extreme cases can now be examined. In the first case assume that the processes A and B do not compete (both are of the same sign), and for simplicity take

$$A \, \Delta t \approx B \, \Delta t \approx \alpha. \tag{2.67}$$

Then the nonsplit scheme (2.63), the split scheme (2.66), and the exact solution (2.66) through second order in the time step give

$$
\begin{aligned}
T^{n+1}/T^n &= 1 + 2\alpha && \textit{nonsplit} \\
T^{n+1}/T^n &= 1 + 2\alpha + \alpha^2 && \textit{operator split} \\
T^{n+1}/T^n &= 1 + 2\alpha + 2\alpha^2 && \textit{second order exact}\,.
\end{aligned}
\tag{2.68}
$$

In the second case, the two processes compete in the sense that $A \approx -B$; for convenience define the net effect of the two processes in one time step by

$$(A + B)\Delta t = \delta \tag{2.69}$$

where δ is small compared with either $|A\Delta t|$ or $|B\Delta t|$. This corresponds to near steady state. Using (2.69) the schemes (2.62) and (2.66), and the exact solution (2.64) become

$$
\begin{aligned}
T^{n+1}/T^n &= 1 + \delta && \textit{nonsplit} \\
T^{n+1}/T^n &= 1 + \delta + A\Delta t\,\delta - (A\Delta t)^2 && \textit{operator split} \\
T^{n+1}/T^n &= 1 + \delta + \delta^2/2 && \textit{second order exact}\,.
\end{aligned}
\tag{2.70}
$$

The results (2.68) show that when the system is not near steady state operator splitting in this example tends to give more nearly accurate results than does the nonsplit scheme. However, (2.70) shows that operator splitting can give very poor results near steady state. For example, in the approach to equilibrium $\delta \rightarrow 0$, reasonable results could only be obtained using a prohibitively small time step whenever $|A\Delta t|$ and $|B\Delta t|$ were large. In general it is poor to split operators into strongly opposing terms.

Problem 2.15

Carry out an analysis similar to the above assuming a fully implicit operator split and a nonsplit scheme for (2.60). Compare the results to the exact solution in the two limits (2.67) and (2.69).

Similar results are found for a fully implicit scheme (as shown in Problem 2.15) where the condition $|A\Delta t|>1$ and $|B\Delta t|>1$ might be encountered in practice.

Operator splitting can result in substantial simplifications in solution method. For example, it may be desired to represent the change in material energy density, which we denote by u_m, as being due to the transport of u_m from point to point by material motion (advection) and by thermal diffusion due to temperature gradients: in one dimensional planar Cartesian coordinates this is described by the partial differential equation

$$\frac{\partial u_m}{\partial t} = -\frac{\partial (u u_m)}{\partial x} + \frac{\partial}{\partial x}\left(K \frac{\partial u_m}{\partial x}\right), \tag{2.71}$$

where K is the thermal conductivity (we can define $u_m = C_V T$ with C_V a constant to relate the diffusion term to a temperature gradient). An operator split approach to solving (2.71) would proceed by finding the change in u_m due to the advection term first, using for example an explicit approximation similar to (2.51), and then finding the further change in u_m due to diffusion, using for example a modification of (2.30) in which all quantities on the right hand side are to be evaluated at the new time (fully implicit diffusion equation). The values of u_m obtained from the first step in this process (advection) are used as the current values to be advanced by the diffusion step. In particular, if $K=K(u_m)$, then u_m represents the energy density following advection. The process can be summarized as shown below, where the input values, equation and output values are shown:

	Input	Output
$\frac{\partial u_m}{\partial t} = -\frac{\partial (u u_m)}{\partial x}$	u_m^n, u^n	u_m^A
$\frac{\partial u_m}{\partial t} = \frac{\partial}{\partial x}\left(K(u_m)\frac{\partial u_m}{\partial x}\right)$	u_m^A	u_m^{n+1}

Problem 2.16

Use operator splitting to solve (2.35) and (2.44) explicitly by taking as one process $A = -T/\tau(T)$ and as the other process $B = -T_{eq}/\tau(T)$; update T between steps and compare the results for $T_O=1.0$, $T_{eq}=0.2$ and $K\Delta t=0.2$ with the results in Figure 2.4. What is the steady state temperature, and how does it depend on Δt? Repeat the analysis starting with process B, and then using process A with updated temperatures.

Another example of operator splitting arises in the solution of the diffusion equation in more than one dimension. In two dimensional Cartesian coordinates, we have

$$\frac{\partial u_m}{\partial t} = -\frac{\partial}{\partial x}\left(K\frac{\partial u_m}{\partial x}\right) + \frac{\partial}{\partial y}\left(K\frac{\partial u_m}{\partial y}\right). \tag{2.72}$$

The gradient along the x direction gives a new energy u'_m which is then used in the solution for the gradient along the y direction to find u_m^{n+1}. An implicit scheme may be used for each step in the scheme.

Operator splitting is not always advisable (in fact situations can usually be found in which any splitting scheme fails). Consider, for example, the emission and absorption of radiation energy density u_R which can be described by

$$\frac{1}{c}\frac{\partial u_R}{\partial t} = k_P a T_e^4 - k_P u_R + \frac{1}{c}\frac{\partial u_R}{\partial t}\Big|_{other} \tag{2.73}$$

where k_P is the Planck mean absorption coefficient, c is the speed of light, and a is the radiation constant. The first term on the right of (2.73) represents the emission of photons by the material, and the second term represents absorption of photons by the material. Operator splitting as in (2.59) could be used to model (2.73), and if either the emission or the absorption term were to dominate the process the method would lead to acceptable results. However, if any region were to approach the steady state $\partial u_R/\partial t=0$ during a calculation, operator splitting these terms would produce inaccurate, or possibly unstable results, which might only be avoided by resorting to unacceptably small time step constraints. The difficulty arises because in steady state the emission and absorption terms in (2.73) may be extremely large and the difference between the two terms needed to accurately model the approach to equilibrium may be quite small. If (2.73) is to be solved for systems which include equilibrium regions where $u_R \approx aT^4$, then the emission and absorption terms should be differenced together. The processes described by $(\partial u_R/\partial t)_{other}$ may often be treated separately from the emission-absorption terms by operator splitting.

Equations (2.71) and (2.72) represent two different aspects of operator splitting. It is used in (2.72) to separately solve for the effect of a single physical process in several directions (equivalently, several dimensions), and in this way can be used to reduce a multidimensional problem to a succession of essentially one dimensional problems. In this type of operator splitting it is sometimes useful to alternate the order in which different directional gradients are solved, taking for example the x gradients first in odd cycles and the y gradients first in even cycles. In this way any tendency to biasing based on the initial choice is partially eliminated.

Equation (2.71) is an example where operator splitting can be used to separate the order in which different physical processes are solved. The first process, advection, could be incorporated with the hydrodynamic equations (of which it is in fact a part). Once the hydrodynamic equations have been solved, then the second process in (2.71), thermal heat conduction, may be treated. This might be done in conjunction with radiative processes.

The finite difference schemes developed in the following chapters may be combined using the method of operator splitting to describe a variety of astrophysical phenomena. Table 2.1 and Table 2.2 give an overview of typical splitting schemes. The first example is a two dimensional Eulerian hydrodynamics scheme including grid motion. It is summarized in Table 2.1, which lists physical processes on the left, and a brief explanation of their function on the right. The order of execution in an operator split approach corresponds to the succession of processes (left column) from top to bottom. With the exception of gravitation, the processes are given in finite difference

Table 2.1

Eulerian Hydrodynamics: Two dimensional (axially symmetric) explicit Eulerian hydrodynamic on a moving grid. The first column represents the primary physics unit (perhaps a subroutine, or a driver calling several subroutines. The second column describes the physical processes which are accomplished by the physics unit. The acceleration step represents the beginning of the computational cycle. The equation of state calculation represents the conclusion of the hydrodynamic cycle.

Physics Unit	Principal Function
Acceleration by external forces	Pressure force and gravitational force; centrifugal acceleration; performed over the entire grid.
Artificial viscous stresses	Acceleration and material heating due to artificial viscous stresses; performed over the entire grid.
Mechanical heating	Work associated with pressure forces; performed over the entire grid.
Scalar transport	Advection of mass, internal energy and angular momentum; radial (axial) gradients are advected first on even (odd) cycles for entire grid.
Momentum transport	Advection of axial and radial momentum; performed over the entire grid.
Gravitation	Solves Poisson's equation for the gravitational potential
Grid motion	Determine new positions of the radial and the axial grid elements based on instantaneous values of the grid velocities.
Time step	Calculates the time step for the next computational cycle.
Equation of state	Obtains the new pressure from the specific energy and mass density.

Table 2.2

Supernovae: One dimensional Lagrangian hydrodynamics and multigroup neutrino radiation transport. The first column gives the chapter where finite difference representations of the topics listed in the second column are discussed. The topics represented in the second column may represent a physics unit (a subroutine, or a driver which calls several other subroutines). The third column summarized the physical processes. The calculation of thermonuclear processes represents the beginning of the computational cycle.

Chapter	Physics Driver	Principal Function
10	Thermonuclear reactions	Nuclear fusion of fuel in the stellar core and in the stellar envelope; nucleosynthesis; material heating associated with fusion, performed over the entire grid.
10	Electron capture	Neutronization of heavy nuclei and material heating; heavy nuclei neutrino radiation emission (all energy groups); over the entire grid.
3	Equation of state	Pressure and temperature calculated from thermal energy, density and new compositions; heat capacity obtained from the energy, temperature and equation of state.
4	Hydrodynamics	Material motion due to pressure gradients and gravitational forces; shock heating and acceleration; over the entire grid. New coordinates are obtained from the velocities.
7	Neutrino Compton scattering	Electron-neutrino scattering; energy exchange and material heating. For all energy groups; over the entire grid.
6	Neutrino transport, emission and absorption	Neutrino spatial diffusion, emission and absorption; changes in composition associated with emission and absorption; radiation material heating; for each energy group over the entire grid.
7	Neutrino radiation hydrodynamics	Radiation acceleration; neutrino spectral shift due to material motion; over the entire grid.

form in Chapter 5. A second example is given in Table 2.2, which describes a one dimensional Lagrangian neutrino radiation hydrodynamics code capable of modeling supernova explosions. In this example the middle column gives the physics driver, which represents an entire process such as hydrodynamics. The right column describes the detailed processes covered by the physics driver. In this example, a finited difference representation for the physics governed by each driver may be found in the chapters denoted in the left hand column.

2.4 INITIAL VALUES AND BOUNDARY CONDITIONS

It was stated in Chapter 1 that the initial value and boundary conditions appropriate to a particular type of second order partial differential equation were related to its classification as hyperbolic, parabolic or elliptic (see Table 1.1). The finite difference methods discussed in the previous sections can be used to establish several of these relationships. Consider, for example, a typical hyperbolic equation (the wave equation)

$$\frac{\partial^2 F}{\partial x^2} - \frac{1}{c^2}\frac{\partial^2 F}{\partial t^2} = 0, \tag{2.74}$$

and the finite difference approximation to it (uniform spatial and temporal zoning will be assumed for simplicity):

$$F_k^{n+1} - 2F_k^n + F_k^{n-1} = \left(\frac{c\,\Delta t}{\Delta x}\right)^2 (F_{k+1}^n - 2F_k^n + F_{k-1}^n)\,. \tag{2.75}$$

This scheme may be rewritten in the form

$$F_k^{n+1} = 2(1-\sigma^2)F_k^n + F_k^{n-1} + \sigma(F_{k+1}^n + F_{k-1}^n)\,, \tag{2.76}$$

where $\sigma \equiv c\ \Delta t\ /\ \Delta x$, which shows that the four values $F^n{}_{k\pm 1}$, $F^n{}_k$ and $F^{n-1}{}_k$ are needed in order to advance F_k to the new time (see Figure 2.8). This four point scheme may be used to show that if $F^n{}_k$ and $F^{n-1}{}_k$ are known for all k, then F_k is uniquely determined for all $t > t^n$.

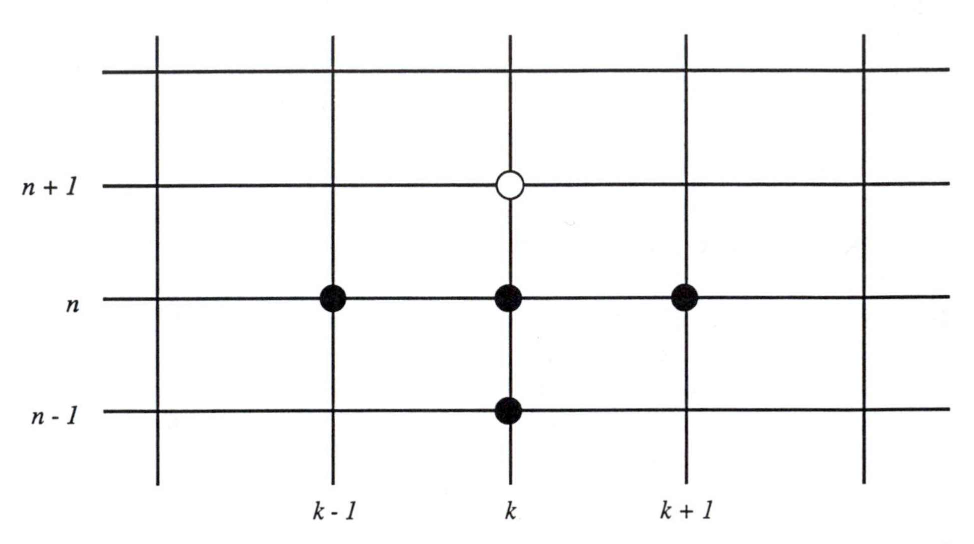

Figure 2.8. Four point scheme used to advance variables from time t^n to t^{n+1} for the wave equations (2.75). Solid points are knowns; open point is to be determined.

Furthermore, the value of F_k is determined by its previous values all of which lie within a limited spatial domain. For example, $F^{n+4}{}_k$ (see

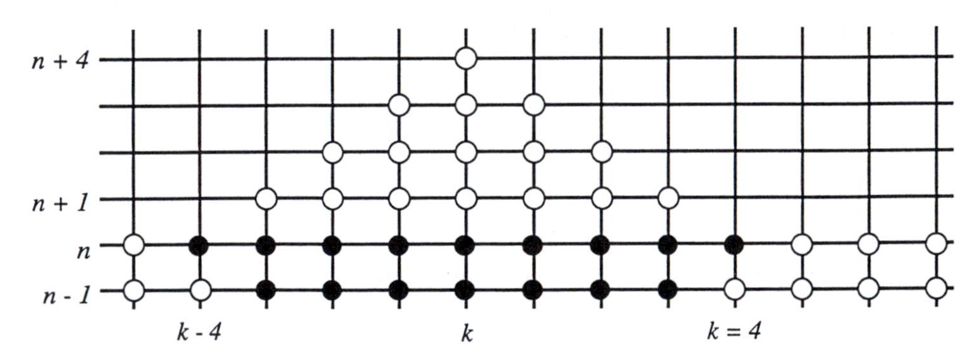

Figure 2.9. Domain of dependence of $F^{n+4}{}_k$ in the wave equation (2.75) assuming that F^n and F^{n-1} are known.

Figure 2.9) is completely determined by the values of $F^n{}_l$ $(k-4 \le l \le k+4)$ and $F^{n-1}{}_m$ $(k-3 \le m \le k+3)$. Specifiying $F^n{}_k$ and $F^{n-1}{}_k$ along a boundary curve is a finite difference representation of Cauchy initial data for the finite difference equation (2.76).

As a second example consider the finite difference representation (2.76) of the wave equation defined on a bounded spatial domain. Figure 2.10 shows a space-time grid whose spatial boundaries are denoted by $k=1$ and $k=K$. The initial Cauchy data ($F^n{}_k$ and $F^{n-1}{}_k$ for $1 \le k \le K$) combined with the Dirichlet boundary conditions ($F^m{}_1$ and $F^m{}_K$ for $n \le m$) uniquely determine all of the F_k for all times in the future. In this case Dirichlet conditions at $k=1$ and $k=K$ are needed to evolve F_1 and F_{K-1} to future times, replacing information which would propagate forward along characteristic curves through the points $k < 1$ and $k > K$ if the domain of Cauchy data were extended.

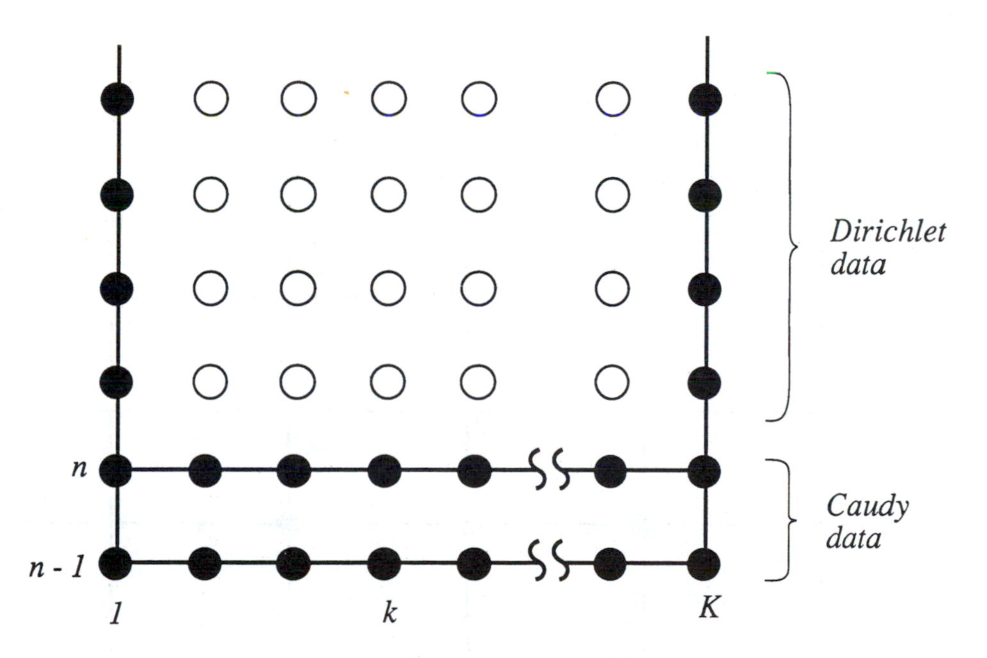

Figure 2.10. Filled circles at times n and $n-1$ give initial (Cauchy) data, and the filled circles at $k=1$ and $k=K$ give boundary (Dirichlet) data for (2.76).

Problem 2.17

Use a simple example to show that a combination of initial and boundary conditions on a closed curve on a space-time grid overspecifies the solution to the wave equation (2.76).

The diffusion equation is a typical example of a parabolic partial differential equation. Assuming a constant diffusion coefficient, the diffusion equation in one spatial dimension is

$$\frac{\partial T}{\partial t} = K \frac{\partial^2 T}{\partial x^2} . \tag{2.77}$$

It may be solved by using the explicit finite difference approximation

$$T_k^{n+1} - T_k^n = \frac{K \Delta t}{\Delta x^2} (T_{k+1}^n - 2\, T_k^n + T_{k-1}^n) ; \tag{2.78}$$

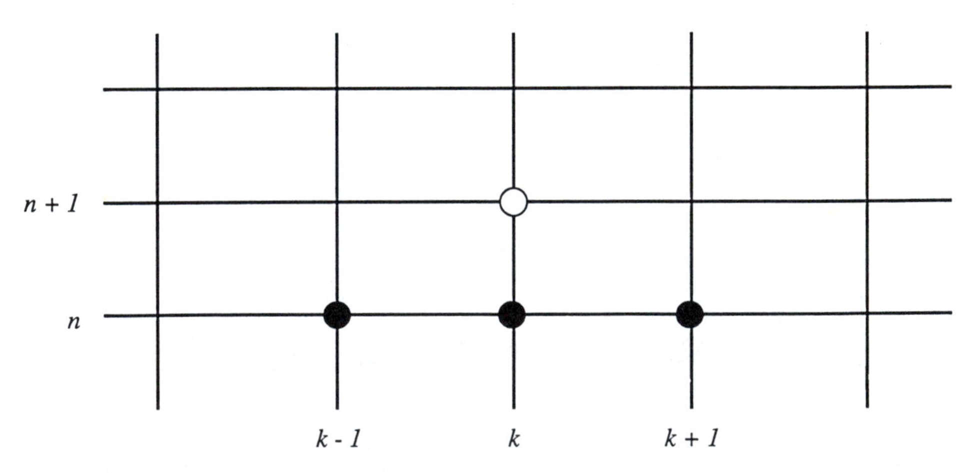

Figure 2.11. Explicit three point difference scheme used to solve the diffusion equation (2.77).

this is equivalent to the three point scheme (Figure 2.11)

$$T_k^{n+1} = \alpha (T_{k+1}^n + T_{k-1}^n) + (1 - 2\alpha) T_k^n ,$$

(2.79)

where $\alpha \equiv K \, \Delta t \, / \, \Delta x^2$ and shows that three values of T^n at time t^n are sufficient to specify $T^{n+1}{}_k$. Thus, for example, Dirichlet or Neumann conditions along an open boundary uniquely determine the solution to (2.79). Figure 2.12 shows an open boundary consisting of the spatial boundaries $k=1$ and $k=K$, and the initial (Dirichlet) conditions along $t = t^n$.

Problem 2.18

A fully implicit finite diference approximation to the diffusion equation (2.77) is given by

$$T_k^{n+1} - T_k^n = \frac{K \Delta t}{\Delta x^2} (T_{k+1}^{n+1} - 2 T_k^{n+1} + T_{k-1}^{n+1}) ,$$

(2.80)

and an acceptable set of boundary and initial conditions are the values $T^o{}_k$ for $1 < k < K$ and $T^n{}_1$ and $T^n{}_K$ for $n \geq 1$ (this initial data corresponds to the solid dots on the grid in Figure 2.12). Show that these data, in combination with the finite difference equation (2.80) determine a solution for future times.

Poisson's equation for the gravitational potential Φ is a typical example of an elliptical partial differential equation. In two spatial dimensions it has the form in Cartesian coordinates

$$\frac{\partial^2 \Phi}{\partial x^2} + \frac{\partial^2 \Phi}{\partial y^2} = 4 \pi G \rho$$

(2.81)

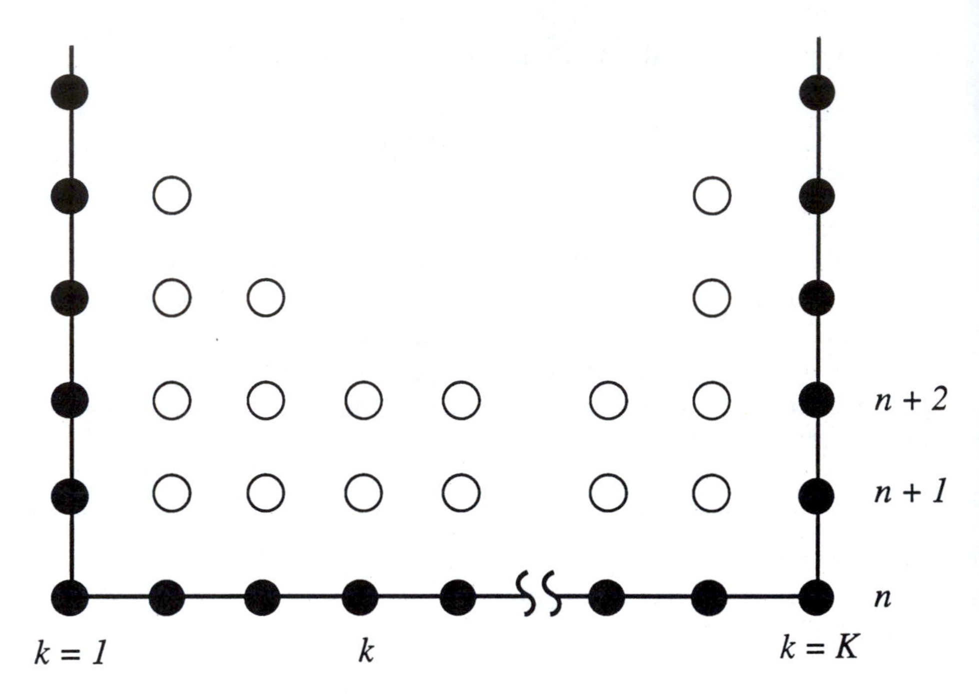

Figure 2.12. Initial conditions at time n and boundary conditions ($k=1$ and $k=K$) along an open boundary for the diffusion equation (2.79).

where ρ is the mass density and G is the gravitational constant. A finite difference representation of this equation is

$$\frac{\Phi_{k+1,j} - 2\Phi_{k,j} + \Phi_{k-1,j}}{\Delta x^2} + \frac{\Phi_{k,j+1} - 2\Phi_{k,j} + \Phi_{k,j-1}}{\Delta y^2} = 4\pi G \rho_{k,j} \,. \tag{2.82}$$

If $\rho_{k,j}$ is assumed to be known throughout the grid, (2.82) is equivalent to the five point scheme on the x-y grid shown in Figure 2.13. It should be noted that the potential is not known initially at any point inside the grid. Using this scheme it is readily shown that Cauchy initial conditions on an open boundary are inadequate to determine Φ throughout the x-y plane, and that a combination of Cauchy and Dirichlet or Neumann conditions along a closed boundary over determine the solution.

Problem 2.19

Use simple examples to prove that: (a) Cauchy conditions on an open boundary under determine the solution to (2.81); (b) a combination of Cauchy and Dirichlet conditions over determine the solution to (2.81).

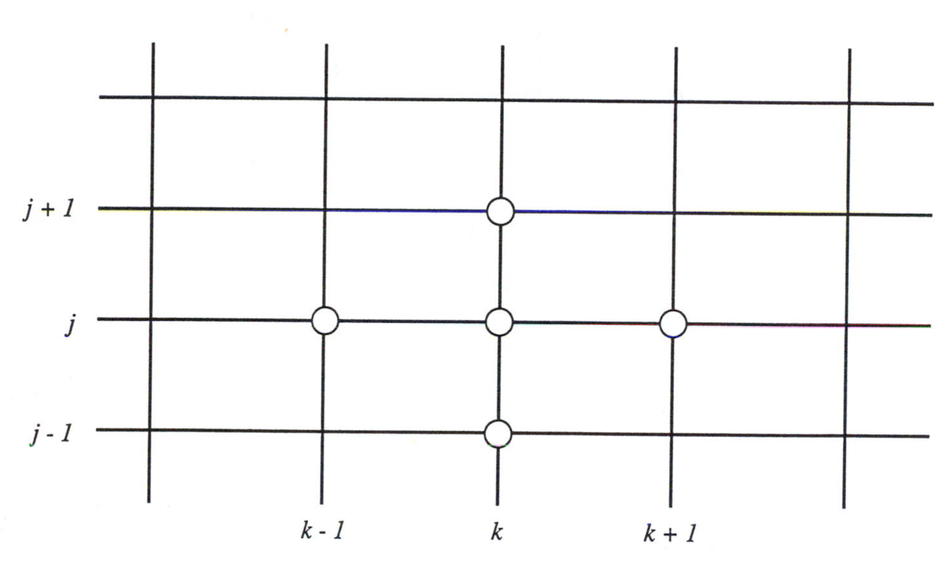

Figure 2.13. A five point scheme used to solve the finite difference representation (2.82) of Poisson's equation. Open circles correspond to the same time.

The appropriate boundary conditions for an elliptical partial differential equation such as (2.81) are Dirichlet or Neumànn conditions (or a linear combination of these) along a closed boundary. This may be established for the corresponding finite difference equation (2.82) for the two dimensional grid shown in Figure 2.14 as follows. Assuming that $2(K-2)$ values are known along the boundaries $k = 1$ and $k = K$, and that $2(J-2)$ values are known along the boundaries $j = 1$ and $j = J$, (2.82) is equivalent to $(K-2)(J-2)$ equations in $(K-2)(J-2)$ unknowns (the interior values of Φ). If the boundary values are properly posed, the system of equations may be solved for the potential at all interior points.

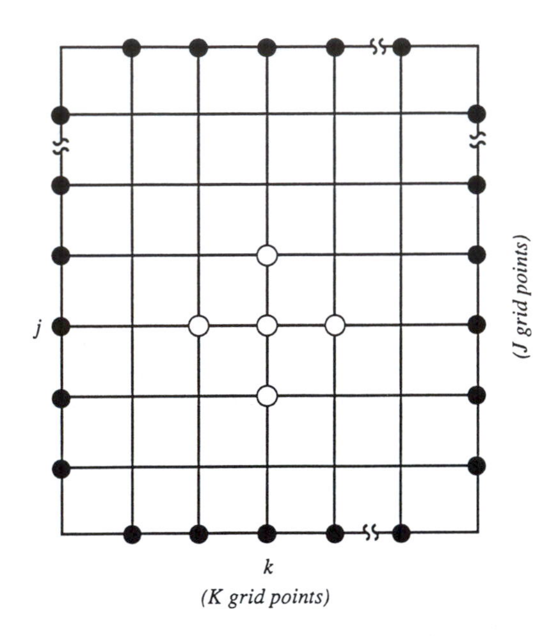

Figure 2.14. Boundary conditions required to solve an elliptical partial differential equation such as (2.81). The filled circles are boundary values, and the open circles are values to be found.

Simple applications of the finite difference methods such as those above may often be used to establish the type of initial and boundary condition that are appropriate to the solution of a given partial differential equation. They may also be used to illustrate other properties of the differential equations (Garabedian 1964). For example, using a finite difference approximation to the wave equation and the stability conditions appropriate to it, it is possible to show that disturbances can be propagated either forward or backward in time (the wave equation contains a form of time reversal invariance). The diffusion equation, on the other hand, does not satisfy time reversal invariance, as can be shown by considering, for example, the stability constraints on (2.79). Backward propagation is unconditionally unstable.

The accuracy of a solution to a set of partial differential equations depends in part on the order of accuracy of the approximations used to construct the solution, as well as on the consistency of the initial values imposed when the problem is generated (Richtmyer and Morton 1967). This poses a special problem for numerical solutions of partial differential equations (which usually does not arise when solving ordinary differential equations) describing complex processes which are introduced

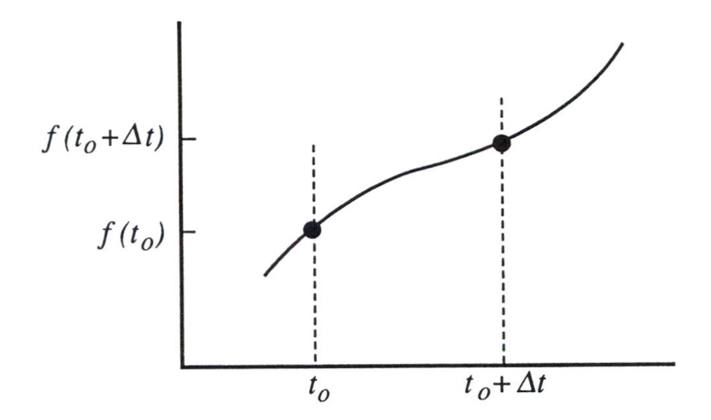

(a)

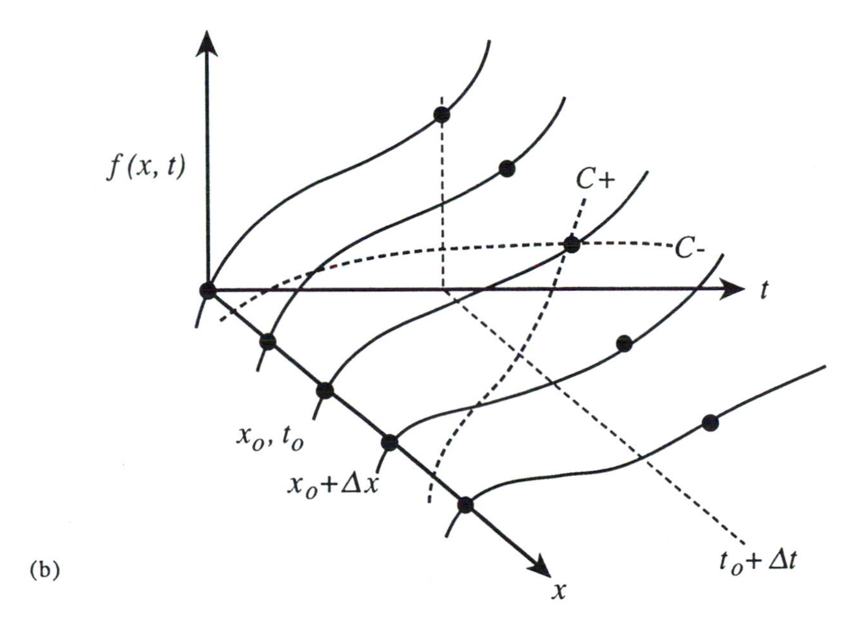

(b)

Figure 2.15. Effect of the accuracy of initial conditions on the accuracy of a numerical solution (a) to an ordinary differential equation, and (b) for a hyperbolic partial differential equation.

through initial conditions. Consider a regular solution $f(t)$ to an ordinary differential equation (Figure 2.15a). Given initial values for $f(t_o)$ and (in most cases) a finite number of its derivatives at t_o, the original differential equation and a Taylor series expansion about t_o may be used to construct an approximate solution at the new time $t_o+\Delta t$ where Δt is not necessarily small. In this case the accuracy of the solution is determined only by the numbers of terms which one is willing to compute. The same is not true for finite difference solutions to partial differential equations. Consider for example a hyperbolic partial differential equation whose solution is $f(x,t)$ (Figure 2.15b). Finite difference methods employ initial values $f(x, t_o)$ which are known at a finite number of points (in this case assumed to be equally spaced along the spatial axis). Given $f(x_o, t_o)$ at x_o, t_o the exact solution $f(x_o,t)$ at $t=t_o+\Delta t$ could be constructed, for example from a Taylor series expansion and the partial differential equation, if $f(x, t_o)$ were known at every point along the x axis lying between the two characteristics $C_\pm$ passing through the point $x_o, t_o+\Delta t$. In a finite difference scheme the interval along the spatial axis lying between the two characteristics may contain the points $x_o-\Delta x$ and $x_o+\Delta x$, but there are still an infinite number of other points in the interval which have not been specified, and upon which the exact solution depends in principle. The accuracy of any answer obtained using finite difference methods will depend in part on the level of consistency with which the initial values have been specified. For smooth initial conditions (say a smooth density distribution initially at rest), the accuracy of the results usually will be determined primarily by the accuracy of the finite difference equations. For more complex initial conditions, particularly in two or more dimensions, the consistency of the initial values may dominate the solution.

Problem 2.20

The mass continuity equation describing arbitrary fluid flow is given by

$$\frac{\partial \rho}{\partial t} + \nabla \cdot \rho \mathbf{v} = 0.$$

An intrinsic property of this equation is that the time rate of change of the total mass Δm of a fluid element is due to the net mass flux crossing the element's surface S:

$$\frac{d\,\Delta m}{dt} = -\int_S \rho \mathbf{v}\cdot dA\ .$$

Construct a finite difference representation of the one-dimensional planar mass continuity equation and use the three principles discussed in Section 2.1 to evaluate it.

Problem 2.21

Some quantities may be transported with material (advected) and may also diffuse through it (examples include magnetic fields or composition). Denoting such a quantity by the variable $\zeta(x, t)$, the partial differential equation for its time rate of change is

$$\frac{\partial \zeta}{\partial t} = -\frac{\partial(\zeta u)}{\partial x} + D\frac{\partial^2 \zeta}{\partial x^2}.$$

Assume that the velocity u and the diffusion coefficient D are constants, and consider the following finite difference representation of the partial differential equation:

$$\frac{\zeta_k^{n+1} - \zeta_k^n}{\Delta t} = -\frac{u}{2\Delta x}(\zeta_{k+1}^n - \zeta_{k-1}^n) + \frac{\zeta_{k+1}^n - 2\zeta_k^n + \zeta_{k-1}^n}{\Delta x^2}.$$

Assume that a steady state solution $\zeta^n{}_k = 0$ exists for $1 \leq k \leq K$, and apply a perturbation $\zeta^{n+1}{}_i = \varepsilon$ to it at a single point i. By imposing the conditions that the solutions be stable and that they not overshoot the physical solution, show that

$$\frac{u\,\Delta x}{D} \leq 1$$

$$\frac{2D\,\Delta t}{\Delta x^2} \leq 1.$$

(Hint: look at $\zeta^{n+1}{}_i$, $\zeta^{n+1}{}_{i+1}$ and $\zeta^{n+1}{}_{i-1}$). The first of these can be thought of as a Reynolds number for a zone. Show that these conditions require that Courant condition also be met:

$$\frac{u\,\Delta t}{\Delta x} \leq 1.$$

Notice that by solving the diffusion and advection terms together the space centered difference for the advection term (which by itself is unconditionally unstable) the latter is stablized. Explain what happens to the solution if $D \rightarrow 0$.

Problem 2.22

A finite difference approximation to the time derivative of a quantity $f(t)$ is said to converge to the differential form if

$$\frac{f^{n+1} - f^n}{\Delta t} = \frac{df}{dt} + O(\Delta t).$$

$$u^{n+1/2} = \frac{r^{n+1} - r^n}{\Delta t^{n+1/2}}$$

$$a^n = \frac{u^{n+1/2} - u^{n-1/2}}{\Delta t^n}$$

where the two time steps are not necessarily equal. Under what condition will the acceleration converge to d^2r / dt^2?

References

P. R. Garabedian, *Partial Differential Equations* (New York: Wiley, 1964).

R. D. Richtmyer and K. W. Morton, *Difference Methods for Initial-Value Problems*, Second Edition (New York: Interscience, 1967).

P. J. Roach, *Computational Fluid Dynamics* (Albuquerque, NM: Hermosa, 1972).

3

PROPERTIES OF MATTER

The systems of partial differential equations which are discussed in the following chapters relate the time evolution of macroscopic material variables such as density, energy and velocity to the spatial matter distribution. The inviscid fluid equations (1.1)-(1.3) discussed in Chapter 1 are one example of such a system of coupled equations. This system of equations is, however, neither mathematically nor physically complete. To complete the description, the physical characteristics of the matter itself must be specified. This requirement introduces the microscopic physics of matter, which tells us how changes in one material variable, such as the pressure, respond to macroscopic changes in other variables such as the density and temperature. In the example of inviscid fluid flow there are five partial differential equations which are given by (1.1)-(1.3). This set of equations contain six dependent variables; these are the velocity (three components), pressure, mass density and specific energy, which will be denoted by $\boldsymbol{v}$, P, ρ, and ε respectively. This system of equations is closed mathematically by specifying an equation of state, for example $P=P(\rho,\varepsilon)$. The additional equations (in the case of inviscid fluid flow), or constitutive relations, contain the material properties needed to completely determine the physical and mathematical behavior of the system.

The material composition may also enter into the equations of state, and may change as a result of reactions between constituents in the system (examples include ionization of atoms, dissociation of molecules, changes in nuclear abundances due to thermonuclear reactions or capture of particles). Therefore an appropriate set of additional variables $\{X_i\}$ specifying the composition of the material may be required as well as the partial differential equations characterizing their rates of change. The equations of state will then depend on the composition as well as on density and energy or temperature: $P(\rho,\varepsilon,X_i)$. In Section 3.1-3.3 expressions for the pressure and energy equations of state which do not depend on composition will be developed. A discussion of some of the processes which may result in a change in composition will be considered at the end of this chapter. The results developed below are of general applicability and can usually be applied to systems with variable composition.

The purpose of this chapter is to discuss some of the ways in which constitutive relations are used in numerical schemes. We shall not discuss techniques for constructing constitutive relations, nor shall we quote specific examples unless they have a direct bearing on the numerical methods. The omission of these matters does not imply that the construction and use of constitutive relations in numerical calculations is unimportant. In fact, the extent to which they are physically complete and consistent often sets the most severe constraint on the credibility of calculations. In simple cases the constitutive relations may be expressed by analytic forms, combinations of analytic forms, or computational algorithms which occur as a part of the numerical program. If the material properties are sufficiently complicated, or if high accuracy is required, tabular arrays may be appropriate. Elementary discussions of equations of state may be found in any standard text on thermodynamics or statistical physics (see for example: Reif 1965; Callen 1961; Landau and Lifshitz 1955) and in many texts on stellar astrophysics (for example: Cox and Guili 1968; Chiu 1968; Bowers and Deeming 1984; Shapiro and Teukolsky 1983).

3.1 EQUATIONS OF STATE FOR HYDRODYNAMICS

The inviscid fluid equations (1.1)-(1.3) can be supplimented by an equation of state whose independent variables are, for example, the mass density, ρ, and the specific internal energy, ε;

$$P = P(\rho, \varepsilon) \ . \tag{3.1}$$

The exact form of (3.1) will depend on the material under study, the density and specific energy range of interest in calculations, and must be determined by microphysics and thermodynamics. Here it is sufficient to assume that a functional dependence (3.1) exists. The inviscid fluid equations represent adiabatic processes, as follows from (1.3), which is just the first law of thermodynamics for isentropic processes. In this case (3.1) can be supplimented when necessary by the additional relation

$$s = s(\rho, \varepsilon) = const. \tag{3.2}$$

where s is the entropy per unit mass. If desired, the values of ρ and ε obtained from each calculational cycle can be inserted into (3.2) to monitor entropy conservation. It should be noted, however, that the entropy does not enter directly into the fluid equations.

The stability of the finite difference approximations to the fluid flow equations are governed in part by the adiabatic sound speed c_s (see Chapter 4), which is defined by

$$c_s^2 = \left(\frac{\partial P}{\partial \rho}\right)_s \ . \tag{3.3}$$

From (3.1) we form the quantity

$$c_s^2 = \left(\frac{\partial P}{\partial \rho}\right)_s = \left(\frac{\partial P}{\partial \rho}\right)_\varepsilon + \left(\frac{\partial P}{\partial \varepsilon}\right)_\rho \left(\frac{\partial \varepsilon}{\partial \rho}\right)_s \ . \tag{3.4}$$

Using the first law of thermodynamics, expressed per unit mass,

$$d\varepsilon = T ds + (P/\rho^2)\, d\rho \ , \tag{3.5}$$

it follows that $(\partial\varepsilon/\partial\rho)_s = P/\rho^2$; substituting this identity into (3.4) gives

$$c_s^2 = \left(\frac{\partial P}{\partial \rho}\right)_\varepsilon + \frac{P}{\rho^2}\left(\frac{\partial P}{\partial \varepsilon}\right)_\rho . \tag{3.6}$$

This expression can be evaluated analytically or numerically to obtain the adiabatic sound speed. It should be noted that the right hand side of (3.6) is expected to be nonnegative for normal fluids. Unstable fluids can have $\partial P/\partial\rho < 0$ and $c_S{}^2 < 0$, but these fluids are not in their thermodynamic ground state. While this is obvious on physical grounds, care must be exercised to guarantee that the equation of state used for computations does not violate this constraint.

For a gamma law equation of state [see (3.8) below] the adiabatic sound speed reduces to the simple expression

$$c_s^2 = \gamma(\gamma-1)\varepsilon .$$

Problem 3.1

The pressure and specific energy of a monatomic ideal gas are given by $P = \rho kT/\mu$ and $\varepsilon = (3/2)kT/\mu$, where μ is the mean molecular weight. Evaluate the adiabatic sound speed for this equation of state.

The fluid equations (1.1)-(1.3) supplimented by the equation of state of the form (3.1) constitute a set of equations which are energy based. It will be noted that the material temperature T does not appear in this representation. If an expression is needed for the temperature it must be obtained from a subsidiary equation of state of the form

$$T = T(\rho, \varepsilon) . \tag{3.7}$$

A simple but useful equation of state having the form (3.1) is the gamma law

$$P = (\gamma - 1)\rho\varepsilon, \tag{3.8}$$

where γ is a constant. For adiabatic processes, (3.8) implies an energy density relation of the form

$$\varepsilon = K_2 \rho^{\gamma-1}, \tag{3.9}$$

which, when combined with (3.8) gives the pressure density relationship

$$P = K_1 \rho^{\gamma}. \tag{3.10}$$

Here K_1 is a constant that depends on the entropy. The simple equation of state (3.8)-(3.10) does not accurately represent the properties of real matter over large ranges of density and temperature. However, for many materials encountered in astrophysics the matter is gas-like, and hence it is possible to find a constant value of γ such that these simple fomulae apply over a limited range of density and temperature with reasonable accuracy. For example, given a complicated equation of state in the general form (3.1) and (3.7), it is often reasonable to construct an effective gamma using the definition

$$\gamma \equiv 1 + \frac{P}{\rho\,\varepsilon}\,.$$

The use of an effective gamma in the equation of state above will usually be reasonable for hydrodynamic calculations where the fractional change in the density and the temperature (or equivalently, the density and specific energy) in a zone is small from one cycle to the next. In such cases γ will not be strictly constant, but its variation during one cycle may be ignored. The assumption of perfect gas behavior within a cycle simplifies the numerics.

3.2 DISSIPATION AND THE ENERGY EQUATION

When nonadiabatic processes are included along with hydrodynamics, additional constitutive relations appear. For example, when thermal conduction accompanies fluid flow, equations (1.1)-(1.3) must be combined with an equation describing thermal energy transport, which introduces the coefficient of thermal conductivity. Similarly, the equation of radiation transport requires opacities, and emission and absorption coefficients. Finally, changes in composition, due for example to ionization, dissociation or nuclear processes, may also occur. In most of these cases the constitutive relations which describe the material are most naturally obtained as functions of the density and temperature rather than as functions of density and thermal energy. The internal energy equation for the system is then most naturally written in the form

$$\frac{d\varepsilon}{dt} = \frac{P}{\rho^2}\frac{d\rho}{dt} + f(\rho, T), \tag{3.11}$$

The first term on the right side of (3.11) is the adiabatic work done on a fluid element, and $f(\rho,T)$ includes all nonadiabatic contributions. For most applications to be discussed in later chapters, the methods of operator splitting (Section 2.3) will be applied to (3.11). Then the adiabatic contributions to the change in energy will be treated with the hydrodynamics (that is we solve (1.1)-(1.3) using the equation of state (3.1)). The nonadiabatic processes may be calculated separately. In operator split form these nonadiabatic effects may be denoted by the expression

$$\frac{d\varepsilon}{dt} = f(\rho,T), \tag{3.12}$$

with additional constitutive relations expressing the coupling coefficients entering into $f(\rho,T)$ as functions of ρ and T. In this approach the hydrodynamics is the only source of changes in mass density and velocity. The processes represented by $f(\rho,T)$ are understood to change only the energy but not the density of the material (for example, $f(\rho,T)$ may include radiation diffusion, emission and absorption, magnetic field diffusion and thermonuclear energy release).

When finite difference approximations to specific processes (3.12) are examined and solution methods sought, we will find that it is desirable to express the energy equation in terms of either the specific energy ε or the temperature, but not both. For Lagrangian calculations (Chapter 4) either approach can in principle be used when formulating the left side of (3.12). In Eulerian calculations, on the other hand, the occurance of advection complicates the situation. In this approach the left side of (3.12) is rewritten in a form which contains terms representing the transport of material energy. If the Eulerian energy equation were expressed in terms of temperature, it is not clear how one would transport it hydrodynamically. Therefore, it is most natural to formulate the left side of (3.12) in terms of internal energy.

One approach might be to express the constitutive relations and the right hand side of (3.12) as functions of ρ and ε. This, however, is usually difficult for any but the simplest equations of state. Another approach, which will be discussed later, is the use of tabular equations of state. The approach which we have found to be most useful in practice uses a pseudo-specific heat C_V, which remains constant during each computational cycle. This is particularly useful when several dissipative processes are treated by the method of operator splitting, since an approximate value of the temperature is easily available at any point in the calculation.

Pseudo-Specific Heat

The left hand side of (3.12) will be rewritten as a function of the temperature T and a pseudo-specific heat C_V

$$\frac{d(\rho C_V T)}{dt} = \frac{d(\rho \varepsilon)}{dt} \tag{3.13}$$

where the pseudo-specific heat is defined by

$$C_V \equiv \varepsilon / T. \tag{3.14}$$

Throughout each computation cycle the pseudo-heat capacity C_V is taken as constant. It is to be emphasized that C_V is not a differential heat capacity in the usual thermodynamic sense, but is defined through (3.14) and an equation of state. It will be assumed, for the purposes of illustration, that an equation of state is available in the form

$$\varepsilon = \varepsilon(\rho, T). \tag{3.15}$$

Thus, given the specific energy (3.15) defines the temperature, and (3.14) is then used to obtain the pseudo-specific heat. The pressure when needed is also assumed to be available in the form

$$P = P(\rho, T)\,. \tag{3.16}$$

Equations of state based on energy and density could also be used.

The energy equation may now be solved in two stages using the method of operator splitting as discussed in Section 2.3. The first splitting solves the adiabatic terms in (3.11). The second splitting solves for the dissipative changes in the energy. In this stage the material velocities and densities are assumed to be constant, and (3.14) is used to rewrite (3.12) in the form

$$\rho C_V \frac{dT}{dt} = f(\rho, T) - \rho T \frac{dC_V}{dt}\,. \tag{3.17}$$

The dissipative energy equation (3.17) may also be solved in two stages using the method of operator splitting by rewriting (3.17) formally as the two step process

$$\rho C_V \frac{dT}{dt} = f(\rho, T) \tag{3.18a}$$

and

$$\rho C_V \frac{dT}{dt} = -\rho T \frac{dC_V}{dt}. \tag{3.18b}$$

The solution of (3.18a) will depend on the structure of $f(\rho,T)$, and will be the subject of later chapters. The solution of (3.18b) is independent of the nature of the dissipative processes, and will be considered next.

Solving for the Pseudo-Specific Heat

The quantity $f(\rho,T)$ appearing on the right side of (3.18a) may represent several processes which are to treated by the method of operator splitting. Then (3.18a) might be of the form (2.58), which would itself be rewritten as in (2.59), and would be solved in successive steps. The result of this process will be an updated value for the temperature, T', and thus for the specific energy $\varepsilon' = C_V{}^n T'$. The final step in the computational cycle involves solving (3.18b). One way to accomplish this would be to develop a finite difference representation for the time rate of change in the pseudo-heat capacity. However, C_V is not related to the equation of state in the usual thermodynamic sense [in fact, $C_V{}^{n+1}$ is only known once T^{n+1} and ε^{n+1} are known; see (3.14)]. Therefore, we seek another approach to solving (3.18b). The solution to the differential equation is obviously

$$C_V T = const. \tag{3.19}$$

The values of the density, temperature and the specific heat, after (3.18a) has been solved, are ρ^{n+1}, T' and $C_V{}^n$, and the new energy ε^{n+1} is

$$\varepsilon^{n+1} = C_V^n\, T'. \tag{3.20}$$

But we also need the values of $C_V{}^{n+1}$ and T^{n+1} corresponding to ε^{n+1}.

Using (3.20) it follows that

$$C_V^{n+1} T^{n+1} = C_V^n T' . \tag{3.21}$$

Thus, if the new temperature can be found, the new value of the heat capacity can be obtained from (3.21). The new temperature is found from the equation of state in the form (3.7):

$$T^{n+1} = T(\rho^{n+1}, \varepsilon^{n+1}) . \tag{3.22}$$

If the equation of state is only available in the form (3.15) as is often the case, then an iterative scheme must be employed. In this case, the current value of the temperature T' is taken as a first guess, $T^{(i)}=T'$, and the equation

$$\Delta\varepsilon \equiv \varepsilon(\rho^{n+1}, T^{(i)}) - \varepsilon^{n+1} \tag{3.23}$$

is evaluated with successive values of $T^{(i)}$. The value of $T^{(i)}$ which reduces $\Delta\varepsilon$ to a sufficiently small fraction of ε is then taken as T^{n+1}. Once the temperature is known, the new pressure can be obtained directly from (3.16). Once the density and specific energy have been obtained, an effective gamma law can be constructed using (3.8), and the energy based equation of state (3.1) may then be used for the next hydrodynamic step. It should be noted that (3.23) works well when the changes in density and temperature over the grid are small. In general, Newton-Raphson iteration works best, but may be more time consuming.

The equation of state of a nearly degenerate gas

$$\varepsilon = \varepsilon_o + \alpha T^2 / E_f = \varepsilon_o + A T^2$$

where E_f is the Fermi energy and α is a constant, may be used to illustrate the procedure above. In this case the equation of state can be inverted to obtain an analytic expression of the form (3.7) so iterations are not necessary. Suppose that the result of the processes described by (3.18a) is the temperature $T' = T^n + \Delta T$, where the superscript n denotes the initial value in a cycle. Then the final value of the specific energy is $\varepsilon^{n+1} = C^n{}_V(T^n + \Delta T)$, and the final value of the temperature can be found by inverting the equation of state in the form $T = [(\varepsilon - \varepsilon_o)/A]^{1/2}$. Inversion yields

$$T^{n+1} = [(\varepsilon^{n+1} - \varepsilon_o)/A]^{1/2} = \{[C^n{}_V(T^n + \Delta T) - \varepsilon_o]/A]\}^{1/2}.$$

Using (3.21) it follows immediately that the new heat capacity is

$$C_V^{n+1} = \frac{A^{1/2} C_V^n (T^n + \Delta T)}{\{C_V^n(T^n + \Delta T) - \varepsilon_o]^{1/2}}.$$

The heat capacity clearly changes with temperature, as expected.

Problem 3.2

Apply the solution method discussed above to the ideal gas equation of state from Problem 3.1, assuming that the temperature following (3.18a) is $T' = T^n + \Delta T$. Find the final temperature, T^{n+1}, heat capacity, C^{n+1}, and the pressure. Explain physically why your results are reasonable.

Problem 3.3

An approximate equation of state for a partially ionized hydrogen gas is given by

$$P = (1 + X)\frac{\rho k T}{m_H}$$

$$\varepsilon = \frac{3}{2}(1 + X)\frac{kT}{m_H} + \frac{\varepsilon_H X}{m_H}$$

where m_H is the mass of the hydrogen atom, $\varepsilon_H = 13.6\ eV$ is its ionization energy and X is the degree of ionization (equal to the ratio of the electron number density to the number density of ions plus neutral atoms) which can be found as the solution to the Saha equation (see Section 3.4) in the form

$$\frac{X^2}{1-X} = 4\,T_{(6)}^{3/2}\,\rho^{-1}\,exp(-0.158\,/\,T_{(6)})\,.$$

In this expression the density is in units of g/cm^{-3}, and the temperature is in units of $10^6\ K$. Set up a subroutine to construct the new temperature, specific energy and heat capacity defined by (3.14) assuming that the input temperature from the solution to (3.18a) is $T' = T^n + \Delta T$.

3.3 Degenerate Matter

White dwarfs, neutron stars and the cores of highly evolved stars contain ions which are nondegenerate, and electrons and other fermions which may be nearly degenerate. In some cases the fermions may be degenerate in part of the system, and nondegenerate on other parts of the system. When this occurs it may be advantageous to use a different formulation of the equation of state for the adiabatic processes than the one discussed in Section 3.1, and to modify the form of the hydrodynamic equations (1.1)-(1.3).

The primary source of difficulty is related to the difference in temperature dependence of the specific energy and the pressure of fermions in nondegenerate and degenerate matter. Figure 3.1 shows $\varepsilon(T)$ schematically for fermions whose chemical potential is denoted by μ. When $kT \ll \mu$, the fermions are essentially degenerate (complete degeneracy holds when $T=0$) and the energy is nearly independent of T, while for $kT \gg \mu$ the fermions are nondegenerate and the temperature dependence of ε will usually be proportional to T^{α}, where α is of the order unity. In the nondegenerate regime (3.14) is a reasonable way of numerically switching from temperature to energy. However, when $kT \ll \mu$, very high precision is

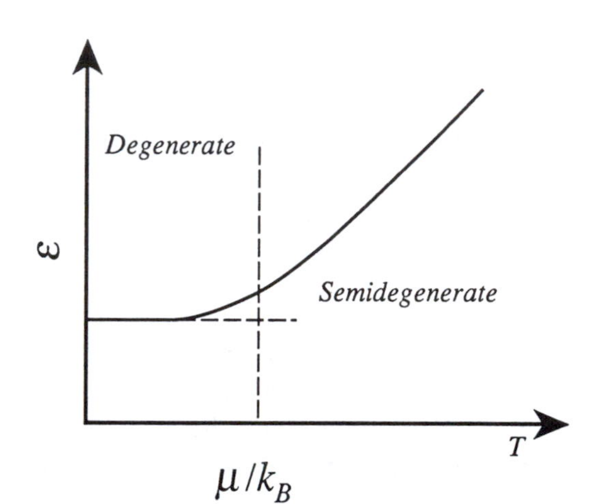

Figure 3.1. Specific energy of a system of fermions versus temperature. In the semidegenerate regime the specific energy varies weakly with temperature. The chemical potential is μ and k_B is Boltzmann's constant.

needed if ε is to be inverted to obtain T. The level of accuracy is limited largely by the accuracy with which a consistent ε and T can be obtained from the equation of state as described in the previous section. Numerical difficulties associated with the temperature dependence of the pressure may also arise.

For a system of fermions P/ρ has the schematic dependence on ρ and s shown in Figure 3.2 ($s_o < s_1 < s_2 < s_3$). The (cold) curve parameterized by s_o corresponds to $T=0$ and the equation of state along it reduces to

$$\varepsilon_c(\rho) \equiv \varepsilon(\rho, 0) \tag{3.23a}$$

$$P_c(\rho) \equiv P(\rho, 0) \tag{3.23b}$$

The equation of state along the cold curve corresponds to complete degeneracy, and gives the specific energy and pressure due to compression. In particular, there are no thermal components in (3.23). The first law of thermodynamics (with constant s) may be used to show that

$$d\varepsilon_c = \frac{P_c\, d\rho}{\rho^2} \tag{3.24}$$

along the cold curve. Thus knowing $\varepsilon_c(\rho)$ and $P_c(\rho)$ is equivalent to (3.24), and the latter need not be solved to find the adiabatic change in the cold energy.

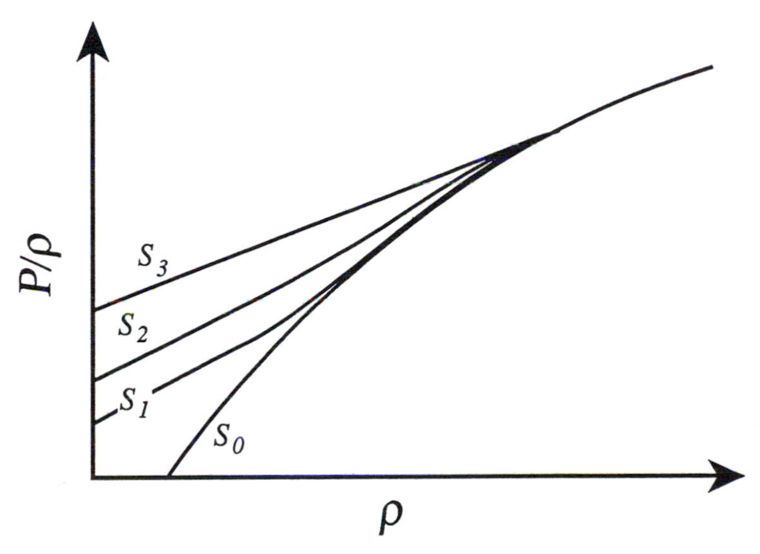

Figure 3.2. Schematic fermion equation of state P/ρ versus ρ. The entropy S is a constant along each curve and $S_o < S_1 < S_2 < S_3$.

The total specific energy $\varepsilon(\rho,T)$ contains compressional energy $\varepsilon_c(\rho)$ and a thermal energy component $\varepsilon_{th}(\rho,T)$ defined by

$$\varepsilon_{th}(\rho,T) \equiv \varepsilon(\rho,T) - \varepsilon_c(\rho). \tag{3.25a}$$

Similarly, a thermal pressure can be defined by the difference

$$P_{th}(\rho,T) \equiv P(\rho,T) - P_c(\rho). \tag{3.25b}$$

By definition, when $T=0$ the thermal energy and pressure vanish. If the definitions of the cold and thermal energy and pressure are substituted into the first law of thermodynamics and (3.24) is used, it follows that

$$\frac{d\varepsilon_{th}}{dt} = T ds + P_{th}(\rho, \varepsilon_{th}) \frac{1}{\rho^2} \frac{d\rho}{dt}. \tag{3.26}$$

This is an equation for the thermal energy, expressed in terms of the independent variables (ρ, ε_{th}). The hydrodynamic cycle increments ρ and ε_{th} from which we may then construct the total energy and total pressure:

$$\varepsilon = \varepsilon_c + \varepsilon_{th}$$
$$P = P_c + P_{th} . \tag{3.27}$$

Problem 3.4

Assume that the cold pressure and the thermal pressure are given by

$$P_c = K\rho^{\gamma_c}$$
$$P_{th} = (\gamma - 1)\rho\varepsilon_{th} . \tag{3.28}$$

Construct the cold energy curve, the total pressure and the total energy. Notice that two effective gammas are used to define the cold and thermal pressure.

The nonadiabatic processes described by (3.12) change the material temperature at constant density. This suggests that the heat capacity be expressed in terms of the thermal energy:

$$C_V \equiv \varepsilon_{th} / T \,, \tag{3.29}$$

which should be compared with (3.14). Using (3.25a) in (3.12)

$$\left(\frac{\partial \varepsilon}{\partial t}\right)_\rho = \left(\frac{\partial \varepsilon_{th}(\rho,\, T)}{\partial t}\right)_\rho + \left(\frac{\partial \varepsilon_c(\rho)}{\partial t}\right)_\rho = \frac{\partial \varepsilon_{th}(\rho, T)}{\partial t}$$

and the energy equation analogous to (3.13) for the thermal energy becomes

$$\rho \frac{d\,\varepsilon_{th}}{dt} = \rho \frac{d\,(C_V\, T)}{dt} = f\,(\rho, T) \,. \tag{3.30}$$

The change in the thermal energy may now be expressed as in (3.17), and may be solved by a slight modification of the method discussed for the total internal energy equation (Section 3.2).

Problem 3.5

Describe how the analysis following (3.17) should be modified when (3.17) is applied to the thermal energy rather than the total energy and the heat capacity is defined as in (3.29).

3.4 THERMODYNAMIC FREE ENERGY

The equations of state (3.15)-(3.16) can be quite complicated for matter in typical astrophysical calculations because of various reactions which change the composition, or because of complications associated with particle-particle interactions. Examples of reactions include: molecular dissociation; excitation and ionization of atoms; nuclear excitation and dissociation; and a variety of reactions which change the composition (nuclear and neutrino interactions, for example). A convenient method of treating some of these processes for an ideal gas uses the thermodynamic free energy per baryon F which, for a gas at temperature T consisting of N species whose mass fractions are X_i, is defined by (Landau and Lifshitz 1958; Callen 1961)

$$F(\rho,T) = \frac{kT}{m_H} \sum_{i=1}^{N} X_i \left[\frac{\varepsilon^o_i}{kT} + \frac{1}{A_i} \ln\left(\frac{X_i \rho \alpha}{g_i A_i^{5/2} T^{3/2}}\right)\right]. \tag{3.31}$$

The mass fractions are defined by

$$X_i \equiv m_i n_i / \rho, \quad \sum_{i=1}^{N} X_i = 1, \tag{3.32}$$

where m_i is the mass (equal approximately to $A_i m_H$) of the ith species, n_i is its number density, and A_i is the atomic or molecular weight. The statistical weight is denoted by g_i, and $\varepsilon^o{}_i$ is the binding energy per nucleon of species i. The advantage of formulating the equations of state in terms of the free energy is that the pressure and specific energy obtained from it using the first law of thermodynamics in the form

$$dF(T, \rho, X_i) = s\, dT + \frac{P}{\rho^2} d\rho + \sum_i \frac{\mu_i}{m_i} dX_i \tag{3.33}$$

and the identities

$$P = \rho^2 \left(\frac{\partial F}{\partial \rho}\right)_T$$

$$\varepsilon = -T^2 \left[\frac{\partial}{\partial T}\left(\frac{F}{T}\right)\right]_\rho$$

$$s = (F - E) / T \tag{3.34}$$

will be thermodynamically consistent, whereas independent approximations for P and ε may not be. Finally, we note that the chemical potential of species i, μ_i, follows from (3.33) and the relation

$$\mu_i = m_i \left(\frac{\partial F}{\partial X_i}\right)_{\rho, T} \qquad i = 1, 2, ...N . \tag{3.35}$$

Problem 3.6

Evaluate the chemical potential for an ideal gas whose free energy is given by (3.31).

3.5 SAHA EQUATION

The composition parameters which are defined in (3.32) and which appear in equations of state, opacities, and reaction rates are generally a function of density and temperature. Their instantaneous values are usually required as input for constitutive relations appearing in the dynamic equations. According to the principles of statistical mechanics (see for example Landau and Lifshitz 1958; Callen 1961), whenever a reaction of the general type

$$N_A A + N_B B \leftrightarrow N_C C + N_D D \tag{3.36}$$

occurs in thermodynamic equilibrium between species of the type A, B, C, and D, the chemical potentials μ of the species satisfy the relation

$$N_A \mu_A + N_B \mu_B = N_C \mu_C + N_D \mu_D . \tag{3.37}$$

The stochiometric coefficients (such as N_A) are integers representing the number of "particles" of each species (such as A) entering into the reaction. The chemical potentials generally depend on the temperature, density and composition. Finally the reaction (3.36) may occur only subject to certain constraints, such as electric charge and baryon number conservation, and equations expressing these requirements must also be written down. Thus an equation (or a series of such equations if more than one reaction occurs) of the type (3.37) and equations for the constraints can be used to solve for the compositions at a given density and temperature.

Ionization Equilibrium

A frequently encountered example of the equilibrium relations (3.36)-(3.37) is the ionization of an atom of atomic number N and atomic weight A which may be written in the more familiar form

$$A_i \leftrightarrow A_{i+1} + e^- \tag{3.38}$$

with the constraint equations

$$n_e = \sum_{i=1}^{N} Z_i n_i \tag{3.39}$$

$$n_A = \sum_{i=0}^{N} n_i \, . \tag{3.40}$$

Here subscript i denotes the i times ionized state of the atom, Z_i is the charge of the ion, n_e denotes the number density of the electrons, n_i the number density of ions in the ionization state i and n_A is the total number density of ions (including neutrals). The equilibrium equation (Saha equation) is obtained from (3.37) and the chemical potentials for an ideal gas, and is easily shown to be

$$\frac{n_{i+1}\, n_e}{n_i} = \frac{g_{i+1}\, g_e}{g_i} \frac{(2\pi m_e kT)^{3/2}}{h^3} e^{-\chi_i / kT} . \tag{3.41}$$

Here χ_i is energy required to remove an electron from the i times ionized ion, $g_e = 2$ is the statistical weight of the electron, and h is Planck's constant. Ionization energies may be found in Lang (1980). For simple ionization

$$g_i = 2J + 1 \tag{3.42}$$

where J is the total angular momentum of the ion. Equations (3.39)-(3.41) represent $N+1$ equations for the $N+1$ unknowns n_i.

Problem 3.7

Derive (3.41), and solve it for the simple case of hydrogen ($\chi = 13.6\ eV$) and an electron pressure $P_e = 10\ dynes/cm^2$. Assume that the electron pressure and electron number density satisfy the ideal gas equation of state $P_e = n_e kT$. Plot the relative concentrations of neutral atoms $n_0/(n_0+n_1)$ and ions $n_1/(n_0+n_1)$ versus temperature in the region of ionization.

Solution of the Saha Equation

Equations (3.39)-(3.41) are a system of nonlinear algebraic equations which must in general be solved by iterative methods. Since this must be done for each zone every cycle of a computation, it can represent a significant fraction of the run time. The following efficient algorithm has been found to work well in practice. It is a simplification of an approach discussed by Zel'dovich and Raizer for equilibrium ionization. Using the known value of n_A, adopt as the first guess for the electron number density for the ionization level i

$$n_e^{(0)} = (i + 1/2)\, n_A \, . \tag{3.43}$$

This guess is used in (3.41) beginning with $i = 0$ and a guess (for example, the old value) for the number density of neutral atoms, $n_0(0)$, to obtain a first guess at the ionization level densities in the form

$$n_{i+1}^{(1)} = \frac{n_i^{(0)}}{n_e^{(0)}} \frac{g_{i+1}}{g_i} C\,(kT)^{3/2} e^{-\chi_i / kT} \, . \tag{3.44}$$

Here $C = 6.0388 \times 10^{21}\ cm^{-3}$ if kT is measured in eV. Using (3.44) an improved estimate of the electron number density is given by

$$n_e^{(1)} = \frac{n_A}{\sum n_i^{(0)}} \sum Z_i\, n_i^{(0)} \, . \tag{3.45}$$

The first iteration for the electron number density (3.45) may now be substituted back into (3.44) to obtain an improved estimate of the ion number densities. A few iterations usually give convergence of n_e and n_i to a few parts in 10^4. Finally, for low temperatures ($C(kT)^{3/2} < 0.4$) where $n_0 + n_1 = n_A$ and $n_e = n_1 << n_0$, the Saha equation may be expanded to yield a solution which does not require iteration:

$$n_1^2 = n_A\, C\,(kT)^{3/2} e^{-\chi_0 / kT} \, . \tag{3.46}$$

Nuclear Photodissociation

Another type of reaction, photodissociation, to which the methods above apply, occurs in models of the late stages of evolution of stellar cores. When material temperatures reach about $8\ MeV$, the thermal energy of nucleons inside nuclei is comparable to their binding energy in the nucleus. At these high temperatures a heavy nucleus

${}_{A}N^{Z}$ of atomic number Z and atomic weight A will photodissociate into alpha particles, He^{4}, and free nucleons. If the heavy nucleus in Fe^{56}, then

$$Fe^{56} \rightarrow 13He^{4} + 4n \tag{3.47}$$

$$He^{4} \rightarrow 2p + 2n \,. \tag{3.48}$$

Applying the methods above to these reactions leads to an equation of the form (3.41) but with some of the number densities of the nuclei raised to high powers. Nevertheless, methods such as the one above can be developed to solve these kinds of problems.

Problem 3.8

Consider the equilibrium of an atomic nucleus ${}_{A}N^{Z}$ of atomic number Z and atomic weight A against photodissociation, and derive the Saha equation for it and the subsequent reaction (3.48). Assume that the reactions conserve total electric charge (all nuclei fully ionized) and total baryon number, and that the nuclei (including free nucleons) are nondegenerate.

3.6 TABULAR EQUATIONS OF STATE

It is not always possible or practical to construct accurate equations of state in analytic form. An alternate approach is to use equations of state in tablular form. In principle, this approach allows for fine resolution of such features as phase transitions, or an accurate representation of regions where thermodynamic quantities change rapidly in one variable, but remain nearly constant with changes in others. Unfortunately, the construction of accurate and thermodynamically consistent tables is often far more difficult than one might guess. A detailed discussion of tabular equations of state lies beyond the scope of this text, but a few introductory observations may be useful. The exercises below will help the reader develop a better understanding of these issues.

To the authors' knowledge no general machinery exists for the construction of equation of state tables. Instead, the approach to constructing a table usually starts with a detailed knowledge of the specific characteristic of the material under consideration. With this in mind, one can make the following more or less general observations regarding their construction.

Clearly, in order to be useful, tabular equations of state must be accurate and thermodynamically consistent. Accuracy can be attained by the application of enough mathematics, and a thorough knowledge of the behavior of the equation of state. The requirement of consistency means that the process of construction itself must be based on fundamental principles of thermodynamics. Experience has shown that the accuracy of the numerical data obtained from a table may sometimes be less important in a calculation than their thermodynamic consistency.

The first step in constructing a table begins by generating enough primary data (the points which go into the table) to span the region of thermodynamic space covered in a calculation. The data must also be sufficiently dense as to resolve specific features of interest, such as phase transitions, regions of ionization, and changes in composition. The primary data must also be thermodynamically consistent. This means, for example, that at each point in the table the primary data satisfies Maxwell's relations (see, for example, Callen 1961), so that the change in thermodynamic potentials will be path independent. This can be often be guaranteed by the method of construction.

Problem 3.9

The van der Waals equation of state is a relatively simple analytic expression for the pressure of a material which undergoes a phase transition (Reif 1965, page 173):

$$P(v,T) = \frac{RT}{v-b} - \frac{a}{v^2} \tag{3.49a}$$

where v is the specific volume of the fluid, and a and b are constants characteristic of the material. The specific energy of the material can be expressed in the form

$$\varepsilon(v,T) = c_V T - \frac{a}{v} + \varepsilon_o \tag{3.49b}$$

where ε_o and the thermodynamic heat capacity $c_V \equiv (d\varepsilon/dT)_V$ are assumed to be constants. The specific entropy of the material, which follows from the first law of thermodynamics, is

$$s(v,T) = c_V \ ln\, T + R\ ln(v - b) + s_o. \tag{3.49c}$$

Here s_o is the reference entropy of the system. The equations of state (3.49a) and (3.49b) will be thermodynamically consistent if they satisfy the Maxwell relation

$$\left(\frac{\partial T}{\partial v}\right)_s = -\left(\frac{\partial P}{\partial s}\right)_v . \tag{3.49d}$$

Show that they do.

Next, discuss the general features of the equation of state and plot its form for a region of T and v which includes the phase transition. Using this as a guide, construct a table of primary data for the pressure equation of state which includes the phase transition region. Use a Maxwell construction to describe the transition region itself (Callen 1961, page 146). Representative values for the coefficients are give below for several materials.

Table 3.1. Coefficients for the van der Waals equation of state for several materials.

Material	a (10^6 atm cm^6)	b (cm^3)
He	0.03415	23.71
A	0.2120	17.10
H_2O	5.468	30.52

The major complication in constructing tabular equations of state lies in the fact that required values seldom fall on the primary points of the table during a calculation. Thus, one or more interpolation schemes are needed to obtain these values from the primary data. Interpolation schemes must be reasonably accurate, but must also produce numerical values which are thermodynamically consistent in

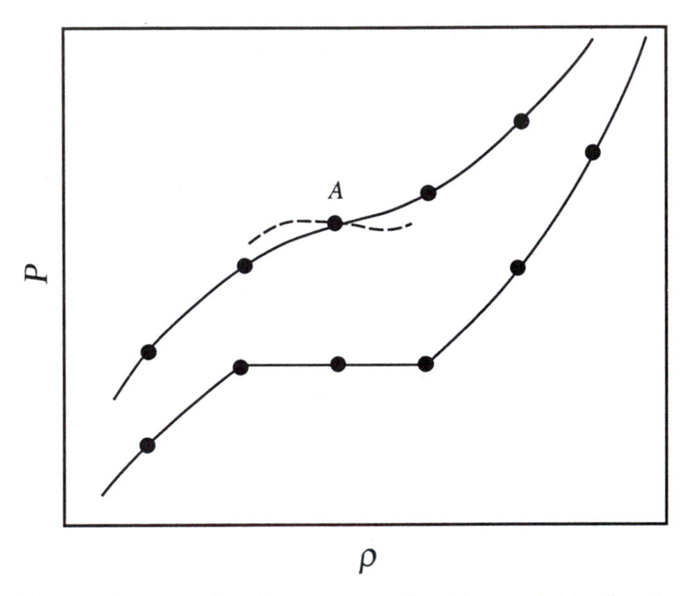

Figure 3.3. Pressure-density curves in the neighborhood of a phase transition shown schematically for a van der Waals gas.

the sense noted above. It should be obvious that thermodynamic consistency of an interpolation scheme is not guaranteed by consistency of the primary data. This is illustrated by the pressure-density curve shown in Figure 3.3. The lower curve contains a phase transition through which the pressure is a constant. The upper curve lies slightly above the critical point, and all points along it represent thermodynamically stable states. The primary data in the table are shown by the points along each curve. The dashed curve passing through point *A* shows schematically a possible interpolation curve. Points to the left of *A* along the interpolation curve may be only slightly below the true curve, but if they lie below point *A* then the interpolated thermodynamic state is unstable. Another source of error is associated with adiabatic processes. If the primary data or an interpolation scheme is not consistent, then the computation of adiabatic processes may lead to irreversible energy dissipation. Such effects may show up as energy nonconservation during the course of a calculation.

A judicious choice of variables may help reduce interpolation errors in tables. For example, the ratios $P/\rho T$ and ε/T are more slowly varying functions than are the original van der Waals equations of state. Constructing tables based on these ratios can give improved accuracy. As illustrated by Problem 3.9 special difficulties arise when interpolation schemes cross phase transition boundaries. An accurate treatment of these regions require special interpolation formulae.

Because of the difficulty associated with their consistent construction, equations of state tables are most useful in regimes where the physics is sufficiently well understood that the table can be constructed once and then used without change for an extended period of time. The use of accurate and consistent tables is usually not practical for regimes where the properties of matter are poorly understood or where the knowledge of the micro-physics underlying the equation of state is changing rapidly.

Problem 3.10

Develop an interpolation scheme for the table of primary data constructed in Problem 3.9. First, consider the bilinear fit to the pressure

$$P(v_i+\Delta v,T_k+\Delta T) = P(v_i,T_k) + C_{Pv}\,\Delta v + C_{PT}\,\Delta T$$

where v_i andT_k represent the primary data point in the table. Here

$$\Delta v \equiv v - v_i \qquad\qquad \Delta T \equiv T - T_i$$

represent the difference between the primary points and arbitrary points between them. Is the interpolated data using the bilinear fit thermodynamically consistent?

References

R. L. Bowers and T. Deeming, *Astrophysics,* Vol. I (Boston: Jones and Bartlett, 1984).

R. L. Bowers and J. R. Wilson, "A Numerical Model for Stellar Core Collapse Calculations," *Ap. J. Suppl.*, **50**, 1982, pg. 115.

H. B. Callen, *Thermodynamics* (New York: Wiley, 1961), pg. 199.

H. Y. Chiu, *Stellar Physics* (Waltham, Mass.: Blaisdell, 1968).

A. N. Cox, "Stellar Absorption Coefficients and Opacities," in *Stellar Evolution,* edited by L. H. Aller and D. B. McLaughlin (Chicago: University of Chicago Press, 1965) pg. 239.

J. P. Cox and R. T. Guili, *Principle of Stellar Structure*, Vol. I (New York: Gordon and Breach, 1968).

L. D. Landau and E. M. Lifshitz, *Statistical Physics* (London: Pergamon Press, 1958).

K. R. Lang, *Astrophysical Formulae* (New York: Springer-Verlag, 1980).

F. Reif, *Statistical and Thermal Physics* (New York: McGraw-Hill, 1965).

S. L. Shapiro and S. A. Teukolsky, *Black Holes, White Dwarfs and Neutron Stars* (New York: John Wiley and Sons, 1983).

Ya. B. Zel'dovich and Yu. P. Raizer, *Physics of Shock Waves and High-Temperature Hydrodynamic Phenomena*, Vol. I, edited by W. D. Hayes and R. F. Probstein (New York: Academic Press, 1967), pg. 201.

4

LAGRANGIAN HYDRODYNAMICS

The fundamental difference between Lagrangian and Eulerian hydrodynamics lies in the choice of coordinates used for each description of the fluid motion. Two sets of coordinates which are frequently adopted will be defined here; the Lagrangian coordinates will be developed in greater detail in this chapter, and the Eulerian coordinates will be discussed in greater detail in Chapter 5 (Courant and Friedrichs 1948). The independent variables in the Lagrangian description consist of labels fixed with the fluid particles themselves. Consequently, the labels move with the fluid throughout its motion. The Eulerian coordinates are simply the spatial coordinates of the fluid. In general the Eulerian coordinates need not be assumed to be fixed (as in the laboratory frame of reference).

Finite difference approximations to the inviscid fluid-flow equations in one spatial dimension can now be developed in Lagrangian form. The use of these methods for supersonic flows where shocks may occur will require special treatment. Lagrangian methods are well suited to one dimensional problems, but can also be extended to higher dimensions. Unfortunately, simple Lagrangian methods in higher dimensions are limited to smooth fluid flow because the Lagrangian mesh moves with the material, and multidimensional flow often leads to complex flow patterns which cannot be followed by the mesh. Multidimensional flow will be discussed in Chapter 5 along with Eulerian methods.

The first part of this chapter considers explicit methods appropriate for problems where the system changes dynamically on time scales comparable to the gravitational free-fall time scale of the system $\tau_G \approx (L^3/MG)^{1/2}$, or on time scales comparable to sound transit time $\tau_s \approx L/c_s$ (where L is a typical scale length of the system, c_s is the sound speed, M is a characteristic mass and G is the gravitational constant). Examples of dynamical processes in astrophysics include protostellar formation, explosive nuclear burning stages in stellar evolution, stellar core collapse and supernova explosions and gravitational collapse to form black holes. An example of an implicit method will then be considered, which is more naturally suited to quasi-dynamic phenomena where the system changes on a Kelvin-Helmholtz time scale $\tau_K \approx E/L$ (for a star, E may be the gravitational potential energy, or the internal thermal energy, and L is the rate at which energy is radiated away from the system), or nuclear burning τ_N time scale. For quasi-static processes τ_K and τ_N $>> \tau_s$ or τ_G.

Newtonian gravitation can be readily incorporated into the Lagrangian finite difference equations for one dimensional problems. However, a relativistic treatment of gravitation and Newtonian gravitation in two or three dimensions will be deferred to Chapter 9.

Finally, we will briefly discuss simple Lagrangian rezoning methods which allow automatic increased mesh resolution without requiring too many zones in the problem at any one time.

Lagrangian Coordinates

Lagrangian coordinates are simply defined as arbitrary labels which we think of as being attached to individual fluid elements throughout their motion (Figure 4.1). This simple concept can be applied in many different ways. For example, let a Lagrangian coordinate ξ represent the location of an element of mass at time $t = 0$. Then the location of the same mass element at time t can be denoted by $r = r(\xi,t)$. Here the spatial coordinate r is a dependent variable which is to be determined as a part of the solution of the fluid equations (note that r is not a Lagrangian coordinate). For example, suppose that the density at the point ξ in a planar system at time $t = 0$ is given by ρ_o;

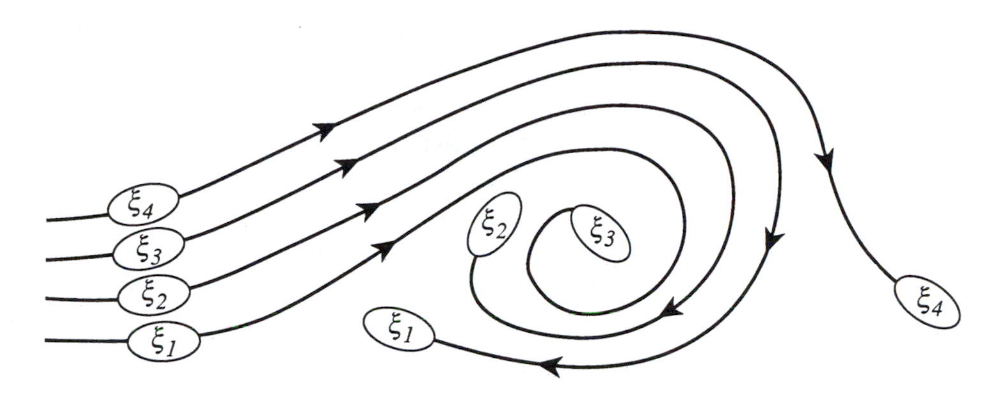

Figure 4.1. Fluid elements carrying Lagrangian labels ξ_i at two times. The solid lines denote fluid stream lines. Arrows show the direction of fluid flow.

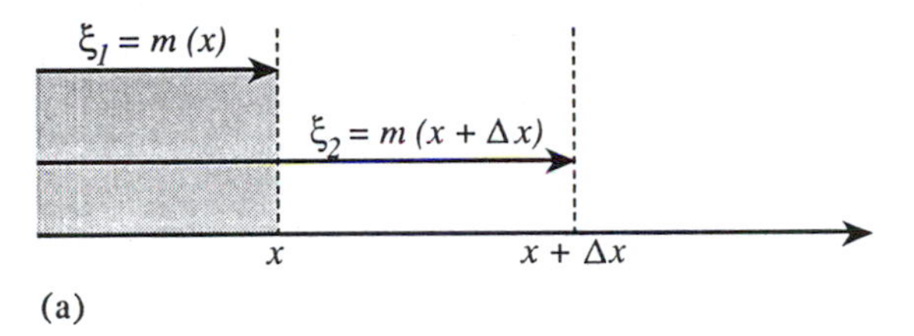

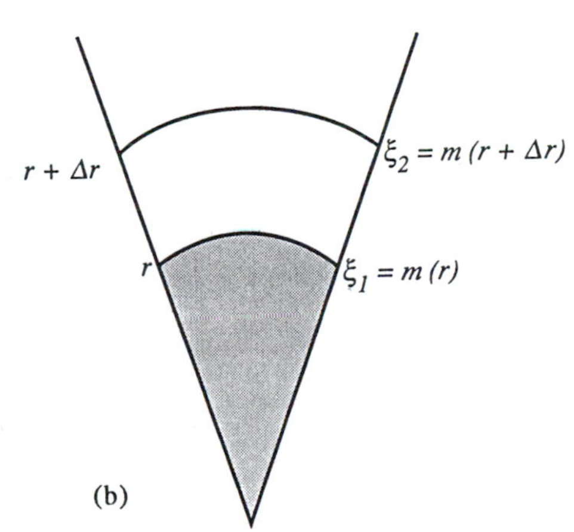

Figure 4.2. (a) Two Lagrangian coordinates ξ_1 and ξ_2 in planar geometry. The mass is measured from the left; the total mass lying to the left of the spatial coordinate x (shown shaded) is $m(x)$. (b) Lagrangian coordinates in one dimensional spherical geometry. The total mass contained inside radius r (shown shaded) is $m(r)$.

then $\rho_0 d\xi = \rho dx$, where ρ is the density at the location x at time t. The Lagrangian coordinate in this case is ξ (Figure 4.2a). The primary advantage of using Lagrangian coordinates is that the mass continuity equation is automatically satisfied (see Problem 4.27).

As another example, consider a one dimensional spherically symmetric distribution of matter whose density is $\rho(r)$. A suitable choice of Lagrangian coordinates for this system would be the total mass interior to radius r: denoting this Lagrangian coordinate by $m(r)$ we have (Figure 4.2b)

$$m(r) = \int_0^r 4\pi\rho(r)\, r^2 dr\,.$$

At any instant t the spatial radius of the mass element is given by $r = r(m, t)$. This choice of Lagrangian coordinates will be used for the remainder of this chapter. In finite difference form, the difference between two consecutive Lagrangian coordinates may then be taken as the width of a computational zone:

$$\Delta m(r) = m(r+\Delta r) - m(r) = \int_r^{r+\Delta r} 4\pi\rho(r)\, r^2 dr\,.$$

Thus the mass element Δm corresponds to all of the material lying between the spatial radii r and $r+\Delta r$. Consequently in the Lagrangian approach to hydrodynamics a fluid element can be defined in such a way that it always contains a specified amount of matter. In particular, there is no exchange of matter between fluid elements. As a result the element's mass remains constant in time (rezoning methods may alter the mass, as will be seen later). The dynamics consist of following each fluid element's motion in space, as determined by interactions with neighboring fluid elements. The Lagrangian approach is the most natural extension of Newtonian mechanics of particles to the fluid regime. In the following section the partial differential equations of fluid mechanics will be developed in Lagrangian form, using the concept of a Lagrangian fluid element.

4.1 HYDRODYNAMIC EQUATIONS

The differential equations describing inviscid fluid flow will first be developed in Lagrangian form. This introduction will establish several concepts and definitions that will be useful later when we discuss finite difference approximations, and when Eulerian methods are considered. At each instant in time, the fluid may be characterized by its density, ρ, pressure, P, specific energy, ε, and velocity, $\boldsymbol{v}$. Consider an infinitesimal, fixed mass of fluid; we will denote its mass by δm, and the volume occupied by this mass (which moves with the mass) by δV. These quantities represent the Lagrangian mass and volume. Next, define a fluid element of fixed mass Δm by

$$\Delta m = \int \delta m = \int \rho \delta V , \tag{4.1}$$

where the integral is over the volume ΔV occupied by the Lagrangian mass Δm.

Lagrangian Time Derivative

Next consider the Lagrangian volume ΔV. It can be shown that the time rate of change of a Lagrangian volume element is given by

$$\frac{d\Delta V}{dt} = \nabla \cdot \boldsymbol{v} \, \Delta V , \tag{4.2}$$

where d/dt has been used to denote the time derivative with respect to a reference frame co-moving with the Lagrangian fluid element of mass Δm. A formal discussion of Lagrangian coordinates is given in Appendix A.

Problem 4.1

Derive (4.2) for the special case of one dimensional Cartesian coordinates, and for one dimensional spherical coordinates. In each case construct the limit of $\Delta V(t+\Delta t)$ and $\Delta V(t)$ as $\Delta V \rightarrow 0$ (see Figure 4.3). See Mihalas and Mihalas (1984, page 58) for a general proof.

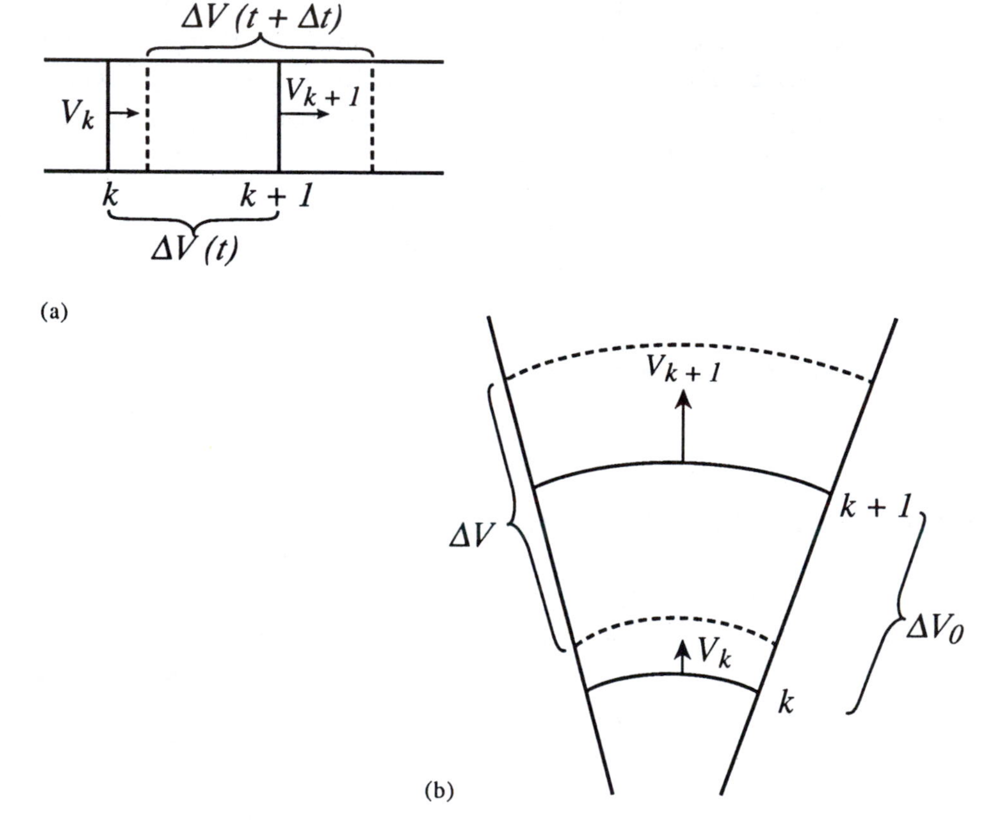

Figure 4.3. (a) Change in volume described by (4.2) in Cartesian coordinates, and (b) in spherical coordinates. Both cases correspond to one dimensional flow.

Mass Continuity Equation

By construction the Lagrangian time derivative of Δm must vanish; therefore using (4.2) we find

$$\frac{d\Delta m}{dt} = \frac{d}{dt}\int \rho\,\delta V = \int \left(\frac{d\rho}{dt}\delta V + \rho\frac{d}{dt}\delta V\right)$$

$$= \int \left(\frac{d\rho}{dt} + \rho\nabla\cdot\mathbf{v}\right)\delta V = 0.$$

Since the volume δV is arbitrary, the integrand must vanish, and we have

$$\frac{d\rho}{dt} + \rho\nabla\cdot\mathbf{v} = 0\,, \tag{4.3}$$

relating the Lagrangian time rate of change of the density to the change in Lagrangian volume. Equation (4.3) is the mass continuity equation in Lagrangian form, and simply expresses the assumption that mass is conserved for any type of fluid motion.

Momentum Equation

The momentum of a Lagrangian fluid element is given by

$$\int \mathbf{v}\ \delta m = \int \rho\mathbf{v}\ \delta V\,,$$

and its Lagrangian time rate of change is equal to the net external force acting on it, which consists of pressure forces and body forces (the latter may be due, for example, to gravitation, magnetic fields or radiation-matter interactions). Pressure differences acting on the surface S of the fluid element produces a net force on it which is given by

$$\int P\ \delta\mathbf{S}\,,$$

where the integral is over the surface of the Lagrangian fluid element; thus

$$\frac{d}{dt}\int \rho \mathbf{v}\ \delta V = -\int P\ \delta S + \int f\ \delta V\ , \tag{4.4}$$

where f denotes the net body force acting on δV. We may apply a generalized version of Gauss's divergence theorem,

$$\int P\ \delta S = \int \nabla P\ \delta V\ , \tag{4.5}$$

to the surface integral in (4.4), and use (4.2) to obtain the momentum equation in Lagrangian form:

$$\rho\,\frac{d\mathbf{v}}{dt} = -\nabla P + f\,. \tag{4.6}$$

Equation (4.6) expresses net momentum conservation for a Lagrangian fluid element, and is equivalent to Newton's second law of motion.

Problem 4.2

Prove (4.5) in one dimension using Cartesian coordinates. Then carry out the steps leading to (4.6).

Internal Energy Equation

Next, we will obtain an expression for the change in energy of the fluid element. Physically this can be expressed as three terms: the change in kinetic energy; adiabatic change in internal energy; and all other sources (here we would include the effects of thermal diffusion, magnetic fields and radiation coupling, for example). Rather than describing the change in total energy, we will work with the change in internal energy, which follows from the first law of thermodynamics (3.5). Taking the Lagrangian time rate of change or the first law gives

$$\frac{d\varepsilon}{dt} = \frac{P}{\rho^2}\frac{d\rho}{dt} + T\frac{ds}{dt} .$$

(4.7)

The term $T(ds/dt)$ represents all dissipative processes. The first term on the right side of (4.7) represents mechanical work due to pressure gradients acting on the fluid element.

Problem 4.3

Consider the adiabatic fluid equations $(ds/dt=0)$ with no external body forces, and show that the first term on the right side of (4.7) is the work due to pressure gradients.

Problem 4.4

Show that the Lagrangian time rate of change of the internal plus kinetic energy of a fluid element Δm is given by

$$\frac{dE}{dt} + \int P\,\mathbf{v}\cdot d\mathbf{S} = \int T\frac{ds}{dt}\,\delta m ,$$

(4.8)

where the total energy of the fluid element is

$$E \equiv \int \rho(v^2/2 + \varepsilon)\,\delta m.$$

Viscosity

The momentum equation (4.6) and the internal energy equation (4.7) must be modified if stresses other than pressure are present. Stresses may arise for mechanical reasons (materials with strength) or because of molecular viscosity (stresses proportional to velocity gradients). In order to include these effects, the pressure must be replaced by a second rank, symmetric tensor which may be written in the form

$$\Pi_{ik} = P\delta_{ik} - \Sigma_{ik} ,$$

where P is the usual isotropic pressure, and Σ_{ik} is a second rank, symmetric tensor. The momentum is then, in tensor form in Cartesian coordinates,

$$\rho \frac{dv^i}{dt} = - \frac{\partial P}{\partial x^i} + \frac{\partial \Sigma_{ik}}{\partial x^k} + f^i , \quad (i = 1, 2, 3) . \tag{4.9}$$

The only stresses that we shall consider here are those associated with viscosity. The general form of Σ_{ik} for an isotropic fluid when it is due solely to gradients in the fluid velocity can be readily found from physical arguments (see Problem 4.5), and is

$$\Sigma_{ik} = \eta \left(\frac{\partial v_k}{\partial x^i} + \frac{\partial v_i}{\partial x^k} - \frac{2}{3}\delta_{ik} \frac{\partial v_l}{\partial x^l}\right) + \zeta\delta_{ik}\left(\frac{\partial v_l}{\partial x^l}\right) . \tag{4.10}$$

The quantity in parenthesis in the first term on the right side of (4.10) is the traceless part of Σ_{ik}, and η is the coefficient of shear viscosity. The last term is proportional [by (4.2)] to the fractional rate of change in the Lagrangian volume of the fluid element. Finally, ζ is the coefficient of bulk viscosity.

Problem 4.5

Derive (4.10) assuming that Σ_{ik} satisfies the following constraints, and discuss each: Σ_{ik} must (a) depend linearly on the velocity gradient; and (b) it must vanish under uniform rotation.

Viscous forces will also do work on the fluid element; therefore (4.7) must be modified to include Σ_{ik}. The result is (in Cartesian coordinates)

$$\frac{d\varepsilon}{dt} = \frac{P}{\rho^2}\frac{d\rho}{dt} + \frac{1}{\rho}\frac{\partial v^i}{\partial x^k}\Sigma_{ik} + T\frac{ds}{dt}. \tag{4.11}$$

The last term on the right side of (4.11) represents entropy changes due to all dissipative effects not directly associated with viscosity.

Problem 4.6

Evaluate Σ_{ik} for one dimensional flow, and discuss its role as an effective pressure in the momentum equation. Also discuss its role in changing the internal energy.

The fluid equations, including viscous and dissipative effects are given by (4.3), (4.9) and (4.11). In the remainder of this chapter we shall ignore all body forces f in (4.9), and all viscous and dissipative terms in the energy equation (artificial viscous stress will be introduced in Section 4.4), and concentrate on finite difference approximations to the adiabatic hydrodynamic equations (Euler's equations):

$$\rho\frac{d\mathbf{v}}{dt} = -\nabla P, \tag{4.12}$$

$$\frac{d\rho}{dt} = -\rho\nabla\cdot\mathbf{v}, \tag{4.13}$$

$$\frac{d\varepsilon}{dt} = \frac{P}{\rho^2}\frac{d\rho}{dt}. \tag{4.14}$$

Finally, the velocity and displacement are related by

$$\frac{dx}{dt} = \mathbf{v} \,.$$

Before considering these equations in detail, we will examine the finite difference approximations describing the propagation of adiabatic sound waves.

4.2 SOUND WAVES

A finite difference representation will be developed below to describe the propagation of small adiabatic perturbations in an ideal fluid. This example introduces the concept of a two-time level finite-difference approximation, and illustrates the importance of temporal and spatial centering on a space-time grid. It is also an example of a coupled system of finite difference equations to which the von Neumann stability analysis of Section 2.2 can be applied. The insight gained from this example will motivate the methods which we shall present for the full hydrodynamic equations in Lagrangian form in the remainder of this chapter, and the Eulerian form of the equations to be discussed in Chapter 5.

The following discussion will be limited to small disturbances in a uniform system, so it is convenient to express all physical variables in terms of their unperturbed (uniform) values, which we denoted by a subscript zero. The perturbation will then represent a small change, which is denoted by a prime. The total (i.e., perturbed) density, pressure and specific energy are then

$$\begin{aligned} \rho &= \rho_o + \rho' \\ P &= P_o + P' \\ \varepsilon &= \varepsilon_o + \varepsilon'. \end{aligned} \tag{4.15}$$

The fluid velocity is $v = v_o + v'$. For simplicity the unperturbed velocity will be assumed to be zero, so that $v = v'$ which is taken to be a small quantity of the same order as the perturbations ρ', P', and ε'. Finally, we restrict attention to one dimensional planar geometry. Substituting (4.15) into (4.12) - (4.14) and dropping terms of second and higher order in the disturbances leads to the following equations describing the perturbations:

$$\rho_o \frac{dv}{dt} = - \frac{\partial P'}{\partial x} \tag{4.16}$$

$$\frac{d\rho'}{dt} = -\rho_o \frac{\partial v}{\partial x} \tag{4.17}$$

$$\frac{d\varepsilon'}{dt} = \frac{P_o}{\rho_o^2} \frac{d\rho'}{dt}. \tag{4.18}$$

These equations constitute a coupled, linear set of three partial differential equations containing four unknowns.

An equation of state is needed to complete the system of equations. Assume that an equation of state of the form (3.1) is available. Then using (4.15) we find, to lowest order in small quantities

$$\begin{aligned} P = P(\rho_o + \rho', \varepsilon_o + \varepsilon') &\approx P_o + \left(\frac{\partial P}{\partial \rho}\right)_\varepsilon d\rho + \left(\frac{\partial P}{\partial \varepsilon}\right)_\rho d\varepsilon \\ &= P_o + \left[\left(\frac{\partial P}{\partial \rho}\right)_\varepsilon + \left(\frac{\partial P}{\partial \varepsilon}\right)_\rho \left(\frac{\partial \varepsilon}{\partial \rho}\right)_s\right] d\rho \\ &\approx P_o + c_s^2 (\rho' - \rho_o) \end{aligned} \tag{4.19}$$

where (3.4) has been used to eliminate the partial derivatives of the pressure. To lowest order, c_S is assumed to be a constant. It should be noted that the solution to (4.18) is trivial; $\varepsilon' = P_o \rho' / \rho_o^2$. Substituting (4.19) into (4.16), taking the time derivative of the result and combining it with the spatial derivative of (4.17) leads to the wave equation for the density perturbations

$$\frac{d^2\rho'}{dt^2} = c_s^2 \frac{\partial^2\rho'}{\partial x^2} . \tag{4.20}$$

In fact a similar equation can be shown to apply to v and to P'; letting ξ represent ρ', P', or v it can be shown that

$$\frac{d^2\xi}{dt^2} = c_s^2 \frac{\partial^2\xi}{\partial x^2} . \tag{4.21}$$

This is an hyperbolic partial differential equation to which, for example, Cauchy boundary conditions apply.

Space-Time Centering

A finite difference representation of (4.21) could be developed involving $\xi^{n+1}{}_k$, $\xi^n{}_k$ and $\xi^{n-1}{}_k$. Such a scheme can be developed which is stable. This type of scheme would require that ρ', P', and v be carried for two time levels for each computational cycle. Another approach that avoids this complication is to return to the first order coupled equations for the perturbations which can be written in the form

$$\frac{d\psi}{dt} = -c_S \frac{\partial\phi}{\partial x} \tag{4.22}$$

$$\frac{d\phi}{dt} = -c_S \frac{\partial\psi}{\partial x} , \tag{4.23}$$

where to simplify notation we have introduced the variables

$$\phi \equiv P' \tag{4.24}$$

$$\psi \equiv \rho_0 c_s v \, . \tag{4.25}$$

We see from (4.22)-(4.23) that spatial gradient of ϕ advances ψ in time, and vice-versa. This suggests a space-time centering for the dependent variables $\phi^n{}_{k+1/2}$ and $\psi^{n-1/2}{}_k$ as shown in Figure 4.4. Referring to (4.24)-(4.25), this implies that the velocity is to be considered a zone-edge centered variable, while the pressure is zone centered. Recalling (3.1) we might also expect to take ρ and ε to be zone centered at the same time level as is the pressure: $\varepsilon(x,t) \to \varepsilon^n{}_{k+1/2}$ and $\rho(x,t) \to \rho^n{}_{k+1/2}$. These associations are clearly reasonable on physical grounds. If a mass element (zone) is interpreted as a basic computational element, the density and specific energy naturally represent its internal properties. Since the dynamics result from imbalances between the internal characteristics of neighboring mass elements or zones, it is also natural to associate a velocity with the boundaries of the elements or zones. If we take the current time to

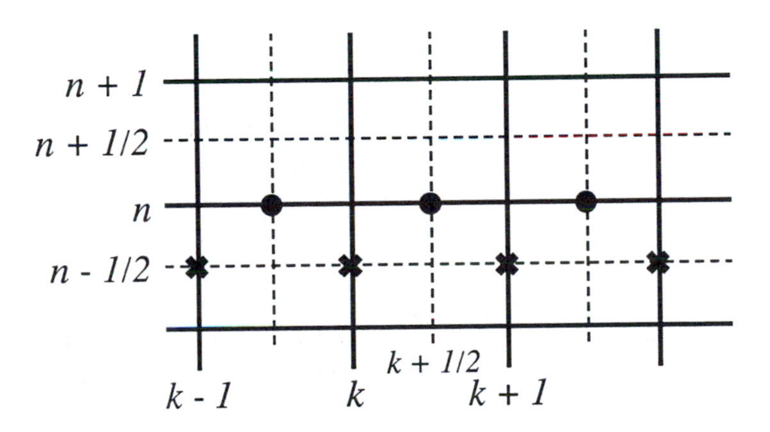

Figure 4.4. Space-time centering of ψ (shown by x) and ϕ (filled circles). The initial conditions for cycle n are $\psi^{n-1/2}{}_k$ and $\phi^n{}_{k+1/2}$ for all k. This method is explicit in time.

be t^n, then (4.22) can be used to advance $\psi^{n-1/2}{}_k$ to $\psi^{n+1/2}{}_k$ using the time step Δt^n. This requires $\phi^n{}_{k+1/2}$ for all spatial grid points. A possible finite difference representation of (4.22) is therefore

$$\frac{\psi_k^{n+1/2} - \psi_k^{n-1/2}}{\Delta t^n} = - c_s \frac{(\phi_{k+1/2}^n - \phi_{k-1/2}^n)}{\Delta x_k} . \tag{4.26}$$

The values $\psi^{n+1/2}{}_k$ may then be used to advance $\phi^n{}_{k+1/2}$ to $\phi^{n+1}{}_{k+1/2}$ using the time step $\Delta t^{n+1/2}$ and the finite difference representation

$$\frac{\phi_{k+1/2}^{n+1} - \phi_{k+1/2}^n}{\Delta t^{n+1/2}} = - c_s \frac{(\psi_{k+1}^{n+1/2} - \psi_k^{n+1/2})}{\Delta x_{k+1/2}} . \tag{4.27}$$

Thus (4.26)-(4.27) represent a possible finite difference approximation to the linearized hydrodynamic equations for adiabatic motion.

Stability

The centering scheme developed above was chosen in part for its symmetry, as suggested by (4.22)-(4.23). The results also are a physically reasonable way of describing finite fluid elements in that the pressure, density and specific energy are zone centered, and the velocity is centered at the zone boundarys. It remains to be seen if the method is numerically stable. The von Newmann stability analysis discussed in Section 2.3 can be applied to the coupled linear system (4.26)-(4.27), by expanding ψ and ϕ in Fourier modes:

$$\psi_k^{n-1/2} = A^n e^{ik\theta} \tag{4.28}$$

$$\phi^n_{k+1/2} = B^n e^{i(k+1/2)\theta} \; . \tag{4.29}$$

We denote current time amplitudes of ψ and ϕ by the same integer power of their initial values A and B [i.e., $A^n = (A)^n$ where A is, for example, the amplitude at $t = 0$], and assume for simplicity that Δx and Δt are constants. Substituting (4.28)-(4.29) into (4.26)-(4.27) yields an amplification matrix (2.55) of the form

$$\mathbf{G} = \begin{bmatrix} 1 & -2i\,\sigma \sin\theta/2 \\ -2i\,\sigma \sin\theta/2 & 1-4\,\sigma^2 \sin^2\theta/2 \end{bmatrix} \tag{4.30}$$

with $V^n \equiv (A^n, B^n)$ and $\sigma \equiv c_s \Delta t/\Delta x$. The stability constraint (2.56) on the eigenvalues λ of $\mathbf{G}$ leads to the condition

$$c_s \Delta t \le \Delta x \; . \tag{4.31}$$

This is conventionally referred to as the Courant condition. It is analogous to the time step constraint (2.54) introduced in the discussion of advection in Chapter 2 (time step constraints which involve a characteristic velocity, such as (4.31) and (2.54), are sometimes referred to collectively as Courant conditions). Physically (4.31) states that a stable solution to the finite difference equations will result as long as the time step Δt is small enough that sound signals cannot cross more than one spatial zone in that time. Thus (4.26)-(4.27) and the time step constraint (4.31) represent an acceptable, explicit solution to the linearized adiabatic hydrodynamic equations.

Problem 4.7

Derive the Courant time step condition (4.31) for the linearized adiabatic hydrodynamic equations.

Next, we note that the mass continuity equation (4.17) can be substituted into the right side of (4.18) to express the internal energy equation in the form

$$\rho_o \frac{d\varepsilon'}{dt} = - P_o \frac{\partial v}{\partial x} .$$

(4.32)

Recall that v is centered at $(k,n+1/2)$ and ε' is centered at $(k+1/2,n)$. If, for notational convenience, we define an energy density variable by

$$\theta \equiv \rho_o \varepsilon',$$

(4.33)

the energy equation can be written as

$$\frac{d\theta}{dt} = - c_S \frac{\partial \psi}{\partial x} .$$

(4.34)

Notice that (4.26)-(4.27) do not contain θ; thus when new values for ψ have been found, they may be used to solve (4.34). Furthermore, since the gradient of ψ advances θ in time, we are again led to a scheme in which the specific energy is zone centered. Thus a finite difference representation of (4.34) which can be used with (4.26)-(4.27) is

$$\frac{\theta^{n+1}_{k+1/2} - \theta^{n}_{k+1/2}}{\Delta t^{n+1/2}} = - c_s \frac{(\psi^{n+1/2}_{k+1} - \psi^{n+1/2}_{k})}{\Delta x_{k+1/2}}$$

(4.35)

Extending the stability analysis to include (4.26)-(4.27) and (4.35) leads again to the Courant condition (4.31). Notice that the gradient used on the right side of (4.27) is the same as appears on the right side of the energy equation. When the Eulerian form of the fluid equations is discussed it will be found that the same flux is used when the density and specific energy are advanced in time.

Implicit Equation for Sound Waves

As a second example of a finite difference approximation to (4.16)-(4.18), consider (4.26) with ϕ^n replaced by ϕ^{n+1}, and (4.27) with $\psi^{n-1/2}$ replaced by ψ^n. The resulting system of equations contains a single set of time levels t^n, connected by the single time step $\Delta t^{n+1/2}$: This scheme,

$$\psi_k^{n+1} = \psi_k^n - \frac{c_S \Delta t^n}{\Delta x_k}\left(\phi_{k+1/2}^{n+1} - \phi_{k-1/2}^{n+1}\right)$$

$$\phi_{k+1/2}^{n+1} = \phi_{k+1/2}^n - \frac{c_S \Delta t^{n+1/2}}{\Delta x_{k+1/2}}\left(\psi_{k+1}^{n+1} - \psi_k^{n+1}\right)$$

is implicit in ϕ and ψ. Substituting (4.28)-(4.29) into the two equations above, and carrying out the stability analysis shows that the finite difference representation is unconditionally stable. The solution of this system of finite difference equations is more involved than is the explicit system. Substituting $\phi^{n+1}{}_{k\pm 1}$ into the right side of the expression for $\psi^{n+1}{}_k$ leads to a relation of the form

$$A_k \psi^{n+1}{}_{k+1} + B_k \psi^{n+1}{}_k + C_k \psi^{n+1}{}_{k-1} - D_k = 0.$$

This is a tridiagonal form and is formally identical to the implicit solution of the diffusion equation discussed in Chapter 2. A similar equation may be obtained for ϕ. In the case of adiabatic sound waves the two implicit equations for ϕ and ψ are linear and are uncoupled.

Problem 4.8

Carry out the von Neumann stability analysis for the implicit equations above, and verify that they are unconditionally stable.

4.3 LAGRANGIAN INVISCID FLUID EQUATIONS

An explicit finite difference approximation to the Lagrangian hydrodynamic equations (4.12)-(4.14) will now be developed and its solution discussed. As was discussed in Chapter 3, the equation of state can be expressed in the form (3.1), even if the underlying physics can only be expresssed in temperature mode.

The Lagrangian hydrodynamic equations were developed in Section 4.1 using the concept of a fluid element whose mass Δm remained constant during the fluid's motion. The time derivatives are taken with respect to a frame of reference co-moving with the fluid, and the spatial gradients are always to be understood to act across the volume element ΔV containing the constant mass Δm. This suggests that Δm is a more natural variable than Δx in representing spatial gradient. The transformation from ∇ to $\partial/\partial m$ can be carried out using (4.1), or in finite difference form $\Delta m = \rho \Delta V$. Consider the planar one dimensional form of (4.12)-(4.14); dividing the momentum equation by ρ , and denoting the mass per unit length by $dm = \rho dx$, leads to the momentum equation expressed in Lagrangian coordinates:

$$\frac{dv}{dt} = -\frac{\partial P}{\partial m} . \tag{4.36}$$

The mass density will be given by

$$V = \frac{1}{\rho} = \frac{\partial x}{\partial m}, \tag{4.37}$$

where V is the specific volume. Since $V = \Delta V / \Delta m$, (4.37) is equivalent to the mass continuity equation [see (4.14) and (4.2)]. The internal energy equation is just (4.14), which we write as [using (4.37)]

$$\frac{d\varepsilon}{dt} = -P\frac{dV}{dt}. \tag{4.38}$$

Finally, the spatial coordinates (which are dependent variables) are given by

$$\frac{dx}{dt} = v. \tag{4.39}$$

Equations (4.36)-(4.39) represent the one dimensional planar Lagrangian hydrodynamics equation; the dependent variables are ρ, P, ε, v and x, and the dependent variables are m and t. The system is closed by assuming an equation of state of the form in equation (3.1).

Problem 4.9

Use the discussion above as a guide to express the one dimensional, spherically symmetric hydrodynamic equations in Lagrangian form. In particular, show that

$$\frac{dv}{dt} = -\,4\pi r^2 \frac{\partial P}{\partial m} \tag{4.40}$$

$$V = \frac{1}{\rho} = 4\pi r^2 \frac{\partial r}{\partial m} \tag{4.41}$$

$$\frac{d\varepsilon}{dt} = -\,P\frac{dV}{dt} \tag{4.42}$$

$$\frac{dr}{dt} = v \tag{4.43}$$

What are the Lagrangian coordinates for this representation of the equations, and how are they related to the spatial coordinates?

Space-Time Centering

The equations governing sound propagation that were discussed in Section 4.2 were obtained by linearizing the full hydrodynamics equations, assuming that the amplitude of the sound waves was small. Physically, (4.36)-(4.39) must also admit sound propagation as a possible solution. This observations suggests that the centering scheme shown in Figure 4.4 for sound waves be tried for the hydrodynamic equations as well. Therefore, the fluid variables will be assigned the following centering:

$$v \rightarrow v^{n+1/2}{}_k \qquad \rho \rightarrow \rho^n{}_{k+1/2}$$

$$P \rightarrow P^n{}_{k+1/2} \qquad \varepsilon \rightarrow \varepsilon^n{}_{k+1/2} \tag{4.44}$$

Two time steps will also be required; these are Δt^n, which is used to advance the velocity, and $\Delta t^{n+1/2}$, which is used to advance the material state variables. The total mass lying, for example, to the left of the spatial coordinates $x^n{}_k$ may be adopted as Lagrangian coordinates ξ_k. Then the difference between two consecutive Lagrangian coordinates is the mass of a zone, and is zone centered:

$$\xi_{k+1} - \xi_k \equiv \Delta m \rightarrow \Delta m_{k+1/2} \,. \tag{4.45}$$

The Lagrangian mass $\Delta m_{k+1/2}$ carries no temporal index since it is assumed to remain constant in time (an exception to this rule arises if the grid is rezoned during a calculation, as we shall see in Section 4.6). The variable centering for the hydrodynamic equations is shown schematically in Figure 4.5.

Solution of the Lagrangian Equations

An explicit finite difference representation of the Lagrangian fluid equations (4.36)-(4.39) will be developed below that is based on the

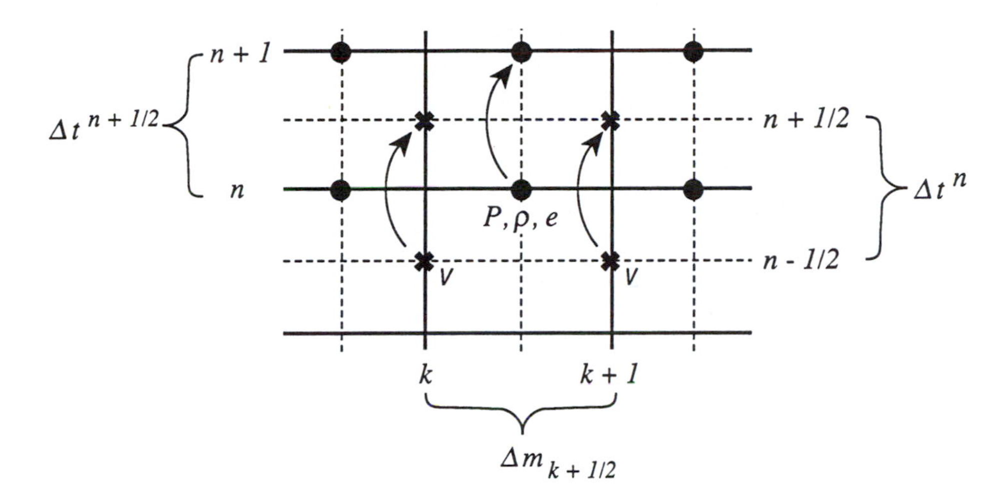

Figure 4.5. Space-time centering of Lagrangian fluid variables. The filled circles represent locations of the intensive variables (density, pressure and internal specific energy). The crosses represent the location of the velocities.

centering (4.44)-(4.45). First, however, an overview of the solution procedure will be outlined. We found in the previous section that the explicit finite difference equations describing sound wave propagation must be solved in a specific order; examining Figure 4.5 suggests that we construct new variables for the hydrodynamic equations at $n+1/2$ and $n+1$ from the current values at times $n-1/2$ and n by the following procedure. First, solve the momentum equation (4.36) for the fluid velocity $v^{n+1/2}{}_k$ using the pressures $P^n{}_{k\pm 1}$. The new spatial coordinates $x^{n+1}{}_k$ are then constructed from (4.39) using $v^{n+1/2}{}_k$. Next, the mass continuity equation (4.37) is solved using the new spatial coordinates $x^{n+1}{}_k$. This step makes explicit use of the assumption that the mass $\Delta m_{k+1/2}$ is fixed. The internal energy equation (4.38) is solved next using the new specific volume. Finally the new energy and density are substituted into the equation of state (3.1) to obtain the pressure. All of the dependent variables are now known at the new time, which completes a computational cycle. If other processes such as radiation transport, magnetic field diffusion or reactions are to be included through operator splitting, they may be solved at this point.

The finite difference equations will be constructed, assuming that the zone masses are not necessarily equal. In most Lagrangian calculations it is desirable to maintain mass zoning which is as nearly uniform as possible, but this is not always possible. The equations will be considered in the order outlined in the paragraph above. Figure 4.6 shows the spatial centering for the pressure and velocity appearing in the momentum equation. The pressure gradient acts across the zone centered at k and bounded by the dashed vertical lines at $k \pm 1/2$. Integrating (4.36) across this zone, and assuming that dv/dt is spatially constant,

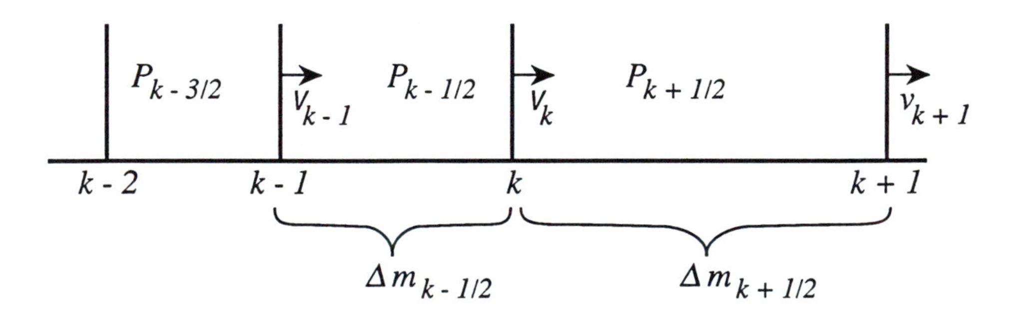

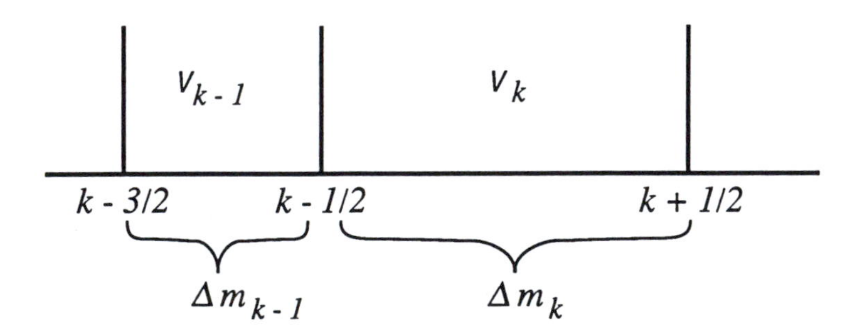

Figure 4.6. Spatial centering for the pressure and velocity appearing in the one dimensional momentum equation. The upper figure (lower figure) shows the mass zones associated with the density $\rho_{k+1/2}$ (ρ_k).

$$\int \frac{dv}{dt}\,\delta m = \frac{(\Delta m_{k-1/2} + \Delta m_{k+1/2})}{2}\left(\frac{dv}{dt}\right)_k = -\int \frac{\partial P}{\partial m}\,\delta m$$

$$= -(P_{k+1/2} - P_{k-1/2})$$

Defining the mass centered at k by

$$\Delta m_k = (1/2)(\Delta m_{k+1/2} + \Delta m_{k-1/2}) \tag{4.46}$$

we have

$$\frac{v_k^{n+1/2} - v_k^{n-1/2}}{\Delta t^n} = -\frac{P_{k+1/2}^n - P_{k-1/2}^n}{\Delta m_k}. \tag{4.47}$$

This finite difference approximation to the momentum equation is first-order accurate in space. It may be noted that (4.46) implies a definition of the Lagrangian coordinates of the staggered zone centered at k. The spatial coordinates of the staggered zone are to be chosen (when needed) so that they divide zone $k+1/2$ into two equal mass halves.

The new spatial coordinates for the original Lagrangian zone boundaries at k are obtained by assuming that they move with the velocity $v^{n+1/2}{}_k$ during the time $\Delta t^{n+1/2}$:

$$x_k^{n+1} = x_k^n + v_k^{n+1/2}\Delta t^{n+1/2}. \tag{4.48}$$

The mass continuity equation simply states that the new Lagrangian zone still contains the mass $\Delta m_{k+1/2}$, from which the new density may be obtained:

$$\frac{1}{\rho_{k+1/2}^{n+1}} = V_{k+1/2}^{n+1} = \frac{x_{k+1}^{n+1} - x_k^{n+1}}{\Delta m_{k+1/2}}. \tag{4.49}$$

Finally, we solve the internal energy equation (4.38). Two approaches will be considered. The simplest method makes use of the fact that (4.36)-(4.39) represents adiabatic motion, and assumes that an effective gamma law equation of state (3.8) can be used to approximate $P = P(\rho, \varepsilon)$. Then the change in internal energy is determined completely by the new density and γ and the adiabatic relation

$$\varepsilon^{n+1}_{k+1/2} = \varepsilon^{n}_{k+1/2} \, (\rho^{n}_{k+1/2} / \rho^{n+1}_{k+1/2})^{1-\gamma} \; . \tag{4.50}$$

Since (4.38) represents adiabatic work done against the pressure, (4.50) is as accurate as the gamma law fit to the actual equation of state.

If an effective gamma law equation of state is not available, the internal energy equation must be solved directly. By substituting the continuity equation into the right side of (4.38) we obtain the equivalent form

$$\frac{d\varepsilon}{dt} = - \, P \frac{\partial v}{\partial m} \; . \tag{4.51}$$

The term $\partial v / \partial m$ is centered at $n+1/2$, and $k+1/2$, and it advances ε from time t^n to time t^{n+1}. A natural centering for the pressure would then be at $n+1/2, k+1/2$, which we define by

$$P^{n+1/2}_{k+1/2} \equiv \frac{1}{2} \left[P(\rho^{n+1}_{k+1/2}, \, \varepsilon^{n+1}_{k+1/2}) + P(\rho^{n}_{k+1/2}, \, \varepsilon^{n}_{k+1/2}) \right] . \tag{4.52}$$

The right side of this expression contains the single unknown $\varepsilon^{n+1}{}_{k+1/2}$, and the incorporation of (4.52) into a finite difference approximation to (4.51) would result in an implicit equation for $\varepsilon^{n+1}{}_{k+1/2}$ which would have to be solved iteratively for all but the simplest equations of state. This complication can be avoided by expanding $P^{n+1/2}{}_{k+1/2}$ in a Taylor series about $\rho^{n}{}_{k+1/2}$ and $\varepsilon^{n}{}_{k+1/2}$; to lowest order

$$P^{n+1/2}_{k+1/2} = P^{n}_{k+1/2} + \frac{1}{2}\left[\left(\frac{\partial P}{\partial \rho}\right)_{\varepsilon} \Delta\rho^{n+1/2}_{k+1/2} + \left(\frac{\partial P}{\partial \varepsilon}\right)_{\rho} \Delta\varepsilon^{n+1/2}_{k+1/2}\right] \tag{4.53}$$

which is to be used with

$$\frac{\Delta\varepsilon^{n+1/2}_{k+1/2}}{\Delta t^{n+1/2}} = \frac{\varepsilon^{n+1}_{k+1/2} - \varepsilon^{n}_{k+1/2}}{\Delta t^{n+1/2}} = -P^{n+1/2}_{k+1/2}\frac{(v^{n+1/2}_{k+1} - v^{n+1/2}_{k})}{\Delta m_{k+1/2}} . \tag{4.54}$$

The change in density at $t^{n+1/2}$ can be approximated by

$$\Delta\rho^{n+1/2}_{k+1/2} = \rho^{n+1}_{k+1/2} - \rho^{n}_{k+1/2} . \tag{4.55}$$

Substituting (4.53) into (4.54), and solving for $\varepsilon^{n+1}{}_{k+1/2}$ gives

$$\varepsilon^{n+1}_{k+1/2} = \varepsilon^{n}_{k+1/2} - \frac{\left[P^{n}_{k+1/2} + \frac{1}{2}\left(\frac{\partial P}{\partial \rho}\right)_{\varepsilon} \Delta\rho^{n+1/2}_{k+1/2}\right]\Delta V \, \Delta t^{n+1/2}}{1 + \frac{1}{2}\left(\frac{\partial P}{\partial \varepsilon}\right)_{\rho} \Delta V \;\; \Delta t^{n+1/2}} \tag{4.56}$$

where

$$\Delta V \equiv V^{n+1}_{k+1/2} - V^{n}_{k+1/2} .$$

The partial derivatives of the pressure must be available at this stage; they are to be evaluated at time t^n. Finally, the pressure for the next computational step is given by

$$P^{n+1}_{k+1/2} = P(\rho^{n+1}_{k+1/2}, \varepsilon^{n+1}_{k+1/2}) . \tag{4.57}$$

Another approach which is more commonly used to time center the pressure is to calculate the mechanical work in several steps, using an updated value of the pressure for the next step. In most cases two steps are adequate In the first step the following equation is solved for the intermediate energy ε':

$$\varepsilon' = \varepsilon^n - P^n \Delta V .$$

The corresponding intermediate pressure is then obtained from the equation of state:

$$P(\varepsilon', \rho^{n+1}) .$$

The final energy is then given by

$$\varepsilon^{n+1} = \varepsilon^n - \frac{1}{2}[P^n + P(\varepsilon', \rho^{n+1})] \, \Delta V.$$

This method requires that the equation of state subroutine be called twice per zone per computational cycle, where as the method involving (4.53) only requires one call to the equation of state subroutine. For complicated equations of state a significant fraction of the computational time per cycle is spent in the equation of state subroutine; the use of (4.53) may be advisable in these cases. However, this expression requires that the partial derivatives of the pressure with respect to density and energy be available in reasonably accurate form. For complicated equations of state this may involve a substantial amount of work. The relative advantages and disadvantages of each method will determine which is best in any given application.

Problem 4.10

What form does (4.56) take for the energy equation if the equation of state is a gamma law (3.8)? Compare the exact form (4.50) and the approximation (4.56) and the two step method for a gamma law equation of state.

To summarize, an explicit finite difference representation for Lagrangian adiabatic hydrodynamics equations in one dimensional planar geometry is given by (4.47)-(4.49), with either (4.50) or (4.56) for the internal energy equation.

Stability

The hydrodynamic equations above are nonlinear, so the von Neumann stability method can not be applied directly to them. If they are linearized, we recover the equations describing the propagation of adiabatic sound waves which were studied in Section 4.2 to which the Courant condition (4.31) applied. There is no reason to assume that the same stability condition is valid for the full nonlinear equations (including an arbitrary equation of state). This is typical of many instances encountered in computational physics. In practice when a stability constraint is known for the linearized form of the equations it is often found that the same constraint can be adapted to the nonlinear equations. This must usually be done empirically. For example, experience has demonstrated that stability of the Lagrangian adiabatic hydrodynamic finite difference equations above is obtained if the Courant time step constraint (4.31) is applied in the form

$$\Delta t^{n+3/2} = \Delta t_C = c_1 \; min \; \{ \Delta x^{n+1}_{k+1/2} / c^{n+1}_{s,k+1/2} \} \; , \tag{4.58}$$

where the adiabatic sound speed is given by (3.4). The sound speed will clearly vary from zone to zone. The time step is to be taken as the minimum over the entire spatial grid. The constant c_1 is only constrained in linear theory to be less than unity; however for the full hydrodynamic equations, nonlinear effects can lead to instabilities for c_1 near unity. In practice it is usually sufficient to set $c_1 \leq 0.5$, although it may be necessary during some stages in a calculation to temporarily lower c_1 below this value. For problems involving subsonic motion it is often possible to set c_1 to a value slightly less than unity.

Time-Step Control

During the course of typical calculations the sound speed in the system may change significantly. Because the Courant stability constraint is based in part on the magnitude of the sound speed, the time step will also vary, perhaps over several orders of magnitude. Since it would be impractical (and indeed unnecessary) to use the smallest time step for the entire calculation, it is desirable to allow Δt^n and $\Delta t^{n+1/2}$ to change from cycle to cycle. Consider for example a typical stellar core collapse calculation which starts with a mass whose central density is a few times 10^9 g/cm^3 and which collapses in about 0.2 $sec.$ to a central density $\rho_b = 2 \times 10^{14}$ g/cm^3. The initial configuration can be calculated with a time step of order 10^{-3} sec, but when the density reaches ρ_b the time step must typically be of order $2 \times 10^{-6} sec$. A calculation of this type requires a few hundred computational cycles, and can be completed in a matter of minutes on large computers if a variable time step is used. If the entire calculation were to be done with a constant Δt it would require about 10^5 computational cycles, and several tens of hours to complete.

In general it will be assumed that the time step can change from cycle to cycle (in practice it will increase during some portions of a calculation, and decrease during others). If $\Delta t^{n+3/2} > \Delta t^{n+1/2}$, then it is usually best not to increase the time step by more than about 20%. This is usually not a severe restriction, since it will allow the time step to increase by about an order of magnitude in every twelve cycles if no other processes limit the time step. A large increase in the time step in one cycle should be avoided, since it may produce anomolous signals which could lead to problems at later times.

The time step Δt^{n+1} must now be specified. This may be done by requiring that the finite difference expression for the acceleration (the left side of the momentum equation) converge to d^2r / dt^2 in the limit as the time steps approach zero (see Problem 2.21). This will occur if the new time step at $n+1$ is defined by

$$\Delta t^{n+1} = \frac{1}{2}\left(\Delta t^{n+3/2} + \Delta t^{n+1/2}\right). \tag{4.59}$$

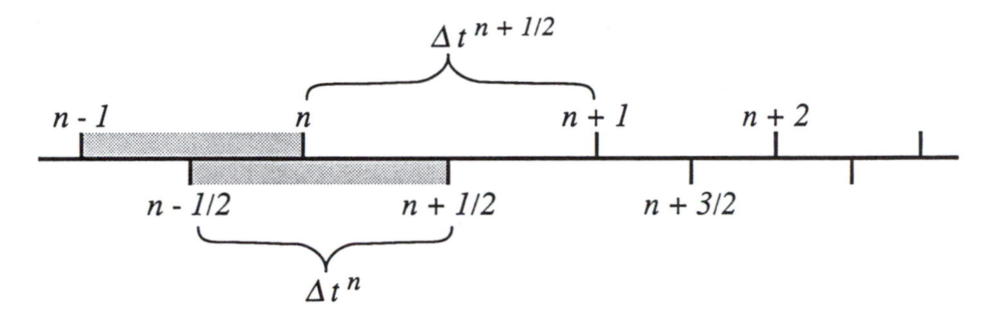

Figure 4.7. Hydrodynamic time steps; the upper portion shows the time steps used to advance the energy, the density and the coordinates. The lower portion shows the time steps used to advance the velocity. The shaded regions represent values for the current cycle.

Figure 4.7 shows the relation between the two time steps in a typical calculation. The new cycle time is defined to be

$$t^{n+1} = t^{n} + \Delta t^{n+1/2} \ .$$

Occasionally the onset of some physical process will result in a drastic decrease (perhaps by several orders of magnitude) in the time step in just a few cycles. This usually offers no difficulties as long as the time step can change rapidly enough so that during each subsequent cycle the calculations are stable.

Momentum and Energy Conservation

Local conservation of the momentum and total energy of an infinitesimal Lagrangian fluid element are consequences of the partial differential equations (4.12)-(4.14). These are fundamental physical properties of the equations which should be preserved by any finite difference representation. It may be recalled that the Lagrangian construct developed above automatically conserves mass. Furthermore, we recall that although (4.14) is an equation for the internal energy, the momentum equation (4.12) may be used in conjunction with it to obtain an equation for the evolution of the total (internal plus kinetic) energy. It follows from this that in the differential limit if momentum and internal energy conservation are satisfied, the total energy conservation will also hold. Unfortunately this is not necessarily true at the finite difference level. Therefore, the issue of energy conservation at the finite difference level must be examined.

We begin by noting that the explicit Lagrangian scheme developed above conserves linear momentum locally. Equation (4.12) states that the rate of change in momentum of a Lagrangian fluid element equals the net force acting on it. By construction the finite difference approximation (4.47) describes this for the finite Lagrangian mass element Δm_k. Global momentum conservation is also guaranteed, as follows from (4.47) by summing over the spatial grid (notice that the sum is over the staggered grid whose mass elements are centered at k):

$$\sum_{k=1}^{K} \frac{(v_k^{n+1/2} - v_k^{n-1/2})}{\Delta t^n} \Delta m_k + \sum_{k=1}^{K} (P^n_{k+1/2} - P^n_{k-1/2}) = 0. \tag{4.60}$$

The total change in momentum in the grid during the time interval Δt^n is given by the left hand term in (4.60), and is just the negative of

$$\sum_{k=1}^{K} (P^n_{k+1/2} - P^n_{k-1/2}) = P^n_{K+1/2} - P^n_{1/2} \,.$$

Since pressure contributions for the interior points cancel exactly, the sum above is exact to machine roundoff. The right side is simply proportional to the net pressure force acting on the boundaries. Thus the net force acting on the system is due to the boundary forces, which is simply a statement of Newton's second law of motion. If the boundary forces vanish, then the total momentum in the grid is conserved.

The partial differential equations also conserve the total energy [kinetic plus internal; see (4.8) with $ds/dt=0$] of an infinitesimal Lagrangian fluid element. The finite difference equations should also preserve this property whenever possible. For simple adiabatic motion, energy is partitioned between internal and kinetic degrees of freedom. Consider a Lagrangian mass $\Delta m_{k+1/2}$; the change in the internal energy density of a Lagrangian fluid element follows directly from (4.54): multiplying by the mass $\Delta m_{k+1/2}$

$$\frac{(\varepsilon_{k+1/2}^{n+1} - \varepsilon_{k+1/2}^{n})}{\Delta t^{n+1/2}}\Delta m_{k+1/2} = - P_{k+1/2}^{n+1/2}(v_{k+1}^{n+1/2} - v_{k}^{n+1/2}) . \tag{4.61}$$

The right side is just the rate at which the pressure does work on the zone boundaries. An expression for the time rate of change in the kinetic energy can be obtained from the momentum equation. Before we can calculate the change in kinetic energy of the zone $k+1/2$ however, a finite difference definition for this quantity must be found. This is nontrivial since the staggered time, staggered space requirements imposed by stability on the finite difference equations results in the velocity and density being defined at different spatial and temporal points (see, for example, Figure 4.5). Since the kinetic energy follows from the momentum change (see Problem 4.4), we should be able to define it using (4.47). However, in order to find a change in kinetic energy which can be compared consistently with (4.61), we seek a definition of $v^2/2$ centered at $k+1/2$ and n. This suggests that we make the following association:

$$v^2/2 \rightarrow \frac{1}{2}v_k^{n+1/2}\, v_k^{n-1/2} \equiv T_k^{\,n} / \Delta m_k. \tag{4.62}$$

In the remainder of this section we will use the symbol T to denote the kinetic energy of a mass element. While the symmetry evident in the definition (4.62) may be comforting, its usefulness (indeed the usefulness of any such finite difference approximation) rests in its consistency with the equations and the physics under investigation. Suppose that we evaluate (4.47) at time $n+1$ and multiply the result by $v^{n+1/2}_k$ to obtain

$$v_k^{n+3/2} v_k^{n+1/2} = (v_k^{n+1/2})^2 - \Delta t^{n+1} \frac{(P^{n+1}_{k+1/2} - P^{n+1}_{k-1/2})}{\Delta m_k} v_k^{n+1/2}.$$

Next, multiply (4.47) by $v^{n+1/2}_k$, and substitute the result into the right hand side of the equation above to obtain, after rearranging terms, the following expression for the change in kinetic energy of a fluid element:

$$\frac{1}{2}(v_k^{n+3/2} v_k^{n+1/2} - v_k^{n+1/2} v_k^{n-1/2})\Delta m_k =$$
$$- v_k^{n+1/2}(\hat{P}^{\,n+1/2}_{k+1/2} - \hat{P}^{\,n+1/2}_{k-1/2})\Delta t^{n+1/2}. \tag{4.63}$$

Here we have defined a pressure at time $n+1/2$ by

$$\hat{P}^{\,n+1/2}_{k+1/2} \equiv \frac{P^{n+1}_{k+1/2}\Delta t^{n+1} + P^{n}_{k+1/2}\Delta t^{n}}{2\Delta t^{n+1/2}}. \tag{4.64}$$

Finally, we define the kinetic energy of $\Delta m_{k+1/2}$ at n, $k+1/2$ by

$$T^{n}_{k+1/2} = \frac{1}{2}(T^{n}_{k+1} + T^{n}_{k}). \tag{4.65}$$

Then (4.63) and (4.65) lead to

$$\frac{(T^{n+1}_{k+1/2} - T^{n}_{k+1/2})}{\Delta t^{n+1/2}} = -\frac{1}{2} v^{n+1/2}_{k+1} (\hat{P}^{n+1/2}_{k+3/2} - \hat{P}^{n+1/2}_{k+1/2})$$
$$- \frac{1}{2} v^{n+1/2}_{k} (\hat{P}^{n+1/2}_{k+1/2} - \hat{P}^{n+1/2}_{k-1/2}).$$
(4.66)

The left side of (4.61) and (4.66) are now centered at the same space-time points, so they may be added to obtain the change in total energy in the time interval $\Delta t^{n+1/2}$. However, the right sides of these equations involve different definitions of the pressure [compare (4.64) and (4.52)] unless $\Delta t^{n+1/2} = \Delta t^{n}$ (constant time step through out the calculation). Making the replacement

$$P^{n+1/2}_{k+1/2} \rightarrow \hat{P}^{n+1/2}_{k+1/2}$$

in (4.61), and adding (4.61) and (4.66) gives

$$\frac{(T^{n+1}_{k+1/2} + \varepsilon^{n+1}_{k+1/2} \Delta m_{k+1/2}) - (T^{n}_{k+1/2} + \varepsilon^{n}_{k+1/2} \Delta m_{k+1/2})}{\Delta t^{n+1/2}}$$
$$+ v^{n+1/2}_{k+1} \hat{P}^{n+1/2}_{k+1} - v^{n+1/2}_{k} \hat{P}^{n+1/2}_{k} = 0.$$
(4.67)

which is a finite difference approximation to the total energy equation (4.8) for an adiabatic system, in which the kinetic energy per unit mass is given by (4.62), and the pressure at $n+1/2$ is given by (4.64). The construction above shows that total energy as defined is conserved only if (4.64) is used in the energy equation (4.54). Unfortunately, this assumes that Δt^{n+1} is known at time t^{n} (see (4.64)). In general, Δt^{n+1} is not determined until after (4.54) has been solved (this leads essentially to an implicit scheme for the hydrodynamics, in which the pressure in (4.54) is defined at the same time as the new energy). It has been suggested above that in practice the change in Δt^{n} and $\Delta t^{n+1/2}$ from cycle to cycle should be small. Under these conditions it is usually reasonable to assume that

$$\hat{P}^{n+1}_{k+1/2} \approx P^{n+1}_{k+1/2} .$$

The finite difference approximation for the momentum equation has been used to show that linear momentum is strictly conserved by the explicit Lagrangian scheme. Equation (4.67), on the other hand, is a derived quantity which does not play a direct role in the solution of the finite difference form of the fluid equations. Therefore the extent to which (4.67) is satisfied gives an indication of the extent to which the scheme conserves total energy. In particular, if (4.67) is not satisfied, it may only mean that the total energy change is not being properly calculated from the instantaneous state of the system rather than implying that the underlying finite difference equations do not conserve energy.

Problem 4.11

Construct a procedure in which (4.64) can be used in the energy equation. Assume that the equation of state is given by the gamma law (3.8).

Spherical Coordinates

The methods used above to obtain a finite difference approximation to the planar Lagrangian hydrodynamic equations may be extended to other one dimensional coordinate systems. A useful system for many stellar problems is spherical coordinates, in which the hydrodynamic equations are given by (4.40)-(4.43). A spherically symmetric spatial grid can be specified initially by the coordinate positions $r^o{}_k$ and the initial state of the material by the velocities $v^o{}_k$, the densities $\rho^o{}_{k+1/2}$, and the specific energies $\varepsilon^o{}_{k+1/2}$ (additional variables will be required to specify such quantities as composition, gravitational potential, radiation fields and magnetic fields if present). The mass interior to a sphere of radius $r^o{}_k$ may be used as Lagrangian coordinates. The zone masses are then given by

$$\Delta m_{k+1/2} = \frac{4\pi}{3}[(r^o_{k+1})^3 - (r^o_k)^3]\rho^o_{k+1/2}.$$

Spatial derivatives are to be expressed in terms of the Lagrangian masses.

First consider the form of the momentum equation in spherical coordinates. Following the discussion leading up to (4.47), we integrate (4.40) over a zone-edge centered mass Δm_k:

$$\int_{\Delta m_k} \frac{dv}{dt}\delta m = \Delta m_k \left(\frac{dv}{dt}\right)_k = \int_{\Delta m_k} \frac{\partial P}{\partial m} 4\pi r^2 \, \delta m.$$

The right hand term may be integrated by parts, assuming that $P_{k+1/2}$ is a constant for $r_k \le r \le r_{k+1}$, to get

$$\int_{\Delta m_k} \frac{\partial P}{\partial m} r^2 \, \delta m = r^2 P \Big|_{k-1/2}^{k+1/2} - \int_{\Delta m_k} P \frac{\partial r^2}{\partial m} \delta m$$

$$= (r^2_{k+1/2} P_{k+1/2} - r^2_{k-1/2} P_{k-1/2}) - P_{k-1/2}(r^2_k - r^2_{k-1/2})$$

$$- P_{k+1/2}(r^2_{k+1/2} - r^2_k) = r^2_k (P_{k+1/2} - P_{k-1/2}).$$

It should be noted that the radii $r_{k\pm 1/2}$, which have not been explicitly defined, do not appear in the final expression for the pressure gradient. Combining the results above gives for the momentum equation in spherical coordinates

$$v_k^{n+1/2} = v_k^{n-1/2} - 4\pi(r_k^n)^2 \frac{(P^n_{k+1/2} - P^n_{k-1/2})}{\Delta m_k} \Delta t^n \tag{4.68}$$

with the zone-edge centered mass Δm_k defined as in (4.46). As expected, the acceleration vanishes if the pressure is uniform $P_{k+1/2} = P_{k-1/2}$. It should be noted that the area appearing in (4.68) is centered at k, rather than at $k \pm 1/2$ where the pressure is defined.

The new coordinates are obtained as for planar coordinates, and are given by

$$r_k^{n+1} = r_k^n + v_k^{n+1/2} \Delta t^{n+1/2}. \tag{4.69}$$

The mass continuity equation (4.41) is solved by dividing the new volume obtained from (4.69) by the (fixed) Lagrangian mass $\Delta m_{k+1/2}$, and is

$$V_{k+1/2}^{n+1} = 1 / \rho_{k+1/2}^{n+1} = \frac{4\pi}{3} \frac{(r_{k+1}^{n+1})^3 - (r_k^{n+1})^3}{\Delta m_{k+1/2}}. \tag{4.70}$$

The specific volume is defined by $V^n{}_{k+1/2}$.

Finally the energy equation (4.42) can be solved in the following manner. In one dimensional spherical coordinates

$$\nabla \cdot \mathbf{v} = \frac{1}{r^2} \frac{\partial r^2 v}{\partial r},$$

and (4.42) may be expressed as

$$\rho \frac{d\varepsilon}{dt} = - \frac{P}{r^2} \frac{\partial r^2 v}{\partial r}.$$

Using (4.41) to rewrite the spatial derivative in terms of Lagrangian coordinates we have

$$\frac{d\varepsilon}{dt} = -4\pi P \frac{\partial r^2 v}{\partial m}.$$

This may now be integrated over $\Delta m_{k+1/2}$, assuming that the pressure $P_{k+1/2}$ is constant throughout the zone; dividing the result by the Lagrangian mass $\Delta m_{k+1/2}$ we find

$$\varepsilon^{n+1}_{k+1/2} = \varepsilon^{n}_{k+1/2} - P^{n+1/2}_{k+1/2} \frac{(A^{n+1}_{k+1} V^{n+1/2}_{k+1} - A^{n+1}_{k} V^{n+1/2}_{k})}{\Delta m_{k+1/2}} \Delta t^{n+1/2}$$

where the area A^n_k is given by

$$A^n_k \equiv 4\pi (r^n_k)^2 .$$

We have specified the pressure centering at time $n+1/2$ as in the planar case, and use the new coordinates and velocities centered at time $n+1$ and $n+1/2$ respectively.

The solution method for the spherical case proceeds exactly as for planar geometry. The Courant stability constraint on the time step is usually sufficient to guarantee stability.

Problem 4.12

Write out a finite difference approximation to the Lagrangian hydrodynamic equations in one dimensional cylindrical coordinates.

Problem 4.13

Do the Lagrangian hydrodynamic equations in spherical coordinates above conserve total energy locally? For simplicity, assume that the time steps and zone masses are constant. Discuss momentum conservation in one dimensional systems exhibiting spherical symmetry.

Boundary Conditions

Boundary conditions must be specified at the boundaries of the computational grid. In spherical geometry an inner $(k=1)$ and an outer $(k=K)$ boundary condition are required. If the inner boundary is located at the origin of coordinates, then the following conditions must be satisfied for all times:

$$r_1 = 0, \qquad v_1 = 0. \tag{4.71}$$

At the outer boundary of a spherical system of radius $R(t)$,

$$r_K = R.$$

The velocity v_K can be specified as a boundary condition, may be determined in the course of the calculation, or may be set by other considerations. For example, at an open boundary the pressure in the boundary zone $P_{K+1/2} = 0$. The inner boundary need not correspond to the origin, but could represent a surface of radius R_i, which could vary with time in a prescribed manner:

$$r_1 = R_i(t), \qquad v_1 = v_i(t). \tag{4.72}$$

Other types of prescribed boundary conditions are also useful in some calculations. A detailed discussion of boundary conditions may be found in Roache (1976).

Problem 4.14

Consider the open boundary condition specified by $P_{K+1/2} = 0$. What happens if the physical boundary is moved inward so that the new boundary corresponds to $P_{L+1/2} = 0$, where $L << K$? Will the two approaches give the same result in all cases?

4.4 SHOCK WAVES

The Lagrangian methods discussed in the previous section are adequate for subsonic flow, but when v approaches the adiabatic sound speed c_s pressure waves may steepen into shock fronts, and the adiabatic hydrodynamic equations are then no longer applicable. A useful frame of reference in which to describe shock waves is one that moves with the shock front's velocity. In this frame the

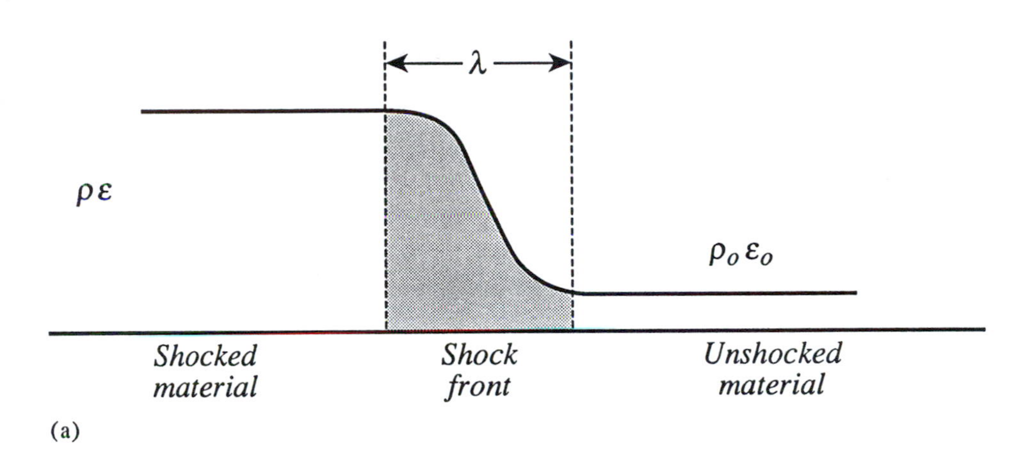

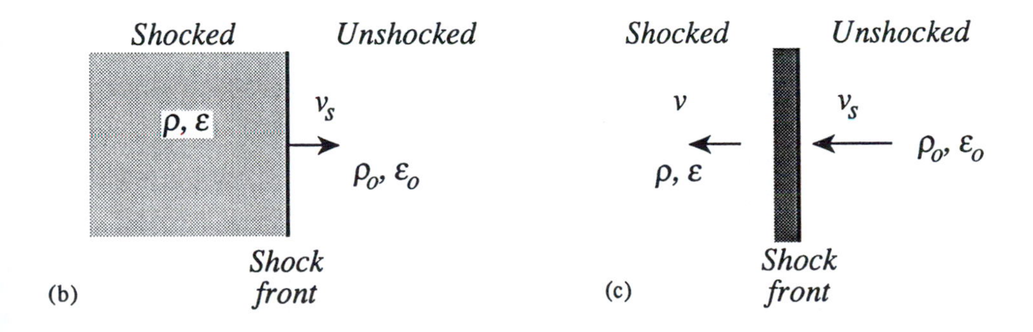

Figure 4.8. Representations of a shock front travelling to the right in a fluid. (a) variation in fluid density and internal energy showing the unshocked fluid into which the front moves, the shock front of thickness l, and the shocked fluid behind it. (b) An ideal shock front (approximated as a discontinuity $l=0$) as seen in the fluid frame. (c) Ideal shock front in a frame of reference at rest relative to the shock.

approaching (upstream) fluid moves with velocity $v \geq c_s$, while the fluid which has passed through the shock front moves subsonically (Figure 4.8). As material flows through the shock front the decrease in its specific kinetic energy is converted into internal energy, and its entropy increases. If the material is ionized and the density is not too low, electrostatic force will tend to couple the electrons and ions so that they move together (no local charge separation will occur) with an average velocity equal to the fluid velocity. After passage through the shock region, a fraction of the fluid's kinetic energy has been converted into thermal energy. Denoting the change in kinetic energy per particle (electron or ion) by $m_\alpha(\Delta v)^2/2$, the resulting temperature of the particle will be of order $T_\alpha \approx m_\alpha(\Delta v)^2/2k_B$. If electron-ion coupling is sufficiently weak, the ratio of ion to electron temperatures will be $T_i / T_e = m_i / m_e$. In many applications the time scale characterizing electron-ion interactions is small enough compared with other time scales of interest that $T_e \approx T_i \approx T$, where T is the material temperature. In these instances the single specific internal energy ε used in the earlier sections is adequate to describe the state of the fluid. The continuous transition region from the unshocked to the shocked fluid has a thickness characterized by particle collisional mean free paths in the fluid. In real fluids thermal conduction and molecular viscosity determine this length scale.

Rankine-Hugoniot Conditions

It is sufficient for many purposes to approximate the shock front by a discontinuity in pressure, internal energy and mass density which moves into the unperturbed fluid with speed v_s (Figure 4.8). Conservation of mass, momentum and energy which must hold across the shock front can be used to relate the density, and specific energy of the shocked matter, and the shock front velocity, to the pressure jump across the shock and to the initial state of the fluid. The results, known as the Rankine-Hugoniot relations, can be written in a number of equivalent forms. Denoting the unshocked fluid variables by subscript zero, one choice of these relations is

$$v = v_o - \left[(P - P_o)(\rho - \rho_o)/\rho\rho_o\right]^{1/2}$$

$$\varepsilon = \varepsilon_o + \frac{1}{2}(P + P_o)(\rho - \rho_o)/\rho\rho_o$$

$$v_s = \frac{\rho v - \rho_o v_o}{\rho - \rho_o},$$

where v_s is the shock speed relative to the laboratory frame. An equation of state of the form (3.1) allows the shocked fluid variable to be fixed in terms of the pressure jump P/P_o across the shock. The Rankine-Hugoniot relations predict the correct entropy jump across the shock front. Another useful representation of these conditions can be expressed in the frame of reference at rest with respect to the shock front (Landau and Lifshitz 1959; Zel'dovich and Raizer 1966; Liberman and Velikovich 1986). Denoting the velocity of the unshocked material relative to the shock front (into which it flows) by u_a and the velocity of the shocked material emerging from the shock front by u_b it is straightforward to show that

$$\frac{V_b}{V_a} = \frac{(\gamma - 1)P_b + (\gamma + 1)P_a}{(\gamma + 1)P_b + (\gamma - 1)P_a},$$

where a (b) denotes the material ahead of (behind) the shock front, and $V \equiv 1/\rho$ denotes the relative volume of the fluid. For an ideal gas, the ratio of the temperature ahead of the shock front to that behind it is clearly

$$\frac{T_b}{T_a} = \frac{P_b}{P_a}\frac{V_b}{V_a}.$$

Finally the velocities of the unshocked and shocked gas are given by

$$u_a^2 = \frac{V_a}{2}\left[(\gamma - 1)P_a + (\gamma + 1)P_b\right]$$

and

$$u_b^2 = \frac{V_a}{2} \frac{[(\gamma+1)P_a + (\gamma-1)P_b]^2}{[(\gamma-1)P_a + (\gamma+1)P_b]}$$

respectively. It should be noted that u_a is the velocity of the shock front in the laboratory frame of reference (when the unshocked fluid is at rest), and u_a is the velocity of the shocked fluid relative to the laboratory frame. These relations may be used to check the accuracy of numerical models of shock wave formation and propagation.

Actually a shock is not a discontinuity; heat conduction and molecular viscosity spread the transition from the unshocked to the shocked material over a distance of several collisional mean free paths, and give rise to the dissipative mechanisms required to increase the material's entropy. These effects follow for example from the viscous equations (4.9)-(4.10). In planar geometry, the momentum equation is (we ignore all but viscous terms for simplicity)

$$\rho \frac{dv}{dt} = \frac{\partial}{\partial x}[(\tfrac{4}{3}\eta + \zeta)\frac{\partial v}{\partial x}]. \tag{4.73}$$

This may be rewritten in Lagrangian coordinates using $dm = \rho\, dx$

$$\frac{dv}{dt} = \frac{\partial}{\partial m}\chi\frac{\partial v}{\partial m}$$

$$\chi \equiv (\tfrac{4}{3}\eta + \zeta)\rho. \tag{4.74}$$

It follows immediately that $\chi \partial v/\partial m$ behaves like a pressure or stress, and acts to change the total momentum of the system only through the boundaries. Multiplying (4.74) by v and integrating over the grid gives

$$\frac{d}{dt}\int \frac{v^2}{2}\delta m = \int v \frac{\partial}{\partial m}\chi\frac{\partial v}{\partial m}\delta m$$

$$= \chi v \frac{\partial v}{\partial m}\Big/_{k=1}^{k=K} - \int \chi\left(\frac{\partial v}{\partial m}\right)^2 \delta m.$$

(4.75)

In many instances the terms evaluated at the grid boundaries will vanish, and then the viscosity acts to reduce the kinetic energy. The viscous change in internal energy in the grid follows from (4.11);

$$\frac{d}{dt}\int \varepsilon \delta m = \int \chi\left(\frac{\partial v}{\partial m}\right)^2 \delta m$$

(4.76)

which is exactly equal (apart from boundary terms) to the reduction in kinetic energy (4.75).

Molecular viscosity spreads shock fronts out over several collisional mean free paths. For matter at ordinary conditions the collisional mean free path is of the order 10^{-5} cm, and modelling a shock transition at these length scales for astrophysical problems would be prohibitive even on modern computers. Despite these limitations, the viscous force term can be successfully modified to model the most important aspects of shock phenomena.

Artificial Viscous Stress

The "viscosity" to be considered below (von Neumann and Richtmyer 1950) is used to construct an artificial viscous stress, which is meant to act when shocks are present, and ideally should have little effect in regions of smooth flow (it is definitely not intended to approximate or represent molecular viscosity). Therefore, χ should be chosen so as to be small when velocity gradients are small, but to be large when velocity gradients are large. Molecular viscosity in an ideal gas is proportional to $\rho v_{th}\lambda$, where v_{th} is the thermal velocity and λ is the collisional mean free path. This form could be adopted in constructing an artificial viscous stress, however it would result in weak shock fronts being spread out over large distances in the fluid. A choice satisfying the requirements above, and having the dimensions of a viscosity, is the artificial scalar viscosity

$$\chi_Q \approx \rho l^2 |\nabla \cdot v| . \tag{4.77}$$

Since the viscosity enters into the momentum equation as a stress, it is conventional to define the artificial viscous stress by

$$Q_q = - \rho l^2 |\nabla \cdot v| \frac{\partial v}{\partial x} = \rho^2 l^2 \frac{dV/dt \, |dV/dt|}{V} . \tag{4.78}$$

The distinction between artificial viscosity, defined for example by (4.77), and artificial viscous stress, defined as in (4.78), is often not made in the literature. It is not uncommon to find the term artificial viscosity used to describe both (4.77) and (4.78). Introducing artificial viscous stress into the momentum equation in planar geometry yields

$$\rho \frac{dv}{dt} = - \frac{\partial P}{\partial x} - \frac{\partial Q_q}{\partial x} . \tag{4.79}$$

In nonplanar geometry an obvious extension of the momentum equation which includes the effects of artificial viscosity involves the introduction of an artificial viscous stress tensor. This could be accomplished in one of several ways. In one dimensional problems, for example, a scalar artificial viscosity such as (4.77) may be used along with the (geometrically correct) expression for a scalar viscosity as in (4.9)-(4.10). This results in a tensor expression for Q_q which replaces (4.78), as will be discussed below.

Finally, the internal energy equation in planar geometry can be constructed in analogy with (4.76) and becomes

$$\frac{d\varepsilon}{dt} = - \frac{(P + Q_q)}{\rho^2} \frac{d\rho}{dt} .$$

(4.80)

The quantity Q_q, which is quadratic in velocity gradients, is a quadratic viscous stress. Other forms (for example linear in the velocity gradients) have often been used as well.

The equations (4.78)-(4.80), and the mass continuity equation, can be solved analytically for a plane, steady-state shock travelling in a fluid described by a gamma law equation of state (Richtmyer and Morton 1967). The solution can be shown to satisfy the Rankine-Hugoniot relations across the shock, and the artificial viscous stress is found to be vanishingly small everywhere except within a distance of order l near the shock front. Thus the analytic model based on (4.78) effectively smears the shock out over a spatial region of extent a few time l. Furthermore the analytic solution demonstrates that the smeared out shock moves into the unperturbed fluid with the same velocity as would the discontinuous shock front. Because the Rankine-Hugoniot relations are also satisfied across it, the artificial viscous stress delivers the correct amount of thermal energy to the fluid behind the shock. The artificial viscous stress (4.78) constitutes a convenient interpolation scheme between the unshocked and shocked fluid. We now consider how to incorporate an artificial viscous stress into the finite difference equations for one dimensional Lagrangian hydrodynamics. It should be noted that the methods developed below are not the only effective way of treating shock wave propagation. Another approach developed by Godunov (1959), and Richtmyer and Morton (1967) is based on solving the one dimensional shock tube problem (the Riemann problem) in each zone, and has been applied in several different numerical schemes (see, for example, Colella and Woodword 1984).

Space-Time Centering

We shall consider first the incorporation of an artificial viscous stress in the one dimensional Lagrangian equations in planar coordinates, and examine the question of centering. From (4.78) it follows that

$Q_q \approx \rho l^2 |\Delta v|(\Delta v)/\Delta x^2$, which is naturally centered at $n-1/2$ and $k+1/2$. Thus we try as a finite difference representation of (4.78) (see Figure 4.9)

$$Q_{k+1/2}^{n-1/2} = -l^2 \rho_{k+1/2}^{n-1/2} \frac{|v_{k+1}^{n-1/2} - v_k^{n-1/2}|(v_{k+1}^{n-1/2} - v_k^{n-1/2})}{(\Delta x_{k+1/2})^2}, \tag{4.81}$$

where

$$\rho_{k+1/2}^{n-1/2} \equiv \frac{1}{2}(\rho_{k+1/2}^{n} + \rho_{k+1/2}^{n-1}).$$

The length scale, l, is to be determined during the calculation. If it is treated as a constant the shock will be spread out over a distance of

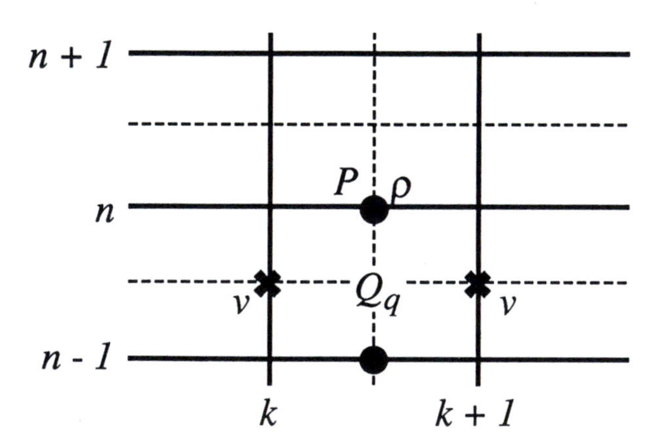

Figure 4.9. Space-time centering for a one dimensional quadratic artificial viscous stress. Also shown are the velocity, pressure and density.

this magnitude. Alternately we may set $l^2 = C_Q\,(\Delta x_{k+1/2})^2$ in which case shocks will be spread out over two to three zones. Typically $C_Q = 2.0$ (see below). In most cases it is desirable to have the shock spread out over several zones. Since the coordinates move dynamically, the zones across which the shock front is spread are in fact relatively thin because the matter there is in compression.

Several constraints are usually imposed on the finite difference representation of Q_q. These include:

$$Q_q \equiv 0 \quad \text{if } \nabla\cdot\mathbf{v}>0$$

$$Q_q \geq 0 \quad \text{if } \nabla\cdot\mathbf{v}<0\,. \tag{4.82}$$

The first condition is usually imposed because shocks only occur in compression, and so there is no *a priori* need for an artificial viscous stress in regions undergoing expansion. For many problems in nonplanar geometry a scalar artificial viscous stress will give quite acceptable results if additional constraints are placed on its use (see

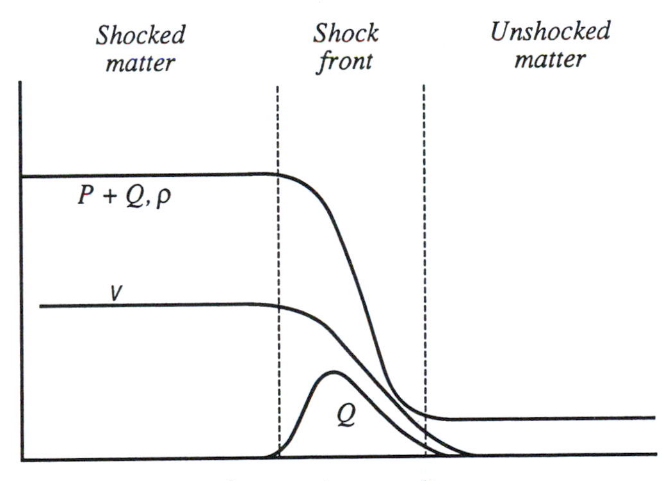

Figure 4.10. Density, pressure plus artificial viscous stress, and artificial viscous stress in a steady shock versus Lagrangian coordinate. Also shown is the fluid velocity, whose gradient is used to establish Q according to (4.81). The shock moves to the right.

below). The behavior of (4.81) with $l \approx \Delta x_{k+1/2}$, is shown schematically in Figure 4.10.

The finite difference approximation for the planar momentum equation including artificial viscous stress is then

$$V_k^{n+1/2} = V_k^{n-1/2} - \frac{(P^n_{k+1/2} + Q^{n-1/2}_{k+1/2} - P^n_{k-1/2} - Q^{n-1/2}_{k-1/2})}{\Delta m_k} \Delta t^n . \tag{4.83}$$

The energy equation (4.80) can be represented in the finite difference form

$$\varepsilon^{n+1}_{k+1/2} = \varepsilon^n_{k+1/2} - \frac{1}{2}(P^{n+1}_{k+1/2} + Q^{n+1/2}_{k+1/2} + P^n_{k+1/2} + Q^{n-1/2}_{k+1/2}) \cdot \frac{(V^{n+1/2}_{k+1} - V^{n+1/2}_k)}{\Delta m_{k+1/2}} \Delta t^{n+1/2}, \tag{4.84}$$

which is just (4.54) with P replaced by an appropriately centered value of $P+Q$. Two important points are to be emphasized. First, the same combination of P and Q enter into the momentum and the energy equations; this is necessary if the finite difference equations are to conserve total energy locally. Second, since $Q^{n+1/2}{}_{k+1/2}$ involves $V^{n+1/2}{}_k$, which is known from the solution of the momentum equation, all we need to do is find an approximation for $P^{n+1}{}_{k+1/2}$ in order to complete the solution of the energy equation (the same method as was used for (4.52) is usually adequate). The heating associated with artificial viscous stress, often termed Q-heating or shock heating, is always positive according to (4.82), as is required for a purely dissipative process.

The artificial viscous stress discussed above works quite well for many shock problems (the question of stability is considered below). Figure 4.11a shows a steady shock in a γ gas calculated with $C_Q = 2.0$, for a pressure jump $P/P_o = 5.0$. The dots represent the value of P/P_o at Lagrangian points in the mesh. Figure 4.11b shows the results of the same calculation with no artificial viscous stress ($C_Q = 0$). The oscillations in pressure about the correct value in the calculation with no artificial viscous stress are not due to instabilities (these will be considered below), but result because the adiabatic equations contain no dissipative mechanism capable of converting the energy jump across the shock into thermal energy. In some applications, a reasonable choice of C_Q will still result in small amplitude oscillations in the pressure behind the shock. These may be removed by adding to Q_q a term linear in the velocity gradient:

$$Q^{n-1/2}_{L,k+1/2} = -C_L(\rho c_s)^{n-1}_{k+1/2}(v^{n-1/2}_{k+1} - v^{n-1/2}_{k}) , \tag{4.85}$$

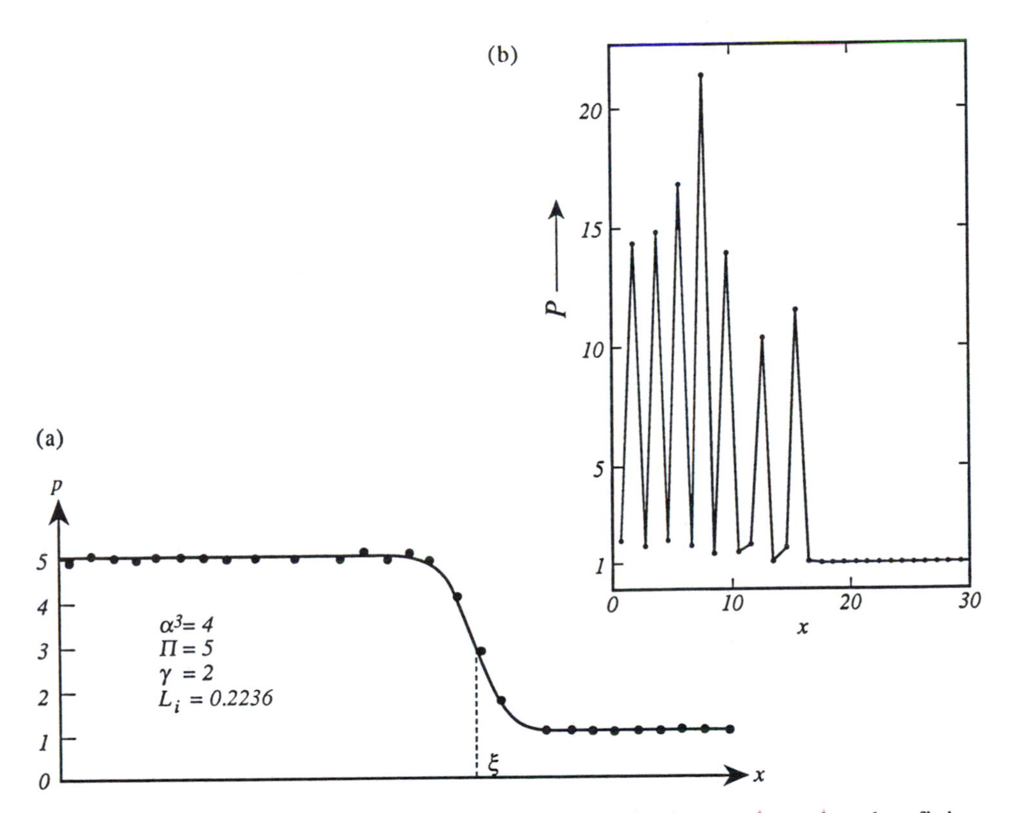

Figure 4.11. Pressure profile in a steady shock as given by the finite difference equations including an artificial viscous stress. (b) Structure of a shock based on the finite difference equations omitting the artificial viscous stress.

where c_s is the local adiabatic sound speed. The linear Q coefficient, C_L, is conventionally constrained by conditions similar to (4.82). The linear artificial viscous stress imposes an additional damping mechanism on a shock, further spreading it out on the grid. Usually $C_L=0$ gives adequate results in astrophysical calculations. When the linear artificial viscous stress imposes an additional damping mechanism on a shock, further spreading it out on the grid. Usually $C_L=0$ gives adequate results in astrophysical calculations. When the linear artificial viscous stress is used, $C_L=0.1$ usually gives sufficient damping.

Stability

The explicit adiabatic Lagrangian hydrodynamic equations are stable for time steps satisfying the Courant condition (4.58). When the artificial viscous stress is small it is reasonable to assume that the hydrodynamic equations will also be stable if the Courant limit is satisfied. However, when the artificial viscous stress is large, this will no longer necessarily be true. Equation (4.83) is nonlinear in the velocity whenever the artificial viscous stress is large; nevertheless a linearized analysis similar to that used for the adiabatic hydrodynamic equations can be useful. Suppose that the momentum equation is written in the form

$$\rho \frac{dv}{dt} = -\frac{\partial Q}{\partial x} = \frac{\partial}{\partial x}\left[\left(\rho l^2 \frac{\partial v}{\partial x}\right) \frac{\partial v}{\partial x}\right] \tag{4.86}$$

where we have used (4.78). For constant density, this is formally identical to the diffusion equation

$$\frac{\partial v}{\partial x} = D \frac{\partial^2 v}{\partial x^2}$$

with the diffusion coefficient [the factor of two arises because (4.86) is quadratic in the velocity gradient]

$$D \equiv 2l^2 \left|\frac{\partial v}{\partial x}\right| . \tag{4.87}$$

Equations (4.81) and (4.83) represent an explicit solution to (4.86), to which the results of Problem 2.8 apply. In particular, the stability limit (2.31) using (4.87) becomes

$$\Delta t_Q \leq \frac{\Delta x^2}{4C_Q|\Delta v/\Delta x|} \rightarrow \frac{C_2 \Delta x_{k+1/2}}{4C_Q|v_{k+1}^{n+1/2} - v_k^{n+1/2}|} . \tag{4.88}$$

This guarantees (with $C_2 \leq 1$) that the explicit diffusion equation is stable for constant coefficients D. Since D is seldom constant, we take $C_2 \approx 0.5$, and adopt as the hydrodynamic time step

$$\Delta t^{n+3/2} = min\{\Delta t_C , \Delta t_Q\}, \tag{4.89}$$

where the minimum is taken over the grid. In practice it is best to calculate the inverse time step, and then take the maximum over the grid, since the denominator in (4.88) may be zero for some zones.

Although rigorous constraints have not been established, the results above are usually adequate in practice. Fortunately, instabilities resulting from artificial viscous stress can usually be recognized in the behavior of the Q near the shock front. Figure 4.12 shows, schematically, the pressure and artificial viscous stress when too large a time step is used in solving (4.83). This should be compared with Figure 4.10, which shows a stable solution.

Problem 4.15

Because the pressure is centered at $n,k+1/2$ in the momentum equation, we could consider centering Q there also. Define Q at $n,k+1/2$ and use the linearized diffusion analogy to show that the resulting explicit analog of (4.86) is unconditionally unstable.

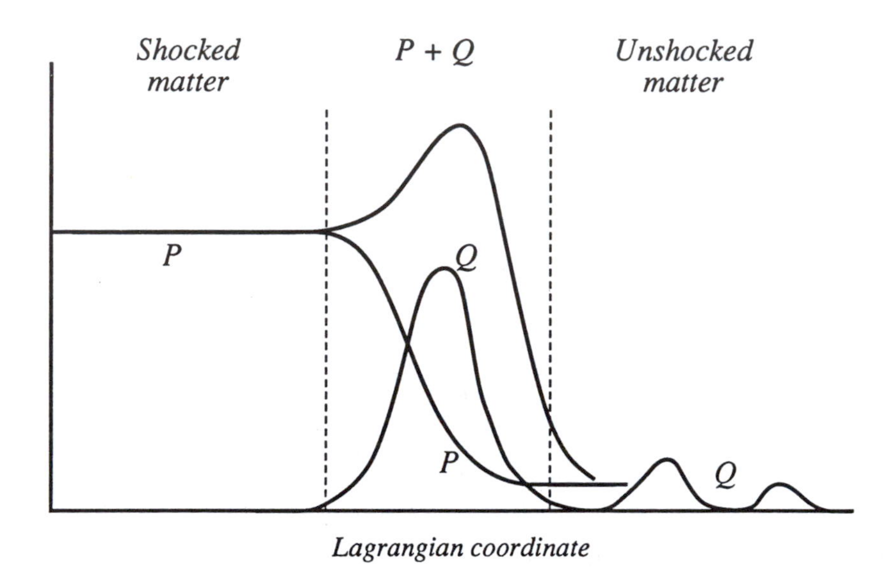

Figure 4.12. Pressure and artificial viscous stress exhibiting an instability associated with the Q. The shock moves to the right.

A relatively simple modification to the artificial viscosity can be made by interpolating the velocities to the zone center, and using the value of Δv there to construct the artificial viscous stress. This method, which can be applied to Lagrangian and Eulerian equations will be discussed in Section 5.13 after monotonic interpolations schemes have been discussed in Chapter 5.

Spherical Coordinates

The analysis following (4.79) assumes planar coordinates and a scalar artificial viscous stress. In nonplanar coordinates it can be argued that the formulation of artificial viscous stress should be done in tensor form, and the results incorporated with the adiabatic equations (4.68)-(4.70a). We shall consider a tensor form for spherical coordinates next. The use of a tensor artificial viscous stress is usually important only when shocks converge at the axis. When this is not expected, the simplier planar stress discussed previously may be adequate.

Starting with the viscous stress tensor written in Cartesian coordinates (4.10), it can be shown that the real viscous acceleration associated with a viscous stress obtained from a scalar viscosity in one dimensional spherical geometry is

$$\rho \frac{dv}{dt} = \frac{1}{r^3} \frac{\partial}{\partial r} [r^3 \chi (\frac{\partial v}{\partial r} - \frac{1}{3} \nabla \cdot \mathbf{v})]. \tag{4.90}$$

The derivation is straightforward but tedious; first all partial derivatives in (4.10) must be replaced by covariant derivatives, and the result evaluated in spherical coordinates assuming that the velocity component in the θ and ϕ directions is zero. The quantity in square braces above represents a tensor viscous stress in one dimensional spherical coordinates. The expression for the real viscous stress acceleration can be used as a guide to construct an artificial viscous stress: define the artificial viscous stress Q_S by

$$\begin{aligned} - r^3 Q_S &\equiv r^3 \chi (\frac{\partial v}{\partial r} - \frac{1}{3} \nabla \cdot \mathbf{v}) \\ &= \frac{2}{3} r^4 \chi \frac{\partial}{\partial r} (\frac{v}{r}). \end{aligned} \tag{4.91}$$

The real viscosity χ must also be replaced by an artificial viscosity. The choice of an artificial viscosity can be motivated by arguments similar to those which lead to (4.77), with

$$\nabla \cdot \mathbf{v} = \frac{1}{r^2} \frac{\partial r^2 v}{\partial r}$$

there and in (4.91). Using the continuity equation, we define the scalar artificial viscosity by

$$\chi = C_Q l^2 |d\rho/dt|. \tag{4.92}$$

The tensor form (4.91) possesses the useful property that Q_S vanishes for homologous contraction where $v \approx -r$. Combining (4.91)-(4.92) we have

$$Q_s = - C_Q l^2 \left|\frac{d\rho}{dt}\right| \left(\frac{\partial v}{\partial r} - \frac{1}{3}\nabla \cdot \mathbf{v}\right). \tag{4.93}$$

Problem 4.16

Show that the artificial viscous stress in (4.90) is dissipative and converts kinetic energy into thermal energy.

The artificial viscosity heating equation in spherical coordinates can be obtained from (4.11) by transforming to spherical coordinates as described above (4.91). The result can be used as a guide in constructing a partial differential equation for the artificial viscous heating. The result is easily shown to be

$$\rho\frac{d\varepsilon}{dt} = \left|\frac{2\chi}{3}\right| \left[r\frac{\partial}{\partial r}\left(\frac{v}{r}\right)\right]^2 = -Q_S r\frac{\partial}{\partial r}\left(\frac{v}{r}\right). \tag{4.94}$$

In the last step the real viscosity has been replace by an artificial viscosity as described above.

Space-Time Centering

There is a greater apparent degree of ambiguity in how to represent the spherical tensor Q in finite difference form than there is for a planar Q. The right side of (4.94) can be written in several forms, as indicated below. A possible choice of centering suggested by the form (4.91) is

$$Q_S \approx r\frac{\partial}{\partial r}\left(\frac{v}{r}\right) = \frac{\partial v}{\partial r} - \frac{v}{r} \rightarrow Q_{k+1/2} \approx \frac{v_{k+1} - v_k}{r_{k+1} - r_k} - \frac{v_{k+1} + v_k}{r_{k+1} + r_k}.$$

The disadvantage of this approximation is that it vanishes for the zone at the origin ($r_1=0$ and $v_1=0$ gives $Q_{3/2}=0$). In effect, we need to decide how to zone center the quantities $\partial v/\partial r$ and v/r in a consistent manner. One choice of centering can be found by examining the volume average of $\nabla\cdot\mathbf{v}$ in spherical coordinates (Whalen 1985):

$$\int_{\Delta V_{k+1/2}} \nabla\cdot\mathbf{v}\,\delta V = \int \frac{1}{r^2}\left(\frac{\partial}{\partial r} r^2 v\right) 4\pi r^2\,dr = 4\pi \int \frac{\partial}{\partial r} r^2 v\,dr$$

and

$$(\nabla\cdot\mathbf{v}) = \frac{A_{k+1} v_{k+1} - A_k v_k}{\Delta V_{k+1/2}},$$

(4.95)

where $A_k \equiv 4\pi r^2{}_k$. Next consider the volume average

$$\int \frac{\partial v}{\partial r}\delta V = 4\pi \int \frac{\partial v}{\partial r} r^2\,dr = 4\pi r^2 v\Big|_k^{k+1} - 8\pi \int rv\,dr$$

$$= A_{k+1} v_{k+1} - A_k v_k - 8\pi\, v_k \int_{r_k}^{r_{k+1/2}} r\,dr$$

$$- 8\pi v_{k+1} \int_{r_{k+1/2}}^{r_{k+1}} r\,dr = 4\pi r^2 (v_{k+1} - v_k)$$

which implies that

$$\left(\frac{\partial v}{\partial r}\right)_{k+1/2} \equiv A_{k+1/2} \frac{(v_{k+1} - v_k)}{\Delta V_{k+1/2}}. \tag{4.96}$$

The area factor $A_{k+1/2}$ will be defined below. Finally, using these results it follows that

$$2\int \frac{v}{r}\delta V = \int \nabla \cdot \mathbf{v}\delta V - \int \frac{\partial v}{\partial r}\, \delta V = \frac{v_{k+1} + v_k}{2}(A_{k+1} - A_k), \tag{4.97}$$

where it is assumed that $A_{k+1/2} = (1/2)(A_{k+1} + A_k)$. Thus (4.96) is a zone-centered velocity gradient and (4.97) is a zone-centered ratio v/r which are consistent with the divergence (4.95). Therefore, a consistent finite difference representation of the spherical artificial viscous stress is given by

$$Q^{n+1/2}_{k+1/2} = \alpha\left[v^{n+1/2}_{k+1}(A^n_{k+1} + 3A^n_k) - v^{n+1/2}_k\,(A^n_k + 3A^n_{k+1})\right]/\Delta V_{k+1/2}\,. \tag{4.98}$$

The factor α could be obtained as an approximation to (4.92). Instead, we require that (4.98) reduce to the planar artificial viscous stress when $\Delta r/r \rightarrow 0$. In this limit all area factors are equal, and comparing (4.98) with (4.81) for $l = \Delta x$ we find

$$C_Q\rho(\Delta v)^2 = 4\alpha\frac{A\Delta v}{\Delta V},$$

or

$$\alpha = \frac{1}{4}C_Q\,\rho^{n+1/2}_{k+1/2}\left(v^{n+1/2}_{k+1} - v^{n+1/2}_k\right)\frac{\Delta V_{k+1/2}}{A_{k+1/2}}. \tag{4.99}$$

Finally, it is common to set

$$C_Q = 0 \quad if \quad v_{k+1}^{n+1/2} > v_k^{n+1/2}, \tag{4.100}$$

which eliminates the artificial viscous stress if the zone is expanding.

Problem 4.17

Consider the artificial viscous stress defined by

$$Q_S \approx \left(\frac{r_{k+1} + r_k}{2\Delta r_{k+1/2}}\right)\left(\frac{v_{k+1}}{r_{k+1}} - \frac{v_k}{r_k}\right).$$

Show that this is a reasonable representation of the differential equation for Q_S and discuss its properties as a representation of the artificial viscous heating.

Having found an expression for $Q^{n+1/2}{}_{k+1/2}$, we determine a finite difference approximation for the viscous acceleration (4.90). Integrating this equation over Δm_k,

$$\begin{aligned}
\int_{\Delta m_k} \frac{dv}{dt}\delta m &= -\int \frac{1}{\rho r^3}\frac{\partial}{\partial r}(r^3 Q)\, 4\pi r^2 \rho\, dr = -4\pi \int \frac{1}{r}\frac{\partial}{\partial r}(r^3 Q)\, dr \\
&= -4\pi r^2 Q\Big|_{k-1/2}^{k+1/2} - 4\pi \int rQ\, dr \\
&= -A_{k+1/2}Q_{k+1/2} + A_{k-1/2}Q_{k-1/2} - \frac{1}{2}Q_{k-1/2}(A_k - A_{k-1/2}) \\
&\quad - \frac{1}{2}Q_{k+1/2}(A_{k+1/2} - A_k),
\end{aligned}$$

which reduces to

$$\frac{v_k^{n+1/2} - v_k^{n-1/2}}{\Delta t^n} = - Q_{k+1/2}^{n-1/2} \frac{(3A_{k+1}^n + A_k^n)}{4\Delta m_k} + Q_{k-1/2}^{n-1/2} \frac{(3A_{k-1}^n + A_k^n)}{4\Delta m_k}. \tag{4.101}$$

As in (4.83), the artificial viscous stress is time centered at $n-1/2$ when the pressure is centered at n (recall the result of Problem 4.15). The momentum equation including artificial viscous stress in one dimensional spherical coordinates is then given by adding (4.101) multiplied by Δt^n to the right hand side of (4.68).

The energy equation (4.94) integrated over the Lagrangian volume $\Delta V_{k+1/2}$, can be rewritten in the following form:

$$\int_{\Delta V_{k+1/2}} \rho \frac{d\varepsilon}{dt}\, \delta V = - \int Q_S \left(\frac{\partial v}{\partial r} - \frac{v}{r}\right) \delta V = - Q_{k+1/2} \int \left(\frac{\partial v}{\partial r} - \frac{v}{r}\right) \delta V$$

$$= - Q_{k+1/2} \left[\frac{A_{k+1} + A_k}{2} (v_{k+1} - v_k) - \frac{A_{k+1} - A_k}{2} \frac{v_{k+1} + v_k}{2} \right];$$

dividing by $\Delta m_{k+1/2}$, the viscous change in internal energy is given by

$$\frac{\varepsilon_{k+1/2}^{n+1} - \varepsilon_{k+1/2}^{n}}{\Delta t^{n+1/2}} = - Q_{k+1/2}^{n} \frac{\left[v_{k+1}^{n+1/2} (A_{k+1}^n + 3A_k^n) - v_k^{n+1/2} (A_k^n + 3A_{k+1}^n) \right]}{4\Delta m_{k+1/2}} \tag{4.102}$$

As in (4.84) for planar geometry, the right side of (4.102) uses the velocity and coordinates centered at times $n+1/2$ and n, respectively. Finally, since the pressure is to be centered at $n+1/2$ [see (4.70a)], the artificial viscous stress may be centered at n :

$$Q_{k+1/2}^n \equiv \frac{1}{2} (Q_{k+1/2}^{n+1/2} + Q_{k+1/2}^{n-1/2}). \tag{4.103}$$

(Recall (4.52). The change in internal energy, including heating associated with artificial viscous stress is obtained by adding the right hand side of (4.102) times $\Delta t^{n+1/2}$ to (4.70a).

The tensor and the scalar artificial viscous stress differ most strongly when shocks occur in portions of the grid where $\Delta r/r$ is large, that is near the origin. Therefore, for many problems which do not involve large pressure gradients near the origin, using the simpler scalar artificial viscous stress is usually adequate. In this case it is sufficient to define $Q^{n-1/2}{}_{k+1/2}$ by (4.81)-(4.82), replace Δx by Δr , and make the replacement $P \rightarrow P+Q$ in (4.68) and (4.70a) to obtain the spherical Lagrangian hydrodynamic equations.

4.5 Rotation in Spherical Systems

Rotation is observed in many stellar systems, including main sequence stars, white dwarfs in binary systems, and evolved stellar objects such as pulsars and compact x-ray sources. Protostellar clouds are also believed to possess rotation. In general these systems involve motions in two or three dimensions, and cannot be modeled by the equations describing one dimensional hydrodynamics. In slowly rotating systems (where, for example, the kinetic energy of rotation is much smaller than the gravitational potential energy) the deviation from sphericity is usually small and some effects of rotation can be approximated using one dimensional models. In these cases rotation is considered to be a small perturbation away from sphericity.

In the lowest order one dimensional approximations rotation appears as a centripetal acceleration term in the momentum equation. In spherically symmetric systems, the momentum equation (4.40) includes on the right side a term a_{rot} representing the centripetal acceleration averaged over angle. Assume that each Lagrangian mass element δm rotates uniformly about the z axis, with an acceleration normal to the z axis $a_R(\theta)$, where θ is the angle between the rotation axis and the radius vector to δm (see Figure 4.13). The total rotational acceleration of the mass shell directed in the radial direction is given by

$$a_{rot} = \frac{1}{4\pi} \int a_R(\theta) \sin\theta \, d\Omega = \frac{1}{2} \int_{-\pi/2}^{\pi/2} \omega^2 r \sin\theta \, d\theta = \frac{\pi \omega^2 r}{4}$$

(4.104)

where ω is the angular velocity of the shell, $a_R(\theta) = r\omega$. The angular momentum of the Lagrangian element δm is $\omega r_\perp{}^2$; integrating over the mass shell Δm_r gives the angular momentum of the shell

$$\int \omega r_\perp^2 \frac{d\Omega}{4\pi} = \frac{\omega r^2}{2} \int \sin^3\theta \, d\theta = \frac{2}{3}\omega r^2 = \frac{2}{3} J(r),$$

(4.105)

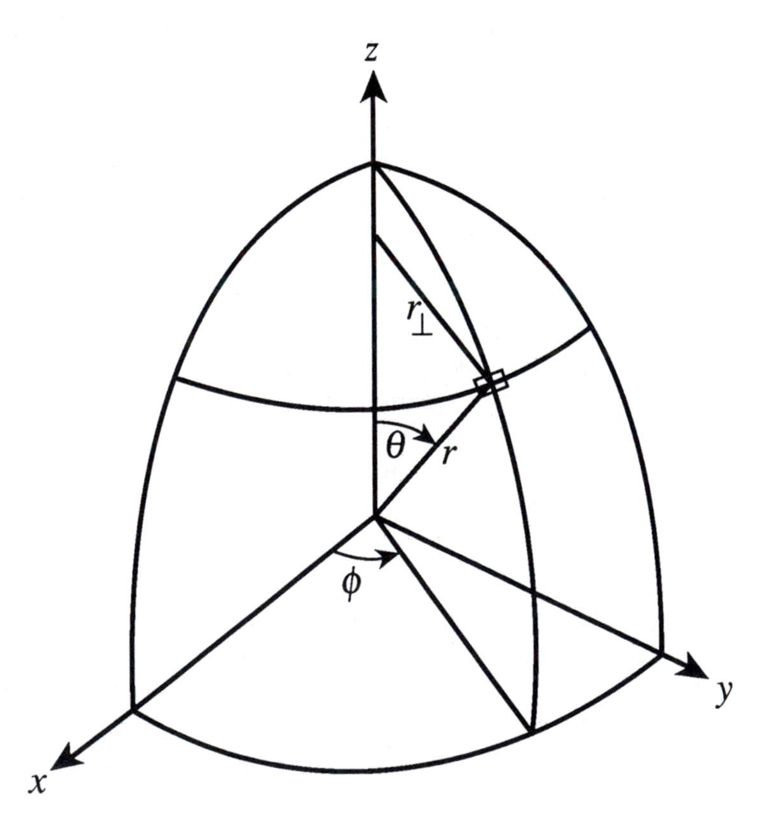

Figure 4.13. Geometry used to calculate the rotational acceleration in a one dimensional system in spherical coordinates.

Note that r is the magnitude of the radius vector from the center of the system. Finally, the rotational energy of the mass shell Δm_r is given by

$$\Delta E_{Rot}(r) = \frac{J(r)^2 \Delta m_r}{3r^2} .$$

(4.106)

Problem 4.18

Derive the rotational energy of a mass shell (4.106). What input should be used to model a $1\ M_{O.}$ star rigidly rotating with total rotational energy $E_{Rot}= 1\times 10^{48}\ ergs$?

The initial rotation model requires values of J_k for each mass shell Δm_k. The corresponding angular velocity ω_k may be found thereafter from (4.105). Then the finite difference approximation

$$a^n_{R,k} = \frac{\pi}{4} \frac{J^2}{(r^n_k)^3} \Delta t^n$$

(4.107)

may be added to the right side of (4.68). The velocity $v^{n+1/2}{}_k$ then contains, in addition to the pressure gradient acceleration, the acceleration due to rotation. The angular momentum J appearing in (4.107) is a constant of the motion.

4.6 NEWTONIAN GRAVITATION

It is relatively easy to add gravitational forces to the one dimensional explicit Lagrangian scheme discussed above. Relativistic theories of gravitation require additional development, as do the finite difference methods applicable to Newtonian gravitation in higher dimensions, and these will be postponed until Chapter 9. Newtonian gravitation acts as a body force f in the momentum equation (4.6) (the mass continuity and internal energy equations remain unchanged, since Newtonian gravitation makes no contribution to either mass or internal energy); it may be written in terms of the gravitational potential Φ:

$$\rho \frac{d\mathbf{v}}{dt} = -\nabla P - \rho\,\nabla\Phi \tag{4.108}$$

where Φ is obtained from the local mass density and Poisson's equation

$$\nabla^2\Phi = 4\pi G\,\rho\,. \tag{4.109}$$

For one dimensional spherically symmetric systems the gravitational potential is easily constructed in terms of the mass $m(r)$ interior to radius r, and these two equations may be combined to give

$$\rho \frac{dv}{dt} = -\frac{\partial P}{\partial r} - \frac{\rho\, m(r)\, G}{r^2}. \tag{4.110}$$

An explicit finite difference representation of the acceleration due to the pressure gradient has already been given (4.68), and we need only add to this the acceleration due to the gravitational force. If v^P includes the change in velocity due to the pressure gradient [as given by (4.68)], then the net change due to the pressure gradient and gravitation can be approximated by

$$v_k^{n+1/2} = v_k^P - \frac{m_k\, G\, \Delta t^n}{r_k^2}, \tag{4.111}$$

where m_k is the mass contained within a sphere of radius r_k, which is

$$m_k = \sum_{i=1}^{k-1} \Delta m_{i+1/2} \ . \tag{4.112}$$

We note that the gravitational acceleration and the pressure gradient acceleration have not been operator split in (4.111).

Problem 4.19
Show that

$$V_k^{n+1/2} = V_k^P - \frac{m_k\, G\, \Delta t^n}{r_k^2 + V_k^{n-1/2}\, r_k\, \Delta t^n} \tag{4.113}$$

is an approximation to the gravitational acceleration which is second order in time.

The effects of gravitation as described by (4.110) may be included in a straightforward way with the finite difference approximations for artificial viscous stress and mechanical work and rotation as developed in previous sections.

4.7 Implicit Hydrodynamics

Explicit hydrodynamic schemes such as the one discussed in the previous sections are suitable for most one-dimensional dynamic phenomena in astrophysics where the system changes significantly on a free-fall or sonic time scale. They could also be used for quasidynamic processes where the evolution of the system occurs over times large compared to τ_S or τ_G. Because of the Courant time step constraint (4.58) modelling phenomena which occur during some stages of stellar evolution using explicit difference schemes could require of order 10^7 computational cycles or more, which is clearly impractical. In calculations of quasidynamic processes the small time scale behavior associated, for example, with sound wave propagation, is not of primary interest, and it may be desirable to employ methods which can accommodate time steps that are very large compared to τ_S. An implicit scheme will be considered next which can be used to describe quasistatic phenomena, dynamic phenomena and situations where both types of processes may be present during the evolution of a system. These schemes are recommended only for problems where a large time step in required.

An example of an implicit finite difference scheme describing adiabatic sound propagation was discussed at the conclusion of section 4.2. That method suggests an approach to an implicit Lagrangian hydrodynamic algorithm.

Problem 4.20

Consider the finite difference approximation for adiabatic sound waves ($0 < \alpha < 1$):

$$\psi_k^{n+1} = \psi_k^n - \sigma\left[\alpha\phi_{k+1/2}^{n+1} + (1-\alpha)\,\phi_{k+1/2}^n - \alpha\phi_{k-1/2}^{n+1} - (1-\alpha)\,\phi_{k-1/2}^n\right]$$

$$\phi_{k+1/2}^{n+1} = \phi_{k+1/2}^n - \sigma\left[\alpha\psi_{k+1}^{n+1} + (1-\alpha)\,\psi_{k+1}^n - \alpha\psi_k^{n+1} - (1-\alpha)\,\psi_k^n\right]$$

(4.114)

where $\sigma \equiv c_S \Delta t / \Delta x$. Under what conditions are these equations unconditionally stable? Discuss the accuracy of the finite difference approximation and its dependence on the value of α.

Finite Difference Equations

Consider the spherically symmetric, one dimensional finite difference equations (4.68)-(4.70a); the discussion above suggests that these may be converted to an implicit scheme by replacing variables evaluated at time t^n on the right hand side of the equations by their new values at time t^{n+1}; this gives a fully implicit scheme. The results of problem 4.20 suggest that we could also replace values at time t^n by a linear combination of old and new values: for example, the linear time average of a variable $A^n{}_k$ could be replaced by

$$\tilde{A}_k \equiv \alpha A_k^{n+1} + (1-\alpha) A_k^n, \quad 0 < \alpha < 1. \tag{4.115}$$

When $\alpha = 1$, a scheme based on (4.115) is fully implicit; for $\alpha = 0$ it is explicit and will require two time levels.

We consider here a specific example which is fully implicit ($\alpha = 1$), and assume that all nonadiabatic effects (except for artificial viscous stresses) are to be separately included by the method of operator splitting (Oliphant 1980; Oliphant and Witte 1987). This assumption is made in order to simplify the solution procedure. Nonsplit schemes can also be developed in one and two spatial dimensions, but these will not be considered here. The radial coordinates and velocities are defined on a staggered space-time grid containing two time levels as was the explicit Lagrangian hydrodynamics scheme discussed earlier;

$$v_k^{n+1/2} = \frac{r_k^{n+1} - r_k^n}{\Delta t^{n+1/2}} . \tag{4.116}$$

The velocity appearing in (4.116) is associated with a time midway between $r^n{}_k$ and $r^{n+1}{}_k$. The mass density of zone $k+1/2$ at time t^{n+1} is given by

$$\rho_{k+1/2}^{n+1} = \frac{3}{4\pi} \frac{\Delta m_{k+1/2}}{[(r_{k+1}^{n+1})^3 - (r_k^{n+1})^3]} , \tag{4.117}$$

where $\Delta m_{k+1/2}$ is the zone mass and may be used as the Lagrangian coordinate.

Implicit Momentum Equation

The implicit momentum equation, including an artificial viscous stress, may be derived from the partial differential equation following the arguments above (4.68), and is given by

$$\frac{v_k^{n+1/2} - v_k^{n-1/2}}{\Delta t^n} = -4\pi \frac{(r_k^{n+1})^2}{\Delta m_k} [P_{k+1/2}^{n+1} - P_{k-1/2}^{n+1} + Q_{k+1/2}^{n+1/2} - Q_{k-1/2}^{n+1/2}] ,$$

(4.118)

which should be compared with the explicit form (4.68). Note in particular that the radius, pressure and artificial viscous stress on the right hand side of the implicit finite difference equation are evaluated at the new time. For quantities such as pressure and radius the new time is t^{n+1}, while for the velocity and artificial viscous stress (which is a function of velocity) it is $t^{n+1/2}$. The form of the artificial viscous stress is, apart from time centering, usually chosen to be identical to the explicit form, and may be expressed in terms of the velocities for zones undergoing compression by

$$Q_{k+1/2}^{n+1/2} = C_Q^2 \, \rho_{k+1/2}^{n+1} / v_{k+1}^{n+1/2} - v_k^{n+1/2} /^2 .$$

(4.119)

The artificial viscous stress is conventionally assumed to vanish in zones which are expanding or which are in homologous motion. Notice that the density in (4.119) is evaluated at the new time.

Implicit Internal Energy Equation

Finally, a representation of the energy equation is needed to complete the system of equations. At this stage either a total energy equation or an internal energy equation could be adopted. Following the analogy with explicit hydrodynamics, the internal energy will be used here. A straightforward generalization of the explicit internal energy equation (which is only first order in time) gives

$$\varepsilon_{k+1/2}^{n+1} - \varepsilon_{k+1/2}^{n} = - \left(P_{k+1/2}^{n+1} + Q_{k+1/2}^{n+1/2}\right) \frac{(A_{k+1}^{n+1} v_{k+1}^{n+1/2} - A_{k}^{n+1} v_{k}^{n+1/2})}{\Delta m_{k+1/2}} \Delta t^{n+1/2}, \tag{4.120}$$

where the area factor is defined in terms of the radius at the new time by

$$A_k^{n+1} = 4\pi (r_k^{n+1})^2. \tag{4.121}$$

In practice the pressure should be time centered for better accuracy. Note that the pressure appearing in the fully implicit representation (4.120) is to be evaluated in terms of the density and specific energy at the new time:

$$P_{k+1/2}^{n+1} = P(\rho_{k+1/2}^{n+1}, \varepsilon_{k+1/2}^{n+1}). \tag{4.122}$$

If the equation of state is a function of ρ and ε , then (4.116)-(4.122) is a set of seven equations in seven variables, and may be reduced to a set of relations defined for each value of k, $(1 \leq k \leq K)$, of the form (Problem 4.21)

$$f_k(r_{k+1}^{n+1}, r_k^{n+1}, r_{k-1}^{n+1}) = 0. \tag{4.123}$$

The solution of this coupled, nonlinear system of algebraic equations gives the new values of $r^{n+1}{}_k$ for the system. Once these new values have been found, the remaining unknowns may be constructed from (4.116)-(4.122). For example, the velocity follows from (4.116), the density from (4.117) and the artificial viscous stress from (4.119). Finally, the internal energy is obtained from (4.122) and (4.120). For simple equations of state, such as the gamma law (3.8), solving for the internal energy is trivial. A solution method for the system of equations (4.123) will be discussed below.

Problem 4.21

Show by direct substitution using (4.116)-(4.122), and by assuming that the equation of state (4.122) is a gamma law (3.8), that the solution for the fully implicit finite difference equations is of the form (4.121).

Internal Energy (Temperature-Based)

In some cases an equation of state of the form (4.122) may not be available and one must work with a temperature-based model:

$$\varepsilon^{n+1}_{k+1/2} = \varepsilon(\rho^{n+1}_{k+1/2}, T^{n+1}_{k+1/2}),$$

$$P^{n+1}_{k+1/2} = P(\rho^{n+1}_{k+1/2}, T^{n+1}_{k+1/2}). \tag{4.124}$$

In general an equation of state of the form (4.124) may be quite complicated, and its use with (4.120) in a fully implicit scheme may lead to a system of equations which are extremely difficult to solve. Therefore, for an equation of state of the form (4.124), a different formulation of the energy equation will be employed (the reasons for this will become apparent later when the solution method for the finite difference equations is discussed). Using (3.15), the first law of thermodynamics (3.5) may be written in the form

$$\left(\frac{\partial \varepsilon}{\partial T}\right)_\rho dT + \left(\frac{\partial \varepsilon}{\partial \rho}\right)_T d\rho = T\, dS + P\frac{d\rho}{\rho^2}. \tag{4.125}$$

Taking the partial derivative of (4.125) at constant temperature gives

$$\left(\frac{\partial \varepsilon}{\partial \rho}\right)_T = T\left(\frac{\partial S}{\partial \rho}\right)_T + \frac{P}{\rho^2}. \tag{4.126}$$

Next, using the Maxwell relation

$$\rho^2\left(\frac{\partial S}{\partial \rho}\right)_T = -\left(\frac{\partial P}{\partial T}\right)_\rho \tag{4.127}$$

to eliminate the derivative of the entropy, (4.125) and (4.126) may be expressed in the final form

$$\left(\frac{\partial \varepsilon}{\partial T}\right)_\rho dT = T\,dS + T\left(\frac{\partial P}{\partial T}\right)_\rho \frac{d\rho}{\rho^2}. \tag{4.128}$$

This is a temperature-based form of the internal energy equation. The nonadiabatic contributions to (4.128), which enter through the entropy term TdS, may be solved separately using the technique of operator splitting. As for the explicit equations, shock heating usually is included with the hydrodynamics. The internal energy equation may then be approximated by the following implicit difference equation, where heating associated with the artificial viscous stress is included in analogy with (4.80):

$$(\varepsilon_T)^n_{k+1/2}\,(T^{n+1}_{k+1/2} - T^n_{k+1/2}) = -\left[T^{n+1}_{k+1/2}\,(P_T)^n_{k+1/2} + Q^{n+1/2}_{k+1/2}\right]\left[\frac{1}{\rho^{n+1}_{k+1/2}} - \frac{1}{\rho^{n}_{k+1/2}}\right]. \tag{4.129}$$

Here ε_T and P_T denote the partial derivatives of the specific energy and pressure with temperature at fixed density. The temperature and artificial viscous stress are evaluated at the new time on the right side. Notice that the temperature derivatives of the specific energy and pressure are evaluated at the old time. This is done to simplify the solution method, and means that not all nonlinear effects in the energy equation are being treated in a fully implicit manner. While it is possible to treat these coefficients in a fully implicit manner, the approximations used here represent a substantial time savings and have been found to work quite well in practice. For temperature-based equations of state, (4.129) replaces (4.120) and (4.124) replaces (4.122), and the solution to the resulting system of equations can in principle be reduced to a set of equations of the form (4.123).

Problem 4.22

Generalize equations (4.125)-(4.127) to include viscous heating, and derive (4.129).

Solution of the Implicit Hydrodynamic Equations

The solution to the system of nonlinear equations (4.123) gives the radius of each Lagrangian zone at the new time, from which all remaining state variables can be constructed. A standard approach for solving coupled nonlinear equations of this type is to use Newton-Raphson iteration (Problem 4.23) and Gaussian elimination (Problem 4.24). Denote the current (initial) value of the Lagrangian radii by $r^n{}_k$. An approximation to the new values at time t^{n+1}, denoted by $r^i{}_k$, will be given by

$$f_k(r^i_{k+1}, r^i_k, r^i_{k-1}) = f_k(r^n_{k+1}, r^n_k, r^n_{k-1})$$

$$+ \left(\frac{\partial f_k}{\partial r^n_{k+1}}\right)_n \Delta r^n_{k+1} + \left(\frac{\partial f_k}{\partial r^n_k}\right)_n \Delta r^n_k + \left(\frac{\partial f_k}{\partial r^n_{k-1}}\right)_n \Delta r^n_{k-1} + \cdots \qquad (4.130)$$

where the partial derivatives are to be evaluated at the initial positions $r^n{}_k$, and the dots represent second and higher order terms.

An approximate solution to (4.123) will then be given by

$$r_k^i = r_k^n + \Delta r_k^n \,, \tag{4.131}$$

where the variations in radius, $\Delta r^n{}_k$, are obtained from the system of equations $(1 \le k \le K)$

$$-\left(\frac{\partial f_k}{\partial r_{k+1}^n}\right)_n \Delta r_{k+1}^n - \left(\frac{\partial f_k}{\partial r_k^n}\right)_n \Delta r_k^n - \left(\frac{\partial f_k}{\partial r_{k-1}^n}\right)_n \Delta r_{k-1}^n = f_k(r_{k+1}^n, r_k^n, r_{k-1}^n) \,, \tag{4.132}$$

which is obtained from (4.130) by equating the left hand side to zero, and dropping terms of second and higher order. Equations (4.132) are a tri-diagonal system which may be solved by means of Gaussian elimination (Problem 4.24). The resulting solution for the new radii (4.131) may then be substituted back into (4.123), which in general will not vanish. Denote its value by

$$f_k(r_{k+1}^i, r_k^i, r_{k-1}^i) = \varepsilon_k^i \,. \tag{4.133a}$$

If the magnitude of all of the errors $\varepsilon^i{}_k$ are small relative to a suitable convergence criterion,

$$|\varepsilon^i{}_k| \le \varepsilon_{conv} \,, \tag{4.133b}$$

then $r^i{}_k$ may be taken as a reasonable solution for the given time step. If the largest of the errors $\varepsilon^i{}_k$ exceeds the convergence criterion as usually happens, then the procedure must be repeated to obtain a better approximation to the radii $r^{n+1}{}_k$. Once the solution is obtained, a new time step is chosen, and the process is repeated. In general, only a few iterations per computational cycle will be needed to satisfy reasonable convergence criteria. However, if the time step for a cycle is too large (compared with the various time scale of the problem) a large number of iterations (in some cases up to a hundred) may be needed to reach convergence. The implicit scheme outlined above has the advantage of permitting calculations with larger time steps than do explicit methods, as well as that of including nonlinear effects which the explicit methods may omit.

Problem 4.23 (Newton-Raphson Method)

Consider a function $f(x)$ of a single independent variable which vanishes for $x = x_o$. An approximation x^{i+1} to the root x_o is given by

$$x^{i+1} = x^{i} - \frac{f(x^{i})}{(\partial f / \partial x)_{x^{i}}}$$

where x^i is the previous guess for the root, and the derivative is evaluated at x^i. Apply this result iteratively to find the positive root of the equation

$$f(x) = e^{-x^{2}} - 0.5$$

starting with initial values $x^1 = 0.5$, $x^1 = 1.8$ and $x^1 = 2.0$. Discuss the implications of the three results.

Problem 4.24

The set of equation (4.132) can be written in the form $(1 \leq k)$

$$A_k \, \Delta r_{k+1} + B_k \, \Delta r_k + C_k \, \Delta r_{k-1} + D_k = 0 , \tag{4.134}$$

where A_k, B_k, C_k and D_k are constant coefficients. Show that the solution to this set of equations is give by the recurrence relations

$$\Delta r_k = \alpha_k \, \Delta r_{k+1} + \beta_k , \tag{4.135}$$

where

$$\alpha_k = \frac{A_k}{B_k - C_k \alpha_{k-1}}, \tag{4.136a}$$

$$\beta_k = \frac{D_k + C_k \beta_{k-1}}{B_k - C_k \alpha_{k-1}} \tag{4.136b}$$

with $1 \leq k$. How must these equations be solved, and how do the boundary conditions enter into the solution?

Implementational Procedure

Two of the most difficult steps in constructing implicit schemes are the elimination of variables to obtain the function (4.123), and the calculation of the partial derivatives which enter into (4.132) which are required to perform the Newton-Raphson interations. Traditionally these two steps are performed analytically, starting with the finite difference equations. The results of Problem 4.21 suggest that these steps may be quite complicated to implement for even the simplest cases. If the algorithm has to be changed (for example, by the addition of a different treatment of artificial viscous stresses, or additional physics packages) the entire process leading to (4.123) and (4.132) must be repeated for each modification. This can make programs based on implicit methods extremely difficult to modify in a dynamic way. We consider next an implementational procedure (Oliphant and Witte 1987; 1988) that has been developed for one dimensional problems to circumvent these difficulties, and which allows implicit schemes to be modified with essentially the same ease as explicit methods. Besides ease of implementation, this method allows new effects, or modifications of existing approximations, to be achieved without perturbing the remainder of the coding. As a result, extensive sets of test problems which may have been run to certify the coding need not all be rerun every time a change in the program is made.

To illustrate the procedure, consider the inviscid fluid equations, and denote the radii at the beginning of the ith iteration by r^i_k. The values at the beginning of a computational cycle are $r^0{}_k = r^n{}_k$. For the $(i+1)$ iteration we define the function

$$f(r^{i+1}_{k+1}, r^{i+1}_{k}, r^{i+1}_{k-1}) = \frac{v^{i+1/2}_k - v^{i-1/2}_k}{\Delta t^n}$$

$$+ 4\pi (r^{i+1}_k)^2 \frac{(P^{i+1}_{k+1/2} - P^{i+1}_{k-1/2})}{\Delta m_k} .$$

(4.137)

Obviously values of $r^{i+1}{}_k$ for which the right hand side of (4.137) vanishes satisfy the momentum equation. We also have

$$v^{i+1/2}_k = \frac{r^{i+1}_k - r^i_k}{\Delta t^{n+1/2}}$$

(4.138)

$$\rho^{i+1}_{k+1/2} = \frac{3}{4\pi} \frac{\Delta m_{k+1/2}}{[(r^{i+1}_{k+1})^3 - (r^{i+1}_k)^3]} .$$

(4.139)

The internal energy equation is

$$\frac{\varepsilon^{i+1}_{k+1/2} - \varepsilon^{i}_{k+1/2}}{\Delta t^{n+1/2}} = -4\pi P^{i+1}_{k+1/2} \frac{[(r^{i+1}_{k+1})^2 v^{i+1/2}_{k+1} - (r^{i+1}_k)^2 v^{i+1/2}_k]}{\Delta m_{k+1/2}} .$$

(4.140a)

If the equation of state is a function of energy and density, then the pressure for the $(i+1)$ iteration is obtained from it:

$$P^{i+1}_{k+1/2} = P(\rho^{i+1}_{k+1/2}, \varepsilon^{i+1}_{k+1/2}) .$$

(4.140b)

For many applications it may be sufficient to linearize the equation of state (4.141b) about $\varepsilon^i{}_{k+1/2}$ and $\rho^i{}_{k+1/2}$. If this is not desired, then an iterative procedure will be required to solve (4.140a) and (4.140b) for the specific energy and the pressure.

We now suppose that a set of radii $r^i{}_k$ are available which satisfy (4.133a), but that not all values of $\varepsilon^i{}_k$ are small enough to consider the solution to have converged [(4.133b) is not satisfied for all radii]. A numerical scheme will be described below which operates on the FORTRAN expressions for (4.137)-(4.140). In this approach the substitutional scheme required to reduce the finite difference equations to the form (4.123) consists of substituting numerical values for the dependent variables into the finite difference equations in sequence. The partial derivatives of $f(r_{k+1}, r_k, r_{k-1})$ appearing in (4.123) are then constructed using numerical derivatives. For example,

$$\frac{\partial f_k}{\partial r_{k+1}} = \frac{f(r_{k+1} + \delta r_{k+1}, r_k, r_{k-1}) - f(r_{k+1}, r_k, r_{k-1})}{\delta r_{k+1}} ,$$

where the δr_k are chosen to be a fraction of the smallest zone width in the problem. The δr_k are not to be confused with the zone widths Δr_k, they are used simply to construct numerical derivatives. The results are used numerically to evaluate the coefficients in (4.134). Gaussian elimination is then performed to obtain new radii $r^{i+2}{}_k$ and to test for convergence. If (4.133b) is not satisfied, then the entire process is then repeated using the new radii. A judicious use of temporary storage arrays and subroutine calls can be developed so that this scheme is fully vectorized. The great advantage of this scheme is that once it has been developed, additional terms may be incorporated into (4.137)-(4.140) simply by modifying the FORTRAN expressions appearing in the program. Furthermore the introduction of additional physics (artificial viscosity, magnetic or radiative acceleration and gravitation) simply require that the appropriate FORTRAN expression be appended to (4.137)-(4.140).

To illustrate how this may be done, consider the inviscid fluid equations (4.137)-(4.140). In the outline below, substitution of variables (such as $r^{i+1}{}_k$) into an equation should be understood to mean setting numerical values into the corresponding FORTRAN arrays. Each iteration in a complete computational cycle consists of the following processes, executed in the order indicated (see Figure 4.14). The initial set of input variables are ${r^{i+1}}_k$ (from the previous iteration or beginning of a computational cycle) and the masses Δm_k and $\Delta m_{k+1/2}$, which do not change in time. In order to solve (4.137) for f_k and its derivatives appearing in (4.132) we need four triplets of radii as shown in the first column of Table 4.1.

Table 4.1

Triplets of radii (first column) used in each of four passes through the implicit hydrodynamic program. The principle result of a pass with each triplet is given in the second column (see Figure 4.14).

Input	Used to construct
(r_{k+1}, r_k, r_{k-1})	$f(r_{k+1}, r_k, r_{k-1})$
$(r_{k+1}+\delta r_{k+1}, r_k, r_{k-1})$	$\partial f_k / \partial r_{k+1}$
$(r_{k+1}, r_k+\delta r_k, r_{k-1})$	$\partial f_k / \partial r_k$
$(r_{k+1}, r_k, r_{k-1}+\delta r_{k-1})$	$\partial f_k / \partial r_{k-1}]$

Step I: Substitute $({r^{i+1}}_{k+1}, {r^{i+1}}_k, {r^{i+1}}_{k-1})$ and $\Delta m_{k+1/2}$ into the right side of (4.138) and (4.139) to obtain ${v^{i+1/2}}_k$ and ${\rho^{i+1}}_{k+1/2}$.

Step II (General Equation of State): (a) Substitute ${\rho^{i+1}}_{k+1/2}$ and an initial guess for ${\varepsilon^{i+1}}_{k+1/2}$ into (4.140b) to obtain ${P^{i+1}}_{k+1/2}$; (b) substitute $\{{r^{i+1}}_k, {v^{i+1/2}}_k\}$, $\Delta m_{k+1/2}$ and the pressure from step IIa into the right side of (4.141a). If the resulting value of ${\varepsilon^{i+1}}_{k+1/2}$ does not agree to within acceptable accuracy with the guess used to evaluate the pressure, increment the initial guess for ${\varepsilon^{i+1}}_{k+1/2}$ and repeat Step IIa; otherwise store ${\varepsilon^{i+1}}_{k+1/2}$.

Step III: The right side of (4.137) may now be evaluated in terms of known quantities to obtain $f({r^{i+1}}_{k+1}, {r^{i+1}}_k, {r^{i+1}}_{k-1}\}$.

Step IV: Using the remaining triplets of radii appearing in the last three rows of Table 4.1 in place of $(r^{i+1}{}_{k+1}, r^{i+1}{}_{k}, r^{i+1}{}_{k-1})$, steps I-III above are repeated and the results used to construct numerical derivatives corresponding to the coefficients in (4.132).

Step V: Arrays containing the coefficients from Step IV are passed to the Gaussian elimination subroutine to obtain the increments $\{\Delta r^{i+1}{}_{k}\}$ and the new estimates $r^{i+2}{}_{k} = r^{i+1}{}_{k} + \Delta r^{i+1}{}_{k}$ are made.

Step VI: If the results have converged, the hydrodynamic cycle is completed. If not, the entire process beginning with Step I is then repeated.

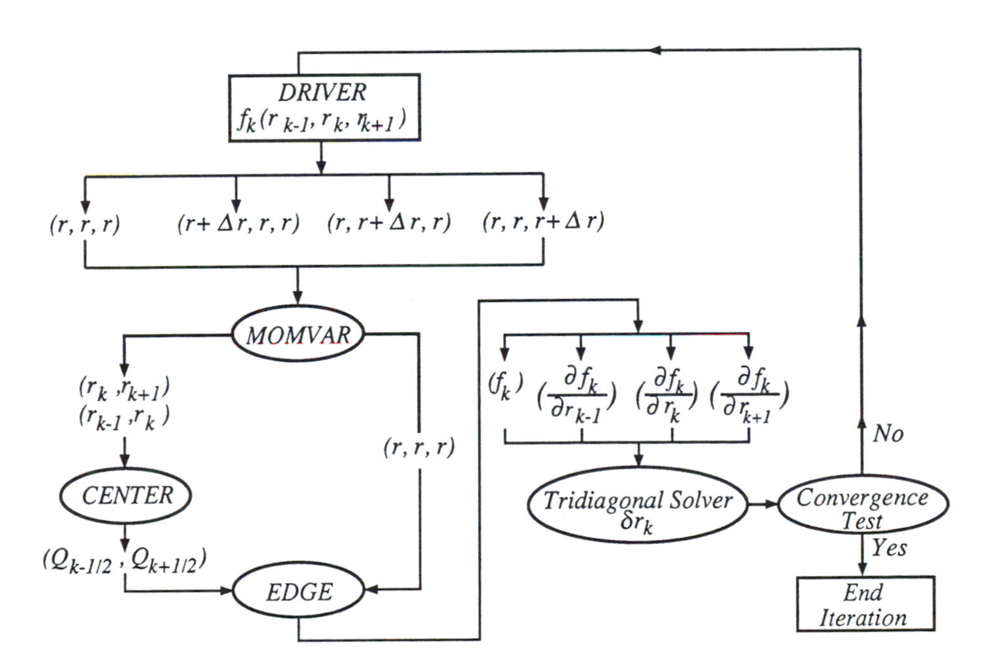

Figure 4.14. Flow chart for the implementational procedure for an implicit hydrodynamic scheme.

Once coding for the substitutional procedure has been constructed, modifications to it may be made in a straightforward way. For example, the artificial viscous stress (4.119) may be added as follows. First, the gradient of the artificial viscous stress is added to the right hand side of (4.137) as in (4.118), and the corresponding artificial viscous heating term in (4.120) is added to the right hand side of (4.140a). Then, the FORTRAN corresponding to the artificial viscous stress is inserted following the FORTRAN expression for (4.138). No other changes are needed in the program. Modifications in the form of the artificial viscous stress may also be trivially implemented by simply replacing the FORTRAN expression corresponding to (4.119) by any other finite difference approximation.

Problem 4.25

Describe how the substitutional procedure discussed above should be modified to include the gravitational acceleration as given by (4.110).

Incorporating Newtonian Gravitation

Newtonian gravitation may be combined with the implicit substitutional scheme described above for the hydrodynamics in several ways. For one dimensional spherically symmetric systems it is sufficient to add an acceleration term to the momentum equation as in (4.110) (see Problem 4.25). Another approach, which is easily extended to the one dimensional relativistic regime, is to describe gravitation in terms of a potential as in (4.108)-(4.109). The addition of Newtonian gravitation to the implicit scheme above will be considered next. This example, which is easily extended to the one dimensional relativistic regime (Schinder, Bludman and Piran 1988), illustrates how additional coupled equations may be combined with the substitutional scheme. We again assume that the basic FORTRAN program describing inviscid flow has been developed (it may also include artificial viscous stresses), and consider the modifications needed to include the potential gradient in the momentum equation and to solve Poisson's equation. This is most easily done using the methods of operator splitting. If Poisson's equation were to be included along with the hydrodynamic equations in nonsplit form, the momentum equation would no longer be tridiagonal in the radii, but would depend on four adjacent values.

The gravitational potential $\Phi^{i+1}{}_{k+1/2}$ is defined as a zone-centered quantity, and is centered at the same time as the other dependent variables. First, the FORTRAN expression for the gravitational acceleration is added to the right side of the momentum equation (4.137)

$$f(r^{i+1}_{k+1}, r^{i+1}_{k}, r^{i+1}_{k-1}) = \frac{V^{i+1/2}_{k} - V^{i-1/2}_{k}}{\Delta t^{n+1/2}}$$

$$+ 4\pi (r^{i+1}_{k})^{2} \frac{(P^{i+1}_{k+1/2} - P^{i+1}_{k-1/2})}{\Delta m_{k}} + \frac{G(\Phi^{n}_{k+1/2} - \Phi^{n}_{k-1/2})}{r^{i+1}_{k+1} - r^{i+1}_{k}} \tag{4.141}$$

where $\Phi^{n}{}_{k+1/2}$ represents the gravitational potential at the beginning of the cycle. The hydrodynamic equations are then solved as described above to obtain new values of $r^{n+1}{}_{k}$. Once the new coordinates are known (and the remaining quantities have been obtained in terms of them) the finite difference representation for Poisson's equation may be solved:

$$(r^{n+1}_{k+1})^{2} \frac{\Phi^{n+1}_{k+3/2} - \Phi^{n+1}_{k+1/2}}{\Delta r^{n+1}_{k+1}} - (r^{n+1}_{k})^{2} \frac{\Phi^{n+1}_{k+1/2} - \Phi^{n+1}_{k-1/2}}{\Delta r^{n+1}_{k}} = G\,\Delta m_{k+1/2} \tag{4.142}$$

where

$$\Delta r^{n+1}{}_{k} = (1/2)\,(r^{n+1}{}_{k+1} - r^{n+1}{}_{k-1}). \tag{4.143}$$

Gaussian elimination may be used to solve (4.142) in terms of the new radii.

4.8 AUTOMATIC REZONING

A major challenge in numerical astrophysics is to maintain the zonal resolution needed to describe physical processes which may arise in different parts of the grid at different stages in a calculation. An obvious (if impractical) solution to this problem would be to carry a very large number of zones throughout each calculation. The increased number of zones would improve the resolution, but each calculation would be slower because more computations would need to be done per cycle and, for explicit methods, because the computational time step would be reduced, thus requiring many more cycles.

Rezoning in the most general sense represents a practical solution to this dilemma. In a typical astrophysical event where dynamic processes are important, the greatest resolution during the early, intermediate and final stages are likely to occur in different spatial regions. For example, the processes controlling stellar evolution may reside either in the stellar core or in the stellar envelope as gravitational contraction and nuclear energy generation alternate as major energy sources. In some evolutionary stages, core nuclear burning dominates the star's energy output, while in other stages relatively thin shell sources located at the core-envelope boundary must be resolved. Protostellar collapse, stellar core collapse and supernova explosions represent typical processes in which the dimensions of the active region may change by several orders of magnitude during the course of a calculation.

One dimensional Lagrangian models of these and similar phenomena are greatly facilitated if zones can be dynamically added to or deleted from the computational grid in such a way as to maintain or increase spatial resolution while maintaining an acceptable computational time step. One of the simplest ways of accomplishing this in one dimensional Lagrangian hydrodynamics can be illustrated by the two simple schemes, the first of which removes a single zone per cycle, and the second of which adds a single zone per cycle.

Consider a zone k which becomes so thin that the hydrodynamic time step in it, $\Delta t^{n+1}{}_k$, satisfies the condition

$$\Delta t^{n+1}{}_k < \xi_D \Delta t^n{}_{min} . \tag{4.144}$$

Here ξ_D is a positive input constant assumed to be less than unity. If the evolution of the system does not depend sensitively on the processes occuring in the vicinity of this zone then it may be combined with its largest neighboring zone. An example where this

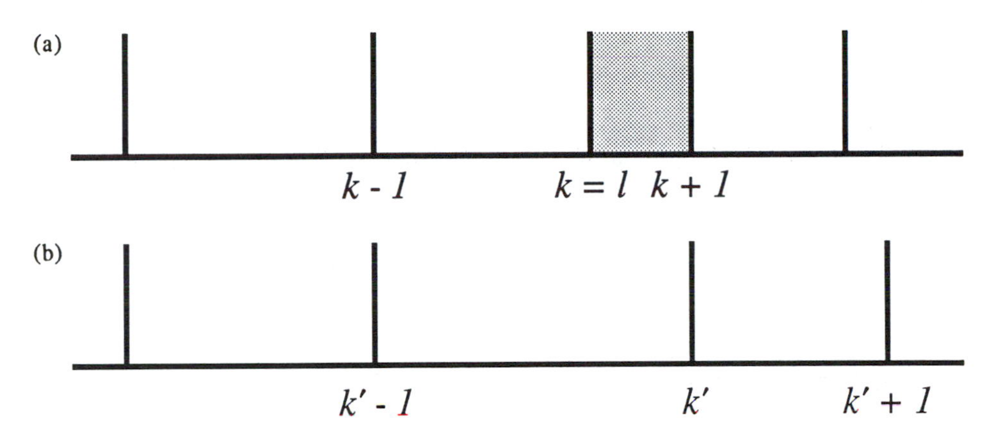

Figure 4.15. Example of dezoning scheme: (a) shows original zoning in which the zone $k=l$ is to be combined with zone $k\text{-}1$; (b) shows the new zoning. The new zone indices are given by $k'=k$ for $k'< l$, and $k' = k+1$ for $k'\geq l$.

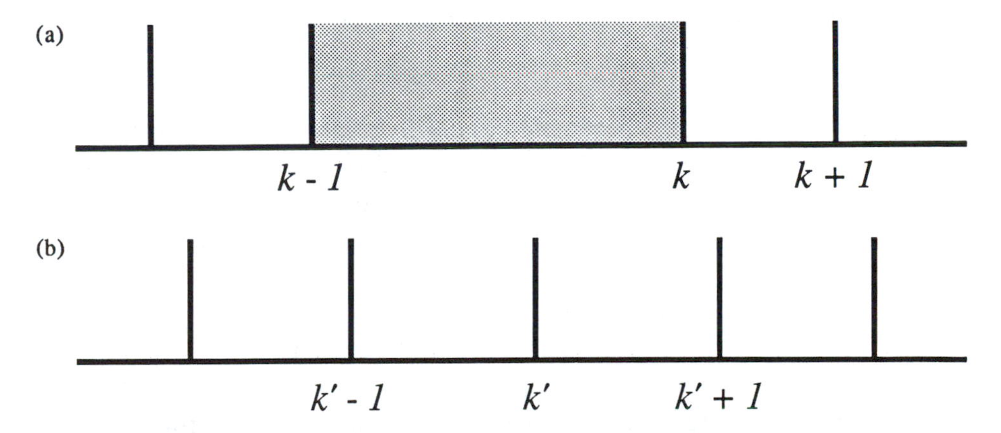

Figure 4.16. Example of rezoning scheme: (a) shows original zoning in which the zone $k\text{-}1$ is to be split into two new zones, as shown in (b).

might be done in the quasistatic core of a star (where the sound speed is greatest) whose outer regions are still dynamic but can be computed with a much larger time step than sound wave propagation in the core. The process of combining two zones, which may be called dezoning, and which is shown schematically in Figure 4.15, reduces the total number of zones in the mesh. To illustrate how this may be done, suppose that zones k and $k-1$ are to be combined into a single new zone which we denote by $k'-1$. The new variables $\rho_{k'-1/2}$, $\varepsilon_{k'-1/2}$, $v_{k'-1}$, and $v_{k'}$ must be determined to complete the dezoning process. There are many ways in which this can be done. The simplest procedure is to require that the rezoning process conserve mass, internal energy and momentum:

$$\Delta m_{k'-1/2} = \Delta m_{k-1/2} + \Delta m_{k+1/2} . \tag{4.145}$$

In addition, the new volume must satisfy

$$\Delta V_{k'-1/2} = \Delta V_{k-1/2} + \Delta V_{k+1/2} . \tag{4.146}$$

The new density $\rho_{k'-1/2}$ follows immediately from (4.144) and (4.145). Conservation of thermal energy leads to the relation

$$\varepsilon_{k'-1/2}\,\Delta V_{k'-1/2} = \varepsilon_{k+1/2}\,\Delta V_{k+1/2} + \varepsilon_{k-1/2}\,\Delta V_{k-1/2} . \tag{4.147}$$

The new velocities may be found from the requirement that the sum of the momenta for the original nodes $k-1$, k and $k+1$ equal the sum of the momenta for the nodes $k'-1$ and k' after dezoning. This gives the single equation for $v_{k'-1}$ and $v_{k'}$

$$\Delta m_{k'-1} v_{k'-1} + \Delta m_{k'} v_{k'} = \Delta m_{k-1} v_{k-1} + \Delta m_k v_k + \Delta m_{k+1} v_{k+1} . \tag{4.148}$$

These constraints are not sufficient to uniquely determine the new velocities. We could also require that dezoning conserve kinetic energy, which would in principle uniquely determine $v_{k'-1}$ and $v_{k'}$ in the differential limit. Because of the difficulty in defining a properly centered kinetic energy in finite difference terms, this approach is not as useful as one might expect. A practical alternative is to assume that the velocities scale in the following way:

$$v_{k'-1} / v_{k-1} = v_{k'} / v_k \equiv K_D \ . \tag{4.149}$$

Equation (4.149) supplies an additional equation which, along with (4.148) may be solved for $v_{k'-1}$ and $v_{k'}$. Once the independent variables have been found, the equation of state may be used to determine the new pressure $P_{k'-1}$. Additional averaging may be required if other physics packages, such as radiation transport, are included. The dezoning scheme based on (4.149) has been used extensively in one dimensional stellar core collapse calculations, and the results have been compared with models which start with a larger number of zone so that dezoning is not necessary. As long as dezoning is not allowed around shock waves, for example, the results are not sensitive to dezoning.

Problem 4.26

The dezoning scheme above can also be used in energy-based Lagrangian codes containing radiation transport. Assume that the material temperature and the internal energy are related by (3.14), and construct expressions for the specific heat $C_{V,\ k'-1}$, and the temperature $T_{k'-1}$ after dezoning.

In some situations it may be desirable to add new zones to a portion of the grid during a computation. For example, if the change in zone thickness during spherical collapse exceeds a prescribed amount, as determined for example from the condition

$$\Delta r_{k+1/2} > \xi_R \, r_k \, , \tag{4.150}$$

then zone k may be split into two new ones (see Figure 4.16, in which the zone is split into two new zones). It is assumed that the input coefficient ξ_R is greater than unity. This process will be termed rezoning. As with dezoning, rezoning is most effective if done in a portion of the grid that is relatively inactive, but which may later involve structure that would be difficult to model with the existing zoning. A simple rezoning prescription can be accomplished by setting the new radial coordinate as follows:

$$r_{k'} = (1/2)(r_k + r_{k-1}) . \tag{4.151}$$

The new zonal volumes follow immediately from (4.151). The simplest approximation for new zonal variables is to assume that all intensive quantities remain unchanged during rezoning. In particular,

$$\rho_{k'-1/2} = \rho_{k'+1/2} = \rho_{k+1/2} ,$$
$$\varepsilon_{k'-1/2} = \varepsilon_{k'+1/2} = \varepsilon_{k+1/2} . \tag{4.152}$$

The zonal masses are then

$$\Delta m_{k'+1/2} = \rho_{k'+1/2} \Delta V_{k'+1/2} . \tag{4.153}$$

The new velocities may be constructed using momentum conservation as described above, with the additional scaling constraint

$$v_{k'-1} / v_{k-1} = v_{k'} / v_k = v_{k'+1} / v_{k+1} = K_R . \tag{4.154}$$

Since ρ and ε do not change, the pressure and composition will also remain unchanged.

The dezoning and rezoning schemes outlined above are too simple to be applied to the entire mesh at one time. Higher order approximations would be required for full mesh rezoning. The methods above can, however, be used to rezone and dezone a few zones per computational cycle, in which case the amount of diffusion introduced is usually minimal, and tends to rapidly damp out as the calculation proceeds. The decision of when, and which zones to dezone or rezone, can be determined dynamically by the code with a minimal amount of logic based, for example, on tests such as (4.144) and (4.150). It is important to anticipate situations where rezoning and dezoning may be needed and to do it only in areas where possible diffusive effects associated with it are small.

Problem 4.27

The mass continuity equation for a one dimensional spherically symmetric system is given in Lagrangian form by

$$\frac{d\rho}{dt} = -\rho \frac{1}{r^2} \frac{\partial (r^2 v)}{\partial r} . \tag{4.155}$$

Show that if the Lagrangian coordinates ξ are defined by

$$\rho_o \, d\xi = 4\pi r^2 \rho \, dr = dm ,$$

that (4.155) is automatically satisfied.

Problem 4.28

Expand the finite difference representation of the Lagrangian momentum equation (4.47) about the space-time point (x_k, t^n) and show that it is second order accurate in space and time if the Lagrangian zones are of constant mass. What happens to the accuracy of the representation if the zones are not of equal mass? In particular, consider the accuracy of the choice (4.46).

References

P. Colella and P. R. Woodward, "Piecewise Parabolic Method (PPM) for Gas-Dynamical Simulations," *J. Comp. Phys.*, **54** (1984): 174.

R. Courant and K. O. Friedrichs, *Supersonic Flow and Shock Waves* (New York: Interscience, 1948).

S. K. Godunov, "A Finite Difference Method for the Numerical Computation and Discontinuous Solutions of the Equations of Fluid Dynamics," *Mat. Sb.*, **47** (1959):271.

L. D. Landau and E. M. Lifshitz, *Fluid Mechanics* (Reading, Mass.: Addison-Wesley, 1959).

M. A. Liberman and A. L. Velikovich, *Physics of Shock Waves in Gases and Plasmas* (Berlin: Springer-Verlag, 1986).

J. von Neumann and R. D. Richtmyer, "A Method for the Numerical Calculations of Hydrodynamical Shocks," *J. Appl. Phys.*, **21** (1950):232.

T. A. Oliphant, "Dynamic and Quasi-Equilibrium Lagrangian MHD in 1-D," J. *Comp. Phys.*, **38** (1980): 406.

T. A. Oliphant and K. H. Witte, *Raven*, Los Alamos National Laboratory Report LA-10826, UC-32, 1987.

---, *Implicit One dimensional Lagrangian Radiation MHD,* Los Alamos National Laboratory Report LA-11887-MS, 1990.

R. D. Richtmyer and K. W. Morton *Difference Methods for Initial-Value Problems*, Second Edition (New York: Interscience Publishers, 1967).

P. J. Roache, *Computational Fluid Dynamics* (Albuquerque, New Mexico: Hermosa, 1976).

P. Schinder, S. Bludman and Piran, *Phys. Rev.*, 1988.

P. Whalen, *One dimensional Covariant Hydrodynamic Equations for Q and Q-free Applications*, Los Alamos National Laboratory preprint, 1985.

Ya. B. Zel'dovich and Yu. P. Raizer, *Physics of Shock Waves and High-Temperature Hydrodynamic Phenomena* (New York: Academic Press, 1966).

5

EULERIAN HYDRODYNAMICS

The simple Lagrangian methods developed in the previous chapter are well suited to one dimensional problems, but fail in higher dimensions when complex fluid flow patterns develop. For example, fluid instabilities, differential rotation and jetting quickly result in distorted flow patterns which cannot be followed by simple Lagrangian mesh methods. Figure 4.1 shows an example where Lagrangian mass elements which are initially neighbors become separated at later times. Modifications have been developed which permit Lagrangian hydrodynamics to follow complicated flow patterns, but these often require complicated and extensive coding. The difficulty with the Lagrangian approach in two or more spatial dimensions is due to the intrinsic inability of the coordinates that are fixed with the fluid particles to retain reasonable geometric and connectivity properties for long periods of time (the discussion of Lagrangian and Eulerian coordinates at the beginning of Chapter 4 should be recalled). As the matter flows, the shape of a typical fluid element will gradually distort and, if the flow is not smooth, will eventually develop a shape which cannot be followed with the initial zoning (the onset of vortex motion in an initially uniform flow is a typical example, as are Rayleigh-Taylor and Kelvin-Helmholtz instability modes). In the Eulerian approach the concept of labels (such as Lagrangian coordinates) fixed with fluid particles (and of fluid elements of fixed mass) is abandoned, and attention is focused instead on a volume in space through which material flows. The method is, therefore, better suited for arbitrary flow in two or three dimensions. Traditional Eulerian hydrodynamics assumes an orthonormal spatial grid which is fixed initially, and does not move during the course of a calculation. In more modern approaches grid motion or grid remapping during the computation can be incorporated with little additional complication in coding. We shall consider Eulerian hydrodynamics in this more general context. In general, the Eulerian approach simply assumes that mass motion relative to the grid can occur. In Chapter 9 numerical methods for one dimensional general relativistic Eulerian hydrodynamics will be developed which are also applicable to special relativistic problems with large gamma.

5.1 EULERIAN REPRESENTATION OF IDEAL FLUIDS

The hydrodynamic equations as seen by a comoving observer are the Lagrangian equations of Section 4.1, in which attention focuses on a fluid element of fixed (Lagrangian) mass. The Eulerian approach focuses on a volume defined with respect to an arbitrary (for example, laboratory) reference frame, and considers how matter flows into and out of it. In the laboratory frame of reference the fluid velocity is $\mathbf{v} = d\mathbf{r}/dt$, and the chain rule of differentiation can be used to show that the Lagrangian time derivative of an arbitrary function $F(\mathbf{r},t)$ is

$$\frac{dF}{dt} = \frac{\partial F}{\partial t} + \mathbf{v} \cdot \nabla F \, . \tag{5.1}$$

Here $\partial F/\partial t$ is understood to be the partial derivative with $\mathbf{r}$ held constant (the Eulerian time derivative). For reasons that will become clear below, it is more convenient to formulate the equations with respect to a frame of reference moving with arbitrary velocity $\mathbf{v}_g$ (this will ultimately be taken to be the grid velocity characterizing a moving mesh). It is to be emphasized that $\mathbf{v}_g$ may be imposed externally, and is in general a function of time.

Consider an arbitrary volume element ΔV defined with respect to a frame of reference moving with velocity $\mathbf{v}_g$. The time rate of change of ΔV, which is due entirely to the motion of the reference frame, can be shown to be

$$\frac{\partial}{\partial t}\int_{\Delta V} dV = \int_{\Delta V} \nabla\cdot \mathbf{v}_g \, dV \, . \tag{5.2}$$

The Eulerian time rate of change of mass in ΔV relative to the moving frame is due to the net mass flux into ΔV: Thus mass conservation requires that

$$\frac{\partial}{\partial t}\int_{\Delta V} \rho \, dV + \int_{\Delta S} \rho(\mathbf{v} - \mathbf{v}_g)\cdot d\mathbf{S} = 0 \tag{5.3}$$

where the mass flux relative to the moving frame of reference is $\rho(\mathbf{v} - \mathbf{v}_g)$. Using (5.2), converting the second term in (5.3) to a volume integral by Gauss' theorem, and noting that ΔV is an arbitrary volume element, we obtain the mass continuity equation in the form

$$\frac{\partial \rho}{\partial t} + \nabla\cdot\rho(\mathbf{v} - \mathbf{v}_g) = -\rho\nabla\cdot\mathbf{v}_g \, . \tag{5.4}$$

For $\mathbf{v}_g = 0$ (the laboratory frame) the standard convective derivative (5.1) is obtained. If $\mathbf{v}_g = \mathbf{v}$ everywhere, then (5.4) reduces to the Lagrangian form of the continuity equation, since in that case $\partial\rho/\partial t$ is the same as $d\rho/dt$. A formal derivation of the transformation to a moving grid can be found in Appendix A.

Problem 5.1

Prove (5.2) for an arbitrary volume element in Cartesian coordinates.

The momentum contained in ΔV is given by

$$\int \rho \mathbf{v} \, dV .$$

It may change as a result of external body forces, f, pressure gradients, or because as matter flows into or out of ΔV momentum is carried with it. Thus, the rate of change of momentum is

$$\frac{\partial}{\partial t} \int_{\Delta V} \rho \mathbf{v} \, dV + \int \rho \mathbf{v} (\mathbf{v} - \mathbf{v}_g) \cdot d\mathbf{S} = -\int P \, d\mathbf{S} + \int_{\Delta V} \mathbf{f} \, dV \tag{5.5}$$

A differential form of the momentum equation is obtained from this expression as follows: evaluate the time derivative on the left hand side, using (5.2) to express the change in volume in terms of the divergence of the grid velocity, and use Gauss' theorem to rewrite the surface intergrals as volume integrals. Since the volume ΔV is arbitrary, the integrand must vanish and we have, for each Cartesian component of the momentum ($k = 1, 2, 3$),

$$\frac{\partial}{\partial t} \rho v^k + \nabla \cdot \rho v^k (\mathbf{v} - \mathbf{v}_g) = -\frac{\partial P}{\partial x^k} + f^k - \rho v^k \nabla \cdot \mathbf{v}_g . \tag{5.6}$$

The body force f may include, for example, magnetic forces, gravitation and radiation-matter coupling.

An equation for the rate of change of the total energy density of the fluid, defined by

$$u_m = \frac{1}{2} \rho \mathbf{v} \cdot \mathbf{v} + \rho \varepsilon ,$$

can also be constructed from (5.4), (5.6) and the first law of thermodynamics. The result, written in conservative form, is

$$\frac{\partial u_m}{\partial t} + \nabla \cdot u_m(\mathbf{v} - \mathbf{v}_g) = - \nabla \cdot (P\ \mathbf{v}) + \mathbf{v} \cdot \boldsymbol{f} - u_m \nabla \cdot \mathbf{v}_g\ ,$$

which expresses the fact that the change in total energy equals the rate at which work is done on the fluid by external forces. Finite difference schemes can be constructed using (5.4), (5.6) and the total energy equation above. The advantage of such an approach is that conservative finite difference representations may be used which guarantee that the total energy is conserved locally as well as globally. Unfortunately, most successful schemes of this type are very complicated. Furthermore, it is not always obvious how one is to include other physics capabilities (such as radiation, gravitation or magnetic fields) with these hydrodynamic methods. An alternate approach which will be followed here is to replace the total energy equation with an expression for the rate of change of material internal energy. At the differential limit there is in fact no real difference between these two approaches; however at the finite difference level subtle differences exist.

The internal energy of a fluid element changes because of work done on it by pressure forces, because of heat flux into it, and because matter which flows into it carries thermal energy as well as mass and momentum. Consider a Lagrangian volume element δV which is small compared with ΔV. The change in internal energy δU of this element is described by the first law of thermodynamics in the comoving frame:

$$\frac{d\delta U}{dt} = T\frac{d\delta S}{dt} - P\frac{d\delta V}{dt} = T\frac{d\delta S}{dt} - P\ \nabla \cdot \mathbf{v}\ \delta V\ . \tag{5.7}$$

Integrating over the Eulerian volume element ΔV , and noting that

$$\delta U = \rho\varepsilon\ \delta V$$

$$\delta S = \rho s\ \delta V$$

we find

$$\int_{\Delta V} \frac{d}{dt} \rho\varepsilon \, \delta V = -\int_{\Delta V} P \nabla \cdot \boldsymbol{v} \, \delta V + \int_{\Delta V} T \frac{d}{dt} \rho s \, \delta V \, . \tag{5.8}$$

The left hand side of this equation is the change in internal energy of ΔV due to processes occuring within ΔV: thus we may write it as

$$\int_{\Delta V} \frac{d}{dt} \rho\varepsilon \, \delta V = \int_{\Delta V} \frac{\partial \rho\varepsilon}{\partial t} \, \delta V + \int_{\Delta V} \rho\varepsilon \, \nabla \cdot \boldsymbol{v}_g \, \delta V. \tag{5.9}$$

Adding to this the flux of thermal energy into ΔV, (5.9) and (5.8) give

$$\int \left(\frac{\partial \rho\varepsilon}{\partial t} + \rho\varepsilon \, \nabla \cdot \boldsymbol{v}_g \right) \delta V + \int \rho\varepsilon \, (\boldsymbol{v} - \boldsymbol{v}_g) \cdot \delta \boldsymbol{S} = - \int P \, \nabla \cdot \boldsymbol{v} \, \delta V + \int T \frac{d\rho s}{dt} \delta V$$

which yields the differential relation

$$\frac{\partial \rho\varepsilon}{\partial t} + \nabla \cdot \rho\varepsilon (\boldsymbol{v} - \boldsymbol{v}_g) = - P \nabla \cdot \boldsymbol{v} - \rho\varepsilon \nabla \cdot \boldsymbol{v}_g + \frac{dq}{dt} \tag{5.10}$$

where the last term on the right,

$$\frac{dq}{dt} \equiv \int T \frac{d\rho s}{dt} \delta V$$

represents all nonadiabatic sources of internal energy, such as thermal diffusion, shock heating and viscosity. Equation (5.10) expresses conservation of internal energy in the moving frame of reference (note that it is not, however, expressed in conservative form). If $\boldsymbol{v} = \boldsymbol{v}_g$ in (5.10), the Lagrangian expression for internal energy is recovered.

The major disadvantage in replacing the total energy equation with the internal energy equation (5.10) is that the latter, because of the term $P \nabla \cdot \boldsymbol{v}$, is not in conservative form and finite difference representations based on it may not conserve total energy to machine round-off. Experience has shown, however, that this limitation can be minimized in practice. Furthermore, the ease with which additional physics can be incorporated into a hydrodynamic scheme based on (5.10) far outweighs the disadvantages inherent in it.

Problem 5.2

Show that the continuity, energy and momentum equations (5.4), (5.6) and (5.10) reduce to the Lagrangian form if $\mathbf{v} = \mathbf{v}_g$. What is the form of the total energy conservation equation analogous to (4.8) for the Eulerian equations on a moving grid?

The Eulerian equations (5.4), (5.6) and (5.10) contain advection terms which do not arise explicitly in the Lagrangian formulation. These terms represent the flux of a quantity (such as mass) across the boundaries of the volume element ΔV, and their presence complicates the numerical treatment of the equations. Finally, because the mass of a volume element is not a constant, the Eulerian equations will be differenced directly in terms of the coordinates.

In the first part of this chapter finite difference approximations to the Eulerian equations will be developed for problems in two spatial dimensions in cylindrical coordinates (reduction to two dimensional Cartesian coordinates is straightforward). Cylindrical coordinates are particularly useful for problems involving axially symmetric rotation. The z axis can be chosen to lie along the rotation axis, and all variables are assumed to be independent of the azimuthal angle ϕ. Table 5.1 gives the axially symmetric inviscid hydrodynamic equations, including the effects of rotation. All variables are functions only of r, z and t. If rotation is absent ($v_\phi=0$) then the angular momentum equation vanishes identically. It should be noted that differential rotation is allowed by the equations.

In two dimensional spherical coodinates r,θ all quantities are assumed to be independent of the azimuthal angle ϕ. The Eulerian hydrodynamic equations for this symmetry, including rotation about the polar axis $\theta=0$ are also included in Table 5.1. If rotation is absent ($v_\phi=0$), the angular momentum equation vanishes identically. Some aspects of these equations will be developed in the Problems. Two dimensional spherical coordinates are an example of curvilinear coordinate systems, and in general are more difficult to treat than the two dimensional axisymmetric system. An approach to curvilinear coordinates will be developed at the end of this chapter.

The only dissipative terms which we shall consider in this chapter are those arising from the treatment of shock waves (artificial viscosity); other dissipative processes, such as radiation or magnetohydrodynamics, will be considered in later chapters.

Problem 5.3

Explain the physical significance of each term in the axisymmetric hydrodynamic equations given in Table 5.1.

Problem 5.4

Rewrite the momentum equations for ρv_ϕ in terms of the angular momentum density: (a) $L=\rho r v_\phi$ for cylindrical coordinates; b) for spherical coordinates with $L=\rho r v_\phi \sin\theta$. What conservation law emerges from the equations in this form?

Advection and Remapping

Finite difference approximations to Eulerian hydrodynamics can be developed along several different lines. A simple example will illustrate two of the most common approaches. Consider a one dimensional, planar grid as shown in Figure 5.1, and suppose that the mass $\Delta m_{k+1/2}$ (shown shaded) undergoes a Lagrangian displacement from time t to time $t+\Delta t$. Since the final mass is also $\Delta m_{k+1/2}$ for Lagrangian a process, the density in zone k at $t+\Delta t$ is easily shown to be

$$\rho_{L,k+1/2} = \Delta m_{k+1/2} / \Delta V_{L,k+1/2} \, , \tag{5.10a}$$

where the subscript L denotes the value of the quantity after a

Lagrangian displacement, and where (in one spatial dimension $\Delta V_{k+1/2} = \Delta x_{k+1/2}$)

$$\Delta V_{L,k+1/2} = x_{k+1} + u_{k+1}\,\Delta t - x_k - u_k\,\Delta t$$

$$= \Delta V_{k+1/2} + (u_{k+1} - u_k)\,\Delta t\,. \tag{5.10b}$$

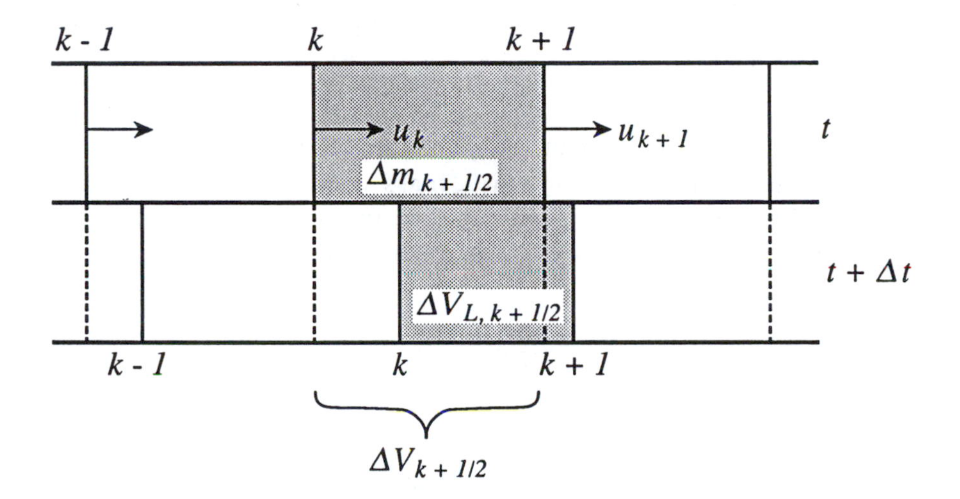

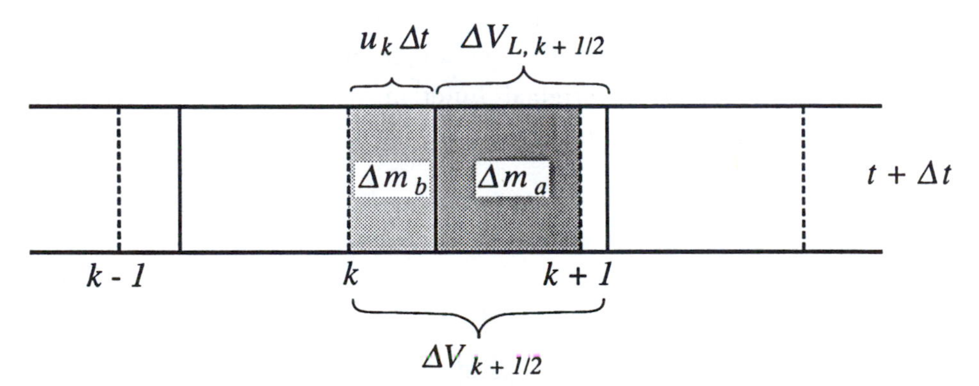

Figure 5.1. Comparison of advection and Lagrangian displacement followed by a remap.

In the following discussion x_k will denote the fixed Eulerian grid coordinates. Combining (5.10a) and (5.10b), and denoting the original density of zone $k+1/2$ at time t by $\rho_{k+1/2}=\Delta m_{k+1/2}/\Delta V_{k+1/2}$, it follows that

$$\rho_{L,k+1/2} = \rho_{k+1/2}(\Delta V_{k+1/2}/\Delta V_{L,k+1/2})$$

which may be rewritten as

$$\rho_{L,k+1/2} - \rho_{k+1/2} = -\rho_{k+1/2}\frac{(u_{k+1}-u_k)\Delta t}{\Delta V_{k+1/2}+(u_{k+1}-u_k)\Delta t}.$$

(5.10c)

To first order in Δt this can be thought of formally as a finite difference approximation for the mass continuity equation in Lagrangian form (the Lagrangian time derivative occurs because the displacements producing (5.10a) are measured in the comoving frame)

$$\frac{d\rho}{dt} = -\rho\frac{\partial u}{\partial x}.$$

(5.10d)

The lower portion of Figure 5.1 shows the Lagrangian zone boundaries (solid) and the original (Eulerian) boundaries (dashed) at the end of the Lagrangian step. If we wish to continue working in the Eulerian grid, the masses must be mapped back onto the original grid. Suppose that both u_k and u_{k+1} are positive, as shown in Figure 5.1. Then the mass in the original Eulerian volume $\Delta V_{k+1/2}$ after the Lagrangian step must be given by

$$\Delta m'_{k+1/2} = \Delta m_a + \Delta m_b ,$$

(5.10e)

where

$$\Delta m_a = \Delta m_{k+1/2}\frac{x_{k+1} - x_k - u_k\Delta t}{\Delta V_{L,k+1/2}},$$

and

$$\Delta m_b = \Delta m_{k-1/2} \frac{u_k \, \Delta t}{\Delta V_{k-1/2} + (u_k - u_{k-1}) \, \Delta t}$$

$$= \Delta m_{k-1/2} \frac{u_k \, \Delta t}{\Delta V_{L,\, k-1/2}} .$$

Combining the results above and dividing (5.10e) by $\Delta V_{k+1/2}$, the Eulerian volume element, we find that

$$\rho'_{k+1/2} - \rho_{L,k+1/2} = -u_k \Delta t \left[\frac{\rho_{L,k+1/2}}{\Delta V_{k+1/2}} - \frac{\rho_{L,k-1/2}}{\Delta V_{k+1/2}} \right] .$$

(5.10f)

To first order in Δt, this can be thought of formally as a finite difference approximation to the equation (the rate of change on the left is the partial derivative taken at fixed spatial coordinates)

$$\frac{\partial \rho}{\partial t} = - u \frac{\partial \rho}{\partial x} .$$

(5.10g)

Finally, combining (5.10f) and (5.10c) into a single step yields

$$\rho'_{k+1/2} - \rho_{k+1/2} = - \frac{\rho_{k+1/2} \, u_{k+1} \, \Delta t}{\Delta V_{L,\, k+1/2}} + \frac{\rho_{k-1/2} \, u_k \, \Delta t}{\Delta V_{L,\, k-1/2}} \frac{\Delta V_{k-1/2}}{\Delta V_{k+1/2}} .$$

(5.10h)

which to first order in Δt is formally a finite difference approximation to the equation

$$\frac{\partial \rho}{\partial t} = - \frac{\partial \rho u}{\partial x} \ . \tag{5.10i}$$

Problem 5.5

Carry out the derivation of (5.10f) and (5.10h) and show that they are first order in Δt representations of (5.10g) and (5.10i) respectively.

The example above suggests that there are two equivalent ways of representing the equation (5.10i) for Eulerian mass motion in finite difference form. In the first way, (5.10i) is solved directly, using for the mass flux into zone $k+1/2$ a quantity proportional to $\rho_{k-1/2} u_k \Delta t$ (this assumes that $u_k > 0$ for all k). This approach is called advection. In the second approach, the Lagrangian step (5.10d) is performed first, followed by a remap step (5.10g). The result of either approach is the same to first order in Δt. In practice, one of the two approaches is usually adopted in first order and then modifications are incorporated to make the scheme yield higher (usually second) order results. This point will be considered later when advection is discussed in detail.

Problem 5.6

The analysis leading from (5.10e) to (5.10i) assumed that u_k and u_{k+1} were positive. What modifications are necessary for arbitrary velocities and how do the results change?

Results similar to those developed above may be obtained to describe the advection of internal energy and momentum for Eulerian hydrodynamics. As for the mass continuity equation, internal energy and momentum advection may be represented in finite difference form by a Lagrangian step followed by a remap step.

5.2 OPERATOR SPLITTING AND FINITE DIFFERENCE APPROXIMATIONS

The two dimensional methods discussed in the remainder of this chapter make extensive use of operator splitting at two levels: first to treat each aspect of the physics (for example hydrodynamics, radiation transport and coupling to matter, and gravitation), and second as a way of handling the multidimensionality of the equations. As the principal example we begin by considering the inviscid equations in cylindrical coordinates without rotation ($v_\phi=0$). In order to simplify notation in subsequent sections, the following change in variable names will be made:

$$u = v_z \qquad v = v_r$$
$$S = \rho v_z \qquad T = \rho v_r . \tag{5.11}$$

We also assume that the Eulerian frame of reference moves with velocity $\mathbf{v}_g$ whose radial and axial components are v_g and u_g respectively. Denoting the grid positions by r_g and z_g , it follows that

$$v_g = d\,r_g\,/dt \qquad u_g = d\,z_g\,/dt.$$

The simplest form of grid motion which we shall consider occurs if $v_g = v_g\,(r_g,t)$ and $u_g = u_g\,(z_g,t)$. The grid is then constrained to move in such a manner as to preserve orthonormality and to remain rectilinear for all times. Other types of grid motion are possible but they involve additional complications which will not be discussed here. In terms of these variables the inviscid momentum equation becomes

$$\frac{\partial T}{\partial t} = -\frac{1}{r}\frac{\partial[rT(v - v_g)]}{\partial r} - \frac{\partial[T(u - u_g)]}{\partial z} - \frac{\partial P}{\partial r} - T\,\nabla\cdot\mathbf{v}_g \tag{5.12}$$

and

$$\frac{\partial S}{\partial t} = -\frac{1}{r}\frac{\partial[rS(v - v_g)]}{\partial r} - \frac{\partial[S(u - u_g)]}{\partial z} - \frac{\partial P}{\partial z} - S\ \nabla\cdot\mathbf{v}_g\ ; \tag{5.13}$$

the mass continuity equation becomes

$$\frac{\partial \rho}{\partial t} = -\frac{1}{r}\frac{\partial[r\rho(v - v_g)]}{\partial r} - \frac{\partial[\rho(u - u_g)]}{\partial z} - \rho\ \nabla\cdot\mathbf{v}_g\ ; \tag{5.14}$$

the internal energy equation becomes

$$\frac{\partial \rho\varepsilon}{\partial t} = -\frac{1}{r}\frac{\partial[r\rho\varepsilon(v - v_g)]}{\partial r} - \frac{\partial[\rho\varepsilon(u - u_g)]}{\partial z} - P\nabla\cdot\mathbf{v} - \rho\varepsilon\ \nabla\cdot\mathbf{v}_g\ . \tag{5.15}$$

The divergence of the grid velocity gives the fractional rate of change of an Eulerian zonal volume, and is given in cylindrical coordinates by

$$\nabla\cdot\mathbf{v}_g = \frac{1}{r}\frac{\partial(rv_g)}{\partial r} + \frac{\partial u_g}{\partial z}\ . \tag{5.16}$$

Finally $\nabla\cdot\mathbf{v}$ satisfies an equation similar in form to (5.16).

Solution of the Eulerian Equations

The explicit Lagrangian hydrodynamic scheme was solved using a space-time centering and solution scheme dictated in part by stability requirements. A similar procedure will be used below which is stable subject to a Courant type time step constraint. The procedure will be outlined first, and then the finite difference equations will be discussed. The hydrodynamic cycle consists of a Lagrangian-like stage, followed by an advection stage. The Lagrangian-like stage begins by calculating the change in momentum caused by pressure gradients (third terms on the right side of (5.12) and (5.13). The mechanical work is obtained next [third term on the right side of (5.15)], using the updated velocities calculated from the momenta. Operator splitting will be used to solve the axial and radial advection terms. In the approach used here all of the transport will be done in one spatial direction, followed by transport in the remaining direction. At the beginning of the advection stage the continuity equation is solved using the first and second terms on the right side of (5.14), and the advection of internal energy is obtained using the first two terms on the right side of (5.15). The first two terms on the right side of (5.12) and (5.13) are then solved to update the momenta. Finally, the last term in each equation due to grid motion (if present) is evaluated. The input for each hydrodynamic cycle will be taken as ρ, ε, S and T. The pressure is to be obtained from the equation of state (3.1), the velocities follow from (5.11) and the grid velocities are arbitrary. At the completion of the hydrodynamic cycle new values have been found for ρ, ε, S and T.

Axisymmetric Grid

For axisymmetric systems described by (5.12)-(5.16), ϕ is an ignorable coordinate. The geometry on a finite difference grid involves several subtle issues, particularly if the zoning is nonuniform. The two dimensional coordinate grid can be defined completely by the positions z_k and r_j of a zone edge (see Figure 5.2a, which shows zone $k+1/2$, $j+1/2$). From these one then defines $\Delta z_{k+1/2} \equiv z_{k+1} - z_k$, $\Delta r_{j+1/2} \equiv r_{j+1} - r_j$, and the zonal volumes

$$\Delta V_{k+1/2,j+1/2} = 2\pi r_{j+1/2} \Delta r_{j+1/2} \Delta z_{k+1/2} , \tag{5.17}$$

where

$$r_{j+1/2} \equiv (1/2)(r_{j+1} + r_j) . \tag{5.18}$$

Although expressions such as $dV = 2\pi r\, dr dz$ are exact in the differential limit, the corresponding finite difference approximations need not be

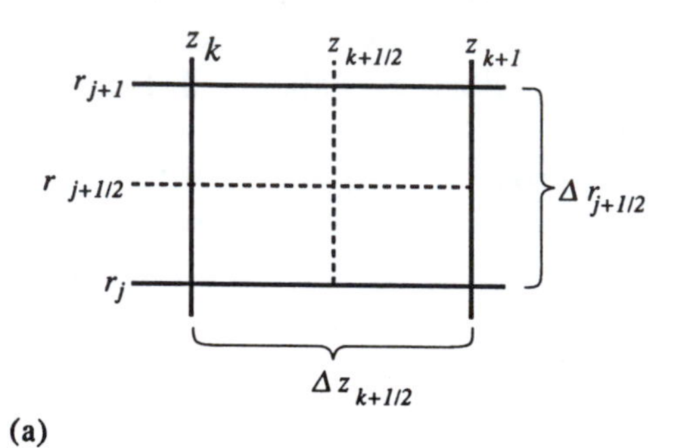

(a)

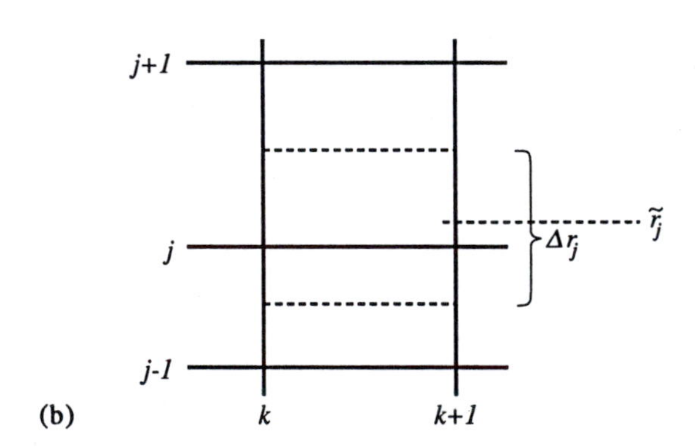

(b)

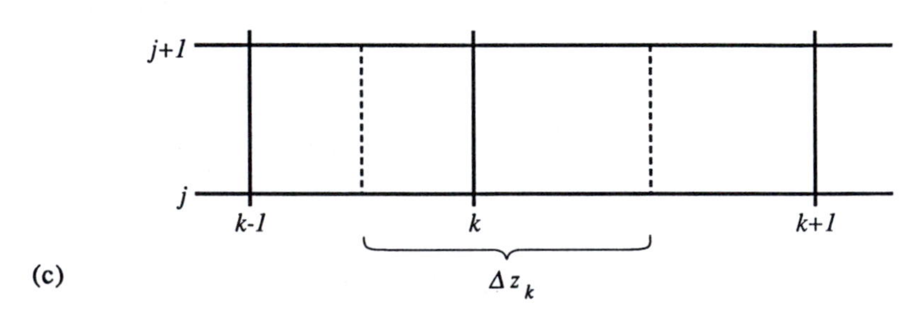

(c)

Figure 5.2. Two dimensional coordinate grid for axially symmetric hydrodynamic equations. The radial and axial coordinates are shown in (a). The radial momentum volume is shown in (b) and the axial momentum volume is shown in (c).

(see Appendix B). It is particularly important to keep this in mind whenever area or volume elements are constructed. In axial coordinates (5.17) is the exact volume if $r_{j+1/2}$ is defined as in (5.18), since

$$\Delta V_{k+1/2,j+1/2} = \pi(r_{j+1}^2 - r_j^2)\, \Delta z_{k+1/2} = 2\pi\frac{(r_{j+1} + r_j)}{2}(r_{j+1} - r_j)\, \Delta z_{k+1/2} \ .$$

Similar arguments show that (see Figure 5.2b)

$$\Delta V_{k+1/2,j} = \pi(r_{j+1/2}^2 - r_{j-1/2}^2)\, \Delta z_{k+1/2}$$

$$= 2\pi \tilde{r}_j\, \Delta r_j\, \Delta z_{k+1/2}$$

is exact if

$$\tilde{r}_j = \frac{1}{2}(r_{j+1/2} + r_{j-1/2}), \tag{5.19a}$$

and

$$\Delta r_j \equiv (r_{j+1/2} - r_{j-1/2}) \ . \tag{5.19b}$$

For uniform zoning the zone edge centered radius (5.19a) is clearly identical to r_j, but when non-uniform zoning is employed the two radii are different. In the differential limit the two radii are identical. The use of two different radii bearing the same radial index is necessary in the finite difference equations to guarantee that consistent volume and surface elements are used. The zone edge centered quantity

$$\Delta z_k \equiv \frac{1}{2}(\Delta z_{k+1/2} + \Delta z_{k-1/2}) \tag{5.20}$$

will be used to define velocity and momentum zones (see Figure 5.2c). In general Δz_k will not equal $\Delta z_{k\pm1/2}$ unless the zoning is uniform. The inconsistent use of quantities such as (5.19a) and r_j does not always produce noticeable errors in simple test problems, but can result in perturbations in fluid variables (which can grow to dominate the calculation), or such phenomena as mass and energy nonconservation.

Space-Time Centering

The two dimensional axisymmetric variables will be centered analogously to the two time level, staggered space-time centering used for the Lagrangian equations, and is shown in Figure 5.3. The intensive variables (ρ, P and ε) will be time centered at t^n, and the velocity and momenta (S, T, u, v, u_g and v_g) will be time centered at $t^{n-1/2}$. The velocity and momentum components may be space centered at the zone edges, or at zone vertices. The use of face centered velocity components for the axially symmetric equations has some advantages because of its simplicity, and may be considered for problems involving pure hydrodynamics without stresses. For these reasons they will be considered first. Eulerian hydrodynamic schemes based on curvilinear coordinates are more

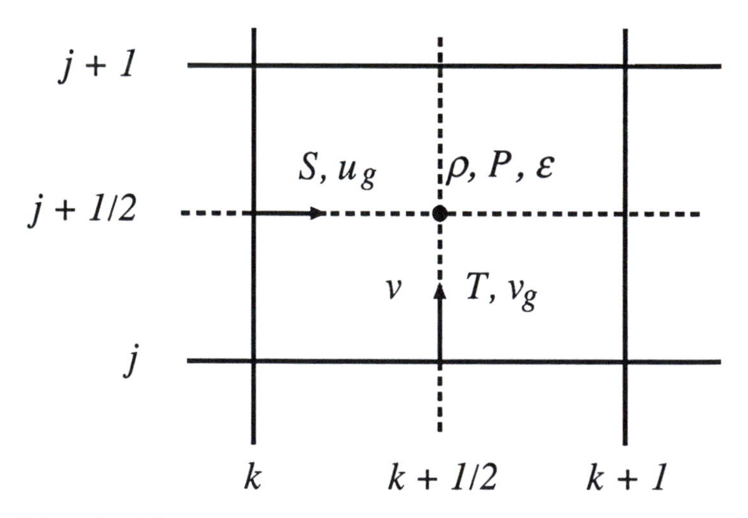

Figure 5.3. Spatial centering of the hydrodynamic variables on a cylindrical grid. Notice that the components of the momentum density and velocity are face centered.

naturally formulated with vertex centered velocities. The latter centering also has advantages, particularly when molecular viscosity or magnetic fields are included. In the discussion below zone edge centering will be adopted, but the approach may be easily modified to allow vertex centering. This choice is particularly convenient for axially symmetric equations in two dimensions. For curvilinear coordinates (to be discussed at the end of the chapter) it is necessary to center both velocity and momentum components at the zone corners. For axisymmetric systems the finite difference variables are

$$
\begin{aligned}
&\rho(z,r,t) \rightarrow \rho^{n}_{k+1/2,j+1/2} \\
&\varepsilon(z,r,t) \rightarrow \varepsilon^{n}_{k+1/2,j+1/2} \\
&P(z,r,t) \rightarrow P^{n}_{k+1/2,j+1/2} \\
&S(z,r,t) \rightarrow S^{n-1/2}_{k,j+1/2} \\
&T(z,r,t) \rightarrow T^{n-1/2}_{k+1/2,j} \\
&u(z,r,t) \rightarrow u^{n-1/2}_{k,j+1/2} \\
&v(z,r,t) \rightarrow v^{n-1/2}_{k+1/2,j} \\
&u_g(z,t) \rightarrow u^{n-1/2}_{g,k} \\
&v_g(z,t) \rightarrow v^{n-1/2}_{g,j} \ .
\end{aligned}
\tag{5.21}
$$

Averages of these variables corresponding to a different spatial centering will be introduced when needed below. As in the Lagrangian case, use of time staggering requires the introduction of two time steps, Δt^{n} and $\Delta t^{n+1/2}$. It should be noted that the grid velocity components are assumed to be known functions of time, and thus are given at times t^{n} and $t^{n+1/2}$. Extensive use of operator splitting will be made in solving the Eulerian hydrodynamic equations. For example, the axial momentum equation (5.13) can be solved in the following order:

$$\frac{\partial S}{\partial t} = -\frac{\partial P}{\partial z}\,, \tag{5.22a}$$

$$\frac{\partial S}{\partial t} = -\frac{\partial[S(u-u_g)]}{\partial z} - \frac{1}{r}\frac{\partial[rS(v-v_g)]}{\partial r}\,, \tag{5.22b}$$

$$\frac{\partial S}{\partial t} = -\,S\,\nabla\cdot\mathbf{v}_g\,. \tag{5.22c}$$

In fact, the second equation, (5.22b), is usually operator split as well, with the change in S due to the axial and radial gradients constructed separately. Since u and S are centered at the same time, it is extremely difficult to correctly time center the nonlinear term $\partial(uS)/\partial z$ in (5.22b). Iterative methods may be used to obtain a consistent centering; however it is usually sufficient simply to center the nonlinear term at the old time. The splitting (5.22a)-(5.22c) complicates the notation for time centering, since three processes combine to give the final value of S (the introduction below of artificial viscosity will add yet another equation to the above). The first step, (5.22a), is a Lagrangian-like process and we might expect to be able to difference it in a manner similar to (4.47). The value of S at time $n+1/2$ including the change due to the pressure gradient will be denoted by superscript P, $S^P{}_{k,j+1/2}$. Similarly, the change in $S_{k,j+1/2}$ at time $n+1/2$ due to advection and grid motion will be denoted by superscripts A and G respectively. If (5.22c) is the final process in the cycle which changes $S_{k,j+1/2}$, then evidently $S^G{}_{k,j+1/2} = S^{n+1/2}{}_{k,j+1/2}$.. Therefore, a hydrodynamic cycle begins with $S^{n-1/2}{}_{k,j+1/2}$, (5.22a) advances it to $S^P{}_{k,j+1/2}$; (5.22b) advances $S^P{}_{k,j+1/2}$ to $S^A{}_{k,j+1/2}$, and (5.22c) advances $S^A{}_{k,j+1/2}$ to $S^{n+1/2}{}_{k,j+1/2}$.

The components of the grid velocity are centered with the fluid velocity components, but are to be associated with an entire grid line; thus

$$u_g \rightarrow u^{n-1/2}_{g,k}$$

$$v_g \rightarrow v^{n-1/2}_{g,j}\,. \tag{5.23}$$

A judicious use of grid motion can greatly enhance zonal resolution during a calculation. For example, a stellar core calculation may result in a substantial fraction of the matter collapsing into a volume of order 10^{-6} times its initial value. By moving portions of the grid inward in a semi-Lagrangian manner, adequate resolution can be maintained for many calculational purposes. Thus, use of a moving

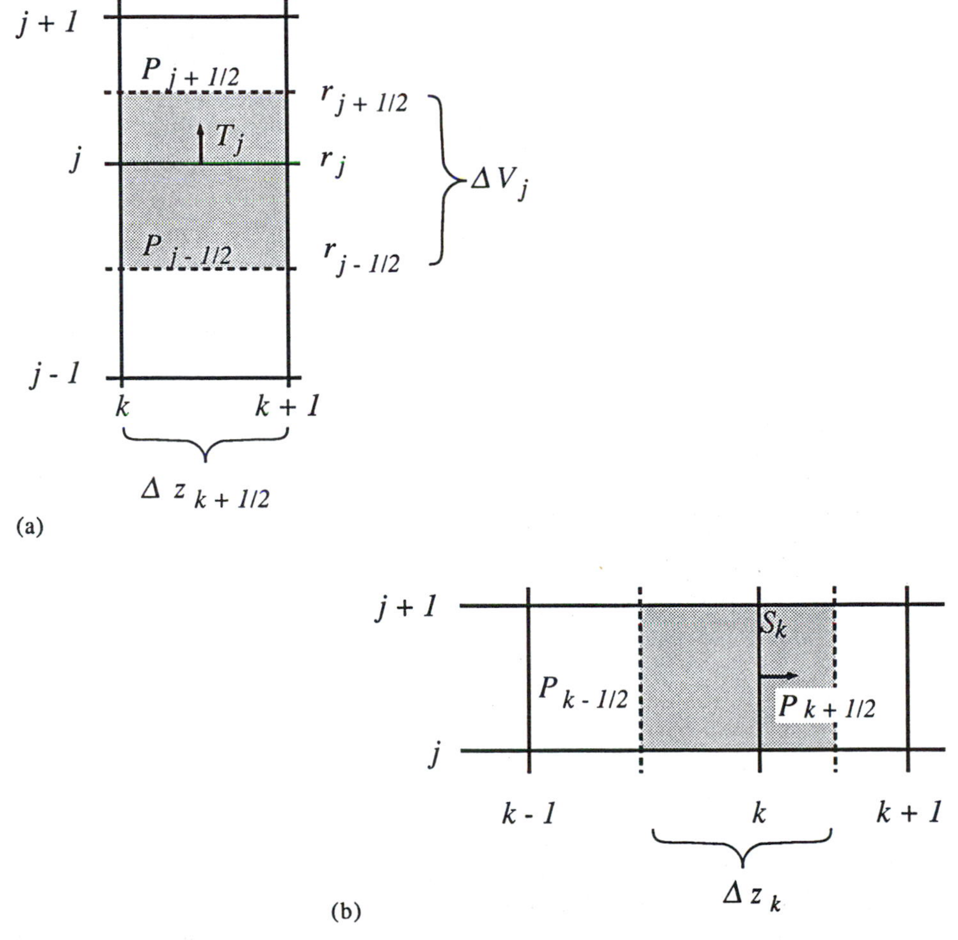

Figure 5.4. (a) Radial momentum zone (shaded) whose volume is ΔV_j used to construct the pressure gradient (5.24). Subscript $k+1/2$ has been deleted for notational convenience. (b) Axial momentum zone (shaded) to be used in constructing the acceleration resulting from the axial pressure gradient.

grid within the Eulerian framework incorporates some of the advantages of a Lagrangian scheme without the Lagrangian limitations to relatively smooth fluid flow. We now consider specific finite difference representations for the various portions of the hydrodynamic cycle outlined above.

5.3 LAGRANGIAN STEP

The Lagrangian-like step starts with a calculation of the change in the momenta due to the components of the pressure gradient in the axial and radial directions. The component in the radial direction will be considered in detail, since it is the most complicated. The acceleration may be integrated over a zone volume centered at $k+1/2,j$ (shown shaded in Figure 5.4a) to obtain

$$\frac{1}{\Delta V_{k+1/2,j}} \int \frac{\partial T}{\partial t}\, dV \equiv \left(\frac{\partial T}{\partial t}\right)_{k+1/2,j} = -\frac{1}{\Delta V_{k+1/2,j}} \int \frac{\partial P}{\partial r}\, dV\,. \tag{5.24}$$

Since all quantities appearing here are centered at $k+1/2$, we shall omit the common index $k+1/2$ for notational simplicity. Thus, the right side of (5.24) can be rewritten as (see Figure 5.4a)

$$\begin{aligned}
\frac{1}{\Delta V_j} \int \frac{\partial P}{\partial r}\, dV &= \frac{2\pi}{\Delta V_j} \int dz \int r \frac{\partial P}{\partial r}\, dr \\
&= \frac{2\pi}{\Delta V_j} \Delta z \left\{ rP\Big|_{j-1/2}^{j+1/2} - \int P\, dr \right\} \\
&= \frac{2\pi}{\Delta V_j} \Delta z \left\{ r_{j+1/2} P_{j+1/2} - r_{j-1/2} P_{j-1/2} - P_{j+1/2}(r_{j+1/2} - r_j) - P_{j-1/2}(r_j - r_{j-1/2}) \right\} \\
&= \frac{2\pi r_j\, \Delta z}{\Delta V_j} (P_{j+1/2} - P_{j-1/2}) = A_j^r \frac{(P_{j+1/2} - P_{j-1/2})}{\Delta V_j}
\end{aligned} \tag{5.25}$$

where

$$A^r_{k+1/2,j} \equiv 2\pi r_j \, \Delta z_{k+1/2} \tag{5.26}$$

is the zonal area whose normal is radially directed. The second line follows using integration by parts. The remaining integral is evaluated by assuming that the pressure is a constant, $P_{j-1/2}$, for $r_{j-1/2} \leq r \leq r_j$ and $P_{j+1/2}$ for $r_j \leq r \leq r_{j+1/2}$. The momenta will also need to be updated to include the effects of advection and grid motion. The new value of T due to the pressure gradient will be denoted by superscript P. For the left side of (5.24) we take

$$\left(\frac{\partial T}{\partial t}\right)_{k+1/2,j} = \frac{T^P_{k+1/2} - T^{n-1/2}_{k+1/2,j}}{\Delta t^n} . \tag{5.27}$$

Combining the results above gives for the change in radial momentum resulting from the pressure gradient in the radial direction

$$T^P_{k+1/2,j} = T^{n-1/2}_{k+1/2,j} - \frac{A^r_{k+1/2,j}\,\Delta t^n}{\Delta V_{k+1/2,j}} (P^n_{k+1/2,j+1/2} - P^n_{k+1/2,j-1/2}) . \tag{5.28}$$

The ratio of area to volume elements could be eliminated to obtain a gradient in terms of Δr_j. The advantage of the form (5.28) will be evident below when advection is discussed. The pressure appearing on the right hand side is obtained from the equation of state

$$P^n_{k+1/2,j+1/2} = P(\rho^n_{k+1/2,j+1/2}\, , \varepsilon^n_{k+1/2,j+1/2}) .$$

The changes in the momenta S and T resulting from ∇P can be made in a single sweep through the r-z grid.

Problem 5.7

The change in axial momentum due to pressure gradients can be constructed as above. Refer to Figure 5.4b and show that

$$S^{P}_{k,j+1/2} = S^{n-1/2}_{k,j+1/2} - A^{z}_{k,j+1/2}\,\Delta t^{n}\,\frac{(P^{n}_{k+1/2,j+1/2} - P^{n}_{k-1/2,j+1/2})}{\Delta V_{k,j+1/2}} \tag{5.29}$$

where

$$\Delta V_{k,j+1/2} \equiv 2\pi r_{j+1/2}\,\Delta r_{j+1/2}\,\Delta z_{k} \tag{5.30}$$

and

$$A^{z}_{k,j+1/2} = 2\pi r_{j+1/2}\,\Delta r_{j+1/2} \tag{5.31}$$

is the zonal area whose normal is axially directed.

Problem 5.8

Set up a finite difference approximation for the momentum change due to pressure gradients in two dimensional spherical coordinates (see Table 5.1). Use centering analogous to (5.21) for all variables except the velocity and momentum; consider these variables to be centered at the node k,j. By using angular momentum instead of linear momentum in the θ direction, the finite difference equations can be constructed to conserve angular momentum explicitly.

Mechanical Work

Following the Lagrangian procedure discussed in Chapter 4, the next step is to calculate the change in internal energy due to mechanical work in (5.15). Thus, consider the differential equation

$$\frac{\partial \rho \varepsilon}{\partial t} = -P\ \nabla \cdot \boldsymbol{v} \tag{5.32}$$

where $\nabla \cdot \boldsymbol{v}$ is, in r-z coordinates,

$$\nabla \cdot \boldsymbol{v} = \frac{1}{r}\frac{\partial r v}{\partial r} + \frac{\partial u}{\partial z}. \tag{5.33}$$

The differencing already introduced for the one dimensional case in spherical coordinates (4.95) will be used here, and is readily extended to two dimensional axial symmetry. First, however, the velocity must be obtained from the momenta S^P and T^P. This may be done using the definitions (5.11), integrated over the appropriate zone volume. For example, integrating $S_{k,j+1/2}$ over the volume $\Delta V_{k,j+1/2}$ (shaded in Figure 5.4b) we obtain (common subscript $j+1/2$ omitted)

$$S = \frac{1}{\Delta V}\int S\, dV = \frac{1}{\Delta V}\int \rho u\, dV = \frac{u}{\Delta V}\int \rho\, dV$$

$$= \frac{u}{\Delta V}\left\{\rho_{k-1/2}\frac{\Delta V_{k-1/2}}{2} + \rho_{k+1/2}\frac{\Delta V_{k+1/2}}{2}\right\}.$$

Defining the edge centered mass density by

$$\rho_{k,j+1/2} = \frac{(\rho_{k+1/2,j+1/2}\,\Delta V_{k+1/2,j+1/2} + \rho_{k-1/2,j+1/2}\Delta V_{k-1/2,j+1/2})}{2\Delta V_{k,j+1/2}}, \tag{5.34}$$

the finite difference representation of $\rho u = S$ can be expressed in the form

$$u_{k,j+1/2} = \frac{S_{k,j+1/2}}{\rho_{k,j+1/2}}. \tag{5.35}$$

Similarly, the radial component of the velocity is obtained from

$$v_{k+1/2,j} = \frac{T_{k+1/2,j}}{\rho_{k+1/2,j}} . \tag{5.36}$$

where the zone edge centered density is

$$\rho_{k+1/2,j} = \frac{(\rho_{k+1/2,j+1/2}\,\Delta V_{k+1/2,j+1/2} + \rho_{k+1/2,j-1/2}\,\Delta V_{k+1/2,j-1/2})}{2\Delta V_{k+1/2,j}} . \tag{5.37}$$

Problem 5.9
Obtain a finite difference representation of $\nabla\cdot\boldsymbol{v}$ in the two dimensional spherical coordinates r,θ. Find the finite difference relation between the velocity components v_r and v_θ and the radial momentum $T=\rho v_r$ and the angular momentum density $L\equiv\rho r v_\phi$. Use the centering from Problem 5.8.

The velocity divergence may be constructed from (5.35) and (5.36) by integrating (5.33) over the volume $\Delta V_{k+1/2,j+1/2}$ (see Figure 5.5):

$$(\nabla\cdot\boldsymbol{v})_{k+1/2,j+1/2} = \frac{A^z_{k+1,j+1/2}\,u^P_{k+1,j+1/2} - A^z_{k,j+1/2}\,u^P_{k,j+1/2}}{\Delta V_{k+1/2,j+1/2}} + \frac{A^r_{k+1/2,j+1}\,v^P_{k+1/2,j+1} - A^r_{k+1/2,j}\,v^P_{k+1/2,j}}{\Delta V_{k+1/2,j+1/2}} . \tag{5.38}$$

The superscript P on the velocity indicates that they are obtained from S^P and T^P. The area factors are defined by (5.26) and (5.31).

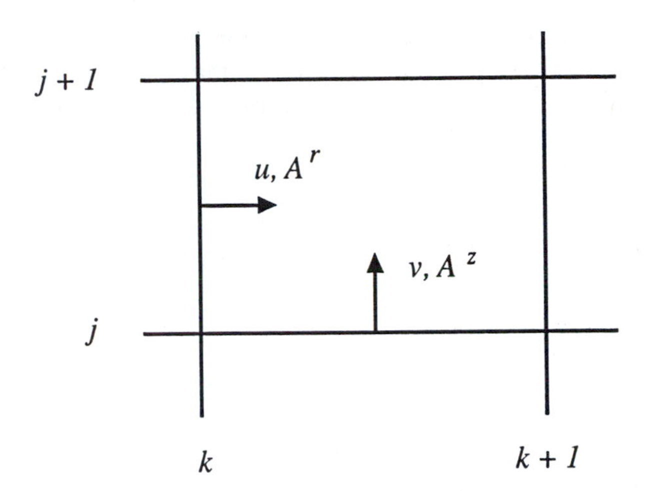

Figure 5.5. Area factors used to construct the divergence of the fluid velocity.

Problem 5.10

The finite difference approximation for $\nabla \cdot \mathbf{v}$ given by (5.38) is a first order accurate time representation of the time rate of change of the volume ΔV [see (4.2)]. Show that (5.38) is second order accurate in time if the radial area factors are defined by

$$(A^r_{k+1/2,j})^{n+1/2} \equiv 2\pi \left(r^n_j + v^P_j \frac{\Delta t^{n+1/2}}{2} \right) \Delta z_{k+1/2} \, .$$

Note that ΔV^{n+1} contains $r^{n+1}{}_j \approx r^n{}_j + v^P{}_j \Delta t^{n+1/2} + O(\Delta t^2)$.

Following the discussion of mechanical work in the one dimensional Lagrangian case ((4.52) to (4.55) in particular) the finite difference representation of (5.32) will use the pressure evaluated midway between the current and the new times as in (4.52). It is now a simple matter to construct a finite difference representation of (5.32) if the equation of state is a gamma law (see Problem 4.10). If the equation of state is not a simple gamma law, or cannot be represented by a smoothly varying gamma law function, then the pressure at the intermediate time may be constructed as in (4.53). Since all quantities are zone centered, the common $k+1/2$ and $j+1/2$ indicies are omitted below. The left side of (5.32) is (superscript C denotes mechanical work update)

$$\Delta(\rho\varepsilon) = \rho^C\varepsilon^C - \rho^n\varepsilon^n = \rho^C\varepsilon^C - \rho^C\varepsilon^n + \rho^C\varepsilon^n - \rho^n\varepsilon^n$$

$$= \rho^C\Delta\varepsilon + \varepsilon^n\Delta\rho \ . \tag{5.39}$$

Mechanical work is in fact a Lagrangian process, so that the change in density must satisfy the equation

$$\Delta(\rho\delta V) = \rho^C\,\Delta(\delta V) + \delta V\,\Delta\rho = 0\,;$$

or, using (4.2) we find

$$\frac{\Delta\rho}{\rho^C} = \frac{\rho^C - \rho^n}{\rho^C} = -\frac{\Delta\delta V}{\delta V} = -\ \nabla\cdot\mathbf{v}\Delta t^{n+1/2}\ , \tag{5.40}$$

which is immediately solved for ρ^C in terms of known quantities. The pressure at $k+1/2,\ j+1/2$ and time $t^{\,n+1/2}$ is given by an expression of the form (4.53), which when substituted into (5.32) can be solved, using (5.39), for $\Delta\varepsilon$ in the form

$$\varepsilon^C - \varepsilon^n = -\frac{\varepsilon^n\,\Delta\rho + \left[P^n + \frac{1}{2}\left(\frac{\partial P}{\partial\rho}\right)_\varepsilon \Delta\rho\right]\,\nabla\cdot\mathbf{v}\ \Delta t^{n+1/2}}{\rho^C + \frac{1}{2}\left(\frac{\partial P}{\partial\varepsilon}\right)_\rho\ \nabla\cdot\mathbf{v}\ \Delta t^{n+1/2}}\ . \tag{5.41}$$

All quantities on the right hand side of (5.41) are evaluated at zone centers $(k+1/2,\ j+1/2)$. The divergence of the velocity is given by (5.38), and $\Delta\rho$ and ρ^C follow from (5.40). The mechanical work may be constructed using a single sweep throughout the grid. The result (5.41) is valid for any orthonormal coordinate system with $\nabla\cdot\mathbf{v}$ evaluated appropriately. In particular, it holds in two dimensional spherical coordinates where

$$\nabla\cdot\mathbf{v} = \frac{1}{r^2}\frac{\partial(r^2 v)}{\partial r} + \frac{1}{r\sin\theta}\frac{\partial(v_\theta \sin\theta)}{\partial\theta} .$$

This completes the Lagrangian stage of the calculation for the inviscid hydrodynamic equations. Artificial viscosity (to be discussed below) can be added at this stage in the cycle.

5.4 Advection

The mass, energy and momentum advection terms appearing in (5.12)-(5.15) constitute a significant complication for the Eulerian hydrodynamic equations. A simple example of advection has already been encountered (see (2.48) and the subsequent discussion).

Before considering these terms in detail, a further comment about advection is in order. Most quantities characterized by an evolution equation will, in an Eulerian formulation, contain advection terms. Thus, when radiation transport, magnetic fields or rotation are included, the energies, magnetic fields and momenta will also be advected. These processes are modelled in finite difference form by constructing fluxes of each quantity, and using the fluxes to calculate the changes in that quantity in each volume element during a time step. At the level of the differential equations, all specific quantities, such as internal energies and specific momenta, move with the mass of the fluid. In particular, there should be no diffusion of intrinsic variables due to the way advection is modelled numerically. Diffusion may occur, such as thermal diffusion, but must be modelled by other means, such as those described in Chapter 6. Not all finite difference schemes achieve this (see Section 5.7 for example). In this section mass advection will be considered in detail, since it is the mass flux that is to be used to advect consistently specific quantities such as the specific energy ε and the specific angular momentum L (see Section 5.7). Terms proportional to $\nabla\cdot\mathbf{v}_g$ will also be omitted from the discussion of energy and momentum advection below, but will be considered with the equations of grid motion at the end of section 5.4.

A comment about the order of approximation of the finite difference schemes must be made at this point. It is well known that simple first order Eulerian advection schemes are extremely diffusive, and in general are not adequate to model most hydrodynamic phenomena. For heuristic reasons we shall introduce the concept of advection using the donor cell (first order) method. A second order method will then be discussed which works well for smooth flow in which the density and other variables change only slightly from point to point. Finally, we consider second order monotonic schemes which work well when steep gradients arise. The latter method is generally the best, and should always be used in practice. The use of strict first order schemes is usually unwarranted and is not recommended.

Mass Advection

The mass advection terms in the continuity equation are typical of the advection of scalar variables in general, so they will be examined first. The results are immediately applicable to internal energy advection, and to the advection of some other quantities as well. We shall continue to develop the two dimensional axially symmetric equations by way of example; the equations in spherically symmetric form will be incorporated into the problems. Consider (5.14), ignoring for now the grid compression term $\nabla \cdot \mathbf{v}_g$:

$$\frac{\partial \rho}{\partial t} = - \frac{\partial [\rho (u - u_g)]}{\partial z} - \frac{1}{r} \frac{\partial [\rho r (v - v_g)]}{\partial r} . \tag{5.42}$$

Integrating the first term on the right side of (5.42) over $\Delta V_{k+1/2,j+1/2}$ (and ignoring for the moment the radial gradient) gives

$$\frac{1}{\Delta V}\int \frac{\partial \rho}{\partial t}\, dV = -\frac{1}{\Delta V}\int \frac{\partial}{\partial z}[\rho(u - u_g)]\, dV$$

$$= \frac{2\pi}{\Delta V}\int r\, dr\, [\hat{\rho}_{k+1}(u_{k+1} - u_{g,k+1}) - \hat{\rho}_k(u_k - u_{g,k})]$$

$$\left(\frac{\partial \rho}{\partial t}\right)_{k+1/2} = -\frac{A^Z_{k+1}\hat{\rho}^Z_{k+1}(u_{k+1} - u_{g,k+1}) - A^Z_k \hat{\rho}^Z_k (u_k - u_{g,k})}{\Delta V_{k+1/2}}$$

(5.43)

(the common index $j+1/2$ has been omitted for simplicity). The mass density at the zone edge will be denoted by

$$\hat{\rho}^Z_k \ .$$

This density can be related to the zone centered density by observing that the advection terms physically represent mass flux into a zone. In particular, if the axial velocity of the fluid relative to the grid $u_k - u_{g,k} > 0$, then mass flows from zone $k-1/2$ into zone $k+1/2$ on the right (Figure 5.6); if the relative velocity $u_k - u_{g,k} < 0$, then mass flows out of zone $k+1/2$ into zone $k-1/2$. This suggests that the densities appearing in (5.43) be defined as follows:

$$\hat{\rho}^Z_k \equiv \frac{1}{2}\rho_{k-1/2}(1 + \alpha^Z_k) - \frac{1}{2}\rho_{k+1/2}(\alpha^Z_k - 1)$$

(5.44)

where

$$\alpha^Z_k \equiv \frac{|u_k - u_{g,k}|}{(u_k - u_{g,k})} \ .$$

(5.45)

The density (5.44) may be substituted into (5.43) giving the change in density associated with axial mass motion. The scheme

(5.43)-(5.45) is known as up-wind, or donor cell differencing, and is first order accurate.

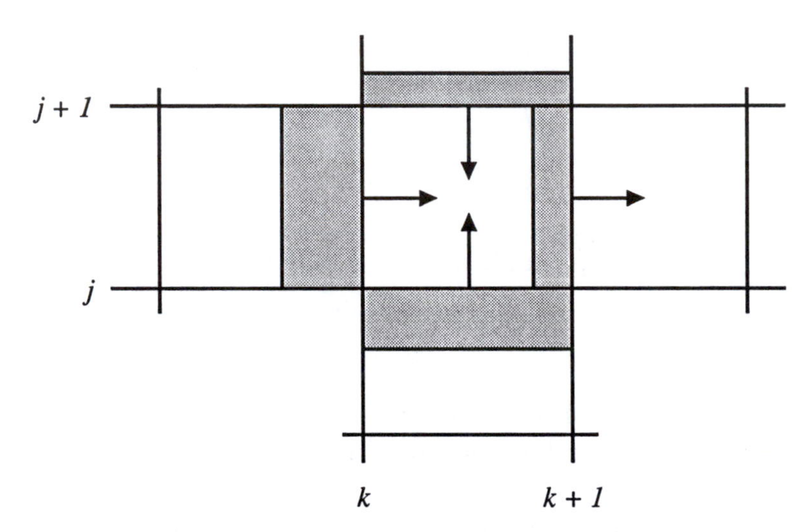

Figure 5.6. Mass flow into zone $k+1/2,\ j+1/2$. In the first order scheme (5.48) the amount of mass from zone $k-1/2,\ j+1/2$ added to the zone $k+1/2,\ j+1/2$ is shown shaded.

The mass flux in the radial direction can be derived in similar fashion. Thus as in (5.44), we may define the zone edge centered density (common index $k+1/2$ omitted)

$$\hat{\rho}_j^R \equiv \frac{1}{2}\rho_{j-1/2}(1 + \alpha_j^R) - \frac{1}{2}\rho_{j+1/2}(\alpha_j^R - 1) \tag{5.46}$$

$$\alpha_j^R \equiv \frac{|v_j - v_{g,j}|}{(v_j - v_{g,j})} . \tag{5.47}$$

Integrating the radial gradient contribution in (5.42) over $\Delta V_{k+1/2,j+1/2}$ yields an expression similar to (5.43). Combining this result with the expressions above, we may express the mass advection equation for two dimensional axially symmetric systems as

$$\frac{\rho^{A}_{k+1/2,j+1/2} - \rho^{n}_{k+1/2,j+1/2}}{\Delta t^{n+1/2}} = - \Big[A^{Z}_{k+1,j+1/2}\, \hat{\rho}^{Z}_{k+1}\, (u^{P}_{k+1,j+1/2} - u_{g,k+1})$$

$$+ A^{Z}_{k,j+1/2}\, \hat{\rho}^{Z}_{k} (u^{P}_{k,j+1/2} - u_{g,k})$$

$$- A^{R}_{k+1/2,j+1}\, \hat{\rho}^{R}_{j+1} (v^{P}_{k+1/2,j+1} - v_{g,j+1})$$

$$+ A^{R}_{k+1/2,j}\, \hat{\rho}^{R}_{j}\, (v^{P}_{k+1/2,j} - v_{g,j})\Big] / \Delta V_{k+1/2,j+1/2}\,.$$

(5.48a)

The fluid velocity components appearing in the right hand terms of (5.48a) include the effects of acceleration due to pressure gradients (and due to artificial viscosity, if present). The first order finite difference approximation (5.48a) has a simple physical interpretation (see Figure 5.6). Consider the simple case $\mathbf{v}_g=0$ and suppose that the material flow is to the right. Then the first term on the right side of (5.48a) gives as the mass carried out of zone $k+1/2,j+1/2$ in the axial direction in a time $\Delta t^{n+1/2}$ the amount

$$(2\pi r_{j+1/2}\Delta r_{j+1/2})(u_{k+1,j+1/2}\Delta t^{n+1/2})\rho_{k+1/2,j+1/2}\,.$$

Thus, the first order scheme assumes uniform density across the donor cell, and transports all material within a distance $u\Delta t$ of the cell boundary into the zone ahead. In keeping with this interpretation, we may define the mass flux at the grid point l (where $l = j, k$) across a surface area whose normal is in the X direction (where $X = R,Z$) by

$$F^{X}_{l} \equiv \hat{\rho}^{X}_{l} (u^{P}_{X,l} - u_{X\,g,l})\,.$$

(5.48b)

The mass fluxes (5.48b) should be stored, since they will be needed below to calculate the transport of specific energy and specific momentum.

A useful test of any advection scheme is its ability to describe properly the translational motion of a sharp pulse. Figure 5.7 shows an initial mass distribution of uniform density ρ which is much larger

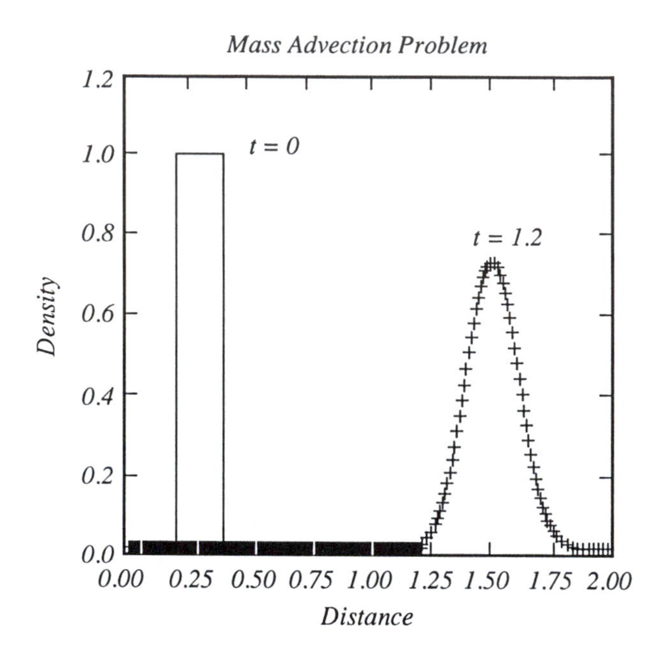

Figure 5.7. Mass advection test problem. The initial density profile ($t = 0$) and the profile after moving a distance *1.25 cm* at time $t = 1.2$ *sec.* using first order advection.

than the surrounding mass density, and which is given a uniform initial velocity (the low density region is intended to simulate a vacuum through which the density pulse travels). In the absence of pressure gradients, the pulse should move through the grid without distortion (see the discussion in Section 2.2). Using the first order scheme above, the density pulse is observed to spread (diffuse) substantially after traversing about six times its original thickness, and to decrease in amplitude by more than twenty percent.

Problem 5.11

The continuity equation (5.42) expresses mass conservation locally. Show that the finite difference representation (5.48) leads to total mass conservation in the entire grid if all boundaries are closed. What happens if one or more of the boundaries are open?

The advection phase of the finite difference representation considered above is first order accurate in spatial derivatives. It is seldom satisfactory to work at this level of accuracy, and definite advantages are gained by treating advection to second order. One way to construct such a scheme (shown in Figure 5.8) is to use in place of (5.44) and (5.46), which corresponds to uniform density across each zone (dashed lines), the density linearly interpolated a distance $(1/2)u\,\Delta t$ back from the zone boundary l (Figure 5.8, which assumes $v_l > 0$):

$$\hat{\rho}_l = \rho_{l-1/2} + \frac{\rho_{l+1/2} - \rho_{l-1/2}}{(1/2)(\Delta x_{l+1/2} + \Delta x_{l-1/2})}\left(\frac{1}{2}\Delta x_{l-1/2} - \frac{1}{2} v_l\,\Delta t\right)$$

$$= \frac{1}{2}\rho_{l-1/2}\frac{(\Delta x_{l+1/2} + v_l \Delta t)}{\Delta x_l} + \frac{1}{2}\rho_{l+1/2}\frac{(\Delta x_{l-1/2} - v_l \Delta t)}{\Delta x_l} \quad . \tag{5.49}$$

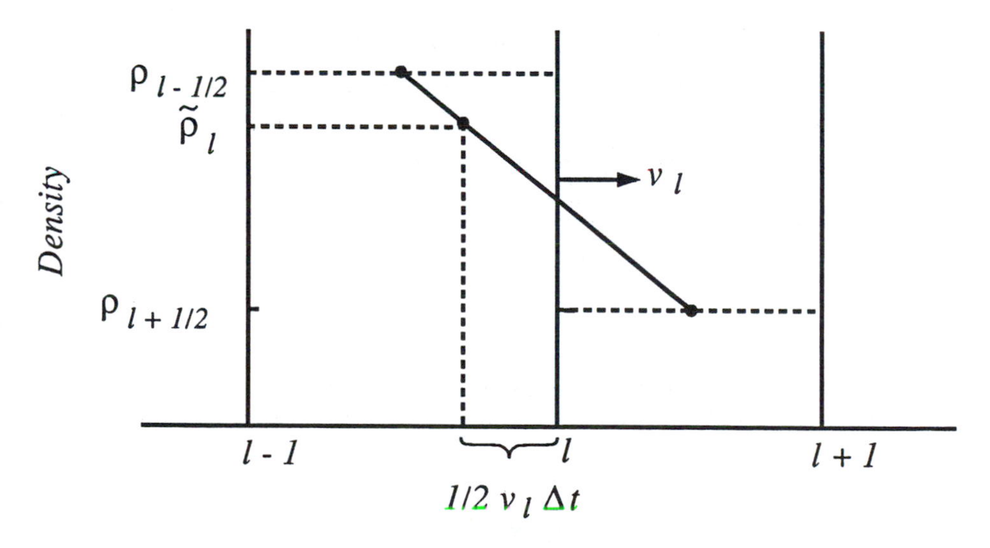

Figure 5.8. Density used to construct a second order approximation for the mass flux across zone boundary l, assuming $v_l > 0$.

This expression automatically takes into account the flow direction through the sign of the velocity component, since if $v_l > 0$ ($v_l < 0$), then (5.49) corresponds to interpolation backward (forward) from the zone boundary l. Setting $l=k$, and replacing Δx by Δz, and v by $u\text{-}u_g$ in (5.49) results in a second order approximation for the density to be used for the axial flow in (5.48) in place of (5.44)-(5.45). This second order scheme corresponds to using a mass flux defined in terms of the fluid velocity relative to the zone boundary and a density interpolated backward (relative to the flow direction) a distance $(1/2)(u\text{-}u_g)\,\Delta t$ from the zone boundary (interpolation on a nonuniform spatial gird is discussed in Appendix C). Finally, substituting $l=j$, and replacing Δx by Δr and v by $v\text{-}v_g$ in (5.49) results in a second order approximation to the density to be used for the radial flow in (5.48) in place of (5.46)-(5.47). We note that this scheme, which is second order in spatial derivatives, is adequate for regions where the density is smooth, but can lead to problems when the density changes significantly across a few zones (as occurs when shock fronts form). A modification to better treat such phenomena will be discussed below.

Equation (5.49) attempts to make the spatial derivatives second order accurate, but it is only first order in the time derivative. A scheme which attempts to make the result second order in both space and time may be constructed by replacing the velocity v_l appearing in (5.49) by the velocity extrapolated backward a distance $v_l\Delta t$, as outlined below. Assume that the fluid flow is to the right (Figure 5.8), and define the transport velocity $v_{T,l}$ by

$$v_{T,l} = v_l - \frac{1}{2} v_l \Delta t \left(\frac{\partial v}{\partial x}\right)_{up}, \tag{5.49a}$$

where the upwind velocity derivative is given by

$$\left(\frac{\partial v}{\partial x}\right)_{up} \equiv \frac{v_{l+1} - v_{l-1}}{2\,\Delta x_l}. \tag{5.49b}$$

If the upwind derivative is of opposite sign to $(v_{l+1} - v_l)/\Delta x_{l+1/2}$, or if any of the velocities v_{l+1}, v_l or v_{l-1} have opposite signs, then

$$\left(\frac{\partial v}{\partial x}\right)_{up} \equiv 0. \tag{5.49c}$$

Equation (5.49a) may be used in place of v_l in (5.49). The results above are readily modified for flow to the left by using upwind derivatives to construct the transport velocity and the derivative.

Problem 5.12
Use (5.49) to construct a second order finite difference representation for the one dimensional continuity equation in planar coordinates

$$\frac{\partial \rho}{\partial t} = -\frac{\partial[\rho(u - u_g)]}{\partial x}. \tag{5.50}$$

Show that the solution is stable as long as the time step is restricted by the relation

$$\Delta t \leq \frac{\Delta x}{|u - u_g|}. \tag{5.51}$$

For simplicity assume a uniform grid and constant fluid and grid velocity.

The first and second order advection schemes above are stable so long as the time step is limited as in (5.51), and the flow is smooth (problems involving nonsmooth flow will be discussed in the section on monotonicity below). For an arbitrary grid, with nonconstant velocities the following constraint is generally sufficient:

$$\Delta t^{\,n+1/2} = C_T \, min \, \{ \frac{\Delta r_{j+1/2}}{|v_{k,j} - v_{g,j}|} , \frac{\Delta z_{k+1/2}}{|u_{k,j} - u_{g,k}|} \} \ . \tag{5.52}$$

The minimum is to be taken over the entire grid. For most problems $C_T=0.5$ works well. For flows which are very subsonic, C_T may approach unity. The value of C_T may be adjusted throughout a calculation as necessary. Physically (5.52), which is a transport time step constraint, specifies that mass can move no further than a fraction C_T of the smallest zonal width in a single cycle. In practice, the inverse of (5.52) is usually calculated so that zones with zero material or grid velocity do not result in division by zero.

Problem 5.13

Construct a finite difference approximation to the mass continuity equation in two dimensional cylindrical coordinates r,θ which is first order accurate in space. Discuss how the approximation can be extended to make it second order accurate.

Energy Advection

The energy advection terms in (5.15) are formally identical with the mass advection terms (replace ρ by $\rho\varepsilon$):

$$\frac{\partial \rho\varepsilon}{\partial t} = - \frac{\partial [\rho\varepsilon(u - u_g)]}{\partial z} - \frac{1}{r}\frac{\partial [r\rho\varepsilon(v - v_g)]}{\partial r} \ . \tag{5.53}$$

This may be expressed in finite difference form in complete analogy with the mass advection terms. In particular, $\rho\varepsilon$ replaces ρ in (5.49), which results in definitions of the internal energy interpolated a distance $u\,\Delta t/2$ behind the zone boundary. The interpolated energy density for axial advection is denoted by

$$(\hat{\rho}\,\hat{\varepsilon})_k^Z$$

and the interpolated energy density for radial advection is denoted by

$$(\hat{\rho}\,\hat{\varepsilon})_j^R .$$

These quantities may be used in (5.48) with $\rho\varepsilon$ replacing ρ on the left side to obtain a second order representation of (5.53). The result is easily shown to be

$$\frac{(\rho\varepsilon)^A_{k+1/2,j+1/2} - (\rho^n\varepsilon^C)_{k+1/2,j+1/2}}{\Delta t^{n+1/2}} = -\{A^Z_{k+1,j+1/2}\,(\hat{\rho}\hat{\varepsilon})^Z_{k+1}\,(u^P_{k+1,j+1/2} - u_{g,k+1})$$

$$+ A^Z_{k,j+1/2}\,(\hat{\rho}\hat{\varepsilon})^Z_k\,(u^P_{k,j+1/2} - u_{g,k})$$

$$- A^R_{k+1/2,j+1}\,(\hat{\rho}\hat{\varepsilon})^R_{j+1}\,(v^P_{k+1/2,j+1} - v_{g,j+1})$$

$$+ A^R_{k+1/2,j}(\hat{\rho}\hat{\varepsilon})^R_j\,(v^P_{k+1/2,j} - v_{g,j})\}\;/\Delta V_{k+1/2,j+1/2}\,. \tag{5.54}$$

The superscript C denotes that the specific energy ε^C contains effects due to mechanical work. Note that the initial value of the energy density involves the initial density ρ^n, and does not include changes due to mass advection. Similarly $(\rho\ \varepsilon)$ is obtained using ρ^n and ε^C. Finally, we note that (5.54) can be rewritten using the mass flux (5.48b), illustrating that the material's specific energy is transported with the mass flux as expected physically. Thus, using the same mass flux for mass advection and specific energy advection guanantees that energy will not diffuse numerically relative to mass.

Problem 5.14

Construct a finite difference approximation to the internal energy equation in two dimensional cylindrical coordinates r,θ.

It can be show by a linearized stability analysis similar to that of Problem 5.12 that the time step constraint (5.52) is sufficient for the advection of internal energy density.

Second order advection as described above works well except in regions of large gradients, such as shock fronts, where it can result in too much mass or energy being removed from a single zone (in extreme cases, the total zonal density may become negative). First order schemes do not in general suffer from these difficulties. A general scheme which circumvents this problem, and which preserves the property of monotonicity, leads to an interpolation procedure which is second order accurate away from discontinuities, and which reverts automatically to first order near them.

Monotonicity

Another way to obtain (5.49) is to fit a quadratic curve to the density in zones $k-1/2$, $k+1/2$ and $k+3/2$ (see Problem 5.15). Although this scheme usually gives improved accuracy compared to first order methods, there will be occasional instances (such as the one shown in Figure 5.9) where the density interpolated a distance $u_k \Delta t/2$ behind the boundary is lower than any of the neighboring densities. There is no physical reason to expect that such a value is meaningful in general. In fact, the undershoot in this example will produce oscillations in the profile in the vicinity of the density jump which would not develop in an analytic solution of the density advection. If the initial profile changes monotonically in space, as does the profile denoted by the dots in Figure 5.9, then mass advection is expected to maintain this monotonic trend. In other words, advection is expected to maintain monotonicity. One way to achieve this is to replace the quadratic fit by a parabolic fit (dashed curve in Figure 5.9) whenever the former leads to physically unacceptable results (Van Leer 1974). In some instances oscillations in density (and in specific energy and pressure) may arise physically near regions where one of these variables changes rapidly. We take as the fundamental requirement of monotonicity that advection maintain monotonic profiles. There are many schemes which will accomplish this. One scheme which has been found to be useful is illustrated in Figure 5.10. In each zone (such as zone $k+1/2$) three slopes can be

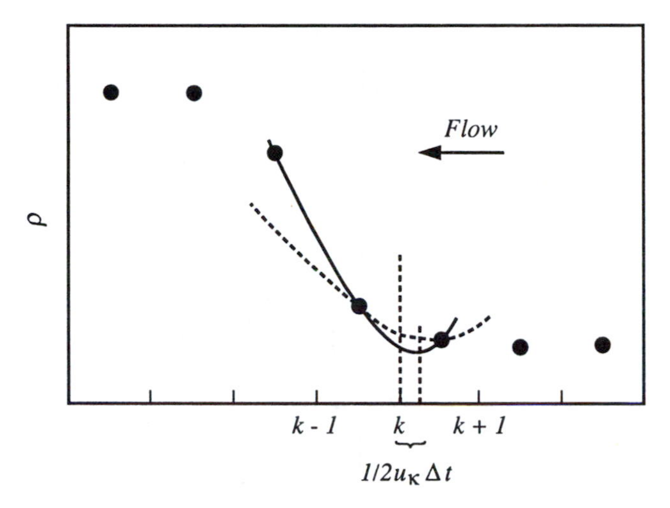

Figure 5.9. A monotonic set of densities (solid dots) versus zone number. The dashed curve is a quadratic fit to the densities at $k+3/2$, $k-1/2$ and $k+1/2$. The dashed curve is a parabola through the two neighboring values.

defined; select the slope of minimum absolute magnitude. If any two of the three slopes have opposite signs, then zero slope is to be used. For $u_k<0$, we interpolate a distance $u_k\Delta t/2$ backward from the zone boundary k. The result is second order accurate in general and maintains monotonicity under advection (near steep gradients, it reduces to first order). This scheme may be implemented as follows. First find the zone centered slope for each zone (consider zone $k+1/2$ shown in Figure 5.10). Using the neighboring densities $\{\rho_{k-1/2}, \rho_{k+1/2}, \rho_{k+3/2}\}$ find

$$\rho_{max} = max\,\{\rho_{k-1/2}, \rho_{k+1/2}, \rho_{k+3/2}\}$$

$$\rho_{min} = min\,\{\rho_{k-1/2}, \rho_{k+1/2}, \rho_{k+3/2}\}\ ,$$

and from them construct

$$S_1 = \frac{min\,\{\rho_{max} - \rho_{k+1/2}\,, \rho_{k+1/2} - \rho_{min}\}}{(\Delta x_{k+1/2}/2)}\ .$$

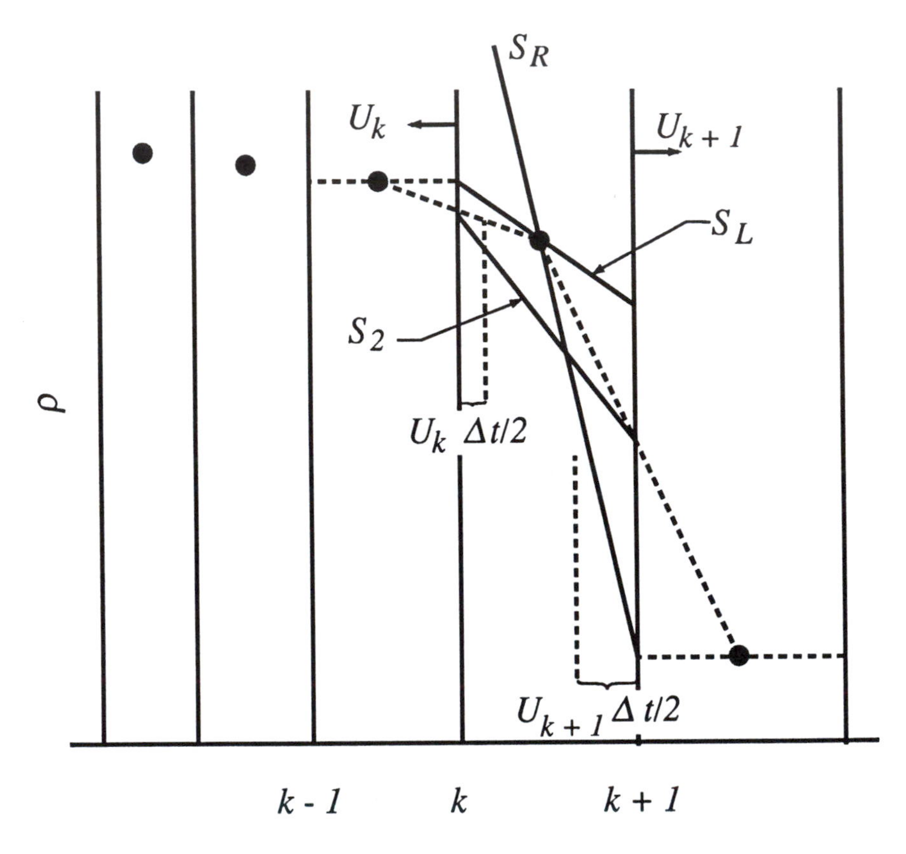

Figure 5.10. A monotonicity scheme for interpolating the density (or other scalar variables).

This quantity represents the absolute value of slope S_L or S_R (shown in Figure 5.10) having the minimum absolute magnitude. Next, using the density interpolated from the zone center to the zone boundary,

$$\rho_k = \frac{\Delta x_{k+1/2}\, \rho_{k-1/2} + \Delta x_{k-1/2}\, \rho_{k+1/2}}{\Delta x_{k+1/2} + \Delta x_{k-1/2}},$$

construct the slope

$$S_2 = \frac{\rho_{k+1} - \rho_k}{\Delta x_{k+1/2}}.$$

Finally, using S_1 and S_2 define the zone centered slope

$$S_{k+1/2} = \frac{S_2}{/S_2/} \min \{ /S_2/ , S_1 \} .$$

If any two of the three slopes S_L , S_R or S_2 shown in Figure 5.10 have opposite signs, then the final slope $S_{k+1/2}$ is set to zero. The second order monotonic value of the density is found using the slope $S_{k+1/2}$ as follows. At each zone boundary k choose the upstream value of the slope. Use it to find a value of the density interplolated a distance $u_k \Delta t/2$ upstream form the boundary. Thus the density used to construct the advected mass is given by

$$\tilde{\rho}_k = \rho_{k-1/2} + \frac{1}{2} S_{k-1/2} (\Delta x_{k-1/2} - u_k \Delta t) \qquad u_k > 0,$$

or

$$\tilde{\rho}_k = \rho_{k+1/2} - \frac{1}{2} S_{k+1/2} (\Delta x_{k+1/2} + u_k \Delta t) \qquad u_k < 0 .$$

This value of the density, and the advection volume $u_k \Delta t/2A_k$, is then used to construct the mass to be advected across the boundary k .

Advection of a square density pulse can be used to test the second order monotonicity scheme considered above. Figure 5.11a shows the initial pulse at time $t = 0$. After travelling a distance comparable to ten times its original thickness, the density pulse appears as shown in Figure 5.11a. For most of the zones the density equals its initial value to machine roundoff. Most of the diffusion which is observed in this test occurs as the pulse moves through a distance comparable to its original thickness. Comparison of this example and the first order results shown in Figure 5.7 clearly exhibit the inadequacy of first order advection in Eulerian schemes.

A second useful test is to advect a Gaussian density pulse through the grid. Figure 5.11b shows an initial pulse whose initial amplitude is $\rho = 1.0$, and Figure 5.11b shows the result using the second order monotonicity scheme above after the pulse has travelled about five times its original width. In this case there is about a four percent reduction in the peak density, but the pulse is still essentially Gaussian. It may be noted that many advection

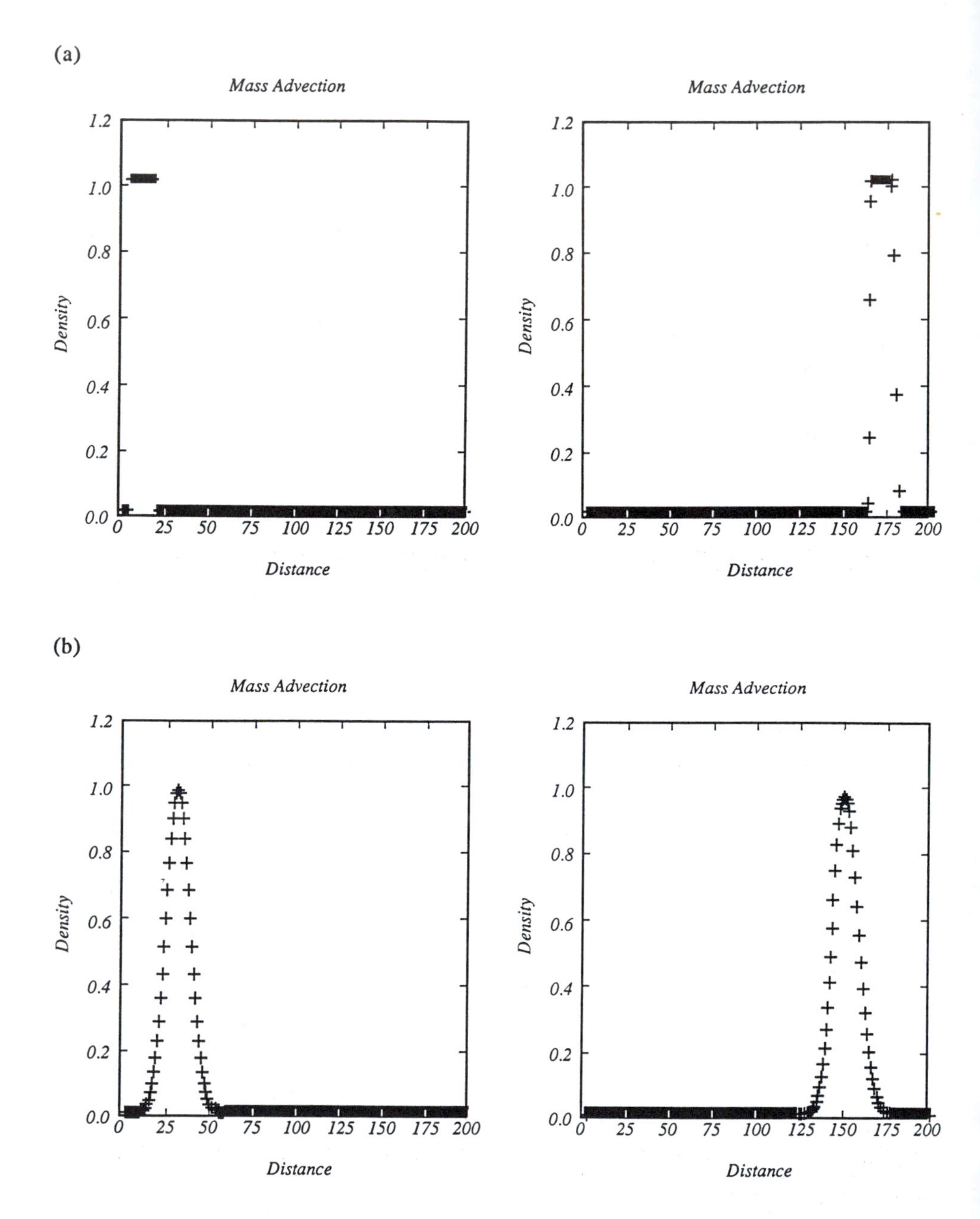

Figure 5.11. Advection of a density pulse using a second order monotonicity scheme. (a) A square distribution at $t = 0$ (left), and following an axial displacement of 1.5 *cm*. (b) A gaussian distribution at t = 0 (left), and following an axial displacement of 1.25 *cm*.

schemes which apparently work well for a square pulse shape do not always work well for a Gaussian pulse. For example, some schemes which retain square pulses also sharpen Gaussian pulses (in some cases turning them into essentially square pulses).

Problem 5.15

Fit the quadratic $\rho(x)=ax^2+bx+c$ to the density (shown schematically by the solid points in Figure 5.9), and show that the result is equivalent to (5.43) with ρ given by (5.49) if ρ is chosen to be $\rho(x_k+u_k\Delta t/2)$. For simplicity, assume uniform spatial zoning.

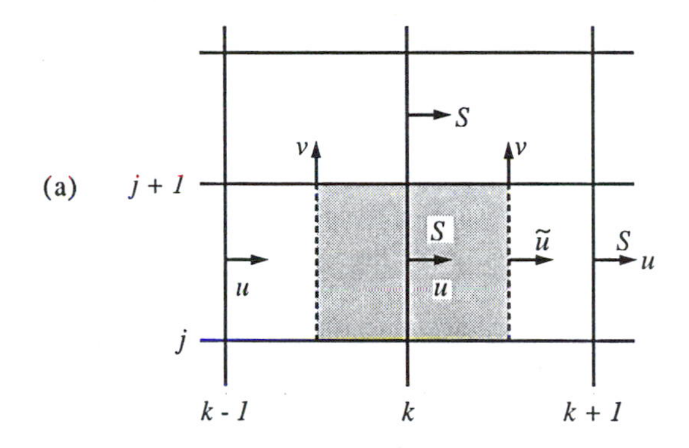

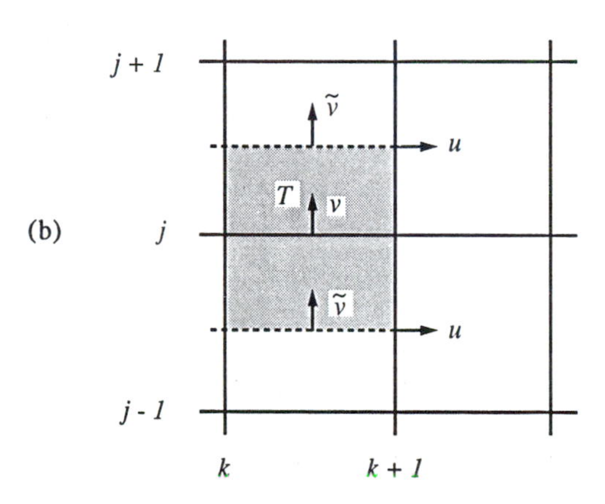

Figure 5.12. (a) Axial advection of the axial component of the momentum density. The axial momentum zone is shaded. (b) Radial advection of the radial component of the momentum density. The radial momentum zone is shaded.

Momentum Advection

The advection of axial momentum $S_{k,j+1/2}$ is described by the first two terms on the right side of the momentum equation (5.13):

$$\frac{\partial S}{\partial t} = - \frac{\partial[S(u - u_g)]}{\partial z} - \frac{1}{r}\frac{\partial[rS(v - v_g)]}{\partial r}$$

(5.55)

Integrating the axial gradient contribution over the volume $\Delta V_{k,j+1/2}$ (shown shaded in Figure 5.12a) gives (common index $j+1/2$ omitted):

$$\left(\frac{\partial S}{\partial t}\right)_k^Z = - \{A^Z_{k+1/2}\, \$^Z_{k+1/2}(u - u_g)_{k+1/2}$$

$$- A^Z_{k-1/2}\, \$^Z_{k-1/2}(u - u_g)_{k-1/2}\}/\Delta V_k \, .$$

(5.56)

The area factors are defined in the obvious way for cylindrical coordinates:

$$A^Z_{k+1/2,j+1/2} = A^Z_{k,j+1/2} \, .$$

Next, the relative velocities and the interpolated momenta appearing in (5.56) must be determined. This is more difficult to accomplish than for mass or energy advection, because none of the quantities (except for the area factors) are properly centered in space. Time centering also complicates the analysis, but this is difficult to accomplish in a completely correct way using explicit methods. The approach used to describe mass advection offers a useful guide. Physically, we expect the fluid's specific momentum to be advected with the mass flux. The mass flux FX_l crossing each zone boundary (5.48b), modified to include a second order monotonic interpolation scheme, is available from the mass advection step. The specific axial momentum flux crossing the area at $k+1/2$ should be proportional to u_k if the fluid's relative axial velocity $(u - u_g)_{k+1/2} > 0$, or to u_{k+1} if $(u - u_g)_{k+1/2} < 0$. We use these observations to construct a second order accurate approximation analogous to (5.44). First, define as in (5.44) the interpolated axial component of the specific momentum by

$$\hat{u}^{Z}_{k+1/2} \equiv \frac{1}{2} u_k (1 + \alpha^{Z}_{k+1/2}) + \frac{1}{2} u_{k+1} (1 - \alpha^{Z}_{k+1/2}) , \tag{5.57}$$

were $\alpha^{Z}{}_{k+1/2}$ is defined as in (5.45). Since no specific momentum flux across $k+1/2$ is expected if the axial mass flux at k and $k+1$ are equal in magnitude but oppositely directed, it is reasonable to define the zone centered mass flux by (common subscript $j+1/2$ omitted)

$$F^{Z}_{k+1/2} = \frac{1}{2} (F^{Z}_{k+1} + F^{Z}_{k}) . \tag{5.58}$$

We then replace (5.56a) with the physically equivalent expression

$$\left(\frac{\partial S}{\partial t}\right)^{Z}_{k} = - \{ A^{Z}_{k+1/2} \hat{u}^{Z}_{k+1/2} F^{Z}_{k+1/2} - A^{Z}_{k-1/2} \hat{u}^{Z}_{k-1/2} F^{Z}_{k-1/2} \} / \Delta V_k . \tag{5.56b}$$

Equations (5.56a) and (5.57)-(5.58) constitute a second order monotonic approximation to the axial transport of axial momentum. The use of the mass fluxes from the mass advection step reduces velocity diffusion. In practice a second order monotonic scheme should be used in place of (5.57).

The contribution to S due to radial gradients (the last term in (5.55)) can be constructed in a similar fashion (see Figure 5.12a). For this term an average radial mass flux is constructed and used to advect axial specific momentum. The result is (common subscript k omitted)

$$\left(\frac{\partial S}{\partial t}\right)^{R}_{j+1} = - \{ A^{R}_{j+1} \hat{u}^{R}_{j+1} F^{R}_{j+1} - A^{R}_{j} \hat{u}^{R}_{j} F^{R}_{j} \} / \Delta V_{j+1/2} \tag{5.59}$$

where the radial mass flux at the top of the momentum zone is

$$F_j^R = \frac{1}{2}\left(F_{j+1/2}^R + F_{j-1/2}^R\right)$$

(5.60)

and the area factor is

$$A_{k,j}^R = 2\pi r_j\, \Delta z_k \,.$$

(5.61)

The specific momentum appearing in (5.59) is defined as in (5.57). The final result, which is a first order approximation to (5.55) is

$$\frac{S_{k,j}^M - S_{k,j}^P}{\Delta t^{n+1/2}} = \left(\frac{\partial S}{\partial t}\right)_{k,j}^Z + \left(\frac{\partial S}{\partial t}\right)_{k,j}^R ,$$

(5.62)

where the momenta appearing on the right side are those obtained from (5.29), containing the effects of acceleration due to pressure gradients. The velocity components are given by (5.35) and (5.36), which use the momenta S^P and T^P, and the density resulting from the solution of the continuity equation:

$$u = S^P/\rho^A \qquad v = T^P/\rho^A .$$

(5.63)

Note that the time step used to advance S due to advection is $\Delta t^{n+1/2}$, the same as is used for mass and internal energy advection. One way to see why $\Delta t^{n+1/2}$ rather than Δt^n is used here is to examine total (kinetic plus internal) energy conservation in Eulerian form. Consider one dimensional planar coordinates with no grid motion: the time rate of change of the total energy is given by

$$\frac{\partial[\rho(\varepsilon + u^2/2)]}{\partial t} + \frac{\partial[\rho u(\varepsilon + u^2/2)]}{\partial x} + \frac{\partial(Pu)}{\partial x} = 0.$$

The second quantity in the advection term, $\rho u\, u^2/2$, which arises from momentum advection, is carried by the same mass flux ρu as is the internal energy. Therefore, the same time step, namely $\Delta t^{n+1/2}$, is used for momentum advection as for mass and energy advection.

Problem 5.16

Derive the Eulerian expression above for total energy conservation for the inviscid hydrodynamic equations in one dimensional planar coordinates.

The advection of radial momentum is described by the first two terms on the right side of (5.12):

$$\frac{\partial T}{\partial t} = -\frac{1}{r}\frac{\partial[rT(v-v_g)]}{\partial r} - \frac{\partial[T(u-u_g)]}{\partial z} . \tag{5.64}$$

A first order finite difference approximation to (5.64) can be constructed along lines similar to those leading to (5.62) by integrating (5.64) over the momentum zone $\Delta V_{k+1/2,j}$; (see Figure 5.12b). Spatial centering complicates the analysis as it does for the advection of axial momentum. An average mass flux is constructed at the surface of the momentum zone, and is used to advect specific radial momentum. The result is easily shown to be

$$\frac{T^M_{k+1/2,j} - T^P_{k+1/2,j}}{\Delta t^{n+1/2}} = \left(\frac{\partial T}{\partial t}\right)^R_{k+1/2,j} + \left(\frac{\partial T}{\partial t}\right)^Z_{k+1/2,j} , \tag{5.65}$$

$$\left(\frac{\partial T}{\partial t}\right)^R_j = -\{A^R_{j+1/2}\,\hat{v}^R_{j+1/2}\,F^R_{j+1/2} - A^R_{j-1/2}\hat{v}^R_{k-1/2}F^R_{k-1/2}\}/\ \Delta V_j . \tag{5.66}$$

$$F^R_{j+1/2} = \frac{1}{2}\,(F^R_{j+1} + F^R_j\,), \tag{5.67}$$

$$\hat{v}^R_{j+1/2} \equiv \frac{1}{2} v_j(1 + \alpha^R_{j+1/2}) + \frac{1}{2} v_{j+1}(1 - \alpha^R_{j+1/2}) . \tag{5.68}$$

The common subscript $k+1/2$ has been omitted from (5.66)-(5.68) for simplicity. The quantity $\alpha^R{}_{j+1/2}$ is defined as in (5.47), and the area factor

$$A^R_{j+1/2} = 2\pi r_{j+1/2} \, \Delta z_{k+1/2} . \tag{5.69}$$

The axial gradient gives (common subscript j omitted)

$$\left(\frac{\partial T}{\partial t}\right)^Z_{k+1/2} = -\{A^Z_{k+1} \hat{v}^Z_{k+1} F^Z_{k+1} - A^Z_k \hat{v}^Z_k F^Z_k\} / \Delta V_{k+1/2}. \tag{5.70}$$

The common subscript j has been omitted for convenience. Also, we define the centered mass flux by

$$F^Z_k = \frac{1}{2}\left(F^Z_{k+1/2} + F^Z_{k-1/2}\right) \tag{5.71}$$

$$A^Z_{k,j} = 2\pi r_j \, \Delta r_j . \tag{5.72}$$

Finally, the specific momentum is defined as in (5.68)

$$\hat{v}^Z_k \equiv \frac{1}{2} v_{k-1/2}(1 + \alpha^Z_k) + \frac{1}{2} v_{k+1/2}(\alpha^Z_k - 1) . \tag{5.73}$$

The common index j has been omitted in (5.73), and $\alpha^Z{}_k$ is defined in analogy with (5.45). Once again, a second order monotonic scheme should be used in place of(5.68).

Problem 5.17

Construct a finite difference representation of momentum advection in two dimensional cylindrical coordinates r,θ.

Problem 5.18

How would you construct a finite difference representation for the two dimensional cylindrically symmetric Eulerian hydrodynamic equations using zone corner centered velocities and momenta? Assume that all variables are centered as in (5.21) except for u,v,S and T, which are all centered at k,j. Use figures similar to Figures 5.3 through 5.4 and Figure 5.12 to illustrate centering and averaging required.

First order (donor cell) momentum advection is extremely diffusive and should be considered only as a motivational example. Substantial improvements in resolution usually result from second order accurate schemes. Higher order schemes may be constructed by replacing expressions such as (5.57) by interpolated values. Straightforward extension of the density interpolation scheme used for mass advection to momentum advection is possible, but other forms of interpolation may be used which have advantages for some problems. Consider for example the density, velocity and momentum near a region of steep composition gradient in a star, as shown schematically in Figure 5.13. Next, consider the radial advection of radial momentum from (5.64) which may be written in the form

$$\frac{\partial T}{\partial t} = -\frac{1}{r}\frac{\partial[rT(v - v_g)]}{\partial r} = -\frac{1}{r}\frac{\partial[r\rho v(v - v_g)]}{\partial r} , \tag{5.74}$$

where $T=\rho v$. A second order finite difference representation of (5.74) can be constructed in many ways. For example, we may interpolate on T, or on v only in the product ρv. Because there is no natural centering for the momentum flux, it is not obvious on which variable above the interpolation should be performed. The physics of the problem illustrated in Figure 5.13 can be used to resolve the ambiguity. Gravitational infall produces a more or less smooth variation in radial velocity (in the absence of shock waves) even though rapid changes in density due to steep composition gradients occur (it can be shown, for example, that the ratio ρ/μ, where μ is the material mean molecular weight, varies smoothly in the absence of shocks and localized energy sources). This suggests that interpolation be carried out on the velocity rather than on the momentum (see Figure 5.13). In fact, an empirical rule, which has been found to work well in practice is to interpolate on the smoothest variable whenever ambiguity arises. A thorough understanding of the important physics being modelled in a set of calculations also helps to identify optimal interpolation schemes.

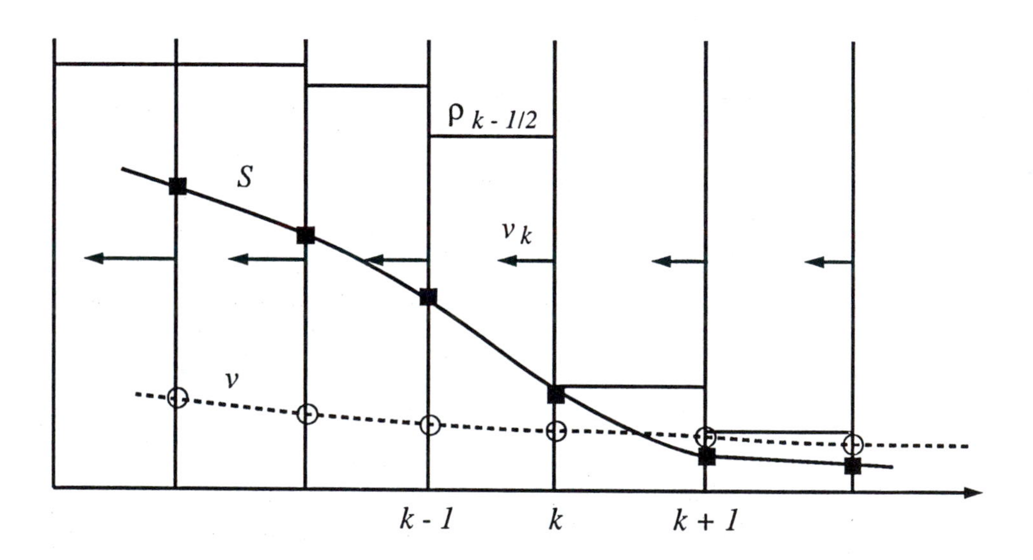

Figure 5.13. Density and velocity for matter undergoing infall in a stellar core near a steep conposition gradient. The solid curve represents the radial momentum density; the dashed curve is radial component of the velocity.

A second order scheme for the radial advection of radial momentum incorporating the arguments above is obtained by replacing (5.68) with the quantity

$$\hat{T}^{R}_{j+1/2} = \rho_{j+1/2} \hat{v}_{j+1/2} \tag{5.75}$$

(common subscript $k+1/2$ omitted) and interpolation is carried out on the velocity. In a similar fashion we denote the remaining interpolated momentum densities by:

$$\hat{T}^{Z}_{k} = \rho_{k} \hat{v}_{k} \tag{5.76}$$

(common subscript j omitted);

$$\hat{S}^{Z}_{k+1/2} = \rho_{k+1/2} \hat{u}_{k+1/2} \tag{5.77}$$

(common subscript $j+1/2$ omitted)

$$\hat{S}^{R}_{j} = \rho_{j} \hat{u}_{j} \tag{5.78}$$

(common subscript k omitted). In (5.75)-(5.78) the interpolated velocity components are denoted by a tilda. In practice they may be represented for example by a second order monotonicity scheme such as the one discussed below.

Monotonic Momentum Advection

A second order monotonic interpolation scheme for momentum advection can be constructed in much the same manner as for density and energy advection. A slightly different form which also works well is shown in Figure 5.14. In this method two slopes s_L and s_R are defined for each momentum zone, based on the velocities u_{k-1}, u_k and u_{k+1} (solid dots in Figure 5.14). The minimum slope (zero if s_L and s_R are of different sign) is adopted, and the velocity is interpolated a distance $u_{k+1/2}\Delta t/2$ upstream along it. It is this value of the velocity which is to be used in (5.77). The quantities

$$\hat{u}_{k,j}\ ,\ \hat{v}_{k+1/2,j+1/2}\ ,\ \text{and}\ \hat{v}_{k,j}$$

may be constructed in similar fashion. If the miminum slope is denoted by s, then it is readily shown that

$$\hat{u}_{k+1/2} = u_k + s\,(\Delta z_k /2 + u_{k+1/2}\Delta t) \tag{5.79}$$

where $u_{k+1/2}$ is defined as in (5.58). Two second order schemes have been presented above; which scheme is used is generally less important than the use of some scheme more accurate than donor cell (first order). The latter should never be used.

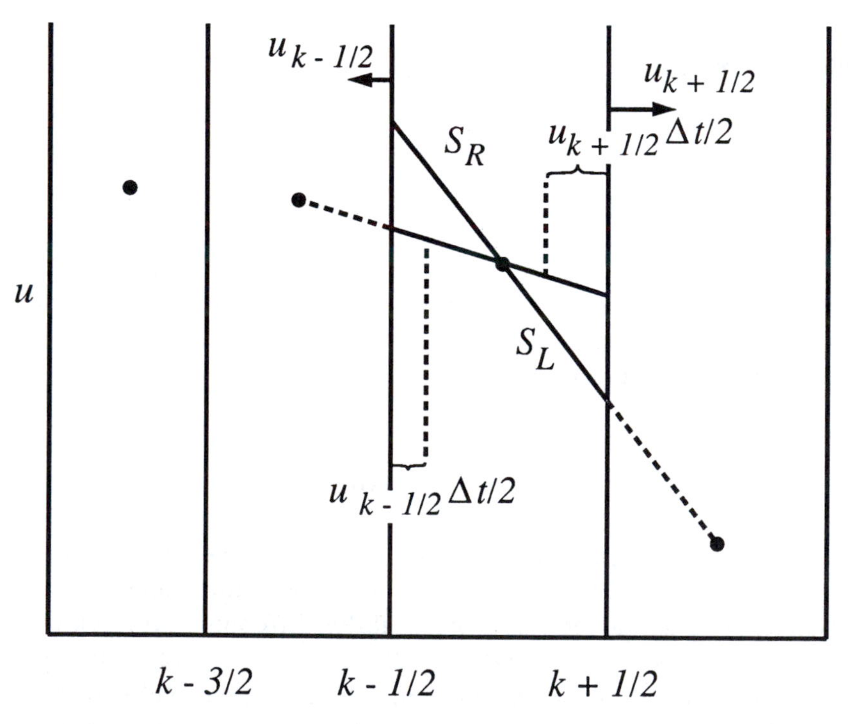

Figure 5.14. Monotonic axial advection of the axial momentum density. Only two slopes S_L and S_R are used, as shown for the momentum zone k. Also shown are the velocities u and the interpolation distance upstream from the zone face.

Problem 5.19

Use the monotonicity scheme of Figure 5.14 to find a second order expression for the interpolated value of $u_{k,j}$ to be used in (5.77).

Problem 5.20

Consider a stellar core (as shown schematically in Figure 5.15) undergoing gravitational infall, whose composition changes rapidly from nondegenerate neutrons (inside) to atoms of the iron group (outside) in the vicinity of zone k. Assume that $\varepsilon = C_V T$ and that C_V is given by the points x. In the absence of shock waves and local energy sources, $T(r)$ will remain smooth as the core evolves. Develop a second order energy advection scheme for this problem.

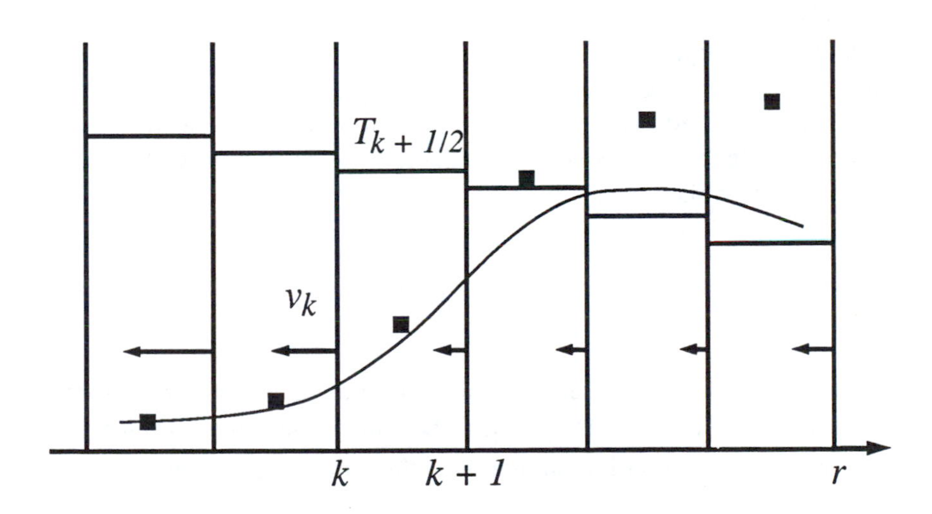

Figure 5.15. Temperature, specific heat (filled squares), and specific energy (solid curve) shown schematically for a stellar core. The matter inside r_k consists predominantly of neutrons, and the matter outside it consists of iron peak nuclei (see Problem 5.20). The fluid is moving to the right (inward).

Small asymmetries may arise when operator splitting is used to solve separately the axial and radial components of advection. Although the effect may be small during each computational cycle, the effect may be cumulative and produce spurious results during the course of a simulation. To compensate for these effects the scalar advection step can be performed with half the timestep in one direction, the entire timestep in the other direction, and concluded with half the timestep in the original direction. This sequence is repeated each computational cycle. A more efficient approach which works well in many applications is to use the full time step for scalar advection along each direction, but to alter the order from one computational cycle to the next. An alternate approach may be used in which the fluxes are constructed in both directions before the density is changed. This, however, requires additional storage.

Grid Motion

Eulerian equations written for a moving grid include terms representing changes in S, T, ρ and $\rho\varepsilon$ which are associated with the rate of change in zonal volume. The change in volume of the zones is described by $\nabla \cdot \mathbf{v}_g$, The changes in fluid variables are all described by equations of essentially the same form. The change in mass density can be used as an example to show how these terms are to be solved. Recalling (5.14), the change in mass density associated with grid motion is

$$\frac{\partial \rho}{\partial t} = - \rho \, \nabla \cdot \mathbf{v}_g \, .$$

The grid velocities are used to calculate the new grid positions and new volume elements $\Delta V'_{k+1/2}$. Denoting the new values of the density by ρ', the equation above has the simple solution

$$\rho' = \rho \, (\Delta V / \, \Delta V') \, .$$

Similar expressions are then used to find new values of the other zone-centered densities. Finally, the momentum density is changed after finding the change in ΔV_k.

This completes the Eulerian inviscid hydrodynamics cycle. At this stage in a calculation the new positions of the grid (if $\mathbf{v}_g$ is nonzero) are calculated from

$$\Delta z^{n+1}_{k+1/2} = \Delta z^{n}_{k+1/2} + (u_{g,k+1} - u_{g,k})\Delta t^{n+1/2}$$

$$r^{n+1}_j = r^n_j + v_{g,j}\, \Delta t^{n+1/2} . \tag{5.80}$$

All other grid related variables may be constructed from (5.80) in preparation for the next computational cycle.

Eulerian Boundary Conditions

A convenient way to impose Eulerian boundary conditions on the finite difference equations is to add a extra strip of zones along the top and bottom of the grid, as well as along each side. These zones do not represent a part of the physical grid, but are used solely for the purpose of defining finite difference representations of the boundary conditions. Thus, for example, the inner most set of density and radial velocity components in a problem will have the values $\rho_{k+1/2,\ 3/2}$ and $v_{k+1/2,\ 2}$, and the inner radius of the grid will be r_2. Similarly, if there are J physical zones in the radial direction the outer most values of the density and radial component of the velocity will be $\rho_{k+1/2,\ J+1/2}$ and $v_{k+1/2,\ J+2}$ respectively.

The simplest boundary conditions that can be imposed on the Eulerian equations represent reflecting boundaries. For a zone centered variable, $q_{k+1/2,j+1/2}$, reflection requires that the normal derivative of q vanish at the boundary. If the left boundary is reflecting, then (for all j)

$$q_{3/2,j+1/2} = q_{5/2,j+1/2} \cdot$$

Other types of boundary conditions can be more difficult to impose in Eulerian form, and are often ill defined. Some examples of Eulerian boundary conditions and the problems which they encounter are given in Problems 5.44 and 5.45.

Problem 5.21

An arbitrary, orthonormal coordinate system x^k is defined by the arc length between two infinitesimally nearby points x^k and $x^k + dx^k$ according to

$$ds^2 = g_{ik}\, dx^k dx^i$$

where repeated indicies are summed, and the metric tensor g_{ik} is diagonal. In this coordinate system physical lengths are measured by

$$dl_i = \sqrt{g_{ii}(dx^i)^2} = \sqrt{g_{ii}}\, dx^i$$

(no summation over i), and the volume element is given by

$$dV = \sqrt{g}\, d^3x$$

where $g = det(g_{ik})$. If $V^k = dx^k/dt$, show that the inviscid hydrodynamic equations may be written in the form

$$\frac{\partial(\rho\sqrt{g})}{\partial t} + \frac{\partial(\rho V^k\sqrt{g})}{\partial x^k} = 0 \tag{5.81}$$

$$\frac{\partial(\rho\varepsilon\sqrt{g})}{\partial t} + \frac{\partial(\rho\varepsilon V^k\sqrt{g})}{\partial x^k} = -P\frac{\partial(V^k\sqrt{g})}{\partial x^k} \tag{5.82}$$

$$\frac{1}{\sqrt{g}}\left[\frac{\partial(S_i\sqrt{g})}{\partial t} + \frac{\partial(V^k S_i\sqrt{g})}{\partial x^k}\right] = -\frac{\partial P}{\partial x^i} - \frac{S_k S_l}{2\rho}\frac{\partial g^{kl}}{\partial x^i} \;. \tag{5.83}$$

The first equation is the mass continuity equation, the second is the internal energy equation and the last is the momentum equation, with $S_i = \rho V_i$ and $V_i = g_{i\,k} V^k$. The last term on the right hand side of the momentum equation (5.83) is analogous to centrifugal force terms encountered in classical mechanics. Notice that in Cartesian coordinates where g_{ik} is constant, these terms vanish.

Evaluate the last term on the right in (5.83) in two dimensional cylindrical coordinates (r,z) and in two dimensional spherical coordinates (r,θ). Discuss the differences. Identify the physical components of the velocity and momentum in the two coordinate systems. What happens if axially symmetric rotation is included in the cylindrical case?

5.5 Shocks in Two Dimensions

The von Neumann artificial viscous stress used in modelling shock propagation which was introduced in Section 4.4 for the one dimensional Lagrangian hydrodynamic equations can be extended to the two dimensional Eulerian equations. An improved, second order monotonically limited form will then be introducted in Section 5.13. In the simplest approach to shock waves in two dimensional geometry the scalar artificial viscosity which was assumed to be proportional to $\Delta x \Delta v \approx [\Delta x^2 (\nabla \cdot \mathbf{v})^2]^{1/2}$ was replaced by the scalar $(\Delta r^2 \Delta v^2 + \Delta z^2 \Delta u^2)^{1/2}$, and a scalar artificial viscous stress was used in the hydrodynamic equations. This was ultimately abandoned for several reasons. The result was observed not to work well when nonuniform zoning occured (as is inevitable in many astrophysical calculations), since the simple scalar above tends to couple effects along orthogonal directions. It is also desirable in some problems to be able to introduce true shear stresses. A natural way to include these extra degrees of freedom is to replace the scalar artificial viscous stress by a tensor artificial viscous stress. Following the analogy with molecular viscosity, the acceleration and energy equations can be modified to include the terms (in Cartesian coordinates)

$$\rho\frac{\partial v^i}{\partial t} = \frac{\partial \Sigma^{ik}}{\partial x^i} \, ,$$

(5.84)

and

$$\rho\frac{\partial \varepsilon}{\partial t} = \Sigma_{ik}\frac{\partial v^i}{\partial x^k} \, .$$

(5.85)

Then the viscous stress tensor Σ_{ik} can be replaced by an artificial viscous stress Q_{ik}, which is in general a tensor. However, no rigorous attempt will be made here to develop a tensor Q_{ij} since in most applications of interest below a simple modification of the scalar form has been found to be adequate. Accordingly, a modification of (4.77)-(4.80) will be considered which allows for shock formation and propagation in two dimensions. Several approximations will be made in constructing finite difference representations for (5.84)-(5.85). A rationale for these approximations may be appreciated if it is remembered that the artificial viscosity was introduced essentially as an interpolation scheme which satisfied the Rankine-Hugoniot jump conditions across a shock front. In this sense, the exact form of Q_{ij} is not expected to be of primary importance. The real justification in fact for employing any artificial stress is that it works. The artificial viscous stress used in the Lagrangian equations is small everywhere except in several zones around the shock front, a property which should be carried over to the multidimensional Eulerian case.

A two dimensional artificial viscous stress will be developed below which is applicable to problems exhibiting axial symmetry. Replacing the stress Σ_{ik} by Q_{ik}, (5.84)-(5.85) are, in cylindrical coordinates:

$$\frac{\partial S}{\partial t} = -\frac{\partial (Q_{zz})}{\partial z} - \frac{1}{r}\frac{\partial (r Q_{zr})}{\partial r}$$

(5.86)

$$\frac{\partial T}{\partial t} = -\frac{\partial (Q_{rz})}{\partial z} - \frac{1}{r}\frac{\partial (rQ_{rr})}{\partial r} \tag{5.87}$$

and

$$\rho\,\frac{\partial \varepsilon}{\partial t} = -Q_{zz}\frac{\partial u}{\partial z} - Q_{rr}\frac{\partial v}{\partial r} - Q_{zr}\frac{\partial u}{\partial r} - Q_{rz}\frac{\partial v}{\partial z}\,. \tag{5.88}$$

Note that although some care has been taken to include curvature effects associated with curvilinear coordinates in (5.86)-(5.87), no such terms appear in the energy equation. Furthermore, no advective terms appear in (5.86)-(5.88). As we shall see, Q_{ij} is determined by the local instantaneous velocity field, and thus is not an intrinsic property of the material. Hence it is not an advected quantity.

It may be recalled that the planar artificial viscous stress discussed in Chapter 4 was proportional to $\rho l^2 |\nabla\cdot\mathbf{v}|\ \partial v/\partial x \approx \rho(\Delta v)^2$. An obvious choice, then, for the components Q_{rr} and Q_{zz} is

$$Q_{rr} \equiv C_Q\,\rho\,(\Delta_r v)^2 \tag{5.89}$$

$$Q_{zz} \equiv C_Q\,\rho\,(\Delta_z u)^2 \tag{5.90}$$

where C_Q is a coefficient which may be adjusted by input, and $\Delta_i \equiv \Delta x^i \partial/\partial x^i$ denotes the change in the velocity component across the zone in the $\mathbf{e}_i$ direction. It is reasonable to expect that these forms will reproduce the behavior of planar shocks in the radial or the axial direction. As in the scalar case, Q_{rr} and Q_{zz} are conventionally taken to be nonzero only for compression [recall the discussion following (4.82)]; thus

$$Q_{zz} = 0 \qquad \text{if } \Delta_z u \geq 0\,,$$

$$Q_{rr} = 0 \qquad \text{if } \Delta_r v \geq 0\,. \tag{5.91}$$

The components (5.89)-(5.90) are usually called compressional artificial viscous stresses.

If Σ^{ik} in (5.84) represents real molecular viscosity, then the terms Σ^{rz} and Σ^{zr} correspond to shear artificial viscous stress, and arise because the gradients $\partial v/\partial z$ and $\partial u/\partial r$ are nonzero. Noting that the energy equation (5.88) contains contributions $Q_{zr}\partial u/\partial r$ and $Q_{rz}\partial v/\partial z$, which should always be positive (viscosity is dissipative and should not decrease thermal energy), an obvious choice for the shear artificial viscous stress components is

$$Q_{rz} = C_S \rho \Delta_z v |\Delta_z v| \tag{5.92}$$

$$Q_{zr} = C_S \rho \Delta_r u |\Delta_r u| \tag{5.93}$$

where C_S is the shear viscosity coefficient. The shear terms may change sign, but will always lead to an increase in the internal energy (5.88). Thus a condition analogous to (5.91) is not required for them. We do not recommend the use of an artificial shear stress unless needed to smooth out very complicated flows.

Space-Time Centering

A finite difference representation for the components of the compressional and shear artificial viscous stress can be constructed from (5.89)-(5.90) and (5.92)-(5.93) in several ways. In this section a relatively standard approach will be taken which illustrates many aspects of the artificial viscosity method. An improved method will be considered later (Section 5.11). First, consider the compressional components; they are naturally given by

$$Q_{zz,k+1/2,j+1/2} = C_Q (u_{k+1,j+1/2} - u_{k,j+1/2})^2 \rho_{k+1/2,\, j+1/2} \tag{5.94}$$

$$Q_{rr,k+1/2,j+1/2} = C_Q(v_{k+1/2,j+1} - v_{k+1/2,j})^2 \rho_{k+1/2,\, j+1/2} \tag{5.95}$$

and are centered as shown in Figure 5.16. The shear components may be similarly defined:

$$Q_{rz,k,j} = C_S\,(v_{k+1/2,j} - v_{k-1/2,j})\,|v_{k+1/2,j} - v_{k-1/2,j}|\,\rho_{k,j} \tag{5.96}$$

$$Q_{zr,k,j} = C_S\,(u_{k,j+1/2} - u_{k,j-1/2})\,|u_{k,j+1/2} - u_{k,j-1/2}|\,\rho_{k,j} \tag{5.97}$$

The time centering of the velocity components (and hence of the artificial viscous stress components) will be considered below. Finally, the constraint (5.91) will be imposed for zones undergoing expansion. The artificial viscous stress Q_{ij} defined by (5.94)-(5.97) is not a tensor, but it has been found to work well in problems using arbitrary zoning. It also naturally incorporates shear effects because (5.96)-(5.97) allow transverse velocity gradients to contribute to the acceleration.

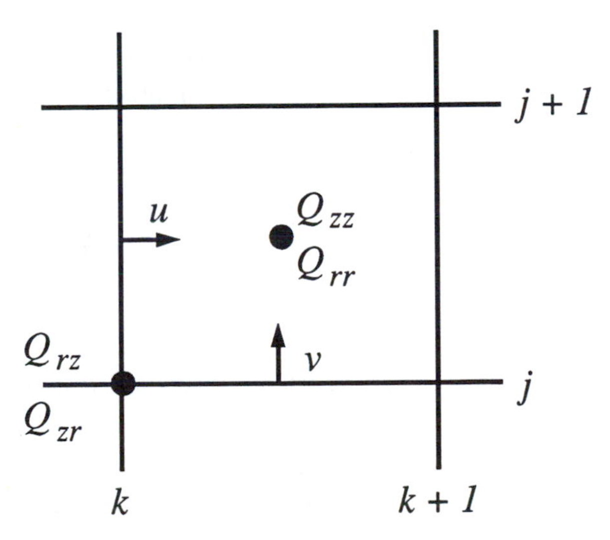

Figure 5.16. Spatial centering of the artificial viscous stress, assuming face centered components for the fluid velocity.

The effects of artificial viscous stress may now be incorporated into the finite difference scheme developed in the first part of this chapter. A convenient way of accomplishing this is to update S and T due to Q_{ij} after the pressure gradient terms (5.28) and (5.29) have been solved, using the updated values u^P and v^P to construct Q_{ij}. The new values of S and T are then used to obtain u^Q and v^Q from (5.35) and (5.36). The artificial viscous heating (5.88) is then done using the same Q_{ij} as was used to solve (5.86) and (5.87), but with the volume rate of change factors based on u^Q and v^Q. We then proceed with the solution of the mechanical work and advection terms as outlined earlier. Thus, the artificial viscosity is included in the Lagrangian stage of the calculation. We now proceed to construct the finite difference equations corresponding to (5.86) and (5.87).

First consider the axial momentum (5.86); integrating over the volume $\Delta V_{k,j+1/2}$ as was done for the pressure gradient in (5.24), the compressional term becomes (common index $j+1/2$ omitted):

$$-\frac{1}{\Delta V_k}\int\frac{\partial[Q^{zz}]}{\partial z}dV = -\frac{A_{k+1/2}Q^{zz}_{k+1/2}-A_{k-1/2}Q^{zz}_{k-1/2}}{\Delta V_k}. \tag{5.98}$$

For notational convenience in the finite difference formulae we have set $Q^{ij}=Q_{ij}$. The shear artificial viscosity term, when integrated over the same volume $\Delta V_{k,j+1/2}$, yields (common index k omitted):

$$-\frac{1}{\Delta V_{j+1/2}}\int\frac{1}{r}\frac{\partial[r\,Q^{zr}]}{\partial r}dV = -\frac{1}{\Delta V_{j+1/2}}\left(A_{j+1}Q^{zr}_{j+1}-A_jQ^{zr}_j\right) \tag{5.99}$$

The density appearing in the artificial viscous stress is defined by

$$\rho_{k,j} = \frac{1}{\Delta z_k}\int\rho\,dz \tag{5.100}$$

and the integral is along the zone boundary defined by fixed r_j, and $z_{k-1/2} \leq z \leq z_{k+1/2}$. This integral actually involves the density of the four zones surrounding the node k,j. Any reasonable choice of $\rho_{k,j}$ will conserve momentum in (5.99), so the ambiguity in (5.100) can be exploited, for example, to conserve total energy. We shall not pursue this point further here, but will simply note that the following prescription works well in practice:

$$\rho_{k,j} = \frac{1}{4}\left\{\rho_{k+1/2,j+1/2} + \rho_{k+1/2,j-1/2} + \rho_{k-1/2,j+1/2} + \rho_{k-1/2,j-1/2}\right\} . \tag{5.101}$$

Problem 5.22

Suggest ways that the average (5.100) might be defined if $\rho_{k,j}$ as given in (5.101) is unacceptable. Under what circumstances would (5.101) not be expected to be a reasonable definition?

The axial momentum change due to artificial viscous stress can be obtained by adding (5.98) and (5.99), multiplying the result by Δt^n and using the final result to update S (common index $j+1/2$ omitted):

$$\frac{S_k^Q - S_k^P}{\Delta t^n} = -\frac{(A_{k+1/2} Q_{k+1/2}^{zz} - A_{k-1/2} Q_{k-1/2}^{zz})}{\Delta V_k} - \frac{(A_{j+1} Q_{j+1}^{zr} - A_j Q_j^{zr})}{\Delta V_k}. \tag{5.102}$$

The momentum is given by (5.29).

The change in radial momentum due to artificial viscous stress follows from (5.87) in essentially the same manner as above. Integrating (5.87) over the volume $\Delta V_{k+1/2,j}$ and defining $\rho_{k,j}$ as in (5.101) it follows that

$$\frac{T^Q_{k+1/2,j} - T^P_{k+1/2,j}}{\Delta t^n} = -\frac{(A_{k+1/2,j+1/2}\,Q^{rr}_{k+1/2,j+1/2} - A_{k+1/2,j-1/2}\,Q^{rr}_{k+1/2,j-1/2})}{\Delta V_{k+1/2,j}}$$

$$- \frac{(A_{k+1,j}\,Q^{rz}_{k+1,j} - A_{k,j}\,Q^{rz}_{k,j})}{\Delta V_{k+1/2,j}}.$$

(5.103)

Problem 5.23

The change in radial momentum in two dimensional cylindrical systems due to molecular viscosity is described by

$$\frac{\partial T}{\partial t} = \frac{1}{r}\frac{\partial}{\partial r}\left(r\eta\frac{\partial v}{\partial r}\right) + \frac{\partial}{\partial z}\left(\eta\frac{\partial v}{\partial z}\right) - \frac{\eta v}{r^2}.$$

(5.104)

Use (4.77) and (4.78) to show that the first two terms on the right of (5.104) lead to (5.87). Construct a finite difference representation for the last term in (5.104) which can be added to (5.103).

The form of the artificial viscous stress developed above spreads shock fronts over about four Eulerian zones. In the Lagrangian formulation the zones near a shock front are compressed and are therefore quite narrow, so that the spatial width of the shock is smaller than for shocks in Eulerian calculations. In the absence of grid motion, strong shocks will be spread out over a relatively large spatial extent in Eulerian hydrodynamics.

Shock Heating

The heating associated with the entropy increase across a shock front is given by (5.88), which must be reduced to finite difference form. Integrate the first term on the right side over the zonal volume $\Delta V_{k+1/2,j+1/2}$:

$$\frac{1}{\Delta V}\int Q^{zz}\frac{\partial u}{\partial z}dV = \frac{1}{\Delta V}Q^{zz}\int\frac{\partial u}{\partial z}dV$$

$$= \frac{Q^{zz}}{\Delta V}(A_{k+1}u_{k+1} - A_k u_k) \; . \tag{5.105}$$

The second term in (5.88) can be evaluated as follows, noting that Q^{rr} is constant in ΔV :

$$\frac{1}{\Delta V}\int Q^{rr}\frac{\partial v}{\partial r}dV = \frac{1}{\Delta V}Q^{rr}\int dz\frac{\partial v}{\partial r}2\pi r\,dr$$

$$= \frac{2\pi Q^{rr}}{\Delta V}\int dz\,\{rv\Big|_j^{j+1} - \int v\,dr\}$$

$$= \frac{2\pi Q^{rr}}{\Delta V}\int dz\,\{r_{j+1}v_{j+1} - r_j v_j - v_j(r_{j+1/2} - r_j) - v_{j+1}(r_{j+1} - r_{j+1/2})\}$$

$$= \frac{2\pi r_{j+1/2}\Delta z_{k+1/2}Q^{rr}_{k+1/2,j+1/2}}{\Delta V_{k+1/2,j+1/2}}(v_{j+1} - v_j)_{k+1/2} \tag{5.106}$$

Equations (5.105)-(5.106) do not use a properly time centered artificial viscous stress. Proper centering can be achieved as illustrated by (4.84) and the accompanying discussion. A simpler approach which works well is to replace the velocity gradient $\partial u / \partial z$ by a time average (common subscript $j+1/2$ omitted)

$$\frac{1}{2}\left(\frac{u_{k+1} - u_k}{\Delta z_{k+1/2}} + \frac{u'_{k+1} - u'_k}{\Delta z_{k+1/2}}\right)$$

where u_k is the velocity obtained from the solution of the pressure gradient (5.29), and u'_k is the velocity obtrained after solving (5.102) for the artificial viscous stress acceleration. A similar time average is constructed for the gradient $\partial v / \partial r$. These expressions are then used in place of the simple time difference in (5.105) and (5.106).

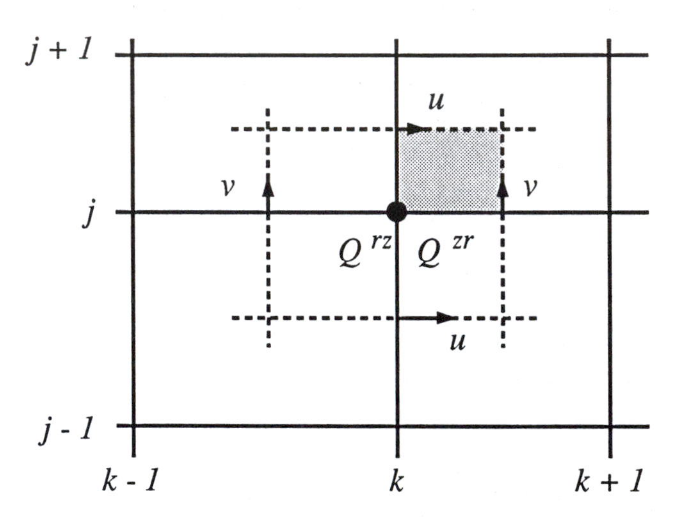

Figure 5.17. Spatial centering of the shear components of the artificial viscous stress used to evaluate shock heating (see Problem 5.24).

The last two terms on the right side of (5.88) are more difficult to evaluate in finite difference form because the specific energy is zone centered and the shear artificial viscous stress is node centered. If the shear terms on the right side of (5.88) are integrated over the nodal volume $\Delta V_{k,j}$ (see Figure 5.17) then an analysis similar to (5.105)-(5.106) can be used to obtain the shear viscous heating of this volume element. However, it is the heating of the volume $\Delta V_{k+1/2,j+1/2}$ which is needed, since the compressional terms (5.105)-(5.106) apply to it. Therefore, the shear heating of $\Delta V_{k,j}$ must be reapportioned among the zones $(k+1/2,\ j+1/2)$, $(k+1/2,\ j-1/2)$, $(k-1/2,\ j+1/2)$, and $(k-1/2,\ j-1/2)$.

Problem 5.24

Evaluate the integral of the shear artificial viscous heating terms in (5.88) over the nodal volume $\Delta V_{k,j}$. Show that

$$\int \rho \frac{\partial \varepsilon}{\partial t} dV = -A^r_{k,j} Q^{zr}_{k,j}(u_{k,j+1/2} - u_{k,j-1/2}) - A^z_{k,j} Q^{rz}_{k,j}(v_{k+1/2,j} - v_{k-1/2,j})$$

(5.107)

where $A^r{}_{k,j} = 2\pi r_j \Delta z_k$ and $A^z{}_{kj} = 2\pi r_j \Delta r_j$ Figure 5.17 shows the spatial centering.

The shear artificial viscous heating may be obtained in finite difference form as described in Problem 5.24, which gives the increase in specific internal energy in $\Delta V_{k,j}$. The fraction of this energy which is to be associated with zone $\Delta V_{k+1/2,j+1/2}$ is just the amount (5.107), multiplied by the ratio of the shaded portion of $\Delta V_{k+1/2,j+1/2}$ to $\Delta V_{k,j}$ itself:

$$\frac{1}{4}\frac{\Delta V_{k+1/2,j+1/2}}{\Delta V_{k,j}} = \frac{r_{j+1/2}\Delta r_{j+1/2}\Delta z_{k+1/2}}{4 r_j \Delta r_j \Delta z_k}$$

Similarly weighted contributions are made to the heating of zone $k+1/2,\ j+1/2$ by terms centered at $(k+1,j)$, $(k+1,j+1)$ and $(k,j+1)$. The third term on the right side of (5.88), integrated over the volume $\Delta V_{k+1/2,j+1/2}$ is

$$\frac{1}{\Delta V_{k+1/2,j+1/2}} \int Q^{zr} \frac{\partial u}{\partial r} dV = \frac{A^r_{k,j} Q^{zr}_{k,j}(u_{j+1/2} - u_{j-1/2})_k}{4\Delta V_{k,j}}$$

$$+ \frac{A^r_{k+1/j} Q^{zr}_{k+1,j}(u_{j+1/2} - u_{j-1/2})_{k+1}}{4\Delta V_{k+1,j}} + \frac{A^r_{k,j+1} Q^{zr}_{k,j+1}(u_{j+3/2} - u_{j+1/2})_k}{4\Delta V_{k,j+1}}$$

$$+ \frac{A^r_{k+1,j+1} Q^{zr}_{k+1,j+1}(u_{j+3/2} - u_{j+1/2})_{k+1}}{4\Delta V_{k+1,j+1}}$$

(5.108)

Integrating the last term in (5.88) over $\Delta V_{k+1/2,j+1/2}$ gives

$$\frac{1}{\Delta V_{k+1/2,j+1/2}}\int Q^{rz}\frac{\partial v}{\partial z}dV = \frac{A^z_{k,j}Q^{rz}_{k,j}(v_{k+1/2} - v_{k-1/2})_j}{4\Delta V_{k,j}}$$

$$+\frac{A^z_{k+1,j}Q^{rz}_{k+1,j}(v_{k+3/2} - v_{k-+1/2})_j}{4\Delta V_{k+1,j}} + \frac{A^z_{k,j+1}Q^{rz}_{k,j+1}(v_{k+1/2} - v_{k-1/2})_{j+1}}{4\Delta V_{k,j+1}}$$

$$+\frac{A^z_{k+1,j+1}Q^{rz}_{k+1,j+1}(v_{k+3/2} - v_{k+1/2})_{j+1}}{4\Delta V_{k+1,j+1}}.$$

(5.109)

Viscous heating of zone $k+1/2,j+1/2$ is given by the negative of the sum of (5.105), (5.106), (5.109) and (5.109), each multiplied by the time step Δt^n. In evaluating the four contributions to $\partial\varepsilon/\partial t$, the velocity components u^Q and v^Q obtained from the viscous acceleration are used. These terms, as noted previously, are also to be used in evaluating the mechanical work.

Shock heating is the result of dissipative processes and, as such, can only lead to an increase in internal energy. The constraint (5.91) on the compressional viscosity guarantees that compressional shock heating will never be negative. The shear artificial viscous stress terms as written above guarantee that these terms also lead only to heating. This results because each term in (5.108) and (5.109) is positive definite (for example, the first term on the right side of (5.108) is, because of (5.93), proportional to $(\Delta_r u)^2/|\Delta_r u|$).

In most calculations values of the artificial viscosity coefficients C_Q between 1.0 and 2.0, and $Q_S=0$ are usually adequate to represent shock propagation in two dimensional axial symmetric problems. For strong shocks, $C_Q = 2.0$ is usually appropriate. The approach developed above also can be extended to other symmetries and coordinate systems. A larger C_Q yields more accurate shock jump (Hugoniot) conditions, but it spreads the shock over more zones. An improved treatment of shocks will be discussed in Section 5.13.

5.6 STABILITY

The stability requirement to be imposed on the explicit Eulerian equations of inviscid hydrodynamics can be understood on simple physical grounds. In the Lagrangian (co-moving) frame stability of the finite difference equations requires that sound waves cross less than one zone width in a time Δt. In the moving grid Eulerian framework the zone boundaries move with speed $\mathbf{v}_g$ and the fluid moves with speed $\mathbf{v}$; the fluid velocity relative to the grid is therefore $\mathbf{v} - \mathbf{v}_g$, and an obvious extension of the Lagrangian stability requirement is that $(|\mathbf{v} - \mathbf{v}_g| \pm c_S)\, \Delta t$ be less than any zone width (see for example, Ricthmyer and Morton (1967) for $\mathbf{v}_g = 0$, which can be easily extended to the case $\mathbf{v}_g$ not equal to zero). For cylindrically symmetric systems in axial coordinated a practical time step constraint is give by

$$\Delta t_c = C_1 \, min\left\{ \frac{\Delta z_{k+1/2}}{|u - u_g|_{k,j+1/2} + c_S} \, , \, \frac{\Delta r_{j+1/2}}{|v - v_g|_{k+1/2,j} + c_S} \right\} \tag{5.110}$$

where c_S is the zone centered adiabatic sound speed given by (3.3). The lack of consistent centering in (5.110) does not cause problems since c_S will usually vary smoothly from zone to zone. For $\mathbf{v} = \mathbf{v}_g$, this reduces to the Courant time step in two dimensions. In practice it is usually sufficient to assign the value $C_1 = 0.5$. For subsonic flow it is often possible to use a value of C_1 only slightly less than unity. When the flow is extremely violent, it may be necessary to set $C_1 = 0.25$. For the Eulerian equations the constraint (5.110) is usually more important in determining accuracy than stability. The use of a time step constraint based on analogy with the Lagrangian finite difference scheme is reasonable since the Eulerian approach involves as its first phase a Lagrangian-like solution of the pressure gradient and mechanical work terms.

As for the Lagrangian formulation with shocks, the presence of artificial viscosity will lead to an additional time step constraint for the Eulerian scheme. The nature of the constraint can be found by an analysis similar to the one used in Chapter 4 for the Lagrangian equations, since the equations for artificial viscosity in Eulerian form also reduce to an explicit diffusion equation to which a time step constraint like (2.31) must also apply. Thus, for example, if these methods are applied to the first two terms in (5.86) we arrive at the time step constraint

$$\Delta t < \frac{1}{4} \Delta z \left(\frac{\rho}{Q_{zz}}\right)^{1/2} .$$

A similar result obtains if only the shear term in (5.86) is included. Proceeding in this way four constraints on Δt are obtained from (5.86)-(5.87), the minimum of which may be used to limit the hydrodynamic time step. A combined form of these constraints in two dimensional axial corrdinates is given by

$$\Delta t = C_2 \; min \left\{ \frac{1}{4} \frac{\rho^{1/2} \Delta z_{k+1/2}}{C_S^{1/2} |\Delta_z v| + C_Q^{1/2} |\Delta_z u|} , \right.$$

$$\left. \frac{1}{4} \frac{\rho^{1/2} \Delta r_{j+1/2}}{C_S^{1/2} |\Delta_r u| + C_Q^{1/2} |\Delta_r v|} \right\} , \tag{5.111}$$

where the shear quantities in the denominator are centered at k,j, and the compressional terms are centered at $k+1/2,j+1/2$.

The actual time step for the hydrodynamics with artificial viscosity should be the minimun of Δt_Q and Δt_C above. As in the Lagrangian approach, if $\Delta t^{n+3/2} > \Delta t^{n+1/2}$, it is usually advisable to restrict the growth in time step to be no more than, say, 20% of the old value. Then

$$\Delta t^{n+3/2} = min \, \{\Delta t_C \, , \Delta t_Q \, , 1.2 \Delta t^{n+1/2}\} \, . \tag{5.112}$$

Finally Δt^{n+1} is given by (4.59) and the new time is given by $t^{n+1} = t^n + \Delta t^{n+1/2}$.

A standard problem used to test shock propagation algorithms for which analytic solutions exist (Landau and Lifshitz 1959) is the

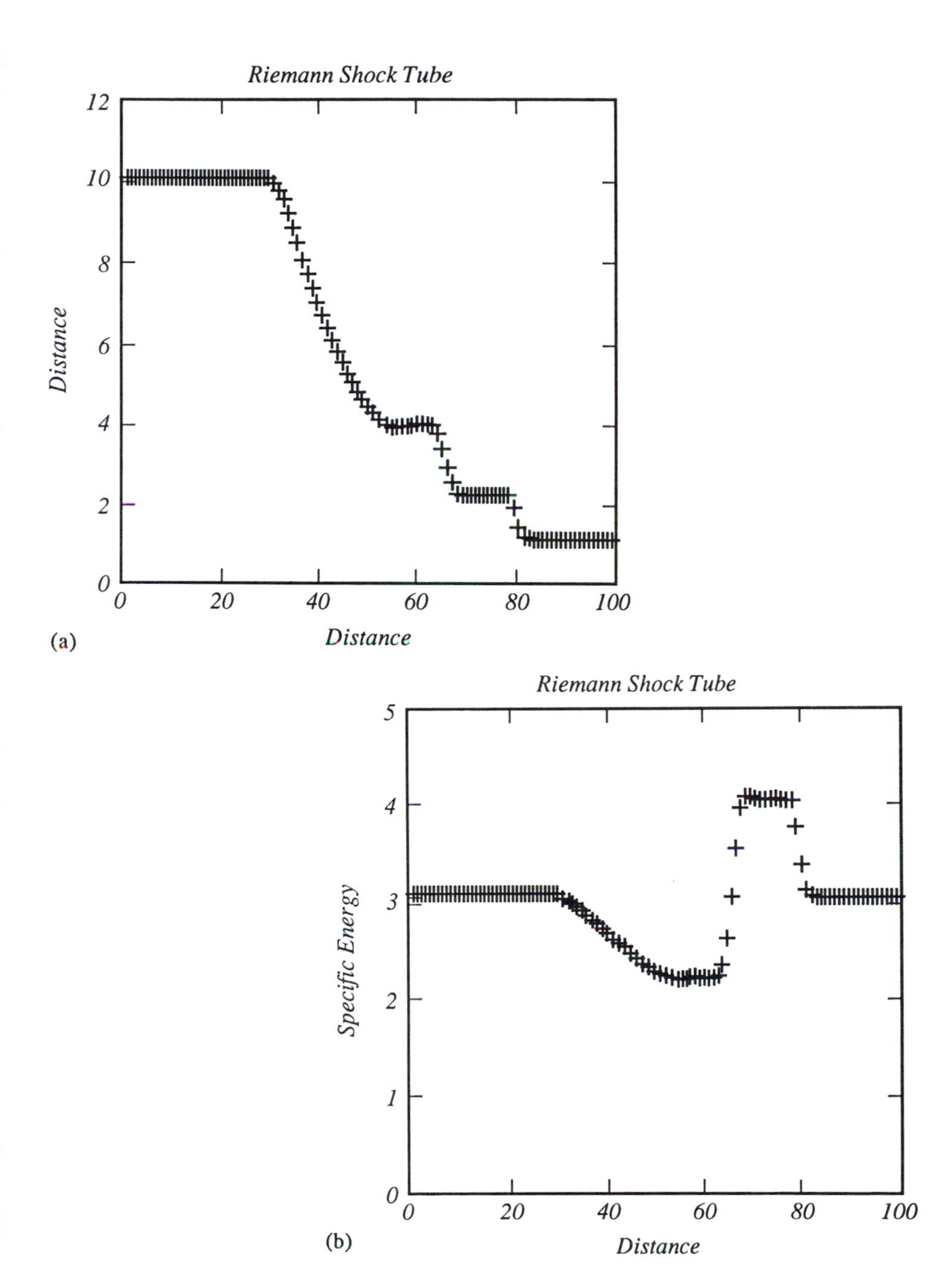

Figure 5.18. Riemann shock tube calculation using second order monotonic advection. (a) Shows the density at t = *15.4 sec.* The initial configuration is shown by the solid line. (b) Shows the specific energy at t = *15.4 sec.*

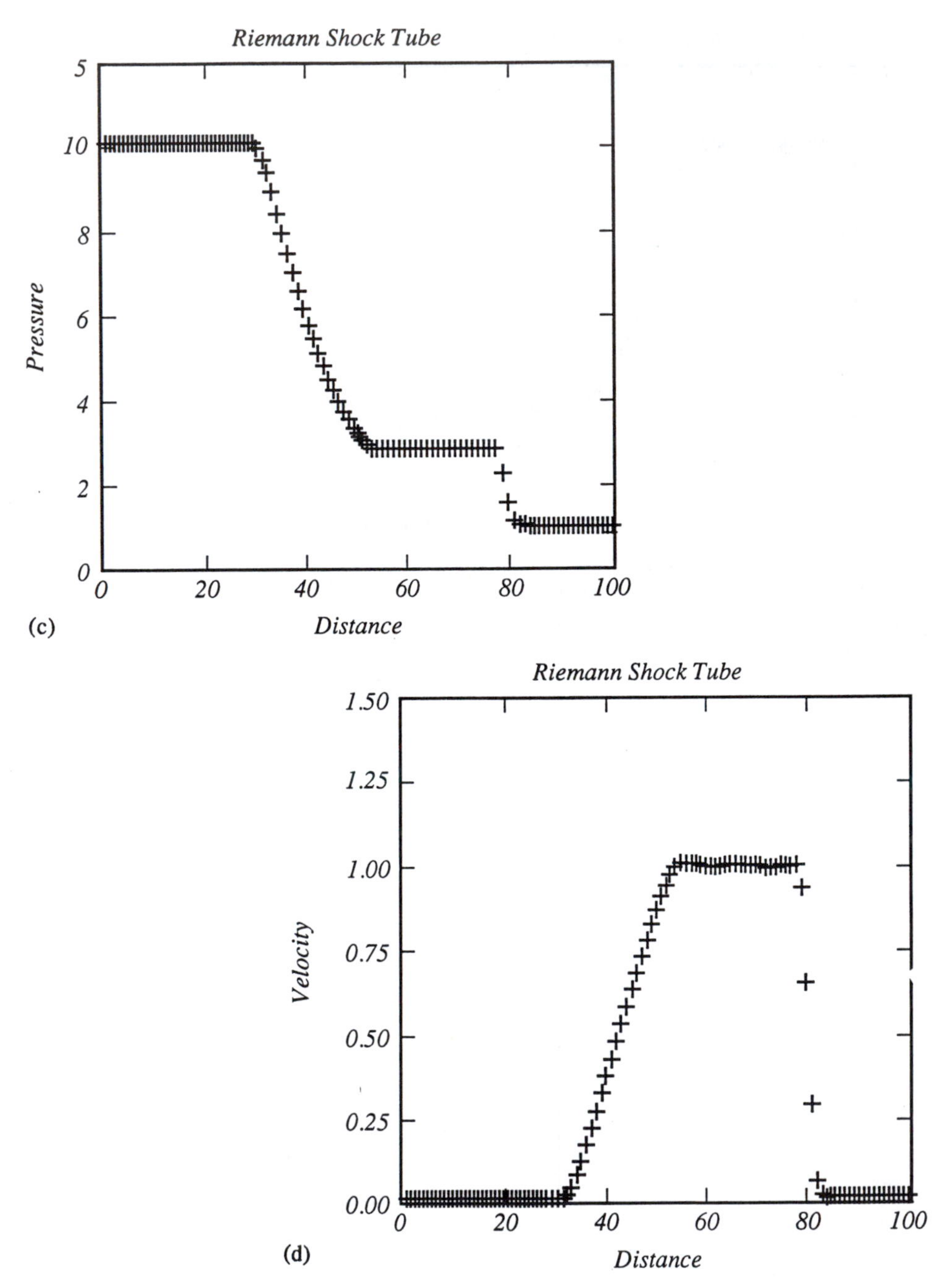

Figure 5.18. Riemann shock tube calculation using second order monotonic advection. (c) Shows the pressure at t = *15.4 sec*. (d) Shows the velocity at t = *15.4 sec*.

shock tube problem, in which a high density gas (density ρ) is placed on one side of a partition in a long tube. The gas on the other side of the partition is at a substantially lower density ρ_o. When the partition is removed, the high density gas expands into the low density gas. As the gas expands, a shock front propagates into the low density region, and a rarefaction moves in the opposite direction into the high density region. Between the shock front and the rarefaction, there is a contact discontinuity which separates the original fluids. Figure 5.18 shows the results of such a calculation after *76.8 sec* using the Eulerian scheme in the axial direction as developed above. The initial densities were chosen such that $\rho_o = 1.0$ and $\rho/\rho_o = 10.0$, and the initial specific internal energies were $\varepsilon = \varepsilon_o = 3.0\ erg/g$. A gamma law equation of state was used with $\gamma = 4/3$. Constant zoning $\Delta z = 1.0\ cm$ was used to cover *100 cm*. The partition was located at *250 cm*. After *76.8 sec* the shock front is at *400 cm*, the contact discontinuity is at *325 cm*,, and the rarefaction is at about *160 cm*. Notice that the shock front is spread out over about three zones, and the contact discontinuity over about seven zones. The second order monotonicity scheme shown in Figure 5.10 was used for the advection of scalar variables and for the momentum.

5.7 AXISYMMETRIC ROTATION

The dynamics of two dimensional, rotating, axially-symmetric systems may be treated by extending the analysis of the previous sections to include the effects of angular velocity v_ϕ, assuming that $v_\phi = v_\phi(r,z,t)$, and that all physical variables are independent of the ϕ coordinate. For simplicity in the discussion below it will be assumed that the grid is fixed; the effects of grid motion on the angular terms may be incorporated in a straightforward manner. Under these assumptions the fluid equations, given in Table 5.1, include an equation for the evolution of the angular momentum, and the addition of a term to the radial momentum equation describing centrifugal forces. Since angular momentum is conserved in the absence of external torques, it is convenient to define the angular momentum density L, and the specific angular momentum L by

$$L = \rho L = \rho r v_\phi = \rho r^2 \omega ,$$

(5.113)

where the angular velocity $\omega = v_\phi / r$. Then the evolution equation for angular momentum may be written as

$$\frac{\partial L}{\partial t} + \frac{1}{r} \frac{\partial [rvL]}{\partial r} + \frac{\partial [uL]}{\partial z} = 0.$$

(5.114)

The change in radial momentum associated with centrifugal forces is described by

$$\frac{\partial T}{\partial t} = \frac{\rho v_\phi^2}{r} = \rho \omega^2 r = \frac{L^2}{\rho r^3} .$$

(5.115)

There are no additional terms due to rotation in the internal energy or mass continuity equations for two dimensional axially symmetric systems.

Space-Time Centering

When axial rotation is included in the fluid dynamics equations, a new variable appears. This may be taken as v_ϕ , ω or the angular momentum density L. For the angular momentum equation we use $L/\rho = L$ as primitive variable; for the centrifugal acceleration the primitive variable will be ω. All three variables are taken to be zone centered (Figure 5.19). For hydrodynamic schemes where the momenta S and T are vertex centered it is natural to center the angular momentum, angular velocity and linear velocity at the vertex also.

The axially symmetric equations including rotation as described by (5.110) and (5.111) do not allow for viscous transfer of angular momentum. Each zone possesses "orbital" angular momentum due to its rotation about the z axis, but does not exhibit intrinsic angular momentum (local rotation cannot occur which produces angular

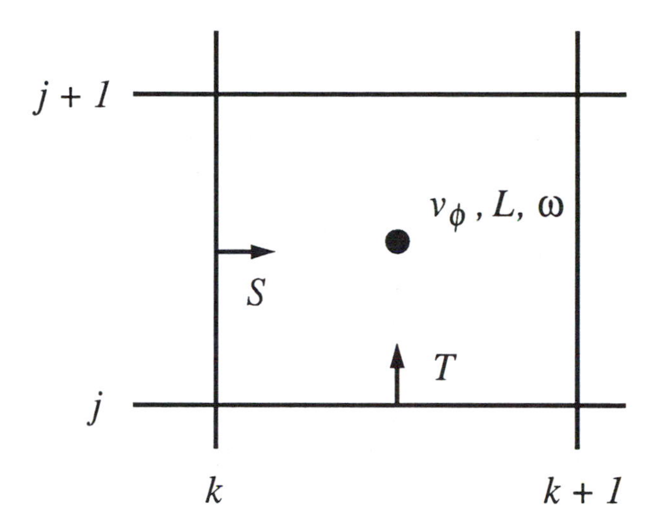

Figure 5.19. Spatial centering of the linear velocity v_ϕ, angular velocity ω, and angular momentum density $L = \rho\, v_\phi\, r = \rho\, \omega\, r^2$.

momentum about an axis parallel to, but offset from the z axis while maintaining two dimensionality). Thus, each zone may be considered to rotate rigidly about the symmetry axis, its angular velocity a constant throughout the zone. The angular momentum per zonal volume is obtained from the volume integral of (5.113):

$$L_{k+1/2,j+1/2} = \frac{1}{\Delta V}\int \rho\omega r^2\, dV = \frac{(\rho\omega)_{k+1/2,j+1/2}}{r_{j+1/2}\Delta r_{j+1/2}} \int r^3 dr$$

$$= \frac{(\rho\omega)_{k+1/2,j+1/2}}{r_{j+1/2}\Delta r_{j+1/2}} \frac{r_{j+1}^4 - r_j^4}{4}$$

$$= (\rho\omega)_{k+1/2,j+1/2} \frac{(r_{j+1}^2 + r_j^2)}{2}$$

where use has been made of the assumption that $\omega_{k+1/2,j+1/2}$ is constant in ΔV (as is ρ). Solving for the angular velocity gives

$$\omega_{k+1/2,j+1/2} = \frac{2L_{k+1/2,j+1/2}}{\rho_{k+1/2,j+1/2}(r_{j+1}^2 + r_j^2)} \; . \qquad (5.116)$$

This may be used to relate the input rotation parameters to the variables in the code. Note that since ω is a constant in ΔV, L and v_ϕ are not. Finally, the specific angular momentum

$$\mathcal{L}_{k+1/2,j+1/2} = (L/\rho)_{k+1/2,j+1/2} \qquad (5.117)$$

is naturally defined from (5.116) and is also zone centered.

Problem 5.25

Suppose that v_ϕ is to be a constant throughout $\Delta V_{k+1/2,j+1/2}$, and show that

$$v_{\phi,k+1/2,j+1/2} = \frac{3(L/\rho)_{k+1/2,j+1/2}\, r_{j+1/2}}{(r_{j+1}^2 + r_{j+1} r_j + r_j^2)} \; . \qquad (5.118)$$

Centrifugal Acceleration

In the rotating frame of reference a fluid element will appear to be accelerated in the outward radial direction as described by (5.115). From (5.113) it is evident that L, ω and v_ϕ cannot each vary in the same way across a zone. Furthermore, the most effective models of angular momentum advection do not result from a single, uniform choice of which angular variable is constant across a zone. Thus, an analysis similar to (5.24)-(5.26) will not be employed here. Instead, observing that edge centered densities $\rho_{k+1/2,j}$ are used to convert from T to v, we may approximate (5.115) by

$$\frac{T'_{k+1/2,j} - T^{n-1/2}_{k+1/2,j}}{\Delta t^n} = \rho^n_{k+1/2,j}(\omega^{n-1/2}_{k+1/2,j})^2 r_j \ , \tag{5.119}$$

where the edge centered angular velocity is chosen to be

$$\omega_{k+1/2,j+1/2} = \frac{2L_{k+1/2,j+1/2}}{\rho_{k+1/2,j+1/2}(r^2_{j+1} + r^2_j)} \ . \tag{5.120}$$

In practice the right side of (5.119) may be evaluated along with the pressure gradient in the radial direction (5.28) in which case T' is just T^P, which then includes changes due to centrifugal acceleration.

Problem 5.26

Derive finite difference representations of the acceleration associated with rotation for: (a) specific angular momentum L constant across a zone; (b) v_ϕ constant across a zone; (c) angular velocity ω constant across a zone.

Angular Momentum Advection

Except for the effects of grid motion (which will be discussed below), the only terms remaining to be evaluated are the two advection terms which determine through (5.114) the time rate of change of the angular momentum. This equation is formally identical to the mass advection equation (5.14) with $v_g=0$. However, effective finite differencing of angular momentum advection is more difficult. As emphasized in Section 5.3 angular momentum must be advected along with mass, which is straightforward with the centering shown in Figure 5.19, and the evolution equation as written in (5.114). In the latter the mass fluxes ρu and ρv are used to carry specific angular momentum.

Diffusion of angular momentum relative to the mass may still occur because of the limited accuracy of the finite difference scheme, and because of differences in the relative smoothness of ρ and L near the rotation axis (Norman, Wilson and Barton 1980). This results because diffusion is significant primarily when second and higher derivative terms are large. Schemes of low order accuracy are often adequate for smooth flow since the diffusive terms are then small. To illustrate this point, consider a rotating protostellar core whose density and angular velocity are initially constant (this is a reasonable approximation during the early stages of protostellar collapse). If ρ is constant, then diffusive terms in the mass equation vanish. However, under these assumptions the specific angular momentum $L=\omega r^2$, and its fractional change with radius $d(\ln L)/dr = 2/r$ are large in the vicinity of the rotation axis. Consequently, diffusive terms associated with angular momentum advection may be large. More important, if the material is inflowing, angular momentum will diffuse inward relative to the rate of mass inflow, and a spurious solution to the fluid flow equations will result (see reference above for example). Similar problems may arise with the linear momenta S and T if gradients in these quantities are very different from the corresponding gradient in ρ. In this case the specific momenta u and v can get out of step with the mass point with which they should be associated. Accuracy can be improved if prior knowledge about how S and T vary in space is available.

A practical solution to this dilemma is to select as a representation of the specific angular momentum an interpolated quantity which exhibits the least variation form zone center to zone edge, since this will tend to minimize diffusion near the axis. The second order scheme described in Section 5.3 for the mass flux can be used with modified zone areas as described below.

The change in angular momentum associated with transport in the radial direction follows upon integration of the first two terms on the left side of (5.114) over $\Delta V_{k+1/2,j+1/2}$:

$$\frac{1}{\Delta V}\int\left(\frac{\partial L}{\partial t}\right)^r dV = -\frac{1}{\Delta V}\int\frac{1}{r}\frac{\partial[r\rho vL]}{\partial r}\,dV$$

$$= -\frac{\hat{A}^r_{j+1}(\rho v)_{j+1}\hat{L}_{j+1} - \hat{A}^r_j(\rho v)_j\hat{L}_j}{\Delta V_{j+1/2}}$$

(5.121)

(common subscript $k+1/2$ omitted). The area factors appearing in (5.121) will be considered below. First however, consider the mass fluxes $(\rho v)_j$. In order to move angular momentum with mass, these factors should be the mass fluxes used to solve the mass continuity equation. In practice, a second order accurate scheme is desirable, and

$$(\rho v)_j = \hat{\rho}_{k+1/2,j}(v_{k+1/2,j} - v_{g,j})$$

(5.122)

where the interpolated density is given by (5.49) and the grid velocity has been included explicitly.

The specific angular momentum should also be interpolated to obtain second order accuracy. An interpolation scheme which has been found to work well in practice exploits assumptions about the subzone structure of L, v_ϕ or ω to reduce diffusion, and chooses the one with the least variation in the problem under consideration, for use in constructing the interpolated specific angular momentum. For each angular variable L, v_ϕ and ω linearly interpolate backward (with respect to the fluid velocity relative to the grid) a distance $(1/2)v_j\Delta t$ from the zone boundary to obtain the values of the variables

$$\hat{L},\ \hat{v}_\phi \ or\ \hat{\omega}.$$

The interpolated angular velocity, for example, is given by [see Figure 5.20, and recall (5.49)]

$$\hat{\omega}_{k+1/2,j} = \frac{1}{2}\omega_{j-1/2}\frac{[\Delta r_{j+1/2} + (v_j - v_{g,j})\Delta t]}{\Delta r_j}$$

$$+ \frac{1}{2}\omega_{j+1/2}\frac{[\Delta r_{j-1/2} - (v_j - v_{g,j})\Delta t]}{\Delta r_j}$$

$$= \omega_j - \frac{(v_j - v_{g,j})\Delta t}{\Delta r_j}\Delta\omega_j \, .$$

(5.123)

A common subscript $k+1/2$ has been omitted on the right side, and the last form makes use of the relations (common index $k+1/2$ omitted)

$$\omega_j \equiv \frac{\omega_{j+1/2}\Delta r_{j-1/2} + \omega_{j-1/2}\Delta r_{j+1/2}}{\Delta r_j}$$

(5.124)

and

$$\Delta\omega_j = \omega_{j+1/2} - \omega_{j-1/2} \, .$$

(5.125)

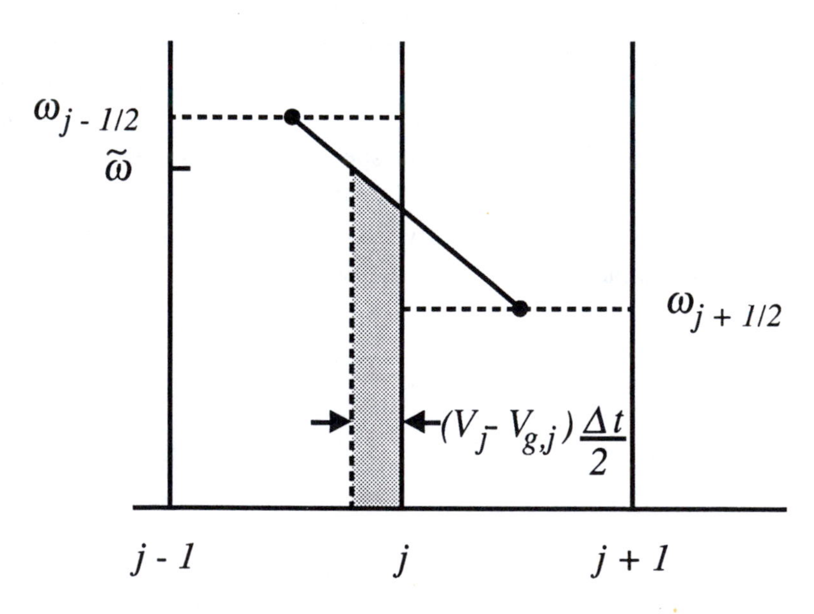

Figure 5.20. Linear interpolation in angular velocity, including grid motion. The interpolated value is second order accurate [compare (5.49) for mass advection].

Formally identical expressions are obtained for the interpolated specific angular momentum and the linear velocity (common subscript $k+1/2$ omitted on the right)

$$\hat{L}_{k+1/2,j} = L_j - \frac{(v_j - v_{g,j})\Delta t^{n+1/2}}{\Delta r_j} \Delta L_j$$

$$\hat{v}_{\phi,k+1/2,j} = v_{\phi,j} - \frac{(v_j - v_{g,j})\Delta t^{n+1/2}}{\Delta r_j} \Delta v_{\phi,j} \; .$$

(5.126)

(5.127)

Next, an expression for the specific angular momentum is constructed using the interpolated linear velocity and angular velocity. This requires a definition of the interpolated radius. A natural choice (which will be used below to construct area factors at time $t^{n+1/2}$) is

$$\hat{r}_j = r_j^{n+1/2} = r_j^n - (v_j - v_{g,j})\Delta t^{n+1/2}/2$$

(5.128)

which is centered in the advection mass (shaded in Figure 5.20). In fact, the interpolated radius given by (5.128) is just the effective radius of the material after a time $\Delta t^{n+1/2}/2$. Three definitions of the interpolated specific angular momentum are now available. Define

$$x_A = \min\{|\Delta\omega/\hat{\omega}|, |\Delta v_\phi/\hat{v}_\phi|, |\Delta L/\hat{L}|\}_{k+1/2,j} \; .$$

(5.129)

Then the advected specific angular momentum is chosen to be

$$\hat{L}_{k+1/2,j} = \hat{v}_{\phi,k+1/2,j}\,\hat{r}_j \quad \text{if } x_A = |\Delta v_\phi/\hat{v}_\phi|$$

$$= \hat{\omega}_{k+1/2,j}\,\hat{r}_j^2 \quad \text{if } x_A = |\Delta\omega/\hat{\omega}|$$

$$= \hat{\mathcal{L}}_{k+1/2,j} \qquad if\ x_A = |\Delta L/\hat{\mathcal{L}}| .$$

(5.130)

The area factors appearing in (5.43) and (5.48) for mass advection were based on the coordinates at time t^n (as were those used for energy and momentum advection). This is usually sufficient for problems which do not involve rotation. When rotation exists and grid motion is employed, improved accuracy results if the area factors are interpolated forward to the time $t^{n+1/2}$. The factors appearing in (5.121), defined at $t^{n+1/2}$ are given by

$$\hat{A}^{r}_{k+1/2,j} \equiv (A^{n+1/2}_{k+1/2,j})^{r} = 2\pi\Delta z^{n+1/2}_{k+1/2} r^{n+1/2}_{j}$$

(5.131)

where the change in Δz is due entirely to grid motion:

$$\Delta z^{n+1/2}_{k+1/2} = z^{n}_{k+1} + \frac{1}{2} u^{n-1/2}_{g,k+1} \Delta t^{n+1/2} - z^{n}_{k} - \frac{1}{2} u^{n-1/2}_{g,k} \Delta t^{n+1/2}$$

$$= \Delta z^{n}_{k+1/2} + \frac{1}{2}(u^{n+1/2}_{g,k+1} - u^{n+1/2}_{g,k})\Delta t^{n+1/2} .$$

(5.132)

The change in r_j results from grid motion and fluid motion, and is given by (5.128). In (5.128) and (5.132) the velocity components are in fact u^Q and v^Q resulting from the pressure gradient and centrifugal acceleration, as well as from artificial viscous stress acceleration.

The axial advection of angular momentum, given by the first and third terms on the left side of (5.114) may be approximated along the lines discussed above. The results are easily shown to be

$$\left(\frac{\partial L}{\partial t}\right)^{z}_{k+1/2,j+1/2} = - \frac{\hat{A}^{z}_{k+1}(\rho u)_{k+1}\hat{\mathcal{L}}_{k+1} - \hat{A}^{z}_{k}(\rho u)_{k}\hat{\mathcal{L}}_{k}}{\Delta V_{k+1/2}}$$

(5.133)

(common subscript $j+1/2$ omitted on the right). The mass fluxes are the same as used for the continuity equation, and the area factors (interpolated to time $t^{n+1/2}$) are

$$(A^{n+1/2}_{k,j+1/2})^z = 2\pi r^{n+1/2}_{j+1/2} \Delta r^{n+1/2}_{j+1/2} \tag{5.134}$$

where

$$r^{n+1/2}_{j+1/2} = r^{n}_{n+1/2} + \frac{1}{4}(v^{n-1/2}_{g,j+1} + v^{n-1/2}_{g,j})\Delta t^{n+1/2}$$

$$\Delta r^{n+1/2}_{j+1/2} = \Delta r^{n}_{j+1/2} + \frac{1}{2}(v^{n-1/2}_{g,j+1} - v^{n-1/2}_{g,j})\Delta t^{n+1/2}. \tag{5.135}$$

Problem 5.27

Derive the axial angular momentum advection approximations (5.133)-(5.135). Finally, explain why in this case

$$\hat{L}_{k,j+1/2} = L_{k,j+1/2} - \frac{(u_{k,j+1/2} - u_{g,k})\Delta t^n}{\Delta z_k}\Delta L_{k,j+1/2}. \tag{5.136}$$

What is the definition of $L_{k,j+1/2}$?

Angular momentum advection is most efficiently computed along with mass and energy advection, since the mass fluxes are available at that stage in the computation. The time step used in (5.121) and (5.133) is $\Delta t^{n+1/2}$ as for the mass advection equation. Finally, if the interpolated area factors (5.131) and (5.134) are used to compute angular momentum flux, they should also be used in the remaining advection equations.

Problem 5.28

Although rotational energy does not appear directly in the finite difference equations developed above, a representation of it is needed for output purposes. Derive expressions for the rotational energy associated with the angular momentum assuming: (a) specific angular momentum L constant across a zone; (b) v_ϕ constant across a zone; (c) angular velocity ω constant across a zone.

Problem 5.29

Show that the time rate of change of the angular momentum is given by

$$\frac{\partial L}{\partial t} + \nabla \cdot [\rho L(\mathbf{v} - \mathbf{v}_g)] = -L\, \nabla \cdot \mathbf{v}_g \tag{5.137}$$

when grid motion is allowed. Express (5.137) in cylindrical coordinates, and discuss the modifications to the finite difference equations discussed above when grid motion is included.

5.8 CURVILINEAR COORDINATES

The hydrodynamic equations discussed in the first part of this chapter exhibit some of the complications associated with finite difference approximations in curvilinear coordinate systems. However, the methods presented there are not adequate to treat general (orthonormal) coordinate dynamics. In this section a modification of the axisymmetric approach will be presented which works well for curvilinear coordinates (the primary restriction is that the coordinates be orthonormal; see Appendix B for a discussion of the geometry of these systems). As a specific example, two dimensional spherical coordinates $x^i = (r, \theta, \phi)$ will be developed. The discussion will emphasis the differences between the r-z coordinate models discussed earlier in this chapter and the requirements for curvilinear systems. Problem 5.21 should be reviewed at this point, since it illustrates several important differences between the two types of coordinates systems.

For purposes of illustration it will be assumed that all quantities depend only on the two independent variables r and θ, and that they are independent of the angle ϕ. In particular, all partial derivatives with respect to ϕ are assumed to vanish identically. The contravariant components of the fluid velocity are defined by

$$V^r = \frac{dr}{dt}, \quad V^\theta = \frac{d\theta}{dt}, \quad V^\phi = \frac{d\phi}{dt} = 0. \tag{5.138}$$

We shall assume that the component V^ϕ vanishes, although this is not required by the assumed symmetry (see Section 5.7 for a discussion of rotation in two dimensional systems). Finally we note that the only nonzero components of the metric tensor are $g_{rr} = 1$, $g_{\theta\theta} = r^2$, and $g_{\phi\phi} = r^2 \sin\theta$. The covariant components are then (summation over repeated indicies is implied)

$$V_k = g_{ki} V^i, \tag{5.139}$$

and if the determinate of the metric tensor is defined by g, then

$$\sqrt{g} = r^2 \sin\theta.$$

The "physical" components of the fluid velocity are given by

$$V_i^P \equiv \sqrt{g_{ii}}\, V^i.$$

We shall denote the two nonzero physical components of the fluid velocity by

$$v = V^P{}_r = V^r \qquad u = V^P{}_\theta = rV^\theta. \tag{5.140}$$

The covariant components of the momentum density of the fluid are defined by $S^i = \rho V^i$ and have nonzero physical components denoted by

$$T = \rho v \qquad S = \rho u. \tag{5.141}$$

Restricting the subsequent discussion to spherical coordinates in no way limits its applicability to other analytic orthonormal coordinate systems. Some modifications are required if a numerical metric is used (the coordinate system obtained from the numerical metric must, however, be orthonormal to keep the equations simple).

Mass Continuity and Internal Energy Equations

The two dimensional hydrodynamic equations in (r,θ) coordinates can be obtained from (5.81)-(5.83). First consider the mass continuity and internal energy equations. The mass continuity equation (5.81) may be written in standard Eulerian form in terms of physical velocity components as

$$\frac{\partial \rho}{\partial t} + \frac{1}{r^2}\frac{\partial}{\partial r}(r^2 \rho v) + \frac{1}{r \sin\theta}\frac{\partial}{\partial \theta}(\rho u \sin\theta) = 0 .$$

The advection terms (the last two terms on the left) may be written out, and parts combined with the time derivative to form the equivalent equation

$$\frac{d\rho}{dt} + \rho\frac{1}{r^2}\frac{\partial(r^2 v)}{\partial r} + \rho\frac{1}{r\sin\theta}\frac{\partial(u\sin\theta)}{\partial\theta} = 0. \tag{5.142a}$$

In this form the time derivative is the usual Lagrangian derivative, and the equation is in standard Lagrangian representation. The rate of change of the internal energy follows from (5.82) and is, in terms of physical velocity components, given in the usual Eulerian form by

$$\frac{\partial \rho\varepsilon}{\partial t} + \frac{1}{r^2}\frac{\partial}{\partial r}(r^2 \rho\varepsilon\, v) + \frac{1}{r \sin\theta}\frac{\partial}{\partial\theta}(\rho\varepsilon\, u \sin\theta) =$$

$$- P\left[\frac{1}{r^2}\frac{\partial}{\partial r}(r^2 v) + \frac{1}{r\sin\theta}\frac{\partial}{\partial\theta}(u \sin\theta)\right].$$

Rewritting this equation in terms of the Lagrangian time derivative we find

$$\frac{d(\rho\varepsilon)}{dt} = - \rho\varepsilon\left[\frac{1}{r^2}\frac{\partial(r^2 v)}{\partial r} + \frac{1}{r\sin\theta}\frac{\partial(u\sin\theta)}{\partial\theta}\right]$$

$$- P\left[\frac{1}{r^2}\frac{\partial}{\partial r}(r^2 v) + \frac{1}{r\sin\theta}\frac{\partial}{\partial\theta}(u\sin\theta)\right].$$

(5.143a)

The two equations expressed in standard Eulerian form above are similar to their counterparts in the cylindrically symmetric two dimensional coordinates (r,z) discussed earlier, and could be solved using the methods developed there. Because a different set of centerings are required (which will be discussed below) in curvilinear coordinates we do not recommend this approach. The reasons for this will become clear when we consider the momentum equation.

Momentum Equation

The momentum equation may also be written in standard Eulerian form in analogy with the equations above using (5.83)

$$\frac{\partial S_i}{\partial t} + \frac{1}{\sqrt{g}}\frac{\partial}{\partial x^k}(S_i V^k \sqrt{g}) = -\frac{\partial P}{\partial x^i} - \frac{S_l S_k}{2\rho}\frac{\partial g^{kl}}{\partial x^k} + f_i$$

If this approach were to be taken it would be necessary to develop a finite difference representation of the noninertial terms which are proportional to the derivatives of the metric tensor. Another approach will be used below which takes advantage of the tensor nature of the momentum equation written in terms of a Lagrangian process and a remap. The momentum equation then has the form

$$\frac{dS_i}{dt} = -\frac{V^k}{\sqrt{g}}\frac{\partial(S_i\sqrt{g})}{\partial x^k} - \frac{S_i}{\sqrt{g}}\frac{\partial(V^k\sqrt{g})}{\partial x^k} - \frac{\partial P}{\partial x^i} - \frac{S_l S_k}{2\rho}\frac{\partial g^{kl}}{\partial x^k} + f_i \tag{5.144a}$$

where the last term on the right represents additional forces, such as those associated with artificial viscous stresses.

The solution of the hydrodynamic equations in curvilinear form will employ the method of operator splitting which is similar to the one used to solve the axisymmteric equations in Section 5.2. The principle differences between that approach and the one employed here is that a true Lagrangian step will be used to solve for the acceleration associated with the pressure gradient in (5.144a) and for the change in internal energy associated with the mechanical work term in (5.143a). The effects of artificial viscous stresses may also be included at this point. In the Lagrangian step the corners of each zone (where the velocities are to be centered) are moved to new locations using the instantaneous velocity and the time step. The resulting zone will no longer be orthonormal. The Lagrangian step is then followed by a remap step in which the distorted Lagrangian zone is mapped back onto an orthonormal Eulerian zone (possibly the original zone). The equations describing the remap (see Problem 5.30) of any scalar density, such as the mass density and the internal energy density can each be written (in terms of physical components) in the form

$$\frac{\partial \rho}{\partial t} = -\left(v\frac{\partial}{\partial r} + \frac{u}{r}\frac{\partial}{\partial \theta}\right)\rho \tag{5.142b}$$

and

$$\frac{\partial \rho\varepsilon}{\partial t} = -\left(v\frac{\partial}{\partial r} + \frac{u}{r}\frac{\partial}{\partial \theta}\right) \rho\varepsilon \,. \tag{5.143b}$$

The remap of a vector density, such as the momentum density, can be written in coordinate form

$$\frac{\partial S_i}{\partial t} = - V^k \frac{\partial S_i}{\partial x^k} + \frac{S_i S_j}{\rho} \Gamma^j_{ik} \,. \tag{5.144b}$$

The Christoffel symbol in the last term on the right side can be expressed in terms of derivatives of the metric tensor (Sokolnikoff 1964). Notice that the Lagrangian form of the momentum equation (5.144a) and the momentum remap both contain velocity terms and terms proportional to the derivative of the metric. The results of the volume remap calculation will be used to remap the mass and all of the scalar quantities that are normally transported by advection in the Eulerian form of the equations.

The space-time centering for two dimensional spherical coordinates will be discussed first, and then the Lagrange and remap steps will be considered. As will become clear below, the overall code structure required to solve the hydrodynamic equations in cylindrical coordinates as discussed in the first part of the chapter can be adopted for curvilinear coordinate applications with minimal changes.

Problem 5.30

Consider the one dimensional planar equations describing a Lagrangian step plus a remap:

Lagrangian Step	**Remap**
$\frac{d\rho}{dt} = - \rho \frac{\partial v}{\partial x}$	$\frac{\partial \rho}{\partial t} = - v \frac{\partial \rho}{\partial x}$

$$\frac{dS}{dt} = -\frac{\partial P}{\partial x} \qquad\qquad \frac{\partial S}{\partial t} = -v\frac{\partial S}{\partial x}$$

$$\frac{d\rho\varepsilon}{dt} = -P\frac{\partial v}{\partial x} \qquad\qquad \frac{\partial \rho\varepsilon}{\partial t} = -v\frac{\partial \rho\varepsilon}{\partial x}$$

(a) Show that the equations describing the Lagrangian step are in fact identical to the Lagrangian equations discussed in Chapter 4. The planar momentum density $S = \rho v$.

(b) Show that if the Eulerian grid is moving with grid velocity v_g, only the remap equations change, and that the mass remap becomes

$$\frac{\partial \rho}{\partial t} = -(v - v_g)\frac{\partial \rho}{\partial x}.$$

Do this by extending the discussion leading to (5.10g) to include grid motion.

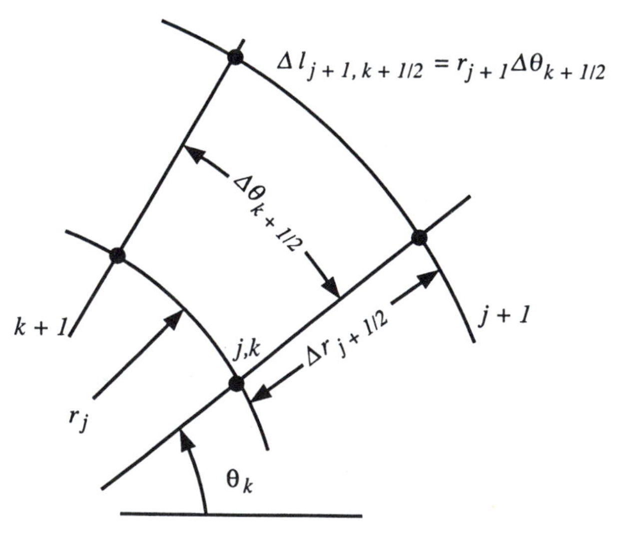

Figure 5.21. The primitive volume element $\Delta V_{j+1/2,k+1/2}$ in spherical coordinates at the beginning of a computational cycle.

Space-Time Centering

The principle issue concerning the centering of fluid variables on a curvilinear grid is the spatial location of the components of the velocity and momenum density. An example of a two dimensional spatial grid in r,θ coordinates is shown in Figure 5.21. The index j will be used to denote the radial variable, and k will denote the angular variable. Thus a point located at (r,θ) shown by the solid dot will be denoted by the indicies (j,k). The azmuthal axis is horizontal, and positive angles are measured away from it as shown. The geometry of lengths, areas and volumes on a curvinilear grid is discussed in Appendix B. The spatial grid can be set up using the primitive variables r_j, and θ_k, from which $\Delta r_{j+1/2}$ and $\Delta\theta_{j+1/2}$ can be constructed in obvious fashion. In general the finite differences $\Delta r_{j+1/2}$ and $\Delta\theta_{j+1/2}$ need not be uniform. In addition to the coordinate differences, several volume and area elements will be needed below; these can be constructed in a straightforward manner, as described in Appendix B. First, consider a zone defined by the coordinate differences $\Delta r_{j+1/2}$ and $\Delta\theta_{j+1/2}$ (see Figure 5.22). The volume of this element is given by

$$\Delta V_{j+1/2,\,k+1/2} = \frac{2\pi}{3}(r_{j+1}^3 - r_j^3)\,(\cos\theta_{k-1} - \cos\theta_k) \tag{5.145a}$$

which is a zone centered quantity. This is the usual zone in which the mass density, specific energy and pressure are centered. The area of the zone face normal to the radial direction at the point $(j,k+1/2)$ which we denote by $A^r{}_{j,k+1/2}$ is

$$A^r_{j,\,k+1/2} = 2\pi r_j^2\,(\cos\theta_k - \cos\theta_{k+1})\,. \tag{5.145b}$$

This area will be used to construct volume changes for radial remap. The area of the zone face normal to the angular direction at the point $(j+1/2,k)$, which will be used for the remap in the angular direction, is easily shown to be

$$A^{\theta}_{j+1/2,k} = 2\pi(r^2_{j+1} - r^2_j)\sin\theta_k. \tag{5.145c}$$

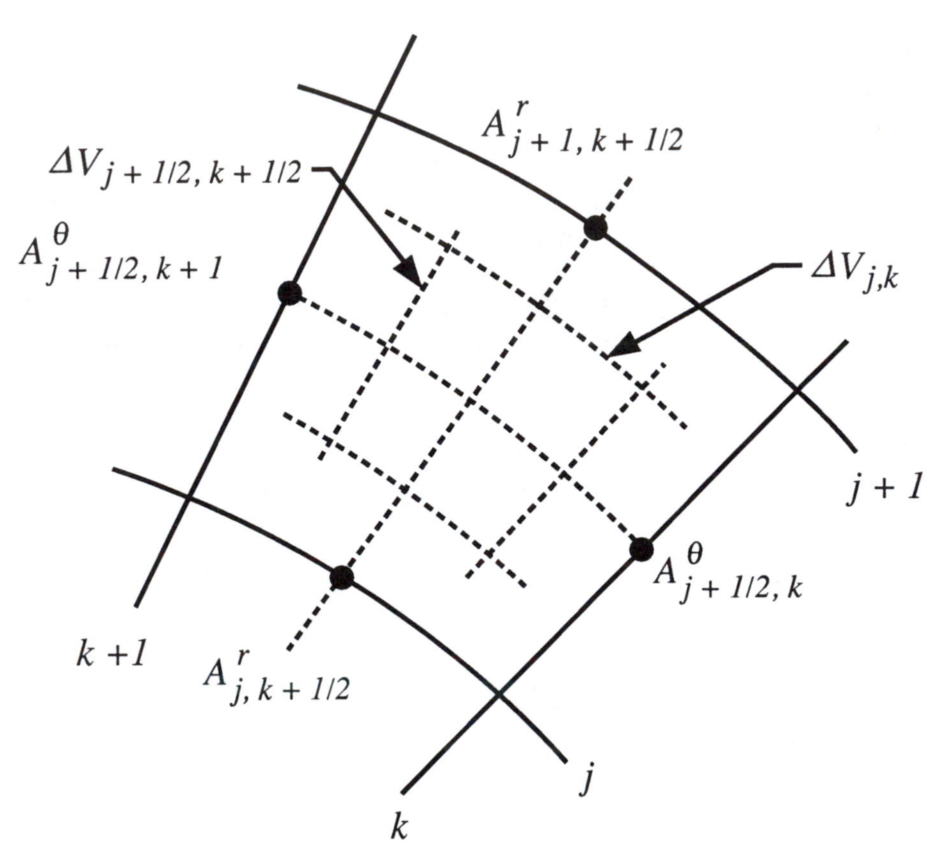

Figure 5.22. Area of the four faces of the volume element at the start of a computational cycle in spherical coordinates.

Expressions will also be needed for the volume centered at the nodes (j,k). It may be recalled that the momentum zones used for the fluid equations in cylindrical coordinates consisted of a fraction of each of the neighboring zones (Figure 5.12). A similar decomposition will be used here. In curvilinear coordinates additional area factors will be needed in the finite difference representations (for example, areas such as $A^r{}_{j,k+1/4}$ and $A^r{}_{j,k+3/4}$ will be needed to represent the two, unequal parts of $A^r{}_{j,k+1/2}$). The momentum volume (the velocity and momentum centering will be discussed below) $\Delta V_{j,k}$ is given in terms of the sum of the volumes of the four neighboring zones:

$$\Delta V_{j,k} \equiv \Delta V_{j+1/2,\,k+1/2} + \Delta V_{j+1/2,\,k-1/2} + \Delta V_{j-1/2,\,k+1/2} + \Delta V_{j-1/2,\,k-1/2} \tag{5.145d}$$

The momentum zone contains one quarter of the mass from each of the surrounding zones $\Delta V_{j+1/2,k+1/2}$. A set of area factors associated with the quarter zones will be needed below. The area factors normal to the radial direction are given by

$$A^{r}_{j,\,k+1/2} = 2\pi r_j^2 (\cos\theta_k - \cos\tilde{\theta}) \tag{5.145e}$$

$$A^{r}_{j,\,k+3/2} = 2\pi r_j^2 (\cos\tilde{\theta} - \cos\theta_{k+1}) \tag{5.145f}$$

where the angle θ is defined by

$$\cos\tilde{\theta} = \frac{1}{2}(\cos\theta_k + \cos\theta_{k+1}) . \tag{5.145g}$$

By construction the sum of (5.145e) and (5.145f) is just $A^{r}{}_{j,k+1/2}$. Finally, the areas normal to the angular direction are

$$A^{\theta}_{j+1/4,\,k} = \pi(\tilde{r}^{\,2} - r_j^2)\sin\theta_k \tag{5.145h}$$

$$A^{\theta}_{j+3/4,\,k} = \pi(r_{j+1}^2 - \tilde{r}^{\,2})\sin\theta_k \tag{5.145i}$$

where

$$\tilde{r}^{\,2} = \frac{1}{2}(r_j^2 + r_j\, r_{j+1} + r_{j+1}^2) . \tag{5.145j}$$

The choice of area factors above is consistent with the construction (5.145d) for the momentum zone.

The spatial centering of the primary fluid variables is shown in Figure 5.23. The most significant difference between the spatial centering for curvilinear coordinates and that used for the axisymmetric equations is the spatial centering of the velocity and the specific momentum, which are chosen to be node centered. This is important in curvilinear coordinate systems since components of vectors can only be compared when they are located at the same point. The pressure, specific energy and density remain zone centered. Finally, the time centering of all variables is independent of the choice of spatial coordinates, and is the same as in Section 5.2 [see, in particular (5.21)].

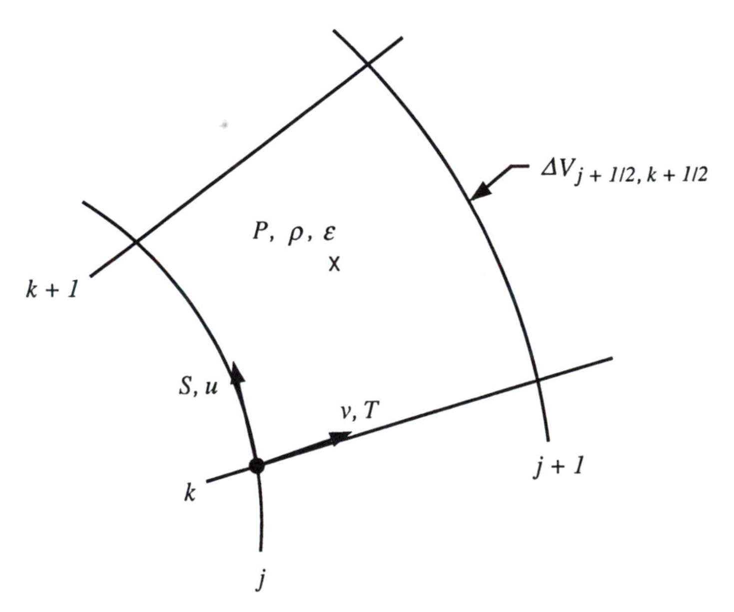

Figure 5.23. Centering of fluid variables on an orthonormal curvilinear grid at the start of a computational cycle. Zone centered variables are located at $j+1/2,\ k+1/2$ and vertex centered variables are located at $j,\ k$.

In principle a two time level representation similar to the one developed earlier should be used with time steps Δt^n and $\Delta t^{n+1/2}$. This representation is necessary for reasons of accuracy and stability. In fact, the stability requirements can usually be achieved in a single time level representation if the order of solution of the terms in the hydrodynamic equations is the same as in the two time level representation. Accuracy will also be maintained as long as the time step does not change by a large fractional amount from cycle to cycle. With these points in mind the following discussion will use a single time step Δt.

Problem 5.31

Carry out a derivation of the area and volume elements listed above for a two dimensional spherical coordinate system. Explain how you would construct a grid with uniform angular zoning.

5.9 Lagrangian Step

The solution to the hydrodynamic equations in curvilinear coordinates can be carried out in the same general way as it was for the axisymmetric equations. The major changes are that the first step is a true Lagrangian step starting from an orthonormal grid at time t^n, which is followed by a remap instead of an advection step. The remap step takes quantities back to an orthonormal grid which may or may not be the same as the original grid. The Lagrangian step constructs the change in specific momentum associated with the pressure gradient. If gravitation is included, the acceleration associated with the gradient in the gravitational potential should also be included here. Next, the change in internal energy associated with mechanical work (and shock heating) is obtained. This ends the Lagrangian step.

The change in the coordinate components of the specific momentum resulting from a pressure gradient is described by

$$\frac{dS_i}{dt} = -\frac{\partial P}{\partial x^i} . \tag{5.146a}$$

where it will be recalled that the left side is the Lagrangian time derivative. Consider the equation for the radial component. The coordinate and physical components of the momentum density are related by $T = g_{rr}S^r = S_r$. Thus (5.146a) can be rewritten in the equivalent form

$$\frac{dT}{dt} = -\frac{\partial P}{\partial r}. \tag{5.146b}$$

A finite difference representation of this equation can be found by integrating it over a node centered (momentum) volume $\Delta V_{j,k}$ given by (5.145d). At the beginning of the Lagrangian step all zones (including the momentum zones) are orthonormal, so integrating (5.146b) over $\Delta V_{j,k}$ is straightforward; the result is easily shown to be

$$\frac{T^{P}_{j,k} - T^{n-1/2}_{j,k}}{\Delta t} = -\frac{A^{r}_{j,k+1/4}(P^{n}_{j+1/2,k+1/2} - P^{n}_{j-1/2,k+1/2})}{\Delta V_{j,k}} - \frac{A^{r}_{j,k-1/4}(P^{n}_{j+1/2,k-1/2} - P^{n}_{j-1/2,k-1/2})}{\Delta V_{j,k}}. \tag{5.147}$$

Because the momentum density is node centered, the pressure gradient in (5.147) requires the four neighboring values of the pressure for its evaluation. Note that all of the area factors are known [see (5.145)] and that each component of the acceleration vanishes if the pressure along that direction is a constant.

Problem 5.32

Integrate (5.146b) over the Lagrangian volume element $\Delta V_{j,k}$ and show that it can be written in the form (5.147). Show that the net change in momentum in the grid described by (5.147) vanishes if the pressure along the boundary is zero. Evaluate the differential limit of (5.147) and show that it reduces to the original differential equation.

Problem 5.33

Express (5.146a) in terms of the physical component of the momentum in the theta direction, and show that the result is

$$\frac{dS}{dt} = -\frac{1}{r}\frac{\partial P}{\partial \theta}. \tag{5.148}$$

Integrate (5.148) over the Lagrangian volume $\Delta V_{j,k}$, and show that it leads to the finite difference representation

$$\frac{S^{P}_{j,k} - S^{n-1/2}_{j,k}}{\Delta t} = -\frac{A^{\theta}_{j+1/4,k}(P^{n}_{k+1/2,j+1/2} - P^{n}_{j+1/2,k-1/2})}{\Delta V_{j,k}}$$

$$- \frac{A^{\theta}_{j-1/4,k}(P^{n}_{j-1/2,k+1/2} - P^{n}_{j-1/2,k-1/2})}{\Delta V_{j,k}}. \tag{5.149}$$

The area factors normal to the theta direction are given by (5.145).

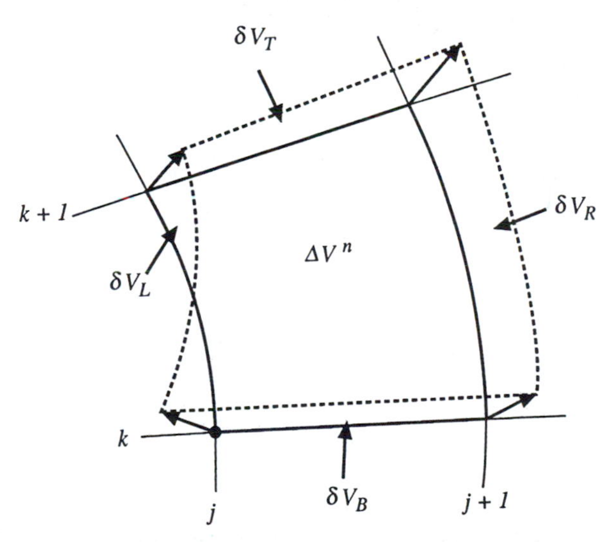

Figure 5.24. Lagrangian displacement of a zone in response to accelerations at the start of a computational cycle. Arrows represent the displacement of vertices after a single time step Δt.

The new components of the momentum density are now used to find the the corresponding velocity components at the zone corners. This is done using relations analogous to (5.35) and (5.36) and the appropriate node centered mass density. The velocity components are then used to calculate the displacement of each zone corner in the time Δt (Figure 5.24). The original orthonormal zone ΔV^n is shown by the solid bounded figure, and the new zone ΔV_L following the Lagrangian displacement by the dashed boundary. Since this is a Lagrangian process the mass of the zone $\Delta m^n{}_{j+1/2,k+1/2}$ remains constant. At this stage the change in volume resulting from the Lagrangian displacement $\Delta V^n{}_{j+1/2,k+1/2} - \Delta V_L$ is obtained. This can be done by finding the volume changes δV_α at each of the four zone boundaries which are shown in Figure 5.24. An approximate construction of these terms which works well in practice may be found as follows.

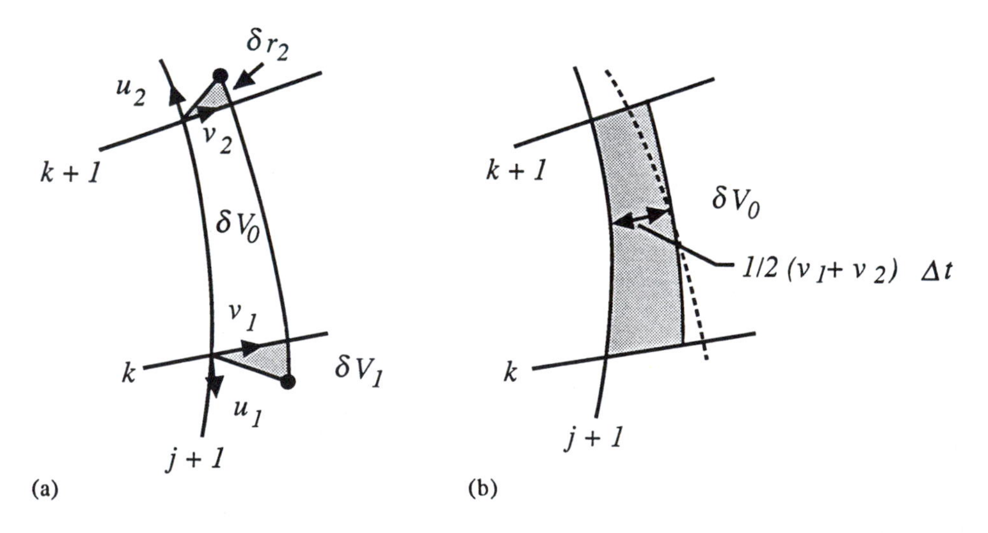

Figure 5.25. (a) Volume change across a zone face at $j,k+1/2$ shown in Figure 5.21. (b) Approximate construction of δV_0 (shaded) assuming that v_1 and v_2 are positive. The filled circles represent two corners of the zone ΔV_L following the Lagrangian displacement.

Consider the volume change δV_R on the right of ΔV^n shown in Figure 5.24. It may be considered to consist of three contributions which are shown in Figure 5.25a. The small corner volumes at the top and bottom (shown shaded) are readily obtained from the velocity components as

$$\delta V_1 = u_1 v_1 (\Delta t)^2 / 2$$

and

$$\delta V_2 = u_2 v_2 (\Delta t)^2 / 2 . \tag{5.150}$$

Notice that these quantities may be of either sign depending on the sign of the velocity components. This feature will be used below when the total volume change is found. Next, the volume change δV_o can be approximated by the volume shown shaded in Figure 5.25b. First, define the displacement

$$\delta r = \frac{1}{2}(\delta r_1 + \delta r_2) \tag{5.151}$$

where $(i = 1, 2)$

$$\delta r_i = v_i \Delta t .$$

The area normal to the radial direction at $j+1$ will be denoted by $A^r{}_{j,k+1/2}(r_j)$. Then the volume δV_o can be approximated by the integral

$$\delta V_o = \int_{r_j}^{r+\delta r} A^r{}_{j,k+1/2}(r_j + \xi)\, d\xi . \tag{5.152}$$

In general ξ is the physical distance measured normal to this area. For some analytic metrics the integral may be evaluated in closed form. If this is not possible, the area may be expanded in a Taylor series for small values of ξ, with the result

$$\delta V_o \approx A(r_j)\,\delta r + \frac{1}{2}\frac{\partial A}{\partial \xi}\delta r^2 + \frac{1}{6}\frac{\partial^2 A}{\partial \xi^2}\delta r^3 + \cdots$$

Once δV_o has been found, the net change in the volume across the zone face at $j+1$ is given by

$$\delta V_R = \delta V_o + \delta V_2 - \delta V_1 \,. \tag{5.153}$$

The sign of the volume changes on the right depends on the direction of motion of the zone corners. The reader may verify that the choice of signs above gives the correct change in δV_o for arbitrary motion of the nodes.

The approximations leading to (5.153) can in principle be improved by going to higher order expansion for δV_o, or (5.150). It is more important, however, to use the results (for any level of approximation) consistently throughout the calculation. The volume changes δV_α are stored for use in the remap stage to be discussed below.

Once all of the volume changes shown in Figure 5.24 have been found the net change in ΔV^n (the subscripts $j+1/2,k+1/2$ are omitted for notational convenience) is given by

$$\Delta V_L = \Delta V^n + \delta V_R + \delta V_T - \delta V_B - \delta V_L \,. \tag{5.154}$$

The signs of the last two terms have again been chosen to guarantee that the proper result be obtained for arbitrary motion of the nodes.

The mass density ρ_L of the zone following the Lagrangian displacement is found using (5.154) and the mass $\Delta m^n{}_{j+1/2,k+1/2}$:

$$\rho_L = \frac{\Delta m_{k+1/2,\,j+1/2}}{\Delta V^n_{j+1/2,\,k+1/2}} \,.$$

At this point in the procedure the acceleration associated with artificial viscous stresses should be added. This can be done as described, for example, in Section 5.6 with appropriate modifications for the velocity centering. We shall assume that this has been done, and will consider the change in internal energy associated with mechanical heating next. An artificial viscous stress in (r,θ) coordinates is considered below.

Mechanical Work

Mechanical heating is described in terms of the physcial components of the fluid variables by the Lagrangian equation

$$\frac{d\rho\varepsilon}{dt} = - P\left[\frac{1}{r^2}\frac{\partial}{\partial r}(r^2 v) + \frac{1}{r\sin\theta}\frac{\partial}{\partial\theta}(u\sin\theta)\right]. \tag{5.155}$$

But the right side is just the pressure times the rate of change of the volume associated with the Lagrangian motion of the fluid, and thus (5.155) can expressed as

$$\Delta(\rho\varepsilon) = \Delta E = -P\frac{\Delta V_L - \Delta V^n}{\Delta V_L}. \tag{5.156}$$

This equation can be solved in essentially the same manner as was (5.32) to construct a finite difference representation of the internal energy equation. If shocks are present, then shock heating should be included at this point. This completes the Lagrangian step of the calculation.

Problem 5.34

A node centered mass density is needed to convert between the momentum density and the velocity. Integrate (5.141) over the momentum volume $\Delta V_{j,k}$, defined by (5.145d) and use the results to show that

$$v_{j,k} = \frac{T_{j,k}}{\rho_{j,k}} \qquad u_{j,k} = \frac{S_{j,k}}{\rho_{j,k}} \tag{5.157a}$$

where the node centered density is defined by

$$4\,\rho_{j,k}\,\Delta V_{j,k} \equiv \rho_{j+1/2,\,k+1/2}\,\Delta V_{j+1/2,\,k+1/2} + \rho_{j+1/2,\,k-1/2}\,\Delta V_{j+1/2,\,k-1/2}$$

$$+\; \rho_{j-1/2,\,k+1/2}\,\Delta V_{j-1/2,\,k+1/2} + \rho_{j-1/2,\,k-1/2}\,\Delta V_{j-1/2,\,k-1/2}\,. \tag{5.157b}$$

Notice that each of the zones contributes one quarter of its mass to the momentum zone $\Delta V_{j,k}$. The results (5.157a) can be used at any stage in the solution as long as the densities, volumes and momenta are mutually consistent. For example ρ_L and ΔV_L may be used to find the momentum density following the pressure acceleration.

Problem 5.35

Construct a finite difference representation of the internal energy equation in two dimensional spherical coordinates using the results above. Assume that the equation of state can be represented by an effective gamma law (3.8).

5.10 Remapping in Curvilinear Coordinates

The remap equations (5.142b)-(5.144b) can be solved in three steps. In the first step, a finite difference representation of the density remap is obtained, and a mass flux across each zone boundary associated with it is constructed. Next, the mass fluxes are used to remap all other scalar variables, such as the specific internal energy (Section 5.4). Finally, the mass fluxes are used to remap the specific momentum. The last step will be considered in detail below, since it is handled in a different manner in curvilinear coordinates.

Mass and Internal Energy Remap

At the conclusion of the Lagrangian step the initial volume element ΔV^n had moved to a new position as shown in Figure 5.24. The mass continuity equation (5.142b) describes the mapping of the mass of this Lagrangian zone onto a new zone. Formally we could try to translate the partial differential equation directly into a finite difference equation. It may be recalled that this approach to the mass continuity equation in conservative form was not very successful. Instead, we found that it was better to construct a finite difference representation which accomplished what the partial differential equation described. A similar approach will be used here. Thus, we consider what the remap process actually does, and base a finite difference representation on it. If the coordinate grid is stationary the new zone is just the initial zone ΔV^n. However, the remap may also be to a new orthonormal zone which does not coincide with ΔV^n. This will occur if the Eulerian grid is moving, and will complicate the results. The approach described below can be easily extended to include these effects. The remap can be done in several ways. The simplest uses two sweeps whose order may be alternated from one cycle to the next. A better method is to sweep in one direction using one half of the time step, then sweep in the orthonormal direction using the full time step, and finally sweep in the original direction with one half of the time step. For simplicity the discussion below that will assume only two sweeps, each using the full time step.

Suppose that the radial sweep is done first. Then the change in volume associated with it is (see Figure 5.24)

$$\Delta V' = \Delta V_L - \delta V_R + \delta V_L \, . \tag{5.158}$$

Next, a second order monotonicity scheme such as the one discussed earlier for advection in cylindrical coordinates is used to find an interpolated density ρ_R and ρ_L in each of the volumes δV_R and δV_L. The mass density in $\Delta V'$ following the radial sweep is then given by

$$\rho' = (\rho^n \Delta V_L - \hat{\rho}_R\, \delta V_R + \hat{\rho}_L \delta V_L) \,/\, \Delta V'. \tag{5.159}$$

The mass associated with the volume change δV_R is $\rho_R \delta V_R$; similar expressions follow for the mass associated with the other volume changes. These quantities are to be saved for use in the momentum remap below. The angular sweep is done next, obtaining the final change in volume from (5.154) to the new volume, which can be written as

$$\Delta V^{n+1} = \Delta V' \; - \; \delta V_T + \delta V_B \,. \tag{5.160}$$

The mass density in the final zone is then given by

$$\rho^{n+1} = (\rho' \Delta V' - \hat{\rho}_T\, \delta V_T + \hat{\rho}_B\, \delta V_B) \,/\, \Delta V^{n+1}. \tag{5.161}$$

By construction, the final zone is again orthonormal, as was assumed at the beginning of the Lagrangian step.

The remap of the remaining scalar variables can be carried out simply by replacing the densities in the (5.160)-(5.161) by the appropriate scalar quantity (see Problem 5.35).

Problem 5.36

Modify (5.160)-(5.161) to describe the remap of an arbitrary scalar variable. Develop a second order monotonic interpolation scheme for the scalar density in the remap volumes δV_α for use in the remap.

Problem 5.37

Carry out an analysis of the volume change δV_L shown in Figure 5.24 similar to the one referring to Figure 5.25. Note that the radial component of the upper and lower node velocities point in opposite directions. In particular, describe the approximation to δV_o in this case.

Momentum Remap in Curvilinear Coordinates

The remap of the momentum density represents a major complication encountered when working in curvilinear coordinates. It is possible to start with the remap terms as expressed in the Lagrange plus remap form (5.144b) and develop a finite difference representation for them using the methods discussed above for the remap of scalar quantities. It would then be necessary to difference the noninertial terms (proportional to the derivative of the metric tensor) appearing on the right side of the momentum equation. Furthermore, the remap terms shown explicitly in (5.144b) and the noninertial terms both result from material motion. Because these two types of terms are, in fact, just different representations of the same effect (motion) they must be differenced in a consistent fashion at the finite difference level. In effect they would be very difficult to solve consistently using the method of operator splitting. Although this might be possible in principle, a much simplier approach will be taken here which can be motivated by the following observations. Rewriting (5.144) in manifest tensor form (where the covariant derivative with respect to x^k is denoted by subscript $;k$

$$\frac{\partial S_i}{\partial t} + \left(S_i v^k\right)_{;k} = - \frac{\partial P}{\partial x^i} + f_i$$

$$\frac{\partial S_i}{\partial t} + V^{;k} S_{i;k} + S_i V^k_{\ ;k} = \frac{d S_i}{d t} + S_i V^k_{\ ;k} .$$

The right left side may be rewritten so that the time derivative is in in the comoving frame. We see that the noninertial terms are a consequence of our choice of coordinates. These terms (which do not appear in the mass continuity equation, for example) arise because as the curvilinear components of a vector quantity are moved from one point to another, the various components are mixed. The Cartesian components of a vector, on the other hand, can be moved freely from point to point without mixing components. This suggests that the remap of the specific momentum be modeled in the following way. As noted in Section 5.4, specific momentum should be transported with mass. Now it may be recalled that the volumes δV_α which were used to remap the volume were also used to remap the mass from the Lagrangian zone back to the final zone ΔV^{n+1}. We should like to use these quantities to remap the specific momentum, thus guaranteeing consistency of the remap process. A straightforward way of doing this is to rotate the curvilinear components of the momentum density and the velocity to a Cartesian frame, remap these variables (which can be moved freely from point to point), and then rotate the resulting Cartesian components back to the original curvilinear coordinate frame.

First consider the vector momentum density at the point j,k (Figure 5.26). Its coordinate components $S^i(\xi)$ in the curvilinear system ξ^i are related to its Cartesian components at the same point $S^k(x)$ by the differential coordinate transformation

$$S^k(x) = \frac{\partial x^k}{\partial \xi^i} S^i(\xi) . \tag{5.162}$$

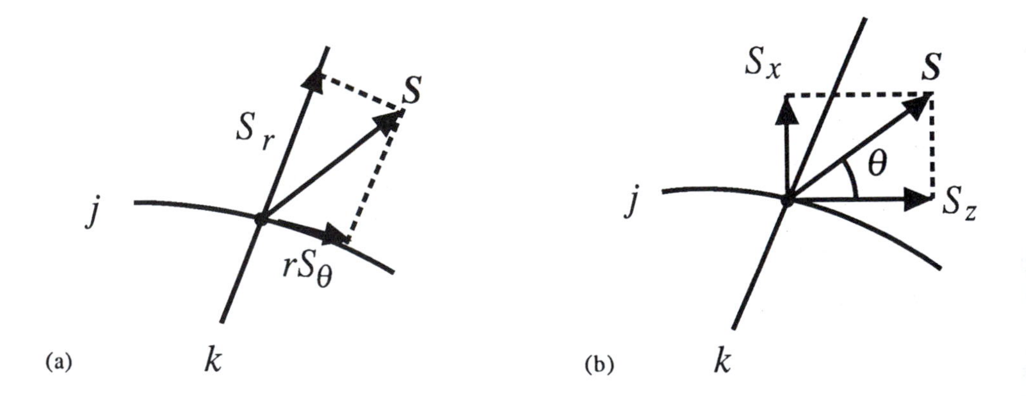

Figure 5.26. (a) Original spherical components of the momentum density. (b) Cartesian components of the momentum density used for momentum advection. A similar decomposition may be made for the fluid velocity.

In terms of r,θ coordinates, and the corresponding Cartesian set z,x, the transformation reduces to

$$S^z = S^r \cos\theta - r\, S^\theta \sin\theta$$

$$S^x = S^r \sin\theta + r\, S^\theta \cos\theta\,.$$

(5.163)

Cartesian components of the velocity vector $V^k(x)$ are given by analogous formulae. We assume that at the start of the momentum remap the components of the momentum and velocity have all been transformed to the Cartesian frame.

Since the Cartesian components of the momentum density and velocity may be moved from point to point freely they may be remapped from ΔV_L to ΔV^{n+1} using the remap masses in a straightforward way. First an effective mass change for the momentum volume is constructed form the remap masses. The change in the momentum volume during the Lagrangian step is shown in Figure 5.27, where the mass changes $\delta M^\theta{}_{j+1/2,k}$ and $\delta M^r{}_{j,k+1/2}$

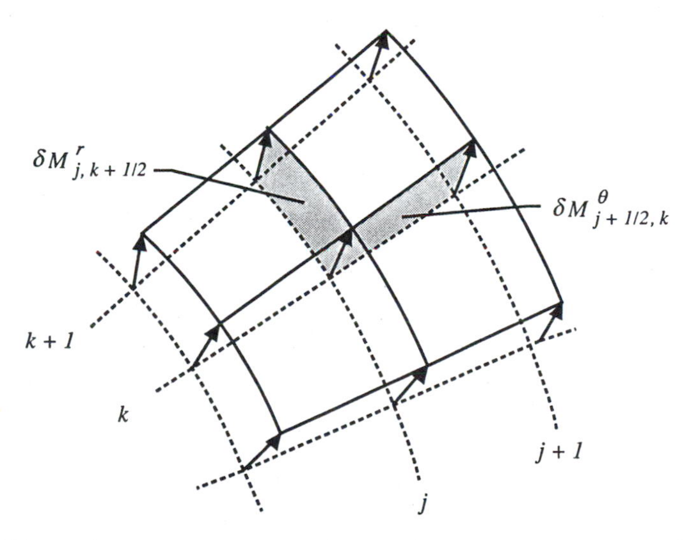

Figure 5.27. Momentum zone used for the momentum remap in spherical coordinates. The initial (orthonormal) zone ΔV^n is shown dashed, and the displaced zone ΔV_L by the solid boundaries. The arrows show the displacement of the verticies after a time Δt. The shaded regions are remap masses obtained from the volume changes and interpolated densities employed for the density remap.

corresponding to the density remap across the left hand and lower boundaries of the zone $\Delta V_{j+1/2,k+1/2}$ are shown shaded. Similar mass changes occur at the other boundaries. These quantities may be used to form the effective mass changes $\Delta M^r{}_{j\pm 1/2,k}$ centered at $j\pm 1/2,k$, which will be used to remap the Cartesian components of the specific momentum:

$$\delta M^r_{j,k+1/2} \frac{A^r_{j,k+1/4}}{A^r_{j,k+1/2}} + \delta M^r_{j,k-1/2} \frac{A^r_{j,k-1/4}}{A^r_{j,k-1/2}} .$$

Finally, the change at $j+1/2,\ k$ is given by the average of the change at j and $j+1$, and we find

$$\Delta M^r_{j+1/2,k} = \delta M^r_{j,k+1/2} + \delta M^r_{j,k-1/2} + \delta M^r_{j+1,k+1/2} + \delta M^r_{j+1,k-1/2}$$

$$\Delta M^r_{j-1/2,k} = \delta M^r_{j,k+1/2} + \delta M^r_{j,k-1/2} + \delta M^r_{j-1,k+1/2} + \delta M^r_{j-1,k-1/2} . \tag{5.164}$$

A similar set of mass can be defined for the angular remap, and will be called $\Delta M^\theta{}_{j,k\pm 1/2}$. The Cartesian components of the specific momentum which is to be associated with each of these effective remap masses are constructed using a second order monotonic interpolation scheme, and are centered at the same points; when interpolated along the radial direction they are denoted by $(i=z,x)$:

$$\hat{V}^{(i)}_{j\pm 1/2,k} \equiv \left[\begin{array}{l} \textit{second order monotonic} \\ \textit{interpolated velocity} \end{array} \right.$$

A similar set of interpolated velocity components may be defined for interpolation in the angular direction. Using the quantities above the components of the total momentum remapped across the outer radial face of the momentum zone are $(i=z,x)$:

$$\hat{V}^{(i)}_{j\pm 1/2,k} \, \Delta M^{r}_{j\pm 1/2,k}$$

(5.165a)

and the components of the total momentum remapped across the angular face are ($i=z,x$)

$$\hat{V}^{(i)}_{j,k\pm 1/2} \, \Delta M^{\theta}_{j,k\pm 1/2} \; .$$

(5.165b)

A finite difference representation of the momentum remap for each Cartesian component of the velocity can now be constructed, and is ($i=z,x$)

$$(V^{(i)})^{n+1}_{j,k} \, \Delta M^{n+1}_{j,k} = (V^{(i)})^{L}_{j,k} \, \Delta M^{L}_{j,k} - (\hat{V}^{(i)} \Delta M^{r})_{j+1/2,k} + (\hat{V}^{(i)} \Delta M^{r})_{j-1/2,k}$$

$$- (\hat{V}^{(i)} \Delta M^{\theta})_{j,k+1/2} + (\hat{V}^{(i)} \Delta M^{\theta})_{j,k-1/2} \; .$$

(5.166)

The first line contains the change in either of the components of the total momentum associated with the radial flow, while the second corresponds to the change in the components associated with angular flow. The subscript L denotes the value of a quantity following the acceleration. Since the mass and volume of the momentum zone are known, a new nodal density may be constructed, and the Cartesian components of the velocity on the right side of (5.166) can be converted to Cartesian components (S^z, S^x) of momentum density. Once these components have been found, the Cartesian components are transformed back to the spherical coordinates system [using the inverse of (5.163)] to find the new values $(S^r)^{n+1}$ and $(S^\theta)^{n+1}$. These components represent the final value of the momentum density, and their construction concludes the hydrodynamic cycle.

The total change in momentum in the computational grid resulting from mass transport can be found by multiplying the finite difference representation (5.166) and its analogue for the angular component, by $\Delta V_{j,k}$ and summing over the grid. The interior terms cancel pair-wise, leaving terms representing the mass flux crossing the outer boundary of the grid. Thus, the transport scheme above conserves momentum by construction.

Boundary Conditions

If $j=2$ represents the origin of coordinates, then the velocity must vanish there, which requires that $v_{2,k} = 0$. The innermost physical momentum zone corresponds to $j=3$, so the change in radial momentum associated with the pressure gradient will not change this value. If the inner boundary does not correspond to the coordinate origin and is closed, then the change in radial velocity will remain zero if

$$P_{3/2,k+1/2} = P_{5/2,k+1/2} \,.$$

Since the radial velocity component remains zero there will be no transport of mass or specific energy across the origin. If the lower grid boundary $k=1$ corresponds to the z axis, then the theta component of the velocity must remain zero. This is achieved by imposing

$$P_{j+1/2,3/2} = P_{j+1/2,5/2} \,.$$

The boundary conditions along the outer portions of the grid depend on the nature of the problem being solved. If $\theta = \pi/2$ represents the outer boundary, and is a plane of symmetry then reflecting boundary conditions are appropriate. In particular

$$P_{j+1/2,K-1/2} = P_{j+1/2,K+1/2}$$

guarantees that the velocity components $u_{j,K}=0$ for all times. This also implies that there will be no flux of mass or specific energy across the boundary. Open boundary conditions may also be imposed along the outer boundaries, or along the inner boundary if it does not represent the origin. The specification of these conditions will depend on the problem under consideration.

Problem 5.38

Express the inviscid fluid equations in curvilinear coordinates on a simple moving grid analogous to the one discussed in the first half of the chapter. Discuss some of the difficulties that would arise if the finite difference representations above were modified to include this type of grid motion?

5.11 Artificial Viscous Stress

Shock wave propagation in curvilinear coordinates can be modeled by the introduction of an artificial viscous stress in much the same manner as described in Section 5.5. The results developed there require modification because of the velocity centering adopted for curvilinear coordinates. Further modifications would also be necessary if a true tensor artificial viscous stress is desired. Finally, the introduction of a monotonically limited artificial viscous stress will be discussed in Section 5.13.

The equations (5.84)-(5.85) describing the artificial viscous force and associated heating can be generalized to an arbitrary coordinate system. A quasitensor approach using coordinate components gives

$$\frac{\partial S_i}{\partial t} = -\frac{1}{\sqrt{g}}\frac{\partial}{\partial x^k}\left(Q_i{}^k\sqrt{g}\right) \tag{5.167}$$

$$\rho\frac{\partial \varepsilon}{\partial t} = Q_i{}^k\frac{\partial v^i}{\partial x^k} \; . \tag{5.168}$$

As noted in Chapter 4, the use of even a quasi-tensor form is seldom necessary in astrophysics, since the differences between tensor and nontensor forms become important only near the origin. In fact, for most applications a simple planar form of the artificial viscous stress gives an adequate respresentation of shock propagation. Therefore, in this section the simple planar form will be chosen which includes only the diagonal components of the stress (off-diagonal, or shear, contributions may be included if desired). Expressing the momentum density and the artificial viscous stress in Lagrangian form in terms of physical components, the artificial viscous stress force may be written in the form

$$\frac{dS_r}{dt} = -\frac{\partial Q_{rr}}{\partial r} + \frac{1}{r}\frac{\partial Q_{r\theta}}{\partial \theta}$$

(5.169a)

$$\frac{dS_\theta}{dt} = -\frac{\partial Q_{\theta r}}{\partial r} + \frac{1}{r}\frac{\partial Q_{\theta\theta}}{\partial \theta} .$$

(5.169b)

For the remainder of this discussion the shear artificial viscous stress components $Q_{r\theta}$ and $Q_{\theta r}$ will be dropped. The interested reader may follow the discussion in Section 5.5 and work out a finite difference representations which includes these terms if needed (Problem 5.39).

The radial (physical) component of the quadratic artificial viscous stress Q_{rr} can be expressed in term of the difference in radial velocities as in (5.89):

$$Q_{rr} \equiv C_q \rho (\Delta_r v)^2.$$

(5.170a)

By analogy, the angular (physical) component of the quadratic artificial viscous stress $Q_{\theta\theta}$ will be given by

$$Q_{\theta\theta} \equiv C_q \rho (\Delta_\theta u)^2.$$

(5.170b)

Constraints may be imposed in the usual way on the components of the artificial viscous stress for expansion. For example, the angular component (5.170b) should be set to zero for zones that are expanding in the angular direction, or which are undergoing homologous compression; for example

$$Q_{\theta\theta} = 0 \qquad if \ \Delta_\theta u \geq 0.$$

(5.170c)

The spatial centering of the two quadratic artificial viscous stresses in the momentum zone is shown in Figure 5.28.

Shock heating will also be expressed in simple planar form. In terms of physical components (as above the shear components are omitted)

$$\rho \frac{\partial \varepsilon}{\partial t} = Q_{rr} \frac{\partial v}{\partial r} + Q_{\theta\theta} \frac{1}{r} \frac{\partial u}{\partial \theta}. \tag{5.171}$$

A finite difference representation of (5.169) and (5.171) for an as yet unspecified representation of the artificial viscosity will be considered next. We will then discuss a monotonicity limited form of the artificial viscosity itself.

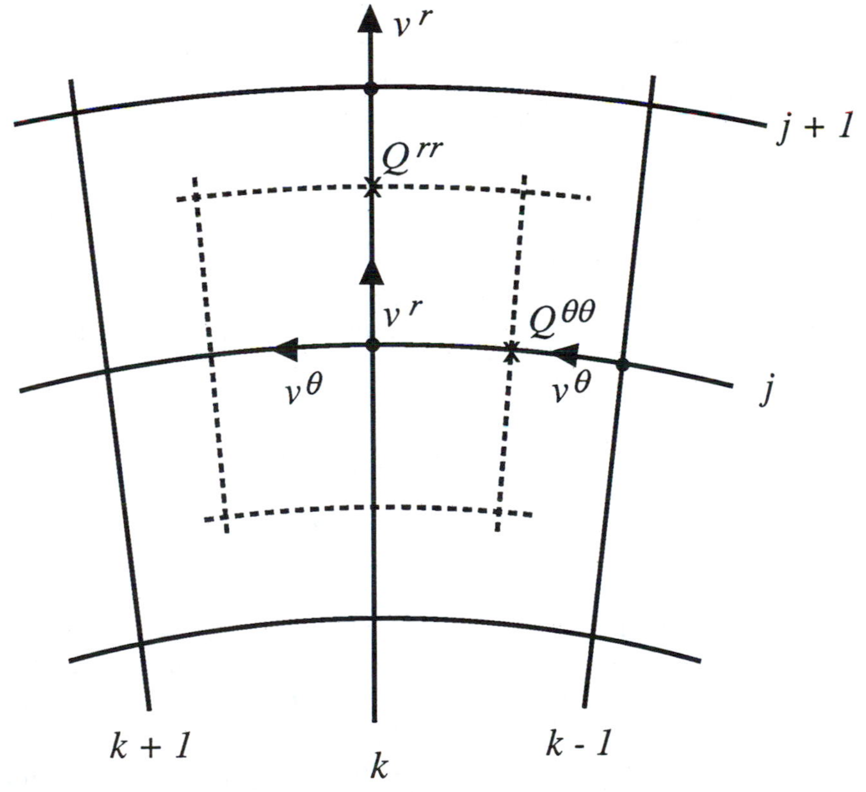

Figure 5.28. Spatial centering of the two diagonal components of the artificial viscous stress in spherical coordinates assuming vertex centered velocities. The momentum zone is bounded by the dashed line.

Artificial Viscous Force

The radial artificial viscous force will be considered first. It could be constructed by integrating (5.169a) over the momentum volume $\Delta V_{j,k}$ in the usual way. Because we have chosen to define the momentum volume as in (5.145d) this approach would result in a spatial centering for Q^{rr} which would be difficult to model (Figure 5.28). A modified form which serves the purpose of an artificial viscous stress is given by

$$\frac{T^{Q}_{j,k} - T^{P}_{j,k}}{\Delta t} = - \frac{A^{r}_{j,k}(Q^{rr}_{j+1/2,k} - Q^{rr}_{j-1/2,k})}{\Delta V_{j,k}} . \tag{5.172}$$

The area factors are defined in analogy with (5.145b), $T^{P}{}_{j,k}$ represents the radial (physical) component of the momentum density obtained from the pressure acceleration, and the volume in the denominator is the value at time t^n. Finally, the physical velocity components used in forming (5.165a) are constructed from (5.153). A practical approximation to (5.172) is to replace the ratio of the area factor to the volume by Δr_j.

An expression for the acceleration associated with the angular artificial viscous stress can be constructed in similar fashion as

$$\frac{S^{Q}_{j,k} - S^{P}_{j,k}}{\Delta t} = - \frac{A^{\theta}_{j,k}(Q^{\theta\theta}_{j,k+1/2} - Q^{\theta\theta}_{j,k-1/2})}{\Delta V_{j,k}} . \tag{5.173}$$

The area factors are defined in the obvious way, and $Q^{\theta\theta}$ uses the velocity components obtained from the pressure acceleration. Finally, the ratio of area to volume elements may in practice be approximated by the angular displacement $r_j\Delta\theta_{k+1/2}$. The viscous acceleration should be implemented following the pressure gradient acceleration.

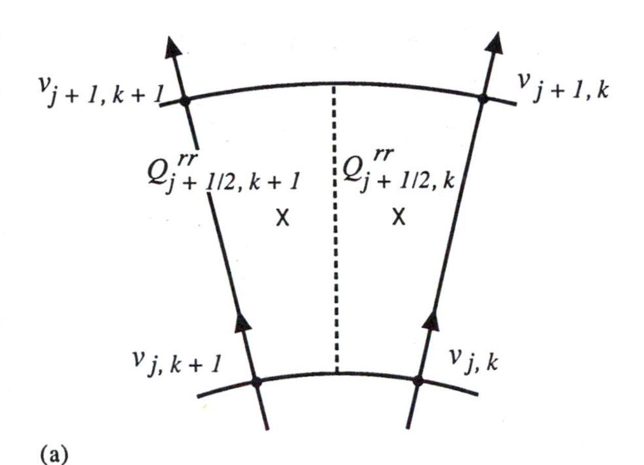

(a)

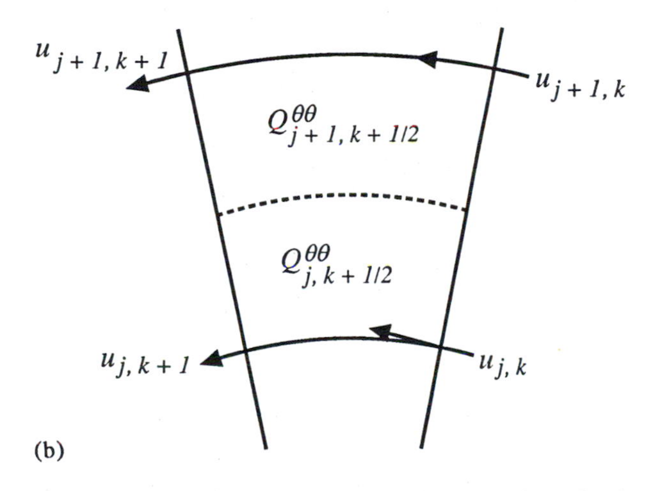

(b)

Figure 5.29. Artificial viscous stresses used in constructing the shock heating of a zone. (a) Shows the radial components used for each portion of the zone; (b) Shows the angular components used for the zone.

Shock Heating

The change in material internal energy described by (5.171) which accompanies shock propagation may be constructed in a number of ways (see Figure 5.29). For example, integrating (5.171) over $\Delta V_{j+1/2,k+1/2}$, the radial stress makes a contribution to the internal energy which can be expressed as

$$\frac{1}{\Delta V_{j+1/2,\,k+1/2}}\int\frac{\partial(\rho\varepsilon)}{\partial t}\,dV = \frac{2\pi}{\Delta V_{j+1/2,\,k+1/2}}\int\sin\theta\, d\theta\int r^2 dr\, Q^{rr}\frac{\partial v}{\partial r}$$

$$= \frac{2\pi}{3}\frac{(r_{j+1}^3 - r_j^3)}{\Delta V_{j+1/2,\,k+1/2}}\Big\{(\cos\theta_k - \cos\theta_{k+1/2})\Big(Q^{rr}\frac{\partial v}{\partial r}\Big)_{j+1/2,\,k}$$

$$+ (\cos\theta_{k+1/2} - \cos\theta_{k+1})\Big(Q^{rr}\frac{\partial v}{\partial r}\Big)_{j+1/2,\,k+1}\Big\}.$$

The first integral has been evaluated by assuming that Q^{rr} is a constant throughout the volume $\Delta V_{j+1/2,k}$, and that $Q^{\theta\theta}$ is a constant throughout $\Delta V_{j,k+1/2}$. The terms on the right side of this expression can be rewritten in a simple form, so that

$$\rho_{j+1/2,\,k+1/2}\frac{(\varepsilon' - \varepsilon^n)_{j+1/2,\,k+1/2}}{\Delta t^n} = \frac{\Delta V_{j+1/2,\,k+1/4}}{\Delta V_{j+1/2,\,k+1/2}}Q^{rr}_{j+1/2,\,k}\frac{|v^{n-1/2}_{j+1,\,k+1} - v^{n-1/2}_{j,\,k+1}|}{\Delta r_{j+1/2}}$$

$$+ \frac{\Delta V_{j+1/2,\,k+3/4}}{\Delta V_{j+1/2,\,k+1/2}}Q^{rr}_{j+1/2,\,k+1}\frac{|v^{n-1/2}_{j+1,\,k+1} - v^{n-1/2}_{j,\,k+1}|}{\Delta r_{j+1/2}}.$$

(5.174)

Heating associated with the angular component of the artificial viscous stress can be obtained in a similar fashion. By construction the artificial viscous stresses are non-negative, and are nonzero in compression only. Thus the terms appearing in (5.174) and the corresponding equation for the angular terms only result in heating. It may be noted that the finite difference representation (5.174) distributes the heating associated with each component of the artificial viscous stress between two neighboring zones.

Problem 5.39

The artificial viscous force can be written in the quasi-tensor form

$$\frac{\partial S_r}{\partial t} = -\frac{1}{r^2}\frac{\partial}{\partial r}(r^2 Q_{rr}) - \frac{1}{r \sin\theta}\frac{\partial}{\partial \theta}(Q_{r\theta} \sin\theta)$$

and

$$\frac{\partial S_\theta}{\partial t} = -\frac{1}{r^3}\frac{\partial}{\partial r}(r^3 Q_{\theta r}) - \frac{1}{r \sin\theta}\frac{\partial}{\partial \theta}(Q_{\theta\theta} \sin\theta).$$

(a) Develop a finite difference representation for the two acceleration terms assuming that the shear stress is zero.

(b) Repeat part (a) but include the shear artificial viscosity. How would you construct the shear stress?

5.12 Time Step Controls and Stability

The Eulerian hydrodynamic equations in curvilinear form have the same type of accuracy and stability time step constraints as do the equations in cylindrical coordinates. A Courant time step constraint based on the pressure and density is necessary for stability, and a transport time step constraint similar to (5.52) is required to maintain accuracy. Finally, the time step must be constrainted in analogy with (5.110) when shocks occur.

Many curvilinear coordinate systems, such as spherical coordinates, contain zones near the origin which may be quite narrow. Since the Courant and transport time steps are proportional to the zone thickness, the time step may be very small here. This may occur even though the fluid velocity there is not large. In many astrophysical problems the fluid motion near the origin may be unimportant (such as in some forms of stellar collapse), but care should be exercised not to use finer zoning than necessary which may unnecessarily limit the effeciency of computations.

Pressure Scaling

In some situations a simple procedure called pressure scaling may be used to avoid small Δt when the time step is limited by the local sound speed, c_S. This method is applicable to regions where

$$c_S >> | v - v_{grid} | , \tag{5.175a}$$

and which are near hydorstatic equilibrium in the sense that the net acceleration

$$\frac{\partial v}{\partial t} = - \frac{1}{\rho} \frac{\partial P}{\partial x} + g << \frac{c_S^2}{l} \tag{5.176b}$$

is small. Where these two conditions are satisfied the acceleration equation may be replaced by

$$\frac{\partial v}{\partial t} = - \alpha \left(\frac{1}{\rho} \frac{\partial P}{\partial x} - g \right) ,$$

with $\alpha < 1$. Explicit finite difference representations of this equation will be stable if the time step satisfies

$$\Delta t < \frac{\Delta x}{c_S{}^*} ,$$

where $c_S{}^* = \alpha c_S$. Since $\alpha < 1$, the time step constraint is reduced and Δt is increased.

The choice of the pressure scaling factor α will depend on the coordinate system. For calculations in spherical coordinates where there is often little fluid motion near the origin the acceleration in the angular direction for small radii will limit the time step. In this case a useful choice for the pressure scaling factor is

$$\alpha_\theta = min \left\{ \frac{r \Delta\theta}{\Delta r} , 1 \right\} , \qquad \alpha_r = 1.$$

This modification is important for spherical coordinate calculations, but it can also be used in other coordinate systems where (5.175) are satisfied.

Independent Time Steps

Pressure scaling may be safe to use when it is known in advance that the region in which scaling occurs remains quasistatic. Another approach is to construct independent time steps in different parts of the grid. Independent time step methods can be developed which are quite general, but programming these methods can be complicated, and the added computation may be time consuming. We consider here a very simple approach for spherical coordinate systems in which the time step is limited near the origin by sound transit time in the angular direction. First, define the radius

$$r_{J_1} \geq \frac{\Delta r_{J_1+1/2}}{\Delta\theta_{k+1/2}}$$

inside of which the Courant condition (based on the sound speed in the angular direction) would limit the time step. We assume, however, that the transport time step constraint $v_\phi \Delta t < r\,\Delta\theta$ is satisfied for $j < J_1$. Next, find the global time step, Δt, outside this region. Then find an independent time step

$$\Delta t_N = \Delta t / 2^N$$

which satisfies the Courant condition for all $j \leq J_1$, and which is used to subcycle the Lagrangian step. The finite difference equations on the outer portion of the grid are solved as usual. Subcycling in the region $j \leq J_1$ is used to compute the Lagrangian step 2^N times. This is done using the independent time step above and the same equations as are used for the global time step. The remap calculation is then done once using the global time step. This method is approximate because the grid produced during each Lagrangian subcycle is nonorthonormal, and expressions such as (5.149) are not strictly correct here. Because the axial flow is assumed to be small in the region where Δt_N is used, very little deviation from orthonormality should occur and the error is expected to be small.

5.13 Monotoniclly Limited Artificial Viscosity

The artificial viscosity (5.170) is constructed using the velocity jump across the shock front. Traditionally this is taken as a simple difference [recall, for example, (5.95) in cylindrical coordinates], and linear terms are seldom used in practice because they can act on regions of smooth flow at large distances from the shock itself. An improvement may be made in the way the artificial viscosity is constructed by extrapolating the velocity from the right and the left zone faces to the zone center, and forming a zone centered velocity difference using extrapolated velocity values. In this scheme a quadratic and linear artificial viscosity are both used. The quadratic term has a small coefficient, and a relatively large coefficient for the linear term. The latter damps out oscillations behind the shock front. The use of interpolation in constructing the velocity jump limits the region over which the linear terms acts. The combination of quadratic plus linear artificial viscosity, and an interpolation scheme produces an effective artificial viscous stress which limits shocks to about two to three zones, while preserving good accuracy in the solution. Finally, the scheme includes a monotonicity constraint to eliminate overshoot in the interpolation procedure.

The construction of the monotonically limited artificial viscosity along either coordinate direction can be accomplished as follows. The first step is to find the minimum and maximum values of the velocities:

$$v_{min} = min\{v_{j+1}, v_j, v_{j-1}\}$$

$$v_{max} = max\{v_{j+1}, v_j, v_{j-1}\}$$

We then define the quantity

$$S_1 = \frac{min\{v_{max} - v_j, v_j - v_{min}\}}{\Delta r_j} . \tag{5.176}$$

Note that the quantity S_1 is always positive. If the set of velocities $\{v_{j-1}, v_j, v_{j+1}\}$ is not monotonic with distance then we set $S_1 = 0$. Next, using a quadratic fit through the three velocities, find the slope S_q at the point j. This slope is easily shown to be (see Appendix C)

$$\frac{S_L \Delta x_{j+1/2} + S_R \Delta x_{j-1/2}}{2 \Delta x_j} = S_q \tag{5.177}$$

where $S_L = (v_j - v_{j-1})/\Delta r_{j-1/2}$ and $S_R = (v_{j+1} - v_j)/\Delta r_{j+1/2}$. Next construct the quantity

$$S_j = S_q \min\{\frac{S_1}{|S_q|}, 1\} \tag{5.178}$$

which has the same sign as S_q, and is the minimum of S_1 and $|S_q|$. S_j is a velocity slope that will be used to extrapolate the velocities at each zone face to the zone center:

$$v_{ext,j+1/2} = v_j - \frac{1}{2} S_j \Delta r_{j+1/2} . \tag{5.179}$$

Note that if the velocities are not monotonic then S_j is zero. The interpolated velocity jump at the center of the zone is given by

$$\{\Delta v_{j+1/2}^{n-1/2}\}_{ext} \equiv (v_{j+1}^{n-1/2} - v_j^{n-1/2}) - \frac{1}{2}(S_j + S_{j+1}) \Delta x_{j+1/2} . \tag{5.180}$$

The zone centered velocity jump is shown in Figure 5.30. If $\Delta v_{j+1/2}$ and $v_j - v_{j-1}$ are of opposite sign, or if $v_j - v_{j-1} = 0$, the velocity jump is set to zero.

The linear and quadratic artificial viscous stress can now be constructed using the monotonically limited velocity jump at the zone center. The linear term is given by

$$Q_{L,j+1/2} = -C_L \rho_{j+1/2} c_{S,j+1/2} \{\Delta v_{j+1/2}\}_{ext} \tag{5.181}$$

where c_S is the local adiabatic sound speed. The quadratic term is given by

$$Q_{q,j+1/2} = C_q \rho_{j+1/2} \{\Delta v_{j+1/2}\}^2_{ext} \, . \tag{5.182}$$

If the zone is expanding Q_q is set to zero. Nominal values for the coefficients in the monotonically limited artificial viscous stress for a gamma law-like equation of state are $C_Q=(\gamma+1)/2$ and $C_L=1.0$.

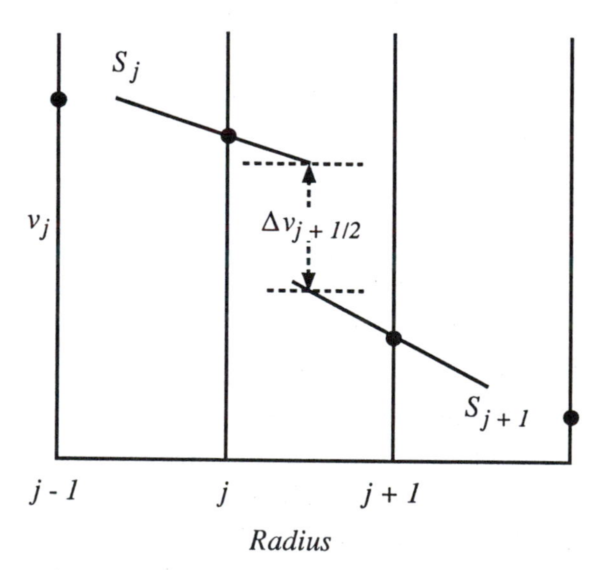

Figure 5.30. Construction of the zone centered velocity jump used to form a monotonically limited artificial viscous stress.

The artificial viscous stress above is planar in form, but it eliminates spurious heating in smooth flows, such as homologous collapse. It is therefore not necessary to introduce the radial factors as was done for the standard von Neumann method. The method can also be applied to Lagrangian methods, such as those discussed in

Chapter 4. The only modification needed in this case is to replace Δr_j in (9.175) by the maximum of $\Delta r_{j+1/2}$ and $\Delta r_{j-1/2}$.

Another modification of the Eulerian artificial viscous stress may be made to counteract some of the diffusion of momentum which occurs in shocks because of advection. A shock front is a discontinuity which moves through the grid in part by advection. Momentum near a velocity discontinuity will diffuse in much the same manner as a density discontinuity diffuses during advection. The diffusion of momentum can lead to a loss of kinetic energy from the problem. This energy should go into thermal energy if energy balance is to be maintained. One way to enforce this is to construct a kinetic energy diffusion corrector Q_D which enters the equations as an additional term in the artificial viscous heating (see Problem 5.41 for a derivation of Q_D in the planar one dimensional model). For two dimensional problems a corrector can be formed along each coordinate direction. The kinetic energy diffusion corrector may then be added to the radial and angular components of the artificial viscous stress discussed above, and the result used in the energy equation (5.166). A value for the coefficient C_D in (5.183) of 0.25 works well in practice. As with all forms of advection, the amount of diffusion increases with increasing fluid velocity relative to the grid. In Lagrangian calculations where the fluid and the grid velocity are the same, a kinetic energy corrector is not necessary.

The monotonically limited artificial viscosity with a kinetic energy diffusion corrector improves overall energy balance, and spreads typical shocks out over only about two zones, thus gives sharper resolution than does the conventional form.

Problem 5.40

Write out a finite difference representation for the artificial viscous stress in two dimensional spherical coordinates. Carry out the derivation of the viscous acceleration (5.173).

Problem 5.41 (Kinetic energy diffusion corrector)

(a) Consider the one dimensional, first order finited difference representation of the momentum advection term in Cartesian coordinates (assuming a uniform space-time grid and fluid flow to the right)

$$\frac{S'_k - S_k}{\Delta t} = -\frac{F_{k+1/2}\,u_k - F_{k-1/2}\,u_{k-1}}{\Delta x}$$

$$F_{k+1/2} \equiv \rho_{k+1/2} u_{k+1/2}$$

and expand it in a Taylor series about a zone edge. Show that, through second order in Δt and Δx

$$\frac{\partial^2 S}{\partial t^2}\frac{\Delta t}{2} = -\frac{\partial(\rho u^2)}{\partial x} + \frac{\partial}{\partial x}\left(\rho u \frac{\partial u}{\partial x}\right)\frac{\Delta x}{2}\,. \tag{5.183}$$

Terms of third order or higher have been dropped, and for simplicity it has been assumed that the space-time grid is uniform.

(b) Consider the quantity

$$Q_D = -\,C_D\,\rho\,/u_{k+1/2}/\,\{\Delta u\}_{ext,\,k+1/2} \tag{5.184a}$$

to be an artificial viscous stress, and show that the increase in thermal energy density associated with momentum diffusion can be modeled by

$$\frac{E'_{k+1/2} - E_{k+1/2}}{\Delta t} = -\,Q_D\frac{\partial u}{\partial x} \tag{5.184b}$$

Explain why the velocity difference $(\Delta u)_{ext,k+1/2}$ should vanish for zones which are not undergoing compression.

Problem 5.42

An important test problem for two dimensional hydrodynamics in spherical coordinates is to send a pressureless slab of matter down the symmetry axis ($\theta = 0$) so that it passes through the coordinate origin. Although a small amount of diffusion is associated with advection, there should be minimal distortion of the structure.

5.14 CONTACT DISCONTINUITIES

Some fluid variables may exhibit sharp gradients with characteristic scale lengths that are very small compared to any practical thickness of Eulerian zones. Examples include composition discontinuities or entropy discontinuities. These examples represent two types of contact discontinuities which occur, either as a result of initial conditions, or as a way of imposing internal boundary conditions. In such cases it may not be sufficient simply to define the contact discontinuity through initial conditions (as it would be in a Lagrangian calculation), since Eulerian diffusion would spread the initial discontinuity out over several zones after a number of computational cycles.

A particularly simple way of treating contact discontinuities will be considered below for a system which contains two different components. The method can be used with either face centered or vertex centered velocity components. For simplicity we shall consider face centered velocity components in cylindrically symmetric systems.

Figure 5.31 shows a typical example of a contact discontinuity that might arise in a model of protostar formation. The contact discontinuity represents a boundary between hot, low density interstellar matter (which may be labeled as component *1*) surrounding a cool, higher density protostellar cloud (labeled as component *2*). If the pressure P_H in the interstellar matter is greater than the pressure P_C in the cloud, then the fluid motion is normal to the contact discontinuity, causing a flow to the lower left. As the protostellar cloud contracts the contact discontinuity will move through the grid, possibly changing shape as it does so. Another example is composition changes which occur at late times in the cores of massive stars. In order to track the contact discontinuity in these cases it is necessary to introduce an additional zonal variable representing the compositions which occupy a zone. Most zones in these examples contain only one component, and may be classed as clean zones. However, the relatively small number of zones defining the contact discontinuity contain both materials. The latter zones may be called mixed zones. The finite difference methods developed in the previous sections apply without modification to each of the clean zones in the grid. In this section we shall consider the modifications necessary to treat the mixed zones which define contact discontinuities.

Consider a typical mixed zone of volume ΔV containing two component, which will be denoted by the index i (where $i=1,2$). The primary purpose of the model presented here is to segregate each component within a zone. This may be done by assigning to each component in a mixed zone the partial volume $\Delta V(i)$ which it occupies. These variables have the same space-time centering as does the zonal volume. Clearly

$$\Delta V^n_{k+1/2,j+1/2} = \sum_i \Delta V^n_{k+1/2,j+1/2}(i) \,. \tag{5.185}$$

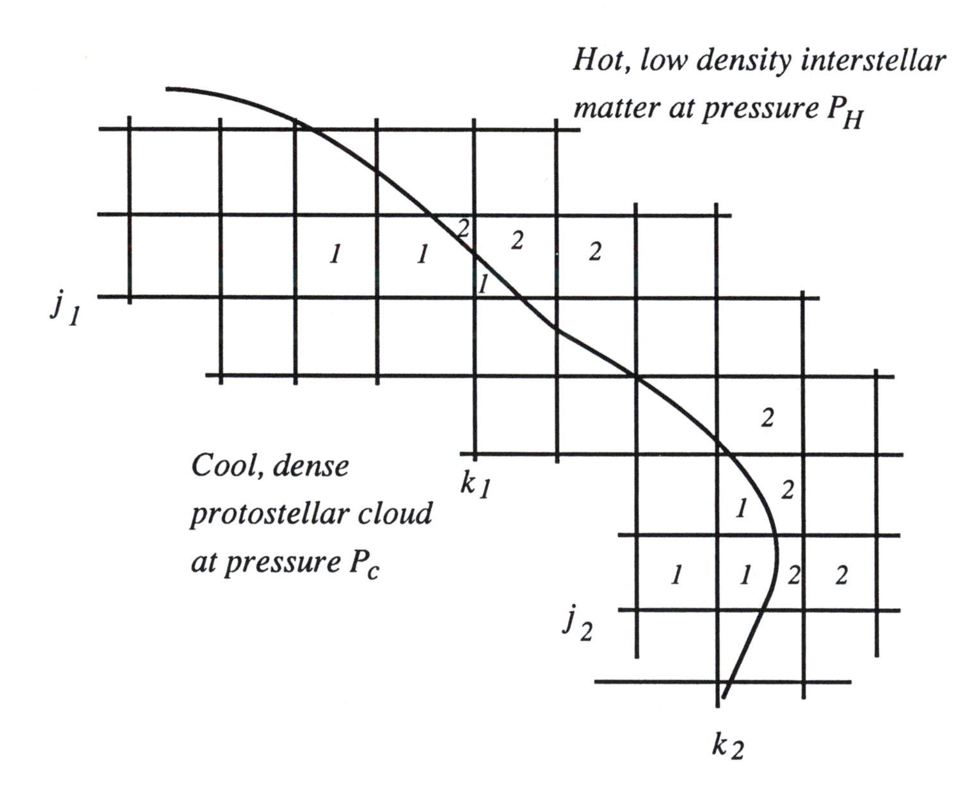

Figure 5.31. Example of a contact discontinuity separating hot, low density interstellar matter from a cool, higher density protostellar cloud. Since $P_H >$ P_C, the contact discontinuity moves to the lower left.

It is also useful to define the volume fractions

$$Vf^{n}_{k+1/2,j+1/2}(i) = \frac{\Delta V^{n}_{k+1/2,j+1/2}(i)}{\Delta V^{n}_{k+1/2,j+1/2}} .$$

(5.186a)

$$\sum_i Vf^{n}_{k+1/2,j+1/2}(i) = 1.$$

(5.186b)

In the absence of grid motion the zonal volume will be a constant in time, but the partial volumes and volume fractions will not. In addition to the partial volumes, we may define the mass of each component in a zone $\Delta m(i)$, which has the same space-time centering as do the total zonal mass, and the density for each composition

$$\rho^{n}_{k+1/2,j+1/2}(i) = \frac{\Delta m^{n}_{k+1/2,j+1/2}(i)}{\Delta V^{n}_{k+1/2,j+1/2}} .$$

(5.187)

Equations (5.187) and (5.186) may be used to show that the average mass density in a mixed zone is given by (common subscripts $k+1/2,j+1/2$ omitted)

$$\langle\rho\rangle = \sum_i \rho(i)\, Vf(i) .$$

(5.188)

Each component will also be characterized by its specific internal energy $\varepsilon(i)$, and pressure $P(i)$. Even though a zone may contain more than one component, the fluid velocity is only defined at the zonal boundaries. The contact discontinuity is not represented explicitly, and no velocity components will be associated directly with it. Furthermore, no attempt will be made to construct an internal boundary between compositions in a zone. Thus the only information which will be used below is the partial volumes, or equivalently the volume fractions, and the masses of each component in a mixed zone. Because of this the method is, in a sense, underdetermined. While this may be thought of as a limitation, it also has advantages in that very little detailed subzonal structure is introduced which could, in some circumstances, begin to dominate the answer. One way in which this inherent ambiguity is partially resolved is to use information about the make up of neighboring zones to spatially order the compositions for the purposes of calculating the transport. The approach taken here has been found in practice to be quite robust. It is also surprisingly forgiving in the sense that if during one time step the transport results in a poor choice, it is usually not reinforced by the subsequent dynamics.

The method of tracking contact discontinuities will be considered in two steps. First, we develop modifications needed for the description of advection, or remapping. The modifications to the Lagrangian step of the hydrodynamic calculation will then be discussed. In practice the Lagrangian step is computed first, followed by the transport (advection or remap) step.

The method of tracking contact discontinuities described below operates alternately on one dimensional strips. This can result in biasing of the results. A practical way around this difficulty is to compute the transport in the axial direction first (using a time step $\Delta t/2$), followed by the radial direction with the full time step, and then finish by computing the transport in the axial direction again, using half the time step $\Delta t/2$.

Transport Near Contact Discontinuities

The first step in computing the transport of components at a contact discontinuity is to use information from the neighboring zones to spatially order the mixed zone. When operator splitting is used to describe a system containing two compositions, four classes of transport will occur (Figure 5.32). The first class, clean to clean transport, has already been discussed. The remaining three classes are clean to mixed transport, mixed to clean transport and mixed to mixed transport. We shall consider the last three cases below. The approach may be applied to the components of either axial or radial flow. It should be noted that as the fluid flows through the grid, zones which were originally clean may be come mixed, while those which were originally mixed may become clean. Examples of this will be evident below.

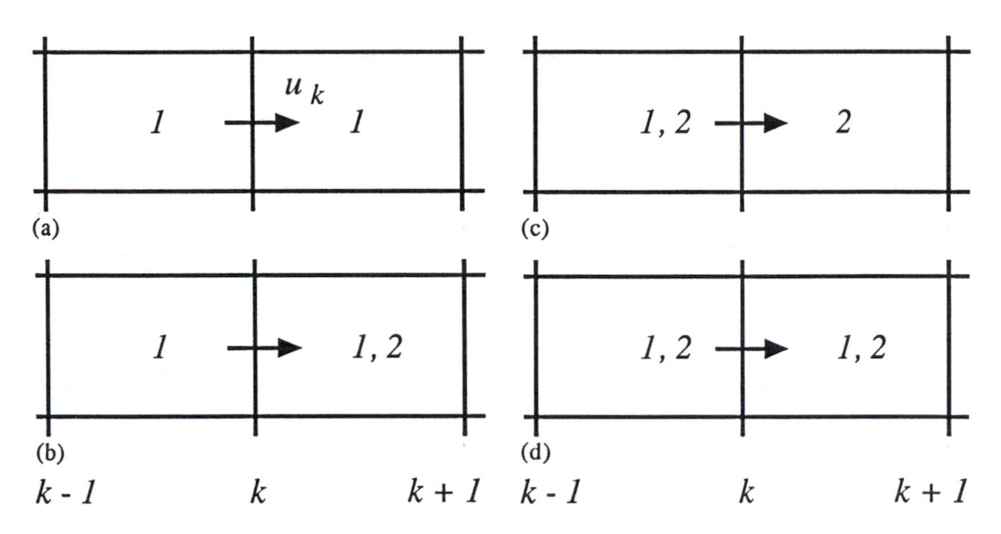

Figure 5.32. The four types of mass transport in a one dimensional sweep near a contact discontinuity: (a) Clean to clean; (b) Clean to mixed; (c) Mixed to clean; and (d) Mixed to mixed. In all cases the fluid velocity at the zone boundary k is directed to the right.

Clean to Mixed Transport

The simplest example of mixed zone transport is clean to mixed transport shown in Figure 5.32b. We shall develop a special case of this type of transport, which illustrates several important features, in which all zones are initially clean and the fluid moves to the right (Figure 5.33). The more general case where the zone ahead contains both components can be constructed by a simple extension of the results below. The contact discontinuity initially corresponds to a zone boundary k . After one time step zone $k+1/2$ will become mixed, while all other zones will remain clean (it may be recalled that the time step constraint derived earlier guarantees that the fluid, and thus the contact discontinuity, moves less than one zone per time step). The change in partial volumes may be constructed by finding the volume of each composition carried across the zone boundaries, and renormalizing the result to the final zonal volume. This may be thought of as a type of normalized remap. The initial conditions in zone $k+1/2$ (the subscript $k+1/2$ will be omitted for notational convenience) are $\Delta V^n(1) = 0$ and $\Delta V^n(2) = \Delta V$, which implies that $Vf^n(1)=0$ and $Vf^n(2)=1$. The area of the zone face at k normal to the flow

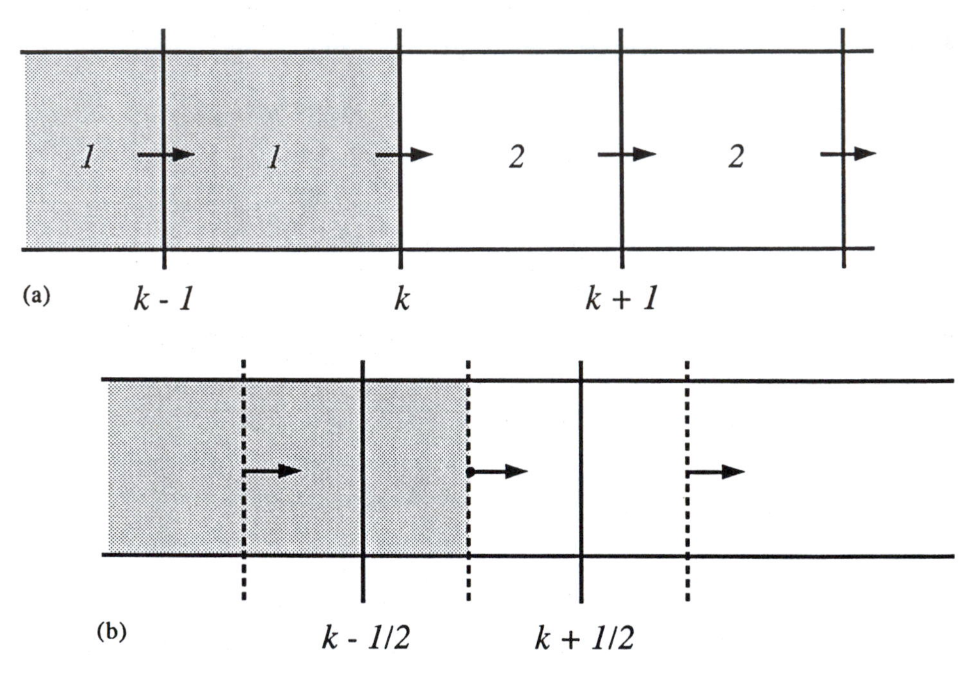

Figure 5.33. (a) The simplest example of clean to mixed transport, in which component *1* moves to the right forming a mixed zone. The fluid velocities are shown at the zone faces. (b) The momentum zone ΔV_k corresponding to part (a). Notice that it contains both components.

will be denoted by A_k. The new partial volumes after a time step Δt are, for the velocities shown,

$$\Delta V'(1) = u_k A_k \Delta t$$

$$\Delta V'(2) = \Delta V^n(2) - u_{k+1} A_{k+1} \Delta t , \tag{5.189}$$

and the (unrenormalized) volume fractions are obtained by dividing (5.189) by the zone volume:

$$Vf'(1) = \frac{u_k A_k \Delta t}{\Delta V}$$

$$Vf'(2) = 1 - \frac{u_{k+1} A_{k+1} \Delta t}{\Delta V} . \tag{5.190}$$

The final (normalized) volume fractions are defined by

$$Vf^{n+1}(i) = \frac{1}{N_V} Vf'(i) \tag{5.191a}$$

$$N_V \equiv \sum_i Vf'(i) . \tag{5.191b}$$

The final partial volumes are given by

$$\Delta V^{n+1}(i) = Vf^{n+1}(i) \, \Delta V^{n+1} . \tag{5.192}$$

The discussion above may be modified to include grid motion (see Problem 5.43). The partial volumes and volume fractions may change due to transport even if ΔV remains constant.

The mass of component 1 in ΔV after one time step is, in this example, just the mass transported across the left boundary:

$$\begin{aligned} \Delta m^{n+1}(1) &= \Delta m^{n}(1) + \rho_k^n(1) \, u_k A_k \, \Delta t \\ &= \rho_k^n(1) \, Vf'(1) \, \Delta V . \end{aligned} \tag{5.193}$$

The upstream value of the density of component i is to be used in general in expressions such as (5.193):

$$\rho_k^n(i) \equiv \begin{cases} \rho_{k-1/2}^n(i) & u_k > 0 \\ \rho_{k+1/2}^n(i) & u_k < 0 \end{cases} . \tag{5.194}$$

Second order schemes are not used to transport materials on either side of a contact discontinuity in this model. The mass of component 2 in the zone after one time step is

$$\Delta m^{n+1}(2) = \Delta m^{n}(2) - \rho^{n}_{k+1}(2)\, u_{k+1} A_{k+1} \Delta t$$
$$= \rho^{n}_{k+1/2}(2)\, Vf^{n}(2)\, \Delta V - \rho^{n}_{k+1}(2)\, [1 - Vf'(2)]\, \Delta V \,. \tag{5.195}$$

Finally, the density of each component in zone $k+1/2$ is

$$\rho^{n+1}(i) = \frac{\Delta m^{n+1}(i)}{\Delta V^{n+1}(i)} \,. \tag{5.196}$$

Notice that this value changes because the total mass of component i changes, and because the partial volume occupied by it has changed. The average mass density of the mixed zone is clearly given by (5.188).

The mass transported across each zone boundary will be needed to calculate the corresponding momentum advection for mixed zones. In this simple example the mass carried across the area A_k is just (5.193),

$$\Delta M_k = \rho_{k-1/2}(1)\, Vf'(1)\, \Delta V \,, \tag{5.197a}$$

while the mass carried across A_{k+1} is, from (5.195)

$$\Delta M_{k+1} = [1 - Vf'(2)]\, \rho_{k+1/2}(2)\, \Delta V \,. \tag{5.197b}$$

It is important to notice that these quantities contain the unrenormalized volume fractions [see (5.191)].

Suppose for illustration we shall assume that all zones to the left of the contact discontinuity have the constant density ρ, and all of them to the right have the constant value ρ_o. Further assume that

$u_k = u_o$ for all zones in the strip (uniform translational motion). Then after a single time step

$$\langle \rho_{k+1/2}^{n+1} \rangle = \rho_o Vf'(2) + \rho Vf'(1),$$

and

$$\rho_{k+1/2}^{n+1}(1) = N_V \rho$$

$$\rho_{k+1/2}^{n+1}(2) = N_V \rho_o . \tag{5.198}$$

But for uniform translation (in this case $A_k = A_{k+1}$) the normalization factor $N_V = 1$. Thus the flow after a time step preserves the densities on either side of the contact discontinuity.

The specific energy is to be transported with mass, using (5.197), in which case the discontinuity in specific energy is preserved under uniform translation (see Problem 5.44)

The transport of mass and energy made specific use of the subzonal information about each composition. The velocity, however, is not associated with individual components of a mixed zone, but are located at the zone boundaries. In particular no velocity is carried (or computed) for the contact discontinuity. Thus the advection of specific momentum is calculated essentially as for clean zones, with the added constraint that it not introduce structure in the velocity field near a contact discontinuity. Returning to the example in Figure 5.33, the density of the momentum zone at the beginning of the time step is given by

$$\rho_k^n = (1/2)(\rho + \rho_o),$$

and the mass carried across the momentum zone faces $A_{k \pm 1/2}$ by mass transport can be approximated by interpolating the masses carried across the zone boundaries:

$$\Delta M_{k+1/2} = \frac{1}{2}(\Delta M_k + \Delta M_{k+1})$$

$$= \frac{1}{2}\Delta V \{\rho Vf'(1) + \rho_o[1 - Vf'(2)]\}$$

$$\Delta M_{k-1/2} = \frac{1}{2}(\Delta M_k + \Delta M_{k-1})$$

$$= \frac{1}{2}\Delta V \left\{\rho\, V f'(1) + \rho \frac{u A \Delta t}{\Delta V}\right\} . \tag{5.199}$$

Using these quantities the change in momentum associated with advection for a zone containing a contact discontinuity may be expressed in the form

$$S_k^{n+1/2}\Delta V = S_k^{n-1/2}\Delta V + u_{k-1/2}\Delta M_{k-1/2} - u_{k+1/2}\Delta M_{k+1/2} . \tag{5.200}$$

The density of the momentum zone at the end of the time step is easily shown to be

$$\rho_k^{n+1} = \frac{1}{2}\left\{\rho(1 + \frac{u_o A \Delta t}{\Delta V}) + \rho_o V f'(2)\right\} . \tag{5.201}$$

Dividing (5.200) by (5.201) gives the new velocity,

$$u_k^{n+1/2} = S_k^{n+1/2} / \rho_k^{n+1} = u_k^{n+1/2} = u_o .$$

The Lagrangian step will also require some modifications for mixed zones. These will be considered below after we have discussed the remaining types of mixed zone transport that can occur in problems involving two components.

Problem 5.43

Generalize (5.189)-(5.192) to the case where the zone boundaries shown in Figure 5.33 are moving with axial grid velocity $u^g{}_k$.

Problem 5.44

Assume that the specific energy for components *1* and *2* in Figure 5.33 are ε and ε_o, respectively. Show that if specific energy is transported with mass, that $\varepsilon^{n+1}(1)=\varepsilon$ and $\varepsilon^{n+1}(2)=\varepsilon_o$ for uniform translation.

For each class of transport we shall first construct the quantity $F_k(i)$, which may be thought of as the volume of component i transported across the zone boundary k. For clean to mixed transport the quantity $F_k(i)$ is evidently given by

$$F_k(1) = u_k \Delta t \, A_k$$
$$F_k(2) = 0 . \tag{5.202}$$

Notice that $F_k(i)$ has the sign of the velocity component at the zone face, and that A_k is the area of the zone normal to the direction of flow.

Mixed to Clean Transport

This case (shown in Figure 5.32c) is more complicated because two components may be transported depending on the ratio $u_k \, A_k \, \Delta t \, / \, \Delta V_d(2)$, where $\Delta V_d(2)$ is the donor cell partial volume of component 2. If the ratio is less than one, only component *2* will be transported and zone $k-1/2$ will remain mixed at the end of the time step. If the ratio exceeds unity, then all of component *2* is transported, followed by a portion of component *1*. In either case we may define

$$F_k(2) = \frac{u_k}{|u_k|} \, min \left\{ |u_k| \, A_k \, \Delta t, \; \Delta V_d(2) \right\}$$

$$F_k(1) = \frac{u_k}{|u_k|} \, max \left\{ 0, \; |u_k| \, A_k \, \Delta t \; - \; \Delta V_d(2) \right\} . \tag{5.203}$$

Here

$$\Delta V_d(i) = \begin{cases} \Delta V_{k+1/2}(i) & u_k < 0 \\ \Delta V_{k-1/2}(i) & u_k > 0 \,. \end{cases}$$

(5.204)

If both materials are transported, the contact discontinuity moves from one zone to the next.

Mixed to Mixed Transport

This case is illustrated by the last example in Figure 5.32. Because both components are common to the two neighboring zones some of each will be transported. The added complication in this instance comes in defining the effective area $A_k(i)$ for each component. One choice, which will be used here, is to linearly interpolate using the volume fractions from each neighboring zone:

$$A_k(i) = \frac{A_k}{2}\left[Vf_{k+1/2}(i) + Vf_{k-1/2}(i)\right] .$$

(5.205a)

By construction the sum of the area factors is

$$A_k = \sum_i A_k(i) .$$

(5.205b)

The quantities $F_k(i)$ may be constructed using the single velocity component u_k for each component of the mixed zone, but using the area factors (5.205a):

$$F_k(i) = u_k \Delta t \, A_k(i) .$$

(5.206)

As for mixed to clean transport, it is possible that the magnitude of

$F_k(i)$ for one of the components could exceed $\Delta V_d(i)$. This is a consequence of our choice for the partial area factors (5.205a). If this occurs then all of the component is to be transported, and the difference can be taken from the remaining component. Another consequence of (5.205a) is that bubbles of one component may occasionally be left behind as the contact discontinuity moves through the grid.

Having constructed the quantities $F_k(i)$, the change in partial volumes for zone $k+1/2$ are given by

$$\Delta V'_{k+1/2}(i) = \Delta V^n_{k+1/2}(i) + F_k(i) - F_{k+1}(i) . \tag{5.207}$$

Notice that since the quantities $F_k(i)$ at each face carry the sign of the velocity, they add correctly in (5.207) for each neighboring zone. Thus the same quantity is used to take partial volume out of one zone and to put it into the next. This guarantees that mass and energy transport are conservative around contact discontinuities. The normalized volume fractions are then defined using (5.191). The unrenormalized partial volumes are to be used in constructing the mass $M_k(i)$ of each composition carried across a zone boundary:

$$M_k(i) = F_k(i)\,\rho_d(i) , \tag{5.208}$$

where the donor cell mass density is defined as in (5.194). Finally, the new masses in a zone containing a contact discontinuity are given by

$$\Delta m^{n+1}_{k+1/2}(i) = \Delta m^n_{k+1/2}(i) + M_k(i) - M_{k+1}(i) . \tag{5.209}$$

The masses (5.208) are used to transport the specific internal energy of each component. The internal energy carried across the zone boundaries are

$$E_{k+1/2}(i) = F_k(i)\,\rho_d(i)\,\varepsilon_d(i) , \tag{5.210}$$

where $\varepsilon_d(i)$ is the donor cell specific energy defined in analogy with (5.194). Then the change in the internal energy is given by

$$E^{n+1}_{k+1/2}(i) = E^{n}_{k+1/2}(i) + E_k(i) - E_{k+1}(i) \, . \tag{5.211}$$

Finally, (5.208) is summed over components, and the result is used to construct the total mass carried across each face of the momenutm zones for use in advecting specific momentum.

Lagrangian Step Near Contact Discontinuities

Because velocities are not associated with components defining a contact discontinuity there are relatively few modifications that need to be made for the Lagrangian step. First, an average pressure is defined in each mixed zone for use in calculating the acceleration. Several methods have been used, such as adjusting the volume fractions so that the component pressures are equal, or by letting each component pressure relax toward an equilibrium value on a sonic time scale. The simplest approach is to define the average pressure as a volume fraction weighted average (the zonal indicies are omitted for simplicity)

$$\langle P \rangle = \sum_i P(i) \, Vf(i) \, , \tag{5.212}$$

where the pressure of an individual components is given by $P(i) = P[\rho(i), \varepsilon(i)]$. This approach has been found to work quite well in practice. The average pressure and the average density (5.188) of a zone containing a contact discontinuity may then be used in the standard finite difference representation of the pressure acceleration.

The pressure acceleration in this scheme does not act directly on the contact discontinuity. They do act indirectly by affecting the rate of mass transfer across zone boundaries, thereby producing motion of the contact discontinuity.

The artificial viscous stress acceleration requires no change for this model of contact discontinuities.

Mechanical heating, on the other hand, requires that the specific energies of each composition be changed. One way to accomplish this is to substitute (5.212) into the expression for the total change in internal energy of the mixed zone:

$$\frac{\partial(\rho\varepsilon\Delta V)}{\partial t} = - P\ \nabla\cdot\mathbf{v}\,\Delta V = - \sum_i P(i)\ Vf(i)\ \nabla\cdot\mathbf{v}\,\Delta V\,.$$

But the total internal energy of the zone is just

$$\rho\varepsilon\,\Delta V = \sum_i \rho(i)\ \varepsilon(i)\ Vf(i)\ \Delta V\,, \tag{5.213}$$

so for each component we have

$$\frac{\partial\,[\rho(i)\,\varepsilon(i)\,]}{\partial t} = - P(i)\ \nabla\cdot\mathbf{v}\,. \tag{5.214}$$

In effect, this scheme apportions the net mechanical work done on the zone among the components, weighted by the fraction of the total internal energy contained by each. The result is particularly simple for a gamma law equation of state, and the result can be represented by finite difference approximations similar to those discussed earlier in this chapter.

Finally we consider the change in internal energy associated with shock propagation. The artifical viscous stress is defined using the velocity difference across the entire zone. If we consider the change arising from the axial component of the artificial viscous stress

$$\frac{\partial(\rho\varepsilon\Delta V)}{\partial t} = - Q_{zz}\frac{\partial u}{\partial z}\Delta V\,.$$

Recalling the observation following (5.214) we may apportion the increase in internal energy from shock heat as follows:

$$\frac{\partial[\rho(i)\,\varepsilon(i)]}{\partial t} = -\frac{\rho(i)\,\varepsilon(i)}{\sum_i \rho(i)\,\varepsilon(i)\,Vf(i)}\left(Q_{zz}\frac{\partial u}{\partial z}\right). \tag{5.215}$$

A similar expression may be constructed for the viscous heating using the radial component of the artificial viscous stress.

Curvilinear Coordinates

The method described above may be applied to contact discontinuities in curvilinear coordinate systems with minor modifications. In the latter case mass transport is modelled as a remap, using remap volumes defined as in Section 5.10. The volume remapped across a zone boundary is then used in place of the advection volumes (such as $u_k A_k \Delta t$) to construct the quantities $F_k(i)$ for clean to mixed and mixed to clean transport. For mixed to mixed transport (5.205a) may be replaced by

$$F_k(i) = \frac{A_k(i)}{A_k}\Delta V_k^R , \tag{5.216}$$

where $\Delta V^R{}_k$ is the remap volume at the zone boundary k.

5.15 A Note on Three Dimensional Methods

The difference between a one dimensional Lagrangian and a one dimensional Eulerian formulation of hydrodyanamics is associated primarily with advection, or remap terms, and constitutes a major increase in complexity. One dimensional and two dimensional Eulerian methods differ primarily because of cross terms in the equations. The extension of two dimensional Eulerian representations to three dimensions, however, is relatively straightforward, since the only new terms which arise are those associated with an additional spatial coordinate, and these may be modeled in standard ways. In any case, the two dimensional finite difference representations discussed above are readily extended to hydrodynamics in three spatial dimensions. Among the principle limitations encountered in three dimensions are those associated with a need for efficiency (typically a three dimensional problem may require of the order of a million spatial zones), and obvious limitations as to the number of full grid size arrays which are to be carried throughout a computation (simple hydrodynamics without gravitation requires a minimum of six such arrays, requiring storage of six million words on a *100x100x100* zone grid).

Smooth particle hydrodynamics is another approach to three dimensional hydrodynamics which has been used widely in recent years. The smooth particle method has the advantage that it does not require an underlying computational grid, and can therefore model complex fluid flow without requiring zones in regions that may be empty during some or all of a calculation. The method is ideal for problems which include large void regions, such as occur in modelling planetary or stellar collisions, or in modelling mass exchange in binary systems. Extensive discussions of the smooth particle hydrodynamics methods and applications can be found in the literature.

Problem 5.45

(a) Specify a set of inflow boundary conditions for a one dimensional Eulerian grid assuming a resevoir containing a gamma law gas whose density and entropy remain finsed at the boundary.

(b) Express the inflow boundary condition of part (a) in finite difference form.

Problem 5.46

Consider a two dimensional grid with a closed left boundary and an open (outflow) right boundary. Discuss what happens within the grid as the mass flows out of the grid. How would the results change if the grid size were doubled in each direction?

Problem 5.47

The rate of change of some variables appears in the Eulerian form

$$\frac{\partial F}{\partial t} + \mathbf{v}\cdot\nabla F = A$$

(material stresses are an example). The operator split form of this equation

$$\frac{\partial F}{\partial t} + \mathbf{v}\cdot\nabla F = 0$$

is often called a remap equation.

(a) Consider the axial remap

$$\frac{\partial F}{\partial t} + u\frac{\partial F}{\partial z} = 0$$

for a zone edge centered variable ($F \rightarrow F_k$) and construct a first order finite difference representation of it. Discuss the stability of the method, and any physical properties that the partial differential equation possesses which should be preserved by the finite difference solution.

(b) Construct a finite difference representation of the axial remap equation above that is second order accurate in spatial derivatives by finding an interpolated value of the gradient located a distance $u_k \Delta t$ behind the zone edge. Show that the result can be written in the form

$$\frac{F'_k - F_k}{\Delta t} = -u_k\left(\frac{\partial F}{\partial x}\right)_B$$

where the prime denotes the new time value of F. The gradient is defined by

$$\left(\frac{\partial F}{\partial x}\right)_B = D_B + \frac{u_k}{|u_k|}\alpha_k\left(D_{k+1/2} - D_{k-1/2}\right)$$

$$D_{k+1/2} = \frac{F_{k+1} - F_k}{\Delta x_{k+1/2}},$$

$$\alpha_k = \frac{\Delta x_B - 2\,|u_k|\Delta t}{2\,\Delta x_k},$$

and the index B, which is $k-1/2$ $(k+1/2)$ if $u_k>0$ $(u_k<0)$, denotes the upstream value.

Problem 5.48

Consider the remap (see Problem 5.47) of a zone centered variable $A_{k+1/2}$, and construct a second order approximation to the equation

$$\frac{A'_{k+1/2} - A_{k-1/2}}{\Delta t} = -\,u_{k+1/2}\left(\frac{\partial A}{\partial x}\right)_B$$

where the zone centered velocity is the average

$$u_{k+1/2} = (1/2)(u_{k+1} + u_k)\,.$$

Show that if the gradient is interpolated a distance $u_{k+1/2}\Delta t$ behind (upstream from) the zone center that the gradient can be written in the form

$$\left(\frac{\partial A}{\partial x}\right)_B = D_B + \beta_{k+1/2}(D_{k+1} - D_k)$$

$$\beta_{k+1/2} = \frac{\Delta x_B - 2\,|u_{k+1/2}|\Delta t}{\Delta x_k + \Delta x_{k+1}},$$

where the index B is k $(k+1)$ if $u_k>0$ $(u_k<0)$.

Problem 5.49

Modify the planar artificial viscous stress equations (5.72)-(5.73) in spherical coordinates to eliminate artificial viscous heating during homologous collapse.

Table 5.1

Two dimensional, inviscid hydrodynamic equations expressed in conservative form in cylindrical (r,z) and spherical (r,θ) coordinates. The velocity components are physical components.

CYLINDRICAL COORDINATES

Momentum Equation

$$\frac{\partial \rho v_r}{\partial t} + \frac{1}{r}\frac{\partial \rho r v_r^2}{\partial r} + \frac{\partial \rho v_r v_z}{\partial z} - \frac{\rho v_\phi^2}{r} = -\frac{\partial P}{\partial r}$$

$$\frac{\partial \rho v_\phi}{\partial t} + \frac{1}{r}\frac{\partial \rho r v_r v_\phi}{\partial r} + \frac{\partial \rho v_z v_\phi}{\partial z} + \frac{\rho v_r v_\phi}{r} = 0$$

$$\frac{\partial \rho v_z}{\partial t} + \frac{1}{r}\frac{\partial \rho r v_r v_z}{\partial r} + \frac{\partial \rho v_z^2}{\partial z} = -\frac{\partial P}{\partial z}$$

Mass Continuity Equation

$$\frac{\partial \rho}{\partial t} + \frac{1}{r}\frac{\partial \rho r v_r}{\partial r} + \frac{\partial \rho v_z}{\partial z} = 0$$

Internal Energy Equation

$$\frac{\partial \rho \varepsilon}{\partial t} + \frac{1}{r}\frac{\partial \rho \varepsilon r v_r}{\partial r} + \frac{\partial \rho \varepsilon v_z}{\partial z} = -P\left(\frac{1}{r}\frac{\partial r v_r}{\partial r} + \frac{\partial v_z}{\partial z}\right)$$

SPHERICAL COORDINATES

Momentum Equation

$$\frac{\partial \rho v_r}{\partial t} + \frac{1}{r^2}\frac{\partial (r^2 \rho v_r^2)}{\partial r} + \frac{1}{r \sin\theta}\frac{\partial (\rho v_r v_\theta \sin\theta)}{\partial \theta} - \frac{\rho (v_\phi^2 + v_\theta^2)}{r} = -\frac{\partial P}{\partial r}$$

$$\frac{\partial \rho v_\theta}{\partial t} + \frac{1}{r^2}\frac{\partial (r^2 \rho v_r v_\theta)}{\partial r} + \frac{1}{r \sin\theta}\frac{\partial (\rho v_\theta^2 \sin\theta)}{\partial \theta} + \frac{\rho v_r v_\theta}{r} - \frac{\rho v_\phi^2 \cot\theta}{r} = -\frac{1}{r}\frac{\partial P}{\partial \theta}$$

$$\frac{\partial \rho v_\phi}{\partial t} + \frac{1}{r^2}\frac{\partial (r^2 \rho v_r v_\phi)}{\partial r} + \frac{1}{r \sin\theta}\frac{\partial (\rho v_\theta v_\phi \sin\theta)}{\partial \theta} + \frac{\rho v_r v_\phi}{r} + \frac{\rho v_\theta v_\phi \cot\theta}{r} = 0$$

Mass Continuity Equation

$$\frac{\partial \rho}{\partial t} + \frac{1}{r^2}\frac{\partial (r^2 \rho v_r)}{\partial r} + \frac{1}{r \sin\theta}\frac{\partial (\rho v_\theta \sin\theta)}{\partial \theta} = 0$$

Internal Energy Equation

$$\frac{\partial \rho\varepsilon}{\partial t} + \frac{1}{r^2}\frac{\partial (r^2 \rho\varepsilon v_r)}{\partial r} + \frac{1}{r \sin\theta}\frac{\partial (\rho\varepsilon v_\theta \sin\theta)}{\partial \theta} = -P\left(\frac{1}{r^2}\frac{\partial (r^2 v_r)}{\partial r} + \frac{1}{r \sin\theta}\frac{\partial (v_\theta \sin\theta)}{\partial \theta}\right)$$

References

L. D. Landau and E. M. *Lifshitz, Fluid Mechanics* (Oxford: Pergamon Press, 1959).

M. L. Norman, J. R. Wilson and R. Barton, *Ap. J.* **239** (1980), pg. 968.

R. D. Richtmyer and K. W. Morton, *Difference Methods for Initial-Value Problems,* Second Edition (New York: Interscience, 1967).

I. S. Sokolnikoff, Tensor Analysis, (Wiley and Sons: New York, 1964).

B. Van Leer, *J. Comp. Phys.* **361** (1980), pg. 361.

RADIATIVE TRANSPORT

The spatial transport of radiation (either in the form of photons or neutrinos) and its interaction with matter plays an important and sometimes dominant role in many astrophysical systems. Matter absorbs, emits and scatters radiation. These processes are accompanied by an exchange of energy and momentum and therefore determine how radiation moves through matter, as well as the extent to which it can affect material motion. In this chapter we will consider the spatial transport of radiation, its emission, absorption, and elastic scattering, but we will ignore material motion. Inelastic scattering, thermal diffusion and radiation forces acting on matter as well as velocity dependent effects will be considered in the next chapter. The equation of radiative transfer and the multigroup diffusion limit will be considered for one dimensional spherically symmetric systems. The equilibrium diffusion limit will be developed for two dimensional axially symmetric systems.

Photons and Neutrinos

Many aspects of the transport of photons and of neutrinos may be modeled by similar equations. Consequently, we shall use both photons and neutrinos as examples in developing the basic finite difference algorithms for radiation transport. Specifically, the transfer equation and the equilibrium diffusion limit will be developed for photons, while the multigroup diffusion limit will be developed for neutrinos. With some modifications the algorithms may be applied to either type of radiation. The primary differences between the two types of radiation are associated with their quantum statistics (photons are bosons, while neutrinos are fermions), and in the type and character of their coupling to matter. A discussion of quantum statistics can be found in standard texts on statistical mechanics (for example, Landau and Lifshitz 1958).

Photons, which are described by Bose-Einstein statistics, can undergo stimulated emission (associated with lasing) and their spectra may exhibit sharp line structure (associated with bound-bound atomic transitions), and absorption edges. Stimulated emission is built into the transport equations, and its occurrence must be incorporated in the development of finite difference approximations. Another important feature associated with bosons is the possible occurrence of a condensed ground state at low temperatures. Most of the other quantum effects enter through the opacities. Since these quantities are similar to an equation of state, we only need their values for a given density and temperature of the material. The details of how these quantities are constructed usually does not affect the finite difference methods in which they are used.

Electrons and neutrinos are examples of particles described by Fermi-Dirac statistics. At low temperatures and high density, fermions become degenerate, and their scattering, absorption or emission by matter may be blocked or inhibited. The presence of blocking factors appear in the transport equations for these particles, as well as in expressions for the opacities. The blocking factors which appear explicitly in the transport equations must be retained, and special care may be needed in representing them in finite difference form. The appearance of blocking factors in the opacities usually does not complicate the finite difference approximations, but may affect the solution methods.

Electron-Ion Collisions and Thermalization

Inelastic collisions between particles (atoms, ions, electrons and photons) in matter are accompanied by exchange of energy between the particles. When particles of essentially the same mass collide, a substantial fraction of the kinetic energy of the faster particle may be transferred to the slower one. When the two masses are very different, relatively little kinetic energy will be exchanged per collision. In general, the rate of energy exchange depends on the nature of the two types of colliding particles (see, for example, Spitzer 1962; Braginskii 1965). For example, the collision time scale for a plasma of electrons and protons can be shown to be roughly

$$\tau_{ee} \approx 5.8 \times 10^{-20} T_e^{3/2} / \rho \quad sec$$

for electron-electon collisions,

$$\tau_{ii} \approx 3.5 \times 10^{-18} T_i^{3/2} / \rho \quad sec$$

for ion-ion collisions, and

$$\tau_{ei} \approx 9.9 \times 10^{-17} T_e^{3/2} / \rho \quad sec$$

for electron-ion collisions, where the temperature is in eV (1.16×10^4 $K = 1\ eV$) and the density is in $g\ cm^{-3}$. In the solar interior the collision times are of order 4×10^{-17} sec for electrons, and 10^{-13} sec for electrons and ions.

Collisions between electrons and photons are also an important source of energy exchange. In a hot, low density hydrogen plasma the dominant mode of energy exchange is Compton scattering. As the temperature of the plasma decreases (or the atomic number of the ions increases) the dominant mode of energy exchange is Bremsstrahlung, followed by bound-free transitions and finally bound-bound transitions. The time scale for these processes can be denoted by $\tau_{e\gamma}$.

The most important radiative processes under typical astrophysical conditions which lead to thermalization are electron-electron, ion-ion, electron-ion and electron-photon collisions, and the characteristic time for these processes usually satisfy the following relations:

$$\tau_{ee} \leq \tau_{ii} << \tau_{ei} \leq \tau_{e\gamma}$$

where the pair of subscripts indicates the colliding particles. If the collisions persist for a time which is long compared with each characteristic time, the material will reach a state of complete thermodynamic equilibrium characterized by a single temperature. An underlying assumption of fluid dynamics (and of the theory of radiation transport which we are considering here) is that τ_{ee} and τ_{ii} are much smaller than the smallest computational times (such as the hydrodynamic time τ_{hyd}) of interest in any problem. This guarantees that the material can be characterized by constitutive relations such as equations of state and opacities, and in particular that the electrons and the ions can each be characterized by a temperature (each species has a Maxwellian, Fermi-Dirac or Bose-Einstein velocity distribution as appropriate, characterized by a parameter T_e or T_i respectively). If $\tau_{ei} > \tau_{hyd}$, then the temperature of the electrons, T_e, and the temperature of the ions, T_i, may be different on time scales comparable to τ_{ei}. When $\tau_{ei} << \tau_{hyd}$, the electron and ion velocity distributions will be characterized by a single parameter which we call simply the material temperature T.

Matter which is not in statistical equilibrium with itself is often said to be a nonLTE system. Shock propagation in matter results in heating of the ions, and can lead to situations where $T_i > T_e$. Diffusion of electric currents in a plasma result in Joule heating of the electrons, and can lead to situations where $T_e > T_i$. Another example results when the collision rates in a plasma are less than the radiation excitation-deexcitation rates, producing nonequilibrium electron occupation levels of the ions. Precise definitions of these terms vary among authors; a more detailed discussion may be found in any standard text on radiation transport (for example, Cox and Guili 1968).

In optically thick matter the radiation spectrum is characterized by the temperature $T_R \approx T$. In optically thin matter, the radiation field may, for many purposes, be characterized by its energy density, through which we may define an effective radiation temperature T_R. In this case the effective radiation temperature T_R will in general differ from the material temperature T.

Radiation and matter in stellar interiors may, to a high degree of accuracy, be characterized by a single temperature, T (a discussion of thermodynamic equilibrium may be found in Cox and Guili, 1968). In some portions of the interstellar medium $T_e \approx T_i$, but they both T_e

and T_i differ from T_R. In HII regions surrounding hot young stars, the electron and ion temperatures are unequal. Unless otherwise stated in this chapter, it will be assumed that $T_e \approx T_i \equiv T$, but that T_R need not equal T.

It is often convenient to describe matter, or matter and radiation, as being in local thermodynamic equilibrium (LTE). A definition of weak LTE assumes that the emission and absorption properties of matter are describable in terms of the material temperature (ionization states are Boltzmannian, as described by the Saha equation, for example). A strong form of LTE further admits a macroscopically nonuniform radiation field (and thus a net radiation flux), but assumes that the radiation field is locally describable by a Planckian distribution characterized by the (radiation) temperature T_R. This may arise in an optically thin material where there is a temperature gradient. Suppose that the material temperature at a point is T; the radiation at that point will have originated elsewhere where the material temperature is T', and because the material is optically thin its spectrum will be characteristic of its point of origin. This may occur, for example, outside of the neutrinosphere during stellar core collapse.

Approximate Models of Radiative Transport

In the next section the classical theory describing radiation transport in a material medium will be summarized. The basic equation in this approach, the radiative transfer equation, describes the propagation of the radiation field. The latter is characterized by its angular dependence and its spectral distribution. This is a complete classical theory, and ideally should be used to describe all classical radiative processes. Unfortunately, this theory is quite complicated, and its implementation in all but one dimensional problems (where four independent variables are needed—the distance, two angle variables and the radiation energy) may be impractical. A representation in one spatial dimension will be considered in Section 6.3. Several levels of approximation to the classical theory of radiative transfer will also be considered which apply in restricted regimes. In all cases it will be assumed that the material is in thermodynamic equilibrium with itself and is characterized by the temperature T.

The first approximation applies when the radiation mean free path is many orders of magnitude smaller than the other physical length scales characterizing the system. It is then reasonable to approximate the full transport theory by an angle-averaged model which retains the spectral resolution of the field. This approach, which is known as multigroup diffusion and is considered in Section 6.4, is not correct in transparent regimes, but does retain a possible non-Planckian feature of the field, and allows the spectrum to evolve dynamically in time. Multigroup diffusion requires, in addition to the independent spatial variables, an energy variable describing the spectrum. This approach is practical in one dimensional problems, and can also be used in some two dimensional cases.

The next level of approximation which will be considered in Section 6.5 arises when the distribution of energy in the radiation spectrum can be ignored. The multigroup equations may then be averaged over radiation spectrum to produce a total radiation energy which is parameterized by the effective radiation temperature T_R. This model describes the diffusion of radiation energy, and its coupling to matter, and can be implemented in multidimensional calculations.

The final approximation which will be considered in Section 6.6 applies when radiation and matter are so strongly coupled that they are characterized by a single temperature (sometimes called radiation conduction).

Operator Splitting

In this chapter it is assumed that all radiation processes can be solved separately from the hydrodynamics using the method of operator splitting. Radiation acceleration will occur, but for many systems it can be included as a part of the hydrodynamics cycle. Thus, all processes that directly affect the motion of the medium will be ignored for now, and the density will be constant in time (spatial variations will, however, be allowed). The only temporal changes in material parameters that will be considered here are changes in thermal energy, and in the case of neutrino transport, changes in material composition which result from neutrino emission and absorption. The finite difference methods developed below may be implemented following the completion of the hydrodynamic computational cycle. This approach is consistent with the philosophy of operator splitting that was introduced in Chapter 2.

6.1 RADIATIVE TRANSFER AND THE DIFFUSION LIMIT

In the absence of interactions with matter, radiation travels in a straight line changing its flux only because of geometric effects. Matter will affect the propagation of radiation in three general ways: emission, absorption and scattering. These processes depend on the nature of the material, and in particular on its thermodynamic state. In the discussions below, it will be assumed that the material is in LTE, having a well defined temperature T which may vary from point to point over distances which are large compared with atomic collisional mean free paths. The radiation interaction processes above will be assumed to depend only on the local material properties. In particular, the effective absorption coefficient, or opacity, κ_a, and effective scattering opacity, κ_s, are functions of frequency or energy as well as of the material temperature, density and composition. We shall assume that these quantities are expressed as functions of the photon frequency or energy:

$$\kappa_a = \kappa_a(\rho,T,\varepsilon) \qquad cm^2 g^{-1}, \tag{6.1a}$$

$$\kappa_s = \kappa_s(\rho,T,\varepsilon) \qquad cm^2 g^{-1}. \tag{6.1b}$$

The opacities may be expressed in terms of single particle cross sections, such as the scattering cross section σ_s. The opacity for a specific process is given by the product of the cross section and the number per unit mass of scatterers or absorbers. The dependence on composition, which may be quite complicated, has not been shown explicitly. In general these quantities may be composition weighted averages of atomic or particle cross sections, with the weighting factors obtained, for example, from the Saha equation.

The partial differential equation of radiation transfer will be reviewed first, and then the diffusion limit will be obtained from it. Space-time centering and finite difference representations will then be developed.

The Radiative Transfer Equation

An arbitrary radiation field whose spectral energy ε and frequency ν are related by $\varepsilon = h\nu$ can be described by a quantity called the specific intensity. For an arbitrary radiation field consisting of photons or neutrinos whose energies lie between ε and $\varepsilon + d\varepsilon$ travelling at speed c, the energy per unit energy interval crossing a unit area in unit time in the direction Ω per unit solid angle is defined in terms of the specific angular intensity

$$I(\mathbf{r}, t, \Omega, \varepsilon) \qquad cm^{-2} s^{-1} sr^{-1} \tag{6.2}$$

(see Figure 6.1). The definition (6.2) differs slightly from the notation used in many texts (where the specific intensity is often denoted by $I_\nu(\mathbf{r}, t, \Omega)$ $erg\ cm^{-2}\ s^{-1}\ Hz^{-1}\ sr^{-1}$). The radiation energy density associated with photons or neutrinos in the energy range ε to $\varepsilon + d\varepsilon$ is defined by integrating the specific intensity over all solid angles

$$E(\mathbf{r}, t, \varepsilon) = \frac{1}{c}\int I(\mathbf{r}, t, \Omega, \varepsilon)\, d\omega \qquad cm^{-3}. \tag{6.3}$$

The element of solid angle, $d\omega$, in spherical coordinates is

$$d\omega = \sin\theta\, d\theta\, d\phi \equiv -d\mu\, d\phi\,. \tag{6.4}$$

In general $d\omega$ will denote the scalar solid angle interval, while Ω is reserved for the unit vector associated with the direction of travel of the photons. The radiant energy flux in the energy range ε to $\varepsilon + d\varepsilon$

is given by the vector quantity

$$\boldsymbol{F}(\boldsymbol{r}, t, \varepsilon) = \int I(\boldsymbol{r}, t, \Omega, \varepsilon)\, \Omega \, d\omega \quad cm^{-2} s^{-1}. \tag{6.5}$$

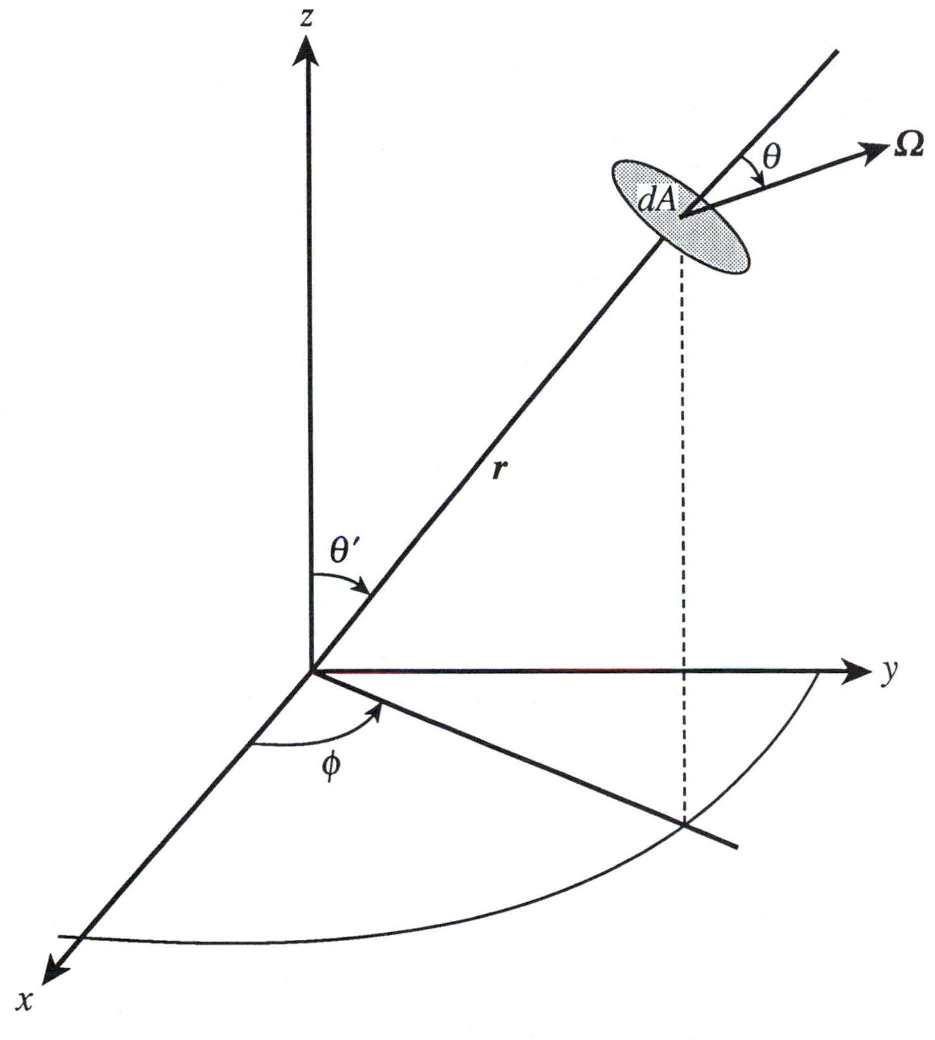

Figure 6.1. Angular geometry defining the specific intensity passing through the area element dA in spherical coordinates. The photons travel in the direction Ω at the point $\boldsymbol{r}$.

The rate of change of the specific angular intensity along a path length ds may be written in the differential form

$$\frac{dI}{dt} = \frac{\partial I}{\partial t} + c\,\Omega \cdot \nabla I = \left(\frac{dI}{dt}\right)_{ea} + \left(\frac{dI}{dt}\right)_{s} . \tag{6.6}$$

The two terms on the right side of (6.6) represent the change in I due to emission and absorption by matter, and to scattering (at this stage we include elastic and inelastic processes).

An extensive discussion of the transfer equation may be found, for example, in Mihalas and Mihalas (1984). The discussion of (6.6) which follows is strictly valid only for isotropic matter which is at rest. The finite difference equations which will be developed from it have, nonetheless, been found to be adequate for many astrophysical applications. Furthermore, they serve as a reasonable starting point for a more nearly rigorous approach to problems which are sensitive to the effects of material motion on the interaction of radiation and matter.

Since the left hand side of (6.6) represents the time rate of change of I along the path ds connecting points P and Q as shown in Figure 6.2, it may be written in the form

$$\frac{dI}{dt} = \lim_{\Delta t \to 0} \frac{I(s+\Delta s, t+\Delta t) - I(s,t)}{\Delta t}$$

$$= \lim_{\Delta t \to 0} \left\{ \frac{I(s+\Delta s, t+\Delta t) - I(s+\Delta s, t)}{\Delta t} + \frac{\Delta s}{\Delta t} \frac{I(s+\Delta s, t) - I(s, t)}{\Delta s} \right\}$$

$$= \frac{\partial I}{\partial t} + c \frac{\partial I}{\partial s} . \tag{6.7}$$

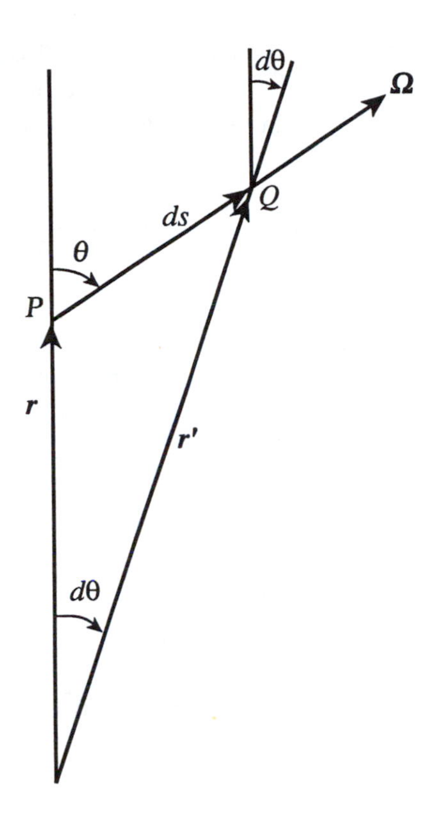

Figure 6.2. Change in angular dependence θ of the specific intensity of a ray travelling nonradially from the point P to the point Q.

The term $I(s+\Delta s, t)/\Delta t$ has been added and subtracted in the second line.

Because of the ubiquity of spherically symmetric systems in astrophysics we will consider the implementation of (6.7) for that case. The equations for planar symmetry may be obtained from the limit $r \to \infty$. For spherically symmetric systems in spherical coordinates $I = I(r,\theta, \varepsilon)$, where θ is the angle between the direction of the photon, Ω, and the radius vector, the path length may be readily found. First write the derivative along the path ds as

$$\frac{\partial I}{\partial s} = \frac{dr}{ds}\frac{\partial I}{\partial r} + \frac{d\theta}{ds}\frac{\partial I}{\partial \theta}.$$

From Figure 6.2 it is easily verified that $dr = \cos\theta\, ds$, and $r\, d\theta = -\sin\theta\, ds$ (the minus sign arises because increasing the path length corresponds to decreasing θ in Figure 6.2), so that

$$\frac{\partial I}{\partial s} = \cos\theta \frac{\partial I}{\partial r} - \frac{\sin\theta}{r}\frac{\partial I}{\partial \theta}.$$

Finally defining the angle variable μ by

$$\mu = \cos\theta \tag{6.8}$$

we obtain

$$\frac{\partial I}{\partial s} = \mu \frac{\partial I}{\partial r} - \frac{(1-\mu^2)}{r}\frac{\partial I}{\partial \theta}$$

$$= \frac{\mu}{r^2}\frac{\partial}{\partial r}(r^2 I) + \frac{1}{r}\frac{\partial}{\partial \mu}[(1-\mu^2)I]. \tag{6.9}$$

The terms on the right side has been written in conservative form, which is a convenient starting point from which to construct finite difference approximations. Combining (6.7)-(6.9) leads to the radiative transfer equation for one dimensional spherical systems:

$$\frac{\partial I}{\partial t} = -\frac{\mu c}{r^2}\frac{\partial}{\partial r}(r^2 I) - \frac{c}{r}\frac{\partial}{\partial \mu}[(1-\mu^2)I]$$

$$+ \left(\frac{dI}{dt}\right)_{ea} + \left(\frac{dI}{dt}\right)_{s}, \tag{6.10}$$

where $I = I(r,t,\mu,\varepsilon)$ (the dependence on r, t, and ε will generally not be exhibited explicitly hereafter).

Problem 6.1

A nearly isotropic radiation field is described by the relation

$$I(\Omega) = I(\mu) = I_o + \mu I_1 \ , \tag{6.11}$$

where μ is given by (6.8), I_0 and I_1 are constants and $I_1 << I_0$. What is the radiation energy density E and the flux $\boldsymbol{F}$? Evaluate the radiation pressure tensor

$$\boldsymbol{P} = \frac{1}{c}\int \Omega\, \Omega\, I(\Omega)\, d\omega \tag{6.12}$$

and show that each component $P_{rad} \equiv (1/3)\ tr\ (\boldsymbol{P})$ satisfies the relation

$$P_{rad} = (1/3)\, E \ . \tag{6.13}$$

The volume element in spherical coordinates, integrated over $d\phi$ is

$$dV = \int d\phi\, r^2 \sin\theta\, d\theta = 2\pi r^2 \sin\theta\, d\theta = 2\pi r^2\, dr\, d\mu \tag{6.14}$$

where $d\mu = -\sin\theta\, d\theta$, and $-1 \leq \mu \leq 1$. Integrating the first term on the right side of (6.10) over dV gives

$$\int \frac{\partial I}{\partial t} dV = -\int \frac{\mu c}{r^2} \frac{\partial}{\partial r}(r^2 I)\, 2\pi r^2\, dr\, d\mu$$

$$= -2\pi c \int_{-1}^{1} \mu\, d\mu\, (r^2 I)\, \Big|_r^{r+\Delta r} , \tag{6.15}$$

where $r + \Delta r$ and r represent the upper and lower radii of ΔV, respectively. Integrating the second term over dV gives

$$-c\int \frac{1}{r}\frac{\partial}{\partial\mu}[(1-\mu^2)I]\ 2\pi r^2\,dr\,d\mu = -2\pi c\int r\,dr\int_{-1}^{1}\frac{\partial}{\partial\mu}[(1-\mu^2)I\ d\mu$$

$$= -2\pi c\int r\,dr\,\left[(1-\mu^2)I\,(\mu)\right]_{-1}^{1} = 0\,.$$

(6.16)

The angular derivative simply leads to a redistribution of radiation among angles, and does not contribute to the net change in radiation energy in a unit volume. Note that at the upper and lower limits of integration the integrand in (6.15) reduces to the total radiation energy $I\cos\theta$ crossing an area $4\pi\, r_k^2$ per unit time.

Problem 6.2

Evaluate (6.15) for the nearly isotropic radiation field which is given by (6.11).

Emission and Absorption (Photons)

The first term on the right side of (6.6) represents the change in I due to local emission and absorption of radiation. If the material is in local thermodynamic equilibrium at temperature T, then it will emit photons at a rate

$$j(\rho,T,\varepsilon) = \rho\,\kappa(\rho,T,\varepsilon)\,B(T,\varepsilon) \qquad cm^{-3}\,s^{-1}$$

(6.17)

where κ is the absorption coefficient (6.1), and $B(T,\varepsilon)$ is the Planck function which can be defined by

$$B(T, \varepsilon) = \frac{2}{h^3 c^2} \frac{\varepsilon^3}{e^{\varepsilon/kT} - 1} \quad cm^{-3} s^{-1}. \tag{6.18a}$$

The integral of the Planck function can be shown (Cox and Guili, 1968) to satisfy

$$\pi \int_0^\infty B(T,\varepsilon)\, d\varepsilon = \sigma T^4, \tag{6.18b}$$

where σ is the Stefan-Boltzmann constant. The relation (6.17) is known as Kirchoff's law, and can be taken as a definition of LTE for matter. In some situations the material may not be in local thermodynamic equilibrium and (6.17) will not hold. Such situations occur, for example, in stellar atmospheres; these complications will not be considered further here (see for example, Mihalas, 1978).

Matter will absorb photons, reducing the radiation intensity by an amount $-\kappa(\rho, T, \varepsilon)\, \rho\, I(\Omega)$. Combining this with (6.17) yields

$$\left(\frac{dI}{dt}\right)_{ea} = \rho\, \kappa(\rho, T, \varepsilon) \left[B(T, \varepsilon) - I(\Omega, \varepsilon) \right]. \tag{6.19}$$

If the radiation field I is in thermodynamic equilibrium with matter, then $(dI/\, dt)_{ea} = 0$, and we let the equilibrium intensity be given by

$$I(\Omega, \varepsilon)_E = B(T, \varepsilon). \tag{6.20}$$

In some applications it is convenient to assume that the radiation field is Planckian but not black body in the sense that it is dilute. In this case

$$I(\Omega, \varepsilon)_{TE} = W \ B(T_R, \varepsilon) \tag{6.21}$$

where the dilution factor W is positive, but less than unity. In general T_R, which may be called a radiation temperature, does not equal the material temperature T. A familiar example of this is sunlight, which is approximately Planckian in spectrum with a temperature of about $6000\ K$, in a room where the air temperature is about $300\ K$.

Scattering of Photons

The second term on the right side of (6.6) describes the change in I due to scattering of radiation in matter. Define the cross section for scattering of a quantum of radiation of energy ε, travelling initially in the direction Ω', into a state with energy ε' travelling in the direction Ω by $\sigma_{\varepsilon\varepsilon'}(\Omega, \Omega')$; then the radiation quanta scattered out of the beam will be proportional to the product of the intensity $I(\Omega, \varepsilon)$, and the scattering cross section $\sigma_{\varepsilon\varepsilon'}(\Omega, \Omega')$ for scattering from (ε, Ω) into (ε', Ω'). The rate of change of $I(\Omega,\varepsilon)$ also depends on the number of quanta already having energy ε' and travelling in the direction Ω'. This is a quantum mechanical effect, which results in a factor proportional to $(1+ c^2 I(\Omega', \varepsilon) / 2h\nu^3)$ thus leading to enhanced scattering into quantum states which are already occupied by photons (see for example, Pomraning, 1973). Photons scatter preferentially into states already occupied by photons because photons are bosons; as we shall see later the plus sign changes to a minus sign for fermions. Let us define the scattering factor as the product of the scattering cross section and the number density of scatterers n_s

$$S_{\varepsilon\varepsilon'}(\Omega, \Omega') \equiv n_s \, \sigma_{\varepsilon\varepsilon'}(\Omega, \Omega') \, .$$

Then combining the factors described above and integrating over final quantum states gives

$$\left(\frac{dI}{dt}\right)_{out} = -\int d\varepsilon' \int d\omega' S_{\varepsilon\varepsilon'}(\Omega, \Omega')\, I(\Omega, \varepsilon) \left[1 + \frac{c^2 I(\Omega', \varepsilon') h^3}{2\, \varepsilon'^3}\right].$$

(6.22)

The stimulated emission factor (second term in square brackets above) makes this expression nonlinear in the specific intensity.

The radiation energy scattered into the beam $I(\Omega, \varepsilon)$ is proportional to: $I(\Omega', \varepsilon')$; the total cross section $n_s \sigma_{\varepsilon\varepsilon'}(\Omega, \Omega')$; the factor ε'/ε ; and the quantum mechanical factor discussed above. Combining these terms leads to an expression for the rate of change of the intensity due to scattering into the beam:

$$\left(\frac{dI}{dt}\right)_{in} = \int d\varepsilon' \int d\omega' S_{\varepsilon\varepsilon'}(\Omega, \Omega')\, I(\Omega', \varepsilon') \frac{\varepsilon}{\varepsilon'} \left[1 + \frac{c^2 I(\Omega, \varepsilon) h^3}{2\, \varepsilon^3}\right].$$

(6.23)

Problem 6.3

The number density of radiation quanta of energy $\varepsilon = h\nu$ travelling in the direction Ω is given by

$$n(\Omega, \varepsilon) = I(\Omega, \varepsilon) / \varepsilon .$$

(6.24)

Use this relation to establish the factor ε'/ε in (6.23).

The change in I due to scattering is given by the sum of (6.22) and (6.23). Elastic (or coherent) scattering results when $\varepsilon = \varepsilon'$, and is characterized by the effective cross section

$$S_{\varepsilon\varepsilon'}(\Omega, \Omega') = n_s \sigma(\Omega, \Omega', \varepsilon)\ \delta(\varepsilon - \varepsilon') .$$

(6.25)

An elastic and isotropic scattering source is characterized by the relation

$$\sigma(\Omega, \Omega', \varepsilon) = \sigma(\varepsilon)\ \delta(\Omega - \Omega'), \tag{6.26}$$

where n_s is the number density of scatterers in the material.

Problem 6.4

Low energy conservative Compton scattering (also called Thomson scattering) may be described by the cross section

$$\sigma_{\varepsilon\varepsilon'}(\Omega, \Omega') = (3/16\pi)\, \sigma(\varepsilon)\ \delta(\varepsilon - \varepsilon')\, [1 + (\Omega{\cdot}\Omega')^2], \tag{6.27}$$

which assumes that $kT << m_e c^2$, and $I_\nu << h\nu^3/c^2$. Evaluate $(dI\,/\,dt)_s$ assuming that the effective cross section is given by (6.27).

Multigroup Diffusion

The radiative transfer equation represents a general description of an arbitrary radiation field (that is one which need not be in thermodynamic equlibirium). For many problems a detailed treatment of transport in angle is not needed, and a description based on the spectral energy density and flux (6.3) and (6.5) are sufficient. Furthermore, in stellar interiors, for example, the radiation field is nearly isotropic and the radiation absorption mean free path λ_a, defined by

$$\lambda_a = [\kappa_a(\rho, T, \varepsilon)\, \rho]^{-1}, \tag{6.28}$$

(where κ_a is the absorption coefficient) satisfies the relation $\lambda_a << l$. Here l is a typical scale length over which the radiation field changes significantly. For matter in thermodynamic equilibrium this may be of order $l^{-1} = d\,(ln\ T\ )\ /\ dr$. In this case radiation transport occurs primarily by diffusion down the material temperature gradient. To illustrate these points, consider the transfer equation for photons in the temperature regime $kT << m_e c^2$ where elastic scattering is described by (6.27). Then (omitting the energy argument)

$$\frac{dI}{dt} = \kappa_a \rho \left[B\ (T) - I\ (\Omega)\right] - \rho\, \kappa_s\, I\ (\Omega)$$

$$+ \frac{3}{16\pi} \rho\, \kappa_s \int d\omega' \left[1 + (\Omega \cdot \Omega\,')^2\right]\ I(\Omega')\ , \tag{6.29}$$

where the left hand side is given by (6.7) in general, and by (6.9) for one dimensional spherically symmetric systems. The low energy Compton scattering cross section σ_s is angle independent. We note, but will not discuss here, that the right side of (6.29) may also include an inelastic scattering term. This will be discussed in the next chapter.

Equation (6.29) may be multiplied by Ω^n, and integrated over solid angle to obtain a series of moment equations which are equivalent to the original transfer equation. For example, the zeroth moment of (6.29), obtained by integrating over $d\omega$, is

$$\frac{1}{c} \int \frac{\partial I}{\partial t} d\Omega + \int \Omega \cdot \nabla\ I\ d\omega = \kappa_a\, \rho \int \left[B\ (T) - I\ (\Omega)\right] d\omega$$

$$- \rho\, \sigma_s \int I(\Omega)\, d\omega + \frac{3\rho\sigma_s}{16\pi} \int d\omega \int d\omega' \left[1 + (\Omega \cdot \Omega\,')^2\right]\ I(\Omega')\ . \tag{6.30}$$

Recalling (6.3) and (6.5), and choosing Ω' to lie along the z axis, (6.30) reduces to the radiation energy equation

$$\frac{\partial E}{\partial t} + \nabla \cdot \boldsymbol{F} = \kappa_a \rho \left[4\pi B(T) - E c \right] . \tag{6.31}$$

The flux is defined by (6.5), and E is given by (6.3). Note that the gain and loss in beam energy due to elastic Compton scattering cancel in the energy equation, as it should.

The first moment is obtained if (6.29) is multiplied by Ω and integrated over solid angle. The result (which ignores third and higher moments) is easily shown to be the radiation momentum equation

$$\frac{1}{c}\frac{\partial \boldsymbol{F}}{\partial t} + c\, \nabla \cdot \boldsymbol{P} = -(\kappa_s + \kappa_a)\rho \boldsymbol{F} , \tag{6.32a}$$

where the pressure tensor $\boldsymbol{P}$ is given by (6.12). The black body term in (6.30) vanishes in the flux equation because it is isotropic and contributes zero net flux. The two terms on the right side of (6.32) arise from scattering out of the beam and absorption by matter. Finally, the term corresponding to scattering into the beam vanishes in the low temperature regime because Compton scattering there is symmetric in angle, and cannot contribute to the net flux. If the cross section has a more complicated angular dependence than is assumed in (6.27) the first moment of the transfer equation becomes (see Problem 6.5)

$$\frac{1}{c}\frac{\partial \boldsymbol{F}}{\partial t} + c\, \nabla \cdot \boldsymbol{P} = -[\kappa_a + \kappa_s (1 - \tilde{\mu})]\, \rho \boldsymbol{F} , \tag{6.32b}$$

where the factor $1 - \mu$ represents a transport correction, and

$$\tilde{\mu} \equiv \frac{\int \sigma(\Omega \cdot \Omega ')\, \bar{\mu}\, d\omega}{\int \sigma\, d\omega} . \tag{6.32c}$$

The transport opacity is then defined by

$$\kappa_{tr} \equiv \kappa_a + \kappa_s[1 - \bar{\mu}] , \tag{6.32d}$$

and the transport mean free path is given by

$$\lambda_{tr} \equiv [\,\kappa_{tr}\,\rho\,]^{-1}. \tag{6.32e}$$

Notice that for the special case of isotropic scattering or Compton scattering the average $\bar{\mu} = 0$, and the transport opacity is just $\kappa_a + \kappa_s$.

Problem 6.5

a. Derive the first two moments of the radiative transfer equation (6.31) and (6.32) from (6.30).

b. Assume that the angular dependence of the cross section is given by

$$\sigma(\Omega\cdot\Omega') = \sigma(\varepsilon)\,(1 + a\Omega\cdot\Omega') .$$

Show that (6.31) remains unchanged, but that the first moment becomes (6.32b). Then show that

$$a = 3\,\bar{\mu} .$$

Additional higher moment equations may also be constructed from (6.29). A characteristic of these equations is that the nth moment equation includes terms involving the $n+1$ moment, and thus a system of n equation for $n+1$ unknowns is obtained. A closed system of equations may be obtained in the diffusion approximation where the specific intensity is given by (6.11), and $\lambda_\varepsilon << l$ as described above [this last condition is equivalent to assuming that the source function is proportional to $B(T, \varepsilon)$]. It then follows that

$$E = \frac{4\pi I_o}{c}$$

(6.33a)

and

$$P = \frac{4\pi I_o}{3c}$$

(6.33b)

(see Problem 6.1). The time derivative in (6.32) is, to order of magnitude, $(1/c)(\partial \mathbf{F} / \partial t) \sim \mathbf{F} / c\tau$, while the other two terms are of order $\mathbf{F} / \lambda_\varepsilon$. Since $c\tau \approx l$ is characteristic of the length scales over which $\mathbf{F}$ changes and $\lambda_\varepsilon << l$, (6.32) reduces in the diffusion approximation to

$$\frac{c}{3}\nabla E \approx - [\kappa_a + \kappa_s (1 - \tilde{\mu})] \rho \mathbf{F} .$$

(6.34)

Substituting (6.34) into (6.31) leads to the multigroup diffusion approximation for the radiation energy

$$\frac{\partial E}{\partial t} = \nabla \cdot D(\varepsilon) \nabla E + \kappa_a \rho [\frac{4\pi B(T)}{c} - E],$$

(6.35)

where the energy dependent diffusion coefficient $D(\varepsilon)$ is defined by

$$D(\varepsilon) = \frac{c/3}{[\kappa_a + \kappa_s(1 - \tilde{\mu})] \rho} = \frac{\lambda_{tr}(\varepsilon) c}{3} .$$

(6.36)

If the radiation field is in thermodynamic equilibrium at a temperature T_R, the energy spectrum will be Planckian

$$E = \frac{4\pi}{c} B(T_R) \,, \tag{6.37a}$$

and the total radiation energy density is given by

$$\int E(\varepsilon)\, d\varepsilon = \frac{8\,\pi^5 k^4}{15\, c^3 h^3}\, T_R^4 = a\, T_R^4. \tag{6.37b}$$

It may should be emphasized that the spectral energy $E(\varepsilon)$ in the multigroup approximation need not be Planckian, but can in principle have any form.

Complete thermodynamic equilibrium between radiation and matter occurs when $T_R = T$; which implies that the emission and absorption cancel, and (6.35) reduces to the thermal diffusion equation with

$$\int E(\varepsilon)\, d\varepsilon = a\, T^4.$$

This is a more restrictive case than (6.37b), and implies that the spectrum is now Planckian.

Energy Diffusion

The multigroup diffusion approximation (6.35) is useful when the radiation field is, in some stage of a process, out of thermodynamic equilibrium and its behavior is sensitive to the details of its spectral distribution. In some astrophysical plasmas (ionized matter surrounding bright, young stars or possibly around active galactic nuclei and quasars) radiation and matter are each approximately in thermodynamic equilibrium, characterized by the temperatures T_R and T respectively. The important point here is that the radiation spectum is Planckian, though not at the material temperature. The radiation energy density is then given by

$$\int_0^\infty E(T_R, \varepsilon)\, d\varepsilon = a T_R^4 \equiv u_R. \tag{6.38}$$

This relation serves to define the effective radiation temperature T_R in terms of the total (local) radiation energy. Integrating (6.35) over all energies results in

$$\frac{\partial a T_R^4}{\partial t} = \nabla \cdot D_R \nabla a T_R^4 + \rho a c \left[\kappa_P (T)\, T^4 - \kappa_P (T, T_R)\, T_R^4 \right] \tag{6.39}$$

where the diffusion coefficient D_R is given by

$$D_R = \frac{c/3}{\kappa_R\, \rho} \tag{6.40}$$

and the generalized Rosseland mean opacity κ_R is defined by the relation

$$\kappa_R(T,T_R)^{-1} \equiv 4\pi \, \frac{\displaystyle\int_0^\infty \frac{\partial B(T_R,\varepsilon)/\partial T_R}{[\kappa_a + \kappa_s(1-\tilde{\mu})]}\, d\varepsilon}{\displaystyle\int_0^\infty \frac{\partial B(T_R,\varepsilon)}{\partial T_R}\, d\varepsilon} . \tag{6.41}$$

The dependence on material temperature arises because the absorption opacity depends in part on the material temperature: $\kappa_a(T,\varepsilon)$. For many astrophysical applications the scattering opacity $\kappa_s(\varepsilon)$ dominates, and thus in problems where scattering dominates the opacity (6.41) depends only on T_R. However, unless the material and radiation temperatures are very different it may be sufficient in this case to set $\kappa_s(T_R) \approx \kappa_s(T)$.

The generalized Planckian mean opacity appearing in the last term on the right of (6.39) is defined, for $T \neq T_R$, by

$$\kappa_P\,(T, T_R) \equiv \frac{\int_0^\infty \kappa_a(T,\varepsilon)\, B(T_R, \varepsilon)\, d\varepsilon}{\int_0^\infty B(T_P, \varepsilon)\, d\varepsilon}. \tag{6.42}$$

The usual Planckian opacity $\kappa_P(T)$ results if $T = T_R$ in (6.42).

Problem 6.6

Carry out the analysis leading to (6.39) and verify that the Rosseland and Planckian mean opacities are given by (6.41) and (6.42). In doing so use (6.38) to define the radiation temperature. Explain physically why κ_R includes κ_s, but κ_P does not.

Problem 6.7

Bremsstrahlung (also called free-free) absorption of photons is proportional to $\kappa(\varepsilon) \approx \varepsilon^{-3} T^{-1/2}$. Ignore scattering and show that the Rosseland mean opacity and the Planckian mean opacity are proportional to $T^{1/2} T_R{}^3$.

The interaction of radiation with matter may produce local accelerations which must be included along with the hydrodynamic equations. This will be discussed in Chapter 7. Emission and absorption will also heat (or cool) matter; in fact, the net energy change in the radiation field due to these processes is obtained by integrating (6.19) over angle and frequency. Recalling (3.13), the change in the material thermal energy resulting from (6.19) is

$$\rho \frac{\partial (C_V T)}{\partial t} = -\int_0^\infty d\varepsilon \int d\omega \left(\frac{\partial I}{\partial t}\right)_{ea} .$$

This will be solved using operator splitting, as discussed below (3.13), by expanding the left hand side and rewritting as

$$\rho\, C_V \frac{\partial T}{\partial t} = -\int d\varepsilon \int d\omega \, \left(\frac{\partial I}{\partial t}\right)_{ea} - \rho T \frac{\partial C_V}{\partial t}.$$

The change in material energy due to the change in temperature will be found first (as in (3.18a)), and then the remaing change due to the change in effective heat capacity can be constructed using (3.18b). Hereafter we consider only the expressions corresponding to (3.18a) corresponding to the equation

$$\rho\, C_V \frac{\partial T}{\partial t} = -\int d\varepsilon \int d\omega \, \left(\frac{\partial I}{\partial t}\right)_{ea}$$

or

$$C_V \frac{\partial T}{\partial t} = -\int d\varepsilon \int d\omega \;\, \kappa_a(\rho, T, \varepsilon) \left[B(T,\varepsilon) - I(\Omega, \varepsilon)\right] , \tag{6.43}$$

where the common density factor cancels. An analysis similar to the the one leading to (6.39) yields the equivalent form

$$C_V \frac{\partial T}{\partial t} = -\, a\, c \left[\, \kappa_P(T)\, T^4 - \kappa_P(T, T_R)\, T_R^4 \,\right] . \tag{6.44}$$

Equations (6.39), and (6.44) represent the radiative diffusion and material coupling equations for matter at rest in the diffusion limit.

Neutrinos

The transfer equation discussed above can be used to model neutrino transport in stellar systems (see, for example, Lund 1985). In many cases it is not necessary to model the angular behavior of the neutrino field in detail, and a multigroup diffusion approximation analogous to (6.35) may be useful. Neutrinos and antineutrinos, which are both fermions, couple to dense matter in somewhat different ways than do photons, which are bosons. Each neutrino species (ν_e, ν_μ, and ν_τ and their antiparticles) may need to be treated differently (a discussion of neutrino astrophysics may be found in: Bowers and Deeming 1984; Shapiro and Teukolsky 1983; Tubbs and Schramm 1975). If the spectral energy density of electron type neutrinos is denoted by $F_\nu(\varepsilon)$ (radiation quantities referring to neutrinos will be denoted by a subscript ν, which should not be confused with frequency) then the total energy density in the neutrino field is given by

$$u_\nu = \int F_\nu(\varepsilon)\, d\varepsilon. \tag{6.45}$$

Since the total number of electrons, electron neutrinos and electron antineutrinos is conserved, it is necessary to keep track of the number density of each species. Therefore, we define the number density of eletron neutrinos by

$$n_\nu = \int F_\nu(\varepsilon)\, \frac{d\varepsilon}{\varepsilon}. \tag{6.46}$$

Besides diffusion, the electron neutrino spectum will change because of emission of neutrinos (associated with electon capture on heavy nuclei and on free proton) and because of absorption by free neutrons. In the multigroup diffusion approximation the emission-absorption terms are linear in F_ν, and the time rate of change of the spectrum may be written as

$$\frac{\partial F_\nu}{\partial t} = \nabla \cdot D_\nu \nabla F_\nu + K_\nu (A_\nu - F_\nu) \tag{6.47}$$

where the energy variable has been omitted. $D_\nu(\varepsilon)$ is the Rosseland mean [defined as in (6.36)] using the transport mean free path, but with the average taken over the equilibrium Fermi-Dirac distribution for neutrinos. This expression includes the quantum effects associated with the blocking factors exactly. $K_\nu(\rho,T,\varepsilon)$ is an effective coupling coefficient which can be expressed in terms of the neutrino opacity of the material κ_ν:

$$K_\nu \equiv \rho\, c\, \kappa_\nu .$$

The latter contains blocking factors characteristic of fermions. $A_\nu(\rho,T,\varepsilon)$, which is the fermion black body energy density, is proportional to the neutrino emissivity of the material. Both K_ν and A_ν depend on the density and temperature of the material, as well as on its composition (see for example: Wilson 1980; Bowers and Wilson 1982; Tubbs and Schramm 1975). The diffusion coefficient is given by an expression analogous to (6.36). Material heating associated with emission and absorption is described by the operator split equation corresponding to (3.18a):

$$\rho\, C_V \frac{\partial T}{\partial t} = -\int \kappa_\nu (A_\nu - F_\nu)\, d\varepsilon \equiv -\int \left(\frac{\partial F_\nu}{\partial t}\right)_c d\varepsilon . \tag{6.48}$$

In addition to material heating or cooling, electron neutrino absorption and emission will result in changes in composition. For example, suppose that the only process leading to the absorption or emission of a ν_e is electron capture on free protons (this occurs in the core of highly evolved stars which are undergoing dynamic collapse):

$$n + \nu_e \leftrightarrow p + e^- .$$

The absorption (emission) of an electron neutrino ν_e increases (decreases) the number of free protons $n_p \equiv Z n_b$, where n_b is the (fixed) number of free baryons per unit volume. Thus, in addition to changing the material energy, emission and absorption will also change the chemical composition of the baryonic fluid. In particular, the change in n_p is given by [see (6.48)]:

$$\frac{dn_p}{dt} = n_b \frac{dZ}{dt} = -\int \left(\frac{\partial F_\nu}{\partial t}\right)_c \frac{d\varepsilon}{\varepsilon} \,. \tag{6.49a}$$

The emission and absorption and transport of antineutrinos may also be important. In this case, expressions analogous to (6.45)-(6.48) arise describing the absorption of an electron antineutrino. An example is absorption by free protons

$$\bar{\nu}_e + p \leftrightarrow n + e^+.$$

The absorption of an electron antineutrino will change the composition by increasing the number density of free neutrons. This is described in analogy with (6.49a) by

$$\frac{dn_n}{dt} = n_b \frac{d(1-Z)}{dt} = -\int \left(\frac{\partial \bar{F}_\nu}{\partial t}\right) \frac{d\varepsilon}{\varepsilon} \,, \tag{6.49b}$$

where $n_n = (1 - Z)\, n_b$ is the number density of free neutrons, and

$$\bar{F}_\nu \equiv \textit{electron antineutrino spectral energy density} \,.$$

The time rate of change of the electron antineutrino spectral energy density due to absorption may be defined in analogy with (6.48). Finally we note that the equations describing emission and absorption of other neutrino species may be developed along similar lines.

Equilibrium Diffusion of Photons

When the electron-radiation coupling in matter is strong ($\tau_{e\gamma}$ small compared to all time scales of interest) the emission and absorption terms in (6.39) cancel at each point and the material and radiation are in thermodynamic equilibrium at the single temperature T. Multiplying (6.44b) by the material density, adding the result to (6.39), and taking the limit $T_R \rightarrow T$ yields the equilibrium energy diffusion equation for photons

$$\rho C_V \frac{\partial T}{\partial t} + \frac{\partial (aT^4)}{\partial t} = \nabla \cdot D_R \nabla aT^4. \tag{6.50a}$$

If we define the total specific heat of the matter plus radiation by

$$C_{tot} = C_V + \frac{4\, aT^3}{\rho} \tag{6.50b}$$

the equilibrium diffusion equation may be rewritten in the more familiar form

$$\rho C_{tot} \frac{\partial T}{\partial t} = \nabla \cdot D_R \nabla aT^4. \tag{6.50c}$$

Using the definition of the diffusion coefficient (6.36), and (6.28) we obtain the alternate form for the thermal diffusion equation

$$\rho C_{tot} \frac{\partial T}{\partial t} = \nabla \cdot \left(\frac{4\, acT^3}{3\, \kappa\rho}\right) \nabla T. \tag{6.50d}$$

In this representation of the thermal diffusion equation the temperature is the primary variable. The right hand side of (6.50d) is just the divergence of the radiant energy flux

$$\mathbf{F}_{eq} \equiv -\frac{4\, acT^3}{3\, \kappa\rho} \nabla T.$$

Finally we note that the luminosity $L(r)$ of a spherical mass distribution is just the flux times the surface area $4\pi r^2$ and is, for a spherical system

$$\nabla T = -\frac{3\kappa\rho}{16\pi ac}\frac{L(r)}{r^2}$$

which will be recognized as one of the equations of hydrostatic stellar structure.

The thermal diffusion equation represents an excellent approximation in stellar interiors. For low mass stars, the second term in (6.50b) can usually be neglected, but for massive stars the radiation contribution can be significant. Finally, because the opacity is usually a strong function of temperature, the diffusion coefficient makes (6.50d) highly nonlinear.

6.2 SPACE-TIME CENTERING

The variables characterizing the radiation field are best centered in time, space, energy and angle to achieve numerical stability and to be compatible with any other physical processes (primarily hydrodynamics). Numerical solutions of the radiative transfer equation require the specific intensity $I(\mathbf{r},t,\Omega,\varepsilon)$, the angular integral of which gives the spectral energy density, and the angular and energy integrals of which give the total radiation energy density. The spatial and temporal variables will be defined on the same space-time grid chosen for the hydrodynamics (see Section 2.1 and Chapter 4). In addition we shall require a set of angular variables and a set of energy groups for the specific intensity at each spatial point. The approach taken here can be used for one dimensional Lagrangian, or higher dimensional Eulerian representations. They are not appropriate for Lagrangian descriptions in more than one dimension.

The radiative transfer equation represents variations in N_S spatial dimensions, N_A angles and N_E energies, and is computationally equivalent to an $D=N_S+N_A+N_E$ dimensional problem. The number of zones required for adequate spatial resolution is usually greater than the number needed for adequate angular resolution, but the occurrence of each independent variable translates into a computational nested loop, and so this distinction is seldom useful in practice. Thus, for example, in two spatial dimensions, $D=5$. In one spatial dimension and spherical symmetry, only one angle is required, and $D=3$. Therefore, representations of the transfer equation will only be considered for one dimensional systems.

A numerical grid will be required to describe the angular and energy dependence of the radiation field. First, a set of discrete angle variables will be introduced to replace the vector Ω (only one angle variable is actually required). Similarly, a discrete set of energy groups will be used to describe the energy dependence;

$$\Omega \rightarrow \mu_i \qquad\qquad \varepsilon \rightarrow \varepsilon_j \, .$$

This suggests a natural centering for the specific intensity in a one dimensional representations which can be denoted by

$$I(\mathbf{r}, t, \Omega, \varepsilon) = I(r, t, \mu, \varepsilon) \rightarrow I^{n}_{i+1/2,\, j+1/2,\, k+1/2} \tag{6.51}$$

Here i, j and k represent the indices of the angular, energy and spatial coordinates respectively. The time centering of the specific intensity is the same as is used for the material energy. Thus, I is centered in the spatial, angular and energy zone. The spatial zones should be the same as those used for either the Lagrangian or Eulerian formulation of the hydrodynamics.

Energy Groups

Now consider the construction of a one dimensional grid defining the energies. We shall refer to the zones in this grid as energy groups. They may be constructed to cover a range ε_1 to ε_J where ε_1 (ε_J) is the minimum (maximum) energy of interest in a given problem. The energy group boundaries will be denoted by ε_j, and the group centered energy by $\varepsilon_{j+1/2}$ (see Figure 6.3). The energy group width is defined by $\Delta\varepsilon_{j+1/2}$. An energy grid is usually constructed by first specifying the group boundaries. The group widths are then fixed,

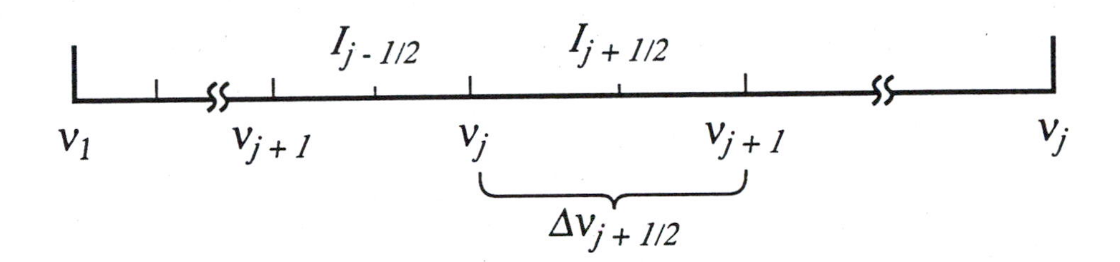

Figure 6.3. Energy groups ν_j, running from a minimum value ν_1 to a maxiamum value ν_J. Note that the group width is not necessarily uniform. The specific intensity is group centered.

but group centered energies are not. The latter may be constructed from ε_j arbitrarly. The ambiguity in how centered energies are defined may often be exploited to satisfy constraints or to improve computational accuracy. For example, the evaluation of (6.18b) on the energy grid involves both $\varepsilon_{j+1/2}$ and $\Delta\varepsilon_{j+1/2}$. In principle, the group centered energies could be chosen such that

$$\pi \sum_{j=1}^{J-1} B(T,\varepsilon_{j+1/2})\, \Delta\varepsilon_{j+1/2} = \sigma T^4$$

exactly. A set of $\varepsilon_{j+1/2}$ satisfying this constraint can be found numerically, and will depend on the temperature of the matter. In practice, it is often sufficient to adopt a simple scheme relating $\varepsilon_{j+1/2}$ to ε_j, adjusted so that the sum over the Planck function is close to $\sigma T^4/\pi$ (see Problem 6.26).

Here we shall consider the simple energy group structure given by

$$\varepsilon_{j+1/2} = \frac{1}{2}(\varepsilon_{j+1} + \varepsilon_j) \tag{6.52}$$

$$\Delta\varepsilon_{j+1/2} = \varepsilon_{j+1} - \varepsilon_j \ . \tag{6.53}$$

In practice ε_j may span several orders of magnitude in energy (for example, in core collapse models of supernovae the neutrino energies of interest may range from one to several hundred MeV), in which case nonuniform energy group zoning is essential. An example of such a structure will be considered when we discuss multigroup methods in Section 6.4. For a reasonably smooth radiation spectrum, ten to twenty energy groups are often adequate to resolve I while maintaining a reasonable time for computations. The situation becomes far more complex if the effects of lines or absorption edges must be included.

The ideal (and simplest) approach to energy group structure is to establish an energy grid at the start of a calculation which is adequate for the entire calculation. It is also possible to rezone the energy grid during the course of the calculation to optimize resolution.

Angular Groups

An independent angular grid can also be constructed following the procedure used for an energy grid. The zones in this grid are called angular groups. The ambiguity in defining $\mu_{i+1/2}$ may be exploited to satisfy constraints on angular integrals (see Problem 6.8). An angular grid which may be used to describe the angular dependence of I in a one dimensional spherically symmetric system is shown in Figure 6.4. The inward direction (inner boundary of the angular grid) is denoted by μ_1 and the outward direction (outer boundary of the angular grid) is denoted by μ_I. The boundaries of the angular zones are μ_i and the zone centered angles may be chosen to be

$$\mu_{i+1/2} = \frac{1}{2}(\mu_{i+1} + \mu_i) \ . \tag{6.54}$$

The angular zone widths are then

$$\Delta\mu_{i+1/2} = \mu_{i+1} - \mu_i \ . \tag{6.55}$$

In general, the angular zoning need not be uniform. Six to eight angular zones may give reasonable results in many one dimensional calculations. The angular zoning for a one dimensional case with eight angular zones is shown schematically in Figure 6.5. Radiation travelling in the direction $\mu_{11/2}$, for example, lies between the cones denoted by μ_5 and μ_6.

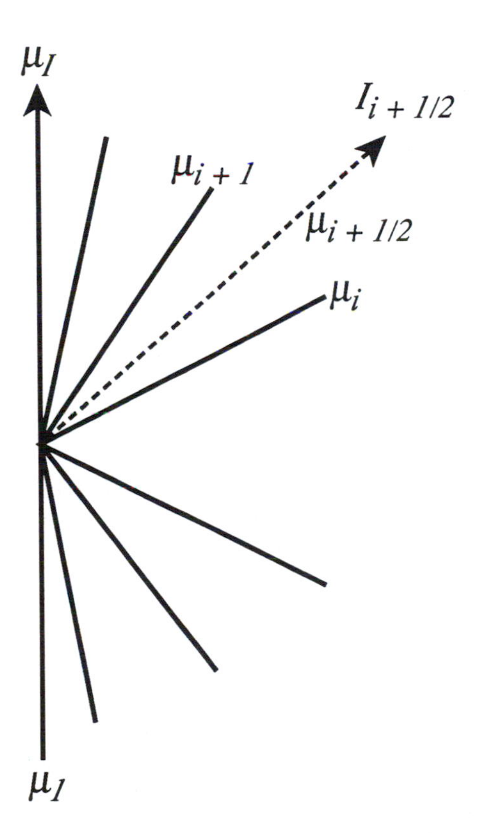

Figure 6.4. Angular groups running from $\mu_1 = -1$ (corresponding to the inward radial direction) to $\mu_I = 1$ (corresponding to the outward radial direction). Note that $\Delta\mu_i$ is not necessarily constant.

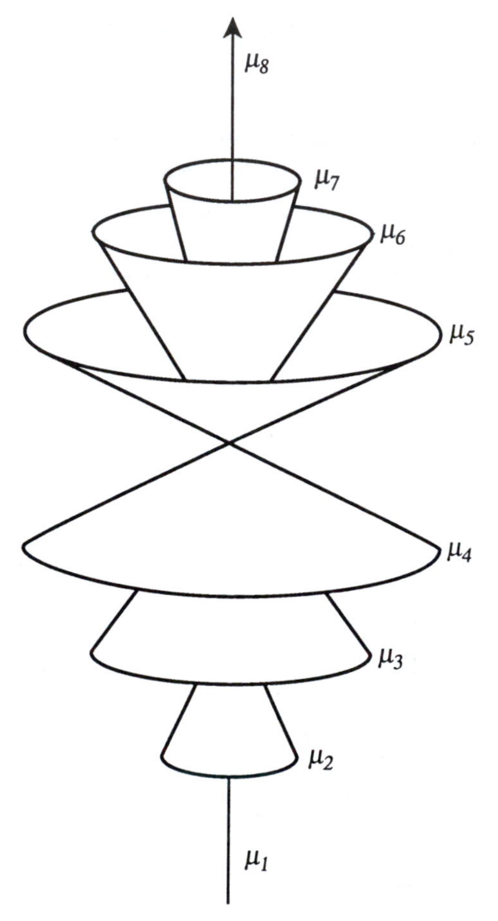

Figure 6.5. Eight angular zones for a one dimensional model of the transfer equation. The first three zones contain photons travelling in the inward direction, and the last three zones correspond to photons travelling in the outward direction.

Problem 6.8

Set up an angular grid μ_i containing six angles with $\Delta\theta_i = \pi/3$, listing the values of μ_i, $\Delta\mu_i$, $\mu_{i+1/2}$, and $\Delta\mu_{i+1/2}$. Evaluate the sums below, and compare the result to the equivalent integral (whose values appear in the right side):

$$\sum_{\mu_i} \mu_{i+1/2}\, \Delta\mu_{i+1/2} = 0\,;$$

(6.56)

$$\sum_{\mu_i} \mu_{i+1/2}\, |\mu_{i+1/2}|\, \Delta\mu_{i+1/2} = \pm \frac{1}{3}$$

(6.57)

where the plus (minus) sign applies for $\mu_i > 0$ ($\mu_i < 0$) in the summation;

$$\sum_{\mu_i} \Delta\mu_{i+1/2} = 2\,.$$

(6.58)

In (6.56) and (6.58) the summations are over all angles ($1 \leq i \leq I$).

Computational models based on the angular averaged multigroup diffusion equation (6.35) for photons, or (6.47) for neutrinos require a finite difference representation for the spectral energy E_ν and the radiation flux F_ν. A natural centering in one spatial dimension, which is consistent with the definitions above, is

$$E_\nu(\mathbf{r}, t, \varepsilon) \rightarrow E^n_{k+1/2,\, j+1/2}$$

$$F_\nu(\mathbf{r}, t, \varepsilon) \rightarrow F^n_{k+1/2,\, j+1/2}\,.$$

(6.59)

This centering is consistent with (6.51), and with the centering for material energy used in the hydrodynamics. It may be recalled that the material temperature and the specific heat introduced in the hydrodynamics was zone centered; thus, in one spatial dimension

$$T \rightarrow T^n_{k+1/2}$$

(6.60a)

$$C_V \rightarrow C^n_{V,\, k+1/2} \; . \tag{6.60b}$$

The temperature, heat capacity and internal energy will then satisfy the relation (3.14). Since ρ and T are zone centered, the absorption coefficients, scattering cross sections and diffusion coefficients are also naturally zone centered.

6.3 RADIATIVE TRANSFER EQUATION

A finite difference representation of the one dimensional transfer equation for photons in spherically symmetric systems (6.29) will be considered using the centering of the previous section. With appropriate redefinitions of the absorption coefficient and scattering cross section, the same methods may be applied to some neutrino transport problems (Lund 1985). The elastic scattering terms in (6.29) connect all angular groups with one another. These terms will be ignored at this time; simple cases will, however, be considered in the next chapter. Inelastic scattering, which connects energy groups and angular groups will be ignored entirely, though an angle average representation will be considered in Chapter 7. In the remainder of this chapter we consider (6.29) with $\kappa_s = 0$.

As noted at the beginning of the chapter, operator splitting is used to treat radiation transport and hydrodynamics separately. The approach discussed below also makes use of operator splitting to decouple the various angular and energy groups. This approximation does not work in all cases (the discussion of operator splitting in Chapter 2 should be recalled). It is most naturally suited to problems where the radiation field is changing dynamically. In these types of problems it has been found to work well. Other methods may be more appropriate for systems that are near equalibrium, or where the change in radiation field is small compared with time scales of other physical processes. When evaluating any operator split method it is necessary to use one's physical understanding of the problem to be solved as a guide to the method's applicability.

Emission and Absorption

First, consider the radiation emission and absorption terms (the spatial and angular gradient terms entering on the left side of (6.29) will be discussed below), and the material coupling equation (6.43): in a one dimensional spherical system these equations are

$$\frac{\partial I(\mu,\varepsilon)}{\partial t} = \kappa_a(\varepsilon)\, \rho\, c \left[B(T,\varepsilon) - I(\mu,\varepsilon) \right] \tag{6.61}$$

$$\rho\, C_V \frac{\partial T}{\partial t} = -\int_0^\infty d\varepsilon\, \kappa_a(\varepsilon)\, \rho\, c \int_{-1}^{1} d\mu \left[B(T,\varepsilon) - I(\mu,\varepsilon) \right]. \tag{6.62}$$

The emission term (first term on the right in (6.62)) implies an order of magnitude rate of change of the internal energy given by

$$\frac{\Delta T}{\Delta t} \approx \frac{a c}{\lambda} \frac{T^4}{\rho\, C_V}.$$

Throughout the central $0.1\ M_{\odot}$ of the sun $T \approx 2 \times 10^7\ K$, $\rho \approx 10^2\ g/cm^3$, and $C_V \approx 3\ k_B / 2\ m_H \approx 10^8\ erg\ K^{-1}\ g^{-1}$. The mean free path at these conditions is $\lambda \approx 1\ cm$ and the rate of change of the temperature as estimated above is of order $10^{15}\ K/sec$. The actual average rate of increase in central temperature as the core exhausts hydrogen and reaches the onset of helium burning ($T \approx 10^8\ K$), which takes about 4×10^9 *years*, is less than $10^8\ K / 10^{17}\ sec \approx 10^{-9}\ K / sec$. The actual heating rate is small because the absorption rate very nearly equals the magnitude of the emission rate. This simple example typifies emission-absorption processes in many astrophysical situations, namely that the net change in I and T result from extremely small differences between large numbers. It is therefore unadvisible to operator split the effects of emission and absorption in (6.61) and (6.62). Hereafter we will consider emission-absorption to be a single process for the purposes of developing finite difference approximations.

Equations (6.61) and (6.62), which are nonlinear in I and T, lead to coupling between all angular and energy groups [note that $\kappa_a = \kappa_a(\rho,T,\varepsilon)$ as well]. A naive and generally unsatisfactory approach to solving these equations would be to use the old values of $I(\mu,\varepsilon)$ and T in (6.62) to obtain a new material temperature, and then to use it in (6.61) to find a new value of $I(\mu,\varepsilon)$. A better approximation is based on the following physical arguments. In the strong coupling limit, the specific intensity rapidly approaches $B(T,\varepsilon)$ at the material temperature, and includes the effects of heating. This suggests that a finite difference representation of (6.61) should be implicit in I, and that $B(T,\varepsilon)$ should be evaluated at the new temperature (implicit in temparature as well). For each angle and energy, the right side of (6.62) can be thought of as representing a partial change in the material temperature. Denoting the partial "temperature" by $\theta(\mu,\varepsilon)$, which is a formal quantity, (6.61) and (6.62) can be decoupled by writing

$$\frac{I' - I}{\Delta t} = \left(\frac{\partial I}{\partial t}\right)_c \tag{6.63}$$

and

$$\Delta\varepsilon\,\Delta\mu\,\left(\frac{\partial I}{\partial t}\right) = -\rho\,C_V\frac{\partial\theta}{\partial t} = \kappa\rho c\left[B(\theta) - I'\right]\Delta\varepsilon\,\Delta\mu \tag{6.64}$$

for each angular and energy group. The partial temperature then has the finite difference representation $\theta(\mu,\varepsilon) \rightarrow \theta_{ij}$ (the spatial index will be omitted since no gradient terms appear in (6.61)-(6.62)) with $\theta_{11} = T^n$, and $\theta_{IJ} = T^{n+1}$. In this approach, the system of equations will be solved $I{\times}J$ times with the partial temperature θ incremented at each step. A temporal index is not used for θ (a superscript could be used to denote how many of the $I{\times}J$ times θ has been changed in a single cycle). For notational convenience, if θ represents the partial temperature at the beginning of one of the $I{\times}J$ angular and energy group calculations, then the same variable with a tilda,

$$\hat{\theta} \equiv \textit{partial temperature},$$

will be used to denote the new value of the partial temperature at the end of that energy and angular group calculation. With these points in mind, the temperature dependence of the Planck function $B(T,\varepsilon)$ on the right side of the coupling term (6.64) may be linearized by expanding it about the initial value θ:

$$B(\tilde{\theta}) = B(\theta) + \frac{\partial B}{\partial \theta}\frac{\partial \theta}{\partial t}\Delta t . \tag{6.65}$$

Substituting (6.65) into (6.64) and solving for $(\partial\theta/\partial t)$, we find

$$\rho C_V \frac{\partial \theta}{\partial t} = -\kappa_{eff}\, \rho c \left[B(\theta) - I'(\mu) \right] \Delta\varepsilon\, \Delta\mu \tag{6.66}$$

where the effective coupling constant is defined by

$$\kappa_{eff} \equiv \frac{\kappa_a}{1 + \dfrac{\kappa_a c}{C_V} \Delta\varepsilon_{j+1/2}\, \Delta\mu_{i+1/2}\, \Delta t \left(\dfrac{\partial B}{\partial \theta}\right)} . \tag{6.67}$$

The denominator in (6.67) is always greater than unity since for photons $\partial B/\partial\theta > 0$. Combining (6.66) and (6.63) leads to the following finite difference representation for the emission and absorption terms in the transfer equation:

$$I^{n+1}_{i+1/2,\, j+1/2} = I^{n}_{i+1/2,\, j+1/2} + \kappa_{eff}\, \rho c \left[B_{j+1/2}(\theta_{ij}) - I^{n+1}_{i+1/2,\, j+1/2} \right] \Delta t^{\, n+1/2} \tag{6.68a}$$

$$\rho C_V \tilde{\theta}_{ij} = \rho C_V \theta_{ij} - \kappa_{eff} \rho c \left[B_{j+1/2}(\theta_{ij}) - I^{n+1}_{i+1/2,\, j+1/2} \right] \Delta \nu_{j+1/2} \, \Delta \mu_{i+1/2} \, \Delta t^{\,n+1/2} .$$

(6.68b)

All spatially varying quantities carry the subsrcipt $k+1/2$ (zone centered), and ρ and C_V are evaluated at time t^n. The transfer equation (6.68a) is implicit, but depends on the current value of θ. Although the Planck function has been linearized using (6.65), addtitional nonlinearities enter through the absorption coefficient. It is often sufficient to evaluate these terms explicitly. Thus quantities in the coupling coefficient which depend on ρ and T use $\rho^n{}_{k+1/2}$ and $T^n{}_{k+1/2}$. Equation (6.68a) may be solved for I^{n+1} for a each angular and energy group, and then the partial coupling (6.68b) solved for a new partial temperature. The next angle or energy group is then solved by setting

$$\theta = \tilde{\theta}$$

in (6.68a) and the process is repeated for that angular or energy group. The final value of θ after I has been incremented for all angles and energy groups is the new material temperature T^{n+1}.

The partial temperature approach may fail if, for example, the change in material temperature associated with two or more energy groups is very large, as one might expect. In practice it is advisable to monitor the change in partial temperatures from one angular or energy group to the next, and to use a small enough time step that the changes are not large relative to the temperature itself. If the swing in partial temperatures becomes unacceptably large during several cycles, and cannot be reduced to an acceptable level for a practical time step, then the approach must be modified. One way to accomplish this is to use the value of T^{n+1} and the mean of the partial temperatures $\langle\theta\rangle$ obtained from the solution of the equations as first guesses in a set of Newton-Raphson iterative schemes. Thus $\langle\theta\rangle$ and T^{n+1} are each used in the complete set of equations above to find two new values for the material temperature. These two values may then be used to form an improved guess for the material temperature with which to continue the iteration. The procedure may be continued until the material temperature converges. The specific intensity obtained for the converged material temperature is then taken as the solution to the transfer equation. Although this sounds expensive, it may be more efficient than reducing the time step.

Problem 6.9

Show that $\partial B / \partial T > B / T$ for photons.

Problem 6.10

Discuss the stability of (6.68a) assuming that κ_{eff} is evaluated at the initial temperature T^n.

An interesting feature of the effective coupling coefficient (6.67) should be noted. Suppose that we define the material thermal energy density at the partial temperature θ by $u_m = \rho C_V \theta$, and the energy emitted per unit frequency and per angle by the material in the form of radiation by

$$\dot{u}_m \Delta t = \kappa_a \rho c \Delta t \Delta\mu \Delta\nu B(\theta) .$$

Then (6.67) may be used to show that

$$\frac{1}{\dot{u}_{eff} \Delta t} \geq \frac{1}{u_m} + \frac{1}{\dot{u}_m \Delta t} .$$

Note that $(du_{eff}/dt) \Delta t$ is given by $(du_m/dt) \Delta t$ with κ replaced by κ_{eff}. In the weak coupling limit ($\kappa_a \to 0$) the energy radiated in a time step is simply $(du_m/dt) \Delta t$, but in the strong coupling limit ($\kappa_a \to \infty$) the energy radiated approaches, but does not exceed, the maximum thermal energy available, u_m. Consequently, the partial temperatures will remain positive throughout the solution of (6.68a) and (6.68b). It should also be noted that the sum over angles and energy groups of (6.68b) recovers the original coupling equation (6.62).

Spatial Gradient Terms

The spatial and angular gradient terms which appear in the radiative transfer equation are given by the first two terms on the right side of (6.10) for a one dimensional spherically symmetric system. These terms are similar in form to advection terms which appear in the hydrodynamic equations, and represent the flow of radiation (moving with constant speed c) through the mesh. The radial gradient term is

$$\left(\frac{\partial I}{\partial t}\right)^R_{adv} = -\frac{\mu c}{r^2}\frac{\partial(r^2 I)}{\partial r} .$$

(6.69)

Integrating over the zone centered volume (6.14) $(r^2 \Delta r)_{k+1/2}$ (the factor $2\pi \ \Delta\mu_i$ cancels from both sides, and will be omitted) for the angular group μ_i yields

$$\left(\frac{\partial I}{\partial t}\right)^R_{adv,\ k+1/2} = -\int \frac{\mu_{i+1/2}\, c}{r^2}\frac{\partial(r^2 I)}{\partial r}\frac{r^2\, dr}{r^2_{k+1/2}\,\Delta r_{k+1/2}}$$

$$= -\frac{\mu_{i+1/2}\, c}{r^2_{k+1/2}\,\Delta r_{k+1/2}}\left\{r^2_{k+1} I_{k+1} - r^2_k I_k\right\} .$$

(6.70)

The angular and energy indices $i+1/2$ and $j+1/2$ have been omitted in the equation above. The angular intensity is a zone centered variable, so a prescription must be found for the intensity I_k at a zone boundary. Physically, radiation is attenuated when it transverses a distance Δx (see Figure 6.6). Therefore, the attenuation experienced by the beam after travelling in a direction $\mu_{i+1/2} > 0$ from a zone center $r_{k-1/2}$ to a zone boundary r_k should be expressed by the approximation

$$I_{i+1/2,\,k} = I_{i+1/2,\,k-1/2}\, exp[-\frac{\tau_k}{2|\mu_{i+1/2}|}] . \tag{6.71}$$

where τ_k is the optical depth of the material, and is defined in integral form by

$$\tau(\varepsilon) = \int_r^{r+\Delta r} \kappa_a(\varepsilon)\, \rho\, dr . \tag{6.72}$$

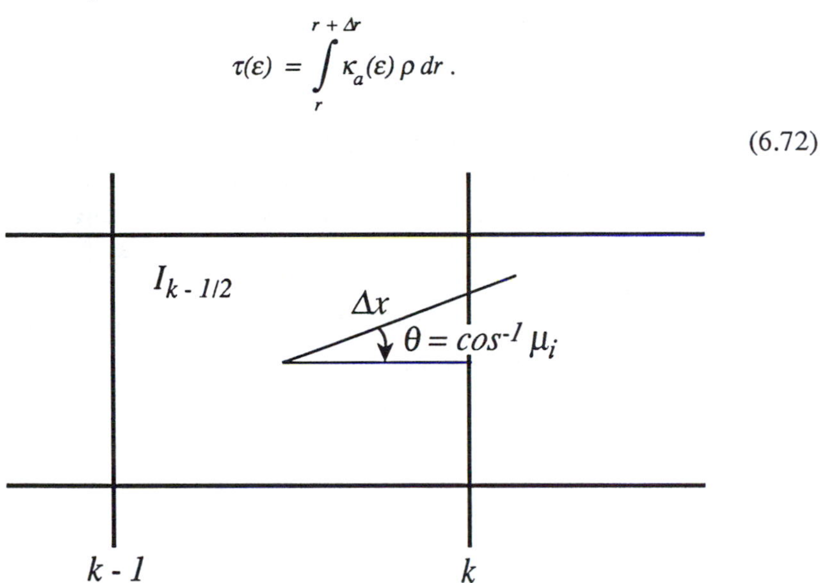

Figure 6.6. Attenuation of the specific intensity as radiation travels from a zone center to a zone boundary due to absorption by matter in a one dimensional, spherically symmetric system. The radial coordinates are denoted by the indices k.

An obvious finite difference representation of the optical depth is the zone centered quantity (common energy and angle subscripts omitted)

$$\tau_{k+1/2} = \rho_{k+1/2}\, \kappa_{k+1/2}\, \Delta r_{k+1/2} , \tag{6.73}$$

which is obtained directly from (6.72) by integrating over the zone. Equation (6.71) requires the optical depth τ_k at the zone boundary. A useful approximation for τ_k is the optical depth average defined by

$$\tau(\varepsilon)_k = \frac{1}{2}\left(\tau(\varepsilon)_{k+1/2} + \tau(\varepsilon)_{k-1/2}\right). \tag{6.74}$$

Over a broad range of density and temperature the absorption coefficient for photons κ_a is approximately proportional to

$$\kappa_a(\varepsilon) \approx \varepsilon^{-3} T^{-1/2}. \tag{6.75}$$

Consequently the mean free path $\lambda(\varepsilon)$ goes as ε^3, and there will be a considerable spread in the spatial extent of the radiation front as it propagates through matter. As a result the change in material temperature due to radiation heating as the front propagates may be spread out over many zones.

For angular groups $\mu_{i+1/2} < 0$, the extrapolation formula analogous to (6.71) is readily shown to be

$$I_{i+1/2,\,k} = I_{i+1/2,\,k+1/2} \exp\left[-\frac{\tau_k}{2|\mu_{i+1/2}|}\right], \tag{6.76}$$

where the zone boundary centered optical depth is again given by (6.73). Note that the extrapolated intensity always uses the value of the zone centered intensity which is upwind with respect to the direction of radiation flow.

In practice a further modification of the extrapolation formulae (6.71) and (6.76) is desirable in the interests of computational speed. In a one dimensional calculation, the solution of the transfer equation requires computational loops over space, angles and energy groups. This is equivalent to a three dimensional problem, and may require long run times. Evaluation of quantities such as (6.76) will need to be carried out at least KXI times per computational cycle, and because the calculation of exponentials is a relatively slow process compared to simple arithmetic operations, a large fraction of the run time may be spent in constructing the values $e^{-\tau/2\mu}$. A significant saving in time will result if an approximation such as

$$e^{-\tau_k/2|\mu_{i+1/2}|} \approx \frac{1}{1+\dfrac{\tau_k}{2|\mu_{i+1/2}|}} \tag{6.77}$$

is used for the exponentials in (6.71) and (6.76). Notice that this expression is positive definite. For forward angles

$$I_{i+1/2,\,k} = \frac{I^{n+1}_{i+1/2,\,k-1/2}}{1+\dfrac{\tau_k}{2|\mu_{i+1/2}|}} \qquad \mu_{i+1/2} > 0\,, \tag{6.78a}$$

and for backward angles

$$I_{i+1/2,\,k} = \frac{I^{n+1}_{i+1/2,\,k+1/2}}{1+\dfrac{\tau_k}{2|\mu_{i+1/2}|}} \qquad \mu_{i+1/2} < 0\,. \tag{6.78b}$$

It may be recalled that the source terms in the transfer equation (6.68) were differenced implicitly. The advection terms in the transfer equation may lead to large changes in the intensity in a single time step and it is also desirable to treat these terms implicitly, as has been indicated by the time index on the right side of (6.78).

Another feature of the extrapolation scheme above is that it reduces to the diffusion limit when $\lambda(\varepsilon) \rightarrow 0$ (see Problem 6.11).

Problem 6.11

The radiation flux is defined by (6.5), which has the finite difference representation (common energy index $j+1/2$ omitted)

$$F_k = c \sum_{\mu_{i+1/2}} \mu_{i+1/2} I_{i+1/2,\,k} \, \Delta\mu_{i+1/2} \, . \tag{6.79}$$

Show that when $\lambda(\varepsilon) \to 0$, the specific intensity is related to the spectral energy density by

$$I \to \frac{c}{2} E \, , \tag{6.80}$$

and that

$$\lim_{\lambda \to 0} F_k \to - \frac{c}{3} \frac{E_{k+1/2} - E_{k-1/2}}{\tau_k}$$

$$= - \frac{\lambda_k c}{3} \frac{E_{k+1/2} - E_{k-1/2}}{\Delta r_k} \, . \tag{6.81}$$

It should be noted that when $\tau >> 1$ the expression (6.79) computes the radiation flux as a difference of large numbers, and as written can be inaccurate.

The advection of the radiation field in the radial direction is given by (6.70), with the zone boundary centered intensity defined according to (6.78). The boundary centered optical depth is given by (6.74), and is related to the absorption coefficient by (6.73).

Angular Gradient Terms

The second term on the right side of (6.10) represents changes in the specific intensity due to changes in the angle of the beam relative to the radial direction as it moves inward or outward in path length. As with the radial gradient term, the angular term represents angular advection associated with the flow of radiation with speed c: this may be written in the form

$$\left(\frac{\partial I}{\partial t}\right)_{adv}^{\mu} = -\frac{c}{r}\frac{\partial}{\partial \mu}\left[(1-\mu^2)\, I(\mu)\right] . \tag{6.82}$$

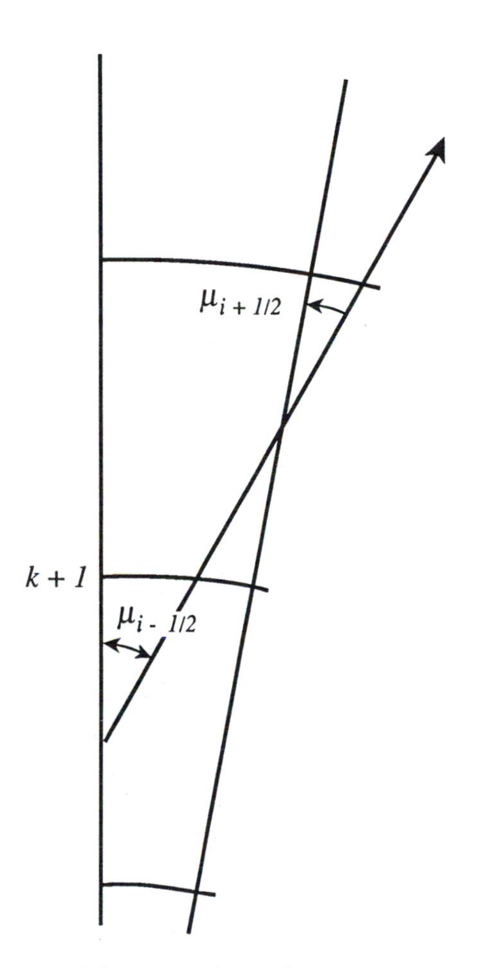

Figure 6.7. The spatial propagation of photons results in their redistribution among angular groups. Shown here, a photon in the angular group $\mu_{i-1/2}$ in the radial zone $k+1/2$ will shift to the angular group $\mu_{i+1/2}$ when it reaches zone $k+3/2$.

Integrating over $d\mu$ and dividing by the angular zone width $\Delta\mu_i$ leads to the finite difference representation (common energy $j+1/2$ index omitted)

$$\left(\frac{\partial I}{\partial t}\right)^{\mu}_{adv,\, i+1/2,\, k+1/2} = -\frac{c}{r_{k+1/2}\,\Delta\mu_{i+1/2}}\left\{(1-\mu_{i+1}^2)I_{i+1,\, k+1/2} - (1-\mu_i^2)I_{i,\, k+1/2}\right\} . \tag{6.83}$$

Equation (6.83) requires a specification of I at the boundaries i and $i+1$ of the angular zone $i+1/2$. The arguments leading to (6.78) apply to the attenuation of I in one dimensional spherical geometry as well, since the transport along the path length which converts a photon initially in the angular group $\mu_{i+1/2}$ into one in the group $\mu_{i-1/2}$ requires that the beam traverse a distance $\Delta x = \Delta r/\mu_i$ (see Figure 6.7). Furthermore, transport from $i+1/2$ to $i-1/2$ corresponds to outward motion of the photons so the extrapolated intensity I_{i+1} arises from attenuation of photons described by $I_{i+1/2}$. Thus, a reasonable interpolation formula is [with use of (6.77)]

$$I_{i+1,\, k+1/2} = \frac{I^{n+1}_{i+1/2,\, k+1/2}}{1 + \dfrac{\tau_k}{2|\mu_{i+1/2}|}} \tag{6.84}$$

(the common energy index $j+1/2$ has been omitted). The advection in the angular direction is given by (6.83) and (6.84), and will be assumed implicit in time.

Solution of the Radiative Transfer Equation

The emission-absorption and advection terms in the transfer equation may now be collected into a finite difference equation for outward moving photons ($\mu_{i+1/2} > 0$) of the form

$$A_+ I^{n+1}_{i+1/2,\, j+1/2,\, k+1/2} - B\, I^{n+1}_{i-1/2,\, j+1/2,\, k+1/2} - C_+ I^{n+1}_{i+1/2,\, j+1/2,\, k-1/2}$$

$$- D = 0 \tag{6.85}$$

where the coefficients are given by:

$$A_+ = 1 + \rho c \Delta t^{n+1/2} \kappa_{eff} + \frac{c / \mu_{i+1/2} / r^2_{k+1} \Delta t^{n+1/2}}{r^2_{k+1/2} \Delta r_{k+1/2} [1 + \frac{\tau_{k+1}}{2 / \mu_{i+1/2} /}]}$$

$$+ \frac{c (1 - \mu^2_{i+1}) \Delta t^{n+1/2}}{r_{j+1/2} \Delta\mu_{i+1/2} [1 + \frac{\tau}{2 / \mu_{i+1/2} /}]} \tag{6.86}$$

$$B = \frac{c (1 - \mu_i^2) \Delta t^{n+1/2}}{r_{j+1/2} \Delta\mu_{i+1/2} [1 + \frac{\tau}{2 / \mu_{i-1/2} /}]} \tag{6.87}$$

$$C_+ = \frac{c / \mu_{i+1/2} / r^2_k \Delta t^{n+1/2}}{r^2_{k+1/2} \Delta r_{k+1/2} [1 + \frac{\tau_k}{2 / \mu_{i+1/2} /}]} \tag{6.88}$$

$$D = I^n + \rho c \kappa_{eff} \Delta t^{n+1/2} B(\theta) . \tag{6.89}$$

The change in partial temperature is described (for either sign of the angle $\mu_{i+1/2}$) by (6.69). For inward moving photons ($\mu_{i+1/2} < 0$) the finite difference representation for the transfer equation can be written in the form

$$A_- I^{n+1}_{i+1/2,\, j+1/2,\, k+1/2} - B\, I^{n+1}_{i-1/2,\, j+1/2,\, k+1/2} - C_- I^{n+1}_{i+1/2,\, j+1/2,\, k+3/2}$$

$$- D = 0 \tag{6.90}$$

where the coefficients are

$$A_- = 1 + \rho c \Delta t^{n+1/2} \kappa_{eff} + \frac{c\,|\mu_{i+1/2}|\, r_k^2 \;\; \Delta t^{n+1/2}}{r^2_{k+1/2}\, \Delta r_{k+1/2} \left[1 + \dfrac{\tau_{k+1}}{2\,|\mu_{i+1/2}|}\right]}$$

$$+ \frac{c\,(1 - \mu^2_{i+1})\, \Delta t^{n+1/2}}{r_{j+1/2}\, \Delta\mu_{i+1/2} \left[1 + \dfrac{\tau}{2\,|\mu_{i+1/2}|}\right]} \tag{6.91}$$

and

$$C_- = \frac{c\,|\mu_{i+1/2}|\, r^2_{k+1}\, \Delta t^{n+1/2}}{r^2_{k+1/2}\, \Delta r_{k+1/2} \left[1 + \dfrac{\tau_{k+1}}{2\,|\mu_{i+1/2}|}\right]} \,. \tag{6.92}$$

All of the variables appearing above are naturally centered on the computational grid unless otherwise noted, and the effective coupling coefficient is defined by (6.67).

The coefficients $A_{\pm}$, B, $C_{\pm}$ and D depend on $\kappa_a(\rho^n{}_{k+1/2}, T^n{}_{k+1/2})$, $\rho^n{}_{k+1/2}$ and I^n, as well as on geometric factors and on the partial temperature θ through $B(\theta)$ and $\partial B / \partial \theta$. All quantities (except the θ dependent factors) are known throughout the grid, and may be set up prior to solution in a given computational cycle. The solution procedure for the one dimensional geometry assumed above involves separate summations over the three indices i, j and k. Examination of (6.85) and (6.90) shows that there is no coupling between different energy groups (because scattering has been ignored, and because emission and absorption have been decoupled using the method of partial temperatures). Therefore, we may solve for the angular and spatial dependence one energy group at a time. For outward (inward) moving photons, the specific intensity at $k+1/2$ depends on the intensity at $k-1/2$ $(k+3/2)$. Thus, for inward motion the solution starts at $k = 1$ and (fixed) $i = 1$, and solve (6.85) outward $1 \leq k \leq K$. The partial temperature is then incremented using (6.69), the next value of i is chosen, and the outward sweep over the grid is repeated. Finally, when all values of k and i corresponding to $\mu_{i+1/2} < 0$ have been used, the energy index j is incremented and the entire procedure is repeated until $j = J$. Note that for $i = 1$, $\mu_1 = -1$ and the coefficient B vanishes identically. For outward moving photons the sum over k begins at the outer boundary, $k = K$, and proceeds inward to $k = 1$. The sum over angular groups begins for the first value of i such that $\mu_i > 0$ and proceeds to $i = 1$ (this is shown schematically in Figure 6.8).

OUTWARD ($\mu_{i+1/2} > 0$)	INWARD ($\mu_{i+1/2} < 0$)
$j = 1$ to J	$j = 1$ to J
i = minimum value for which $\mu > 0$ to I	$i = 1$ to maximum value for which $\mu < 0$
(Increment partial temperatures θ)	
$k = 1$ to K	$k = K$ to 1
(Sweep through grid to find new specific intensity)	

Figure 6.8. Order of computation of specific intensity for photons moving outward (left column) or inward (right column) in one dimensional spherical geometry. The uppermost entry corresponds out the outermost loop. The indices for loops over energy groups, angles and spatial zones are denoted by j, μ, and k respectively.

The method presented above can be extended to planar or cylindrical coordinates for one dimensional geometry. Extension to more than one spatial dimension (which would involve a specific intensity dependent on four or more independent variables—at least two spatial coordinates and two angle coordinates, plus the energy) would lead to very long computational run times. For example, using six angles in two spatial dimensions would be roughly equivalent to running six three-dimensional calculations. In one dimensional modelling, including radiative transfer, hydrodynamics and most other physics options, the actual run time will usually be dominated by the solution of the transfer equation. For problems involving more than one spatial dimension, it may be wise to explore the applicability of multigroup diffusion or possibly simple energy diffusion methods to see if they can adequately model the physics.

6.4 Multigroup Diffusion

The multigroup diffusion approximation to the transfer equation which will be considered now represents a significant reduction in complexity and in execution time because all angle dependent terms are averaged out. The resulting equations then depend only on time, the spatial coordinate r, and the radiation spectral energy $\varepsilon = h\nu$. This represents a substantial savings in time in one dimensional calculations, and can even be practical for two dimensional calculations. For photons the multigroup diffusion approximation is given by (6.35) and (6.44). In this section we shall consider the finite difference equations for multigroup neutrino diffusion described by (6.47)-(6.49). The approach can easily be modified to describe multigroup transport of photons.

The radiative transfer equation is first order in spatial derivatives, while the diffusion equation involves second order terms. In general, it is advisable to solve the spatial gradient and emission-absorption terms together (recall the discussion at the beginning of section 6.3). The diffusion coefficient D_ν, the coupling coefficient K_ν and the emission term A_ν for neutrinos differ from the corresponding quantities appearing in the multigroup equation for photons primarily because photons are bosons whose number is not conserved, while neutrinos are fermions whose number is constrained to satisfy lepton number conservation (for examples and further discussion see, for example: Tubbs and Schramm 1975; Bowers and Wilson 1982; Bowers and Deeming 1984). These details do not significantly affect the development of a finite difference representation for (6.47)-(6.49), and so will not be considered further here. We note only that for many processes the neutrino opacity depends on the energy roughly as ε^2.

The emission and absorption terms in the evolution equation (6.47) are linear in F_ν. The neutrino mean free path appearing in the diffusion coefficient contains neutrino blocking factors, which are proportional to F_ν, and thus the overall equation is, in general, nonlinear. Iterative methods may be used to solve the full nonlinear equations. In the discussions below we assume that the mean free path has been constructed using the initial spectrum ($F_\nu{}^n$), and thus that the resulting equations are linear.

The multigroup equations to be solved for electron neutrinos are given by (6.47)-(6.49); they include spatial transport, emission and absorption, material coupling and the composition changes associated with neutrino emission and absorption. A modification of the partial temperature approach used in the previous section to decouple photon energy groups in the transfer equation can be used to decouple neutrino energy groups in the multigroup approximation. We will consider here the set of equations, analogous to (6.63)-(6.64), in one dimensional spherical geometry. In the equations below the subscript ν denotes neutrinos, and should not be confused with frequency; all equations are expressed below as functions of the neutrino energy ε. The evolution equation for the neutrino spectrum becomes

$$\frac{\partial F_\nu}{\partial t} = \frac{1}{r^2}\frac{\partial}{\partial r}\left(r^2 D_\nu \frac{\partial F_\nu}{\partial r}\right) + \left(\frac{\partial F_\nu}{\partial t}\right)_c . \tag{6.93}$$

The partial temperature method introduced for photons can be extended to the material coupling equation represented by (6.48). If the last term on the right of (6.93) is interpreted as the change in the neutrino spectrum associated with emission and absorption within a single energy group, then for each of these energy groups (6.48) may be approximated by

$$\left(\frac{\partial F_\nu}{\partial t}\right)_c \Delta\varepsilon = -\rho C_V \frac{\partial T}{\partial t} = K_\nu \left[A_\nu(T, Z) - F_\nu\right] \Delta\varepsilon . \tag{6.94}$$

Here the quantity $\rho C_V\, \partial T/\partial t$ is to be understood not as the total change in the material energy, but only as the change associated with the emission and absorption of neutrinos whose energies lie in the range ε to $\varepsilon+\Delta\varepsilon$. In a similar fashion the change in composition described by (6.49a) can be considered energy group by energy group, and the change in Z associated with the emission and absorption of neutrinos whose energies lie in the range ε to $\varepsilon+\Delta\varepsilon$ can be approximated by

$$n_b \frac{\partial Z}{\partial t} = -\left(\frac{\partial F_\nu}{\partial t}\right)_c \frac{\Delta\varepsilon}{\varepsilon} . \tag{6.95}$$

The first equation, (6.93) is the diffusion equation and includes the coupling term to matter; the second equation, (6.94) is the material heating term for each energy group, and the third equation (6.95) represents the change in electric charge due to electron neutrino emission and absorption. For the angle averaged multigroup equations, T will be used for the partial temperatures as well as for the initial or final temperature. By analogy with the partial temperature decomposition, (6.95) may be considered a partial charge decomposition of the neutrino radiation-matter coupling. In the discussions below the neutrino energy will be considered as the independent variable, though it will often be omitted for notational convenience. The method can be applied to other neutrino species (see Problem 6.27, for example). In particular, when electon and positron neutrinos are present simultaneously, the change in charge is given by

$$n_b \frac{\partial Z}{\partial t} = -\left(\frac{\partial F_\nu}{\partial t} - \frac{\partial \bar{F}_\nu}{\partial t}\right) \frac{\Delta\varepsilon}{\varepsilon} .$$

If the two terms on the right side are of comparable magnitude then operator splitting should not be used to solve for the change in charge (see Problem 6.27).

It is convenient to define the neutrino optical depth $\tau_\nu(\varepsilon)$ for neutrinos whose energy lies in the range ε to $\varepsilon+\Delta\varepsilon$, which is obtained from the opacity $\kappa_\nu(\rho,\varepsilon,T)$ in the usual way:

$$\tau_\nu(\varepsilon) \equiv \int \kappa_\nu(\varepsilon)\, \rho \, dr \, .$$

Clearly an expression similar to the above may be defined for each neutrino and antineutrino species under consideration.

Neutrino Emission and Absorption

As indicated in (6.94) the neutrino emission term $A_\nu(T, Z)$ is a function of temperature and composition. The equation for the time rate of change of the spectral energy F_ν given by (6.93) will be differenced implicitly (see below). Arguments similar to those following (6.62) apply to the temperature and composition dependence in (6.94), namely that the emission term should be evaluated at the new values of temperature and charge, T^{n+1} and Z^{n+1}. Linearizing about the initial values (as was done in 6.65) for photons), the emission term may be written as

$$A_\nu(T^{n+1}, Z^{n+1}) = A(T^n, Z^n) + (\frac{\partial A}{\partial T})_Z \frac{\partial T}{\partial t} \Delta t + (\frac{\partial A}{\partial Z})_T \frac{\partial Z}{\partial t} \Delta t$$

$$= A - [(\frac{\partial A}{\partial T})_Z \frac{\Delta \varepsilon}{\rho C_V} + (\frac{\partial A}{\partial Z})_T \frac{\Delta \varepsilon}{n_b \varepsilon}](\frac{\partial F}{\partial t})_c \Delta t$$

$$\equiv A - \frac{\delta A}{\delta F} (\frac{\partial F}{\partial t}) \Delta t$$

(6.96)

where (6.94) and (6.95) have been used to eliminate the time rate of change of the temperature and charge, and all energy group indices have been omitted on the right hand side. Replacing $A_\nu(T,Z)$ on the right side of (6.94) by (6.96), solving for $(\partial F_\nu/\partial t)_c$ and substituting the result into (6.93) yields the relation

$$\frac{\partial F_\nu}{\partial t} = \nabla \cdot D_\nu \nabla F_\nu + K_{\nu,\,eff}\,(A_\nu - F_\nu)$$

(6.97)

where the effective coupling coefficient $K_{\nu,eff}$ is defined by

$$K_{\nu,\,eff} \equiv \frac{K_\nu}{1 + K_\nu\,(\delta A_\nu / \delta F_\nu)\,\Delta t}\,.$$

(6.98)

Equation (6.97) is analogous to the transfer equation whose finite difference form is given by (6.68) (for photons, $A_\nu \rightarrow B_\nu(\theta)$ and $(\partial B_\nu/\partial Z)_T = 0$). The effective coupling coefficient for photons is such that $\kappa_{\nu,eff} \leq \kappa_\nu$ for all T and ρ (recall the discussion following (6.67)). The quantity $\delta A_\nu/\delta F_\nu$ for neutrinos is more complicated (primarily because the neutrino distribution function contains a chemical potential which depends on ρ and T), and care should be taken in writing a finite difference approximation to (6.98) which does not change sign for numerical reasons.

Spatial Diffusion of Neutrinos

We first consider the change in the neutrino spectrum $F_\nu(\varepsilon)$ associated with diffusion (the independent variable ε will be omitted below for notational convenience). The change in F_ν due to spatial diffusion and to emission and absorption of neutrinos can be expressed in finite difference form by integrating (6.93) over the zonal volume $\Delta V_{k+1/2}$:

$$\frac{F^{n+1}_{k+1/2} - F^{n}_{k+1/2}}{\Delta t^{n+1/2}} = \frac{1}{\Delta V_{k+1/2}} \int \frac{1}{r^2}\frac{\partial}{\partial r}\left(r^2 D_\nu \frac{\partial F_\nu}{\partial r}\right) 4\pi r^2 dr$$

$$= \frac{4\pi r_{k+1}^2 D_{\nu,\,k+1} \left(\frac{\partial F_\nu}{\partial r}\right)_{k+1} - 4\pi r_k^2 D_{\nu,\,k} \left(\frac{\partial F_\nu}{\partial r}\right)_k}{\Delta V_{k+1/2}} \, . \tag{6.99}$$

Expressing the radial gradient of F_ν in finite difference form leads immediately to the diffusion equation. Explicit differencing of this equation often leads to severe time step limitations, so it is usually advisable to use an implicit scheme similar to the one discussed in Problem 4.8 and 4.24. A time centered average can be used which would yield second order accuracy. However, the fully implicit time difference has been found to be more robust and will be used below. Since (6.97) including the emission and absorption terms is linear in F_ν, the entire equation may be solved implicitly. When the details of neutrino emission and absorption processes are considered, particularly in very dense matter, the presence of blocking factors will result in nonlinearities in (6.97). This point is discussed by Pomraning for photons (Pomraning 1973; for neutrinos see Bowers and Wilson 1982). For the linear (or linearized) equations the following finite difference representation may be used (common energy subscript $j+1/2$ omitted):

$$F_{k+1/2}^{n+1} - F_{k+1/2}^{n} = \frac{4\pi r_{k+1}^2 D_{k+1}}{\Delta V_{k+1/2}} \frac{(F_{k+3/2}^{n+1} - F_{k+1/2}^{n+1})}{\Delta r_{k+1}} \Delta t^{n+1/2}$$

$$- \frac{4\pi r_k^2 \; D_k}{\Delta V_{k+1/2}} \frac{(F_{k+1/2}^{n+1} - F_{k-1/2}^{n+1})}{\Delta r_k} \Delta t^{n+1/2}$$

$$+ K_{eff,\,k+1/2} (A_{k+1/2} - F_{k+1/2}^{n+1}) \, \Delta t^{n+1/2} . \tag{6.100}$$

The centering and finite difference representation of the diffusion coefficient remains to be specified. Although scattering has not been included as a source term in (6.97) its effect on the mean free path may be introduced as in (6.36). The mean free path, or equivalently the absorption coefficient is naturally zone edge centered because it enters into the flux of energy entering or leaving a zone. The zone edge centered optical depth used in approximating the transfer equation (6.74) may be used as a guide in centering the diffusion coefficient for multigroup diffusion. Thus it would seem reasonable to set (recall that $\kappa_\nu(\varepsilon)\rho = \lambda_\nu(\varepsilon)^{-1}$)

$$\frac{\Delta r_k}{\lambda_{\nu,\,k}} \equiv \frac{1}{2}\left(\frac{\Delta r_{k+1/2}}{\lambda_{\nu,\,k+1/2}} + \frac{\Delta r_{k-1/2}}{\lambda_{\nu,\,k-1/2}}\right). \tag{6.101}$$

Problem 6.12

The flux of neutrino energy in the energy group ε is given by $F_\nu(\varepsilon) = D_\nu(\partial F_\nu/\partial r)$. Show that if the flux from $k-1/2$ to k equals the flux from k to $k+1/2$, then

$$F_\nu = \frac{F_{\nu,\,k+1/2} - F_{\nu,\,k-1/2}}{\dfrac{\Delta r_{k+1/2}}{2\,\lambda_{k+1/2}} + \dfrac{\Delta r_{k-1/2}}{2\,\lambda_{k-1/2}}},$$

which is equivalent to (6.101).

The zone edge centered mean free path (6.101) has been found to work well in multigroup simulations of neutrino flow in optically thick matter. However, when $\lambda_\nu(\varepsilon) > l$, where l is a representative scale length over which physical variables change, (6.100) with a diffusion coefficient based on (6.101) may predict radiation front propagation in excess of the speed of light. When in a numerically well resolved calculation $l > \Delta r$ and $\lambda_\nu(\varepsilon) > l$ leading to $\lambda_\nu(\varepsilon) >> \Delta r$ the material is optically thin to radiation of energy ε. In this case radiation should free-stream with speed c, and the energy lost from a zone of radial width $\Delta r_{k+1/2}$ should be at most $(1/3)F_{k+1/2}\, c\Delta t\, /\, \Delta r_{k+1/2}$. A finite difference representation of the diffusion coefficient which reduces to (6.101) when F_ν is smoothly varying in space, and which approaches (but does not exceed) the free streaming limit as $\lambda_\nu(\varepsilon) \rightarrow 0$ is given by

$$D_{\nu,k} = \frac{c/3}{\dfrac{1}{\lambda_{\nu,k}} + \dfrac{C_L |F_{\nu,k+1/2} - F_{\nu,k-1/2}|}{\Delta r_k \dfrac{(F_{\nu,k+1/2} + F_{\nu,k-1/2})}{2}}} . \tag{6.102}$$

The coefficient C_L may be taken to be $1/3$. For some problems greater accuracy can be obtained by replacing C_L by a function of $\Delta F/F$ so as to better match the true flow through the region where $\tau_\nu \approx 1$. An example for multigroup neutrino transport is given in Bowers and Wilson (1982). The second term in the denominator of (6.102), which is a finite difference representation of $|\partial \ln F_\nu / \partial r|$, is known as a flux limiter. Its introduction into the diffusion approximation is a simple way of interpolating between the optically thick and the optically thin regions. It can clearly not be expected to accurately represent the physics when $\tau_\nu \approx 1$. Nevertheless, flux limited diffusion has been found empirically to be a useful way of allowing for both limits $\tau >> 1$ and $\tau << 1$ without resorting to the time and cost of solving the transfer equation. Further discussions of flux limiters may be found in the literature (Mihalas and Mihalas 1984; Alme and Wilson 1974; and Levermore and Pomraning 1981).

Problem 6.13

Use order of magnitude arguments to show that if (6.102) is used in the diffusion equation, the free streaming limit results when $\lambda_\nu \rightarrow \infty$ if $C_L = 1/3$.

Solution of the Implicit Diffusion Equation

The implicit finite difference approximation to the diffusion approximation including emission and absorption may be solved by the method of Gaussian elimination. Define the coefficient

$$\mathcal{D}_k \equiv \frac{4\pi r_k^2}{\Delta z_k} D_k \, \Delta t^{n+1/2} ;$$

(6.103)

then the multigroup diffusion equation (including emission and absorption) may be rewritten in the form

$$\alpha_{k+1/2} F_{k+3/2}^{n+1} - \beta_{k+1/2} F_{k+1/2}^{n+1} + \gamma_{k+1/2} F_{k-1/2}^{n+1} + \delta_{k+1/2} = 0,$$

(6.104)

with the coefficients

$$\alpha_{k+1/2} \equiv \frac{\mathcal{D}_{k+1}}{\Delta V_{k+1/2}} \qquad \gamma_{k+1/2} \equiv \frac{\mathcal{D}_k}{\Delta V_{k+1/2}} ,$$

(6.105)

$$\beta_{k+1/2} = 1 + \frac{\mathcal{D}_{k+1}}{\Delta V_{k+1/2}} + \frac{\mathcal{D}_k}{\Delta V_{k+1/2}} + K_{eff} \, \Delta t^{n+1/2} ,$$

(6.106)

$$\delta_{k+1/2} = F_{k+1/2}^{n} + K_{eff} A_{k+1/2} \, \Delta t^{n+1/2} .$$

(6.107)

In (6.103)-(6.107) the energy index $j+1/2$ has been omitted. The implicit equation (6.104) can be solved using the method of Gaussian elimination. The solution for F^{n+1} at $k-1/2$ can be expressed in terms of the value at $k+1/2$:

$$F^{n+1}_{k-1/2} = A_{k-1/2} F^{n+1}_{k+1/2} + B_{k-1/2} \, . \tag{6.108}$$

This expression may be substituted into (6.104) and the resulting equation rewritten in the form of the original solution (6.108):

$$F^{n+1}_{k-1/2} = \frac{\alpha_{k-3/2}}{\beta_{k-3/2} - \gamma_{k-3/2} A_{k-3/2}} F^{n+1}_{k+1/2} + \frac{\delta_{k-3/2} + \gamma_{k-3/2} B_{k-3/2}}{\beta_{k-3/2} - \gamma_{k-3/2} A_{k-3/2}} \, .$$

Direct comparison of this result with (6.108) leads immediately to the identification of the coefficients $A_{k+1/2}$ and $B_{k+1/2}$ in terms of the coefficients (6.105)-(6.107):

$$A_{k+1/2} \equiv \frac{\alpha_{k-1/2}}{\beta_{k-1/2} - \gamma_{k-1/2} A_{k-1/2}} \, , \tag{6.109}$$

$$B_{k+1/2} \equiv \frac{\delta_{k-1/2} + \gamma_{k-1/2} B_{k-1/2}}{\beta_{k-1/2} - \gamma_{k-1/2} A_{k-1/2}} \, . \tag{6.110}$$

The coefficients $A_{k+1/2}$ and $B_{k+1/2}$ depend on $A_{k-1/2}$ and $B_{k-1/2}$, and must therefore be constructed sequentially, beginning with $A_{1/2}$ and $B_{1/2}$ and moving outward to $k = K$. The values of $A_{1/2}$ and $B_{1/2}$ are determined by the boundary conditions at $k = 1/2$. Once they have been constructed, (6.108) may be used to find $F^{n+1}{}_{k-1/2}$ in terms of the value ahead, $F^{n+1}{}_{k+1/2}$. Thus, solution of the implicit difference equation (6.104) requires a forward sweep throught the grid to construct the coefficients (6.109)-(6.110), followed by a backward sweep to find the new spectra using (6.108). Finally the entire procedure must be carried out for each energy group until the entire spectrum has been calculated.

Boundary Conditions

Boundary conditions on F_ν at $r = 0$ and at $r = R$, the outer radius of the system, are also required to completely specify the solution to (6.108)-(6.110). For example, we may impose the condition that no point source exist at the center of the system

$$\frac{\partial F_\nu(r,t)}{\partial r}\Big/_{r=0} = 0\,, \tag{6.111a}$$

which leads immediately to the finite difference representation (for each energy group j)

$$F_{1/2,\,j} = F_{3/2,\,j}\,. \tag{6.111b}$$

When this boundary condition is applied to (6.108) it follows that $A_{1/2} = 1$ and $B_{1/2} = 0$. Thus a solution begins at the center, and is carried outward to the boundary of the system. A suitable boundary condition at the outer radius of the grid must also be imposed. For example, if there is no incoming radiation flux, we may set

$$F_\nu(R,t) = 0, \tag{6.112a}$$

which has the equivalent finite difference representation (for each energy group j)

$$F_{K+1/2,\,j} = 0. \tag{6.112b}$$

Other choices of the boundary conditions may also be made depending on the problem to be solved (see Problem 6.14).

Problem 6.14

Modify the boundary conditions above to model a stellar mantle whose inner radius R_o is fixed, but whose outer radius $R(t)$ is a function of time. Assume that the stellar core ($r < R_o$) supplies a neutrino flux at the base of the mantle given by $F_\nu(R_o,t)$. Describe how these boundary conditions are used to solve (6.108)-(6.110).

Problem 6.15

(a). The coefficients (6.105)-(6.106) are all positive (assuming that $F^n{}_{k+1/2,\ j+1/2} > 0$, which is expected on physical grounds), and $\beta_{k+1/2} > \alpha_{k+1/2} + \gamma_{k+1/2}$. Show that $A_{k+1/2} \leq 1$ for all k. What limits, if any, apply to $B_{k+1/2}$?

(b). Starting with the diffusion equation in the form (6.104), show that

$$F^{n+1}_{k+1/2} = A_{k+1/2}\, F^{n+1}_{k-1/2} + B_{k+1/2}\ .$$

is also a solution. How are the coefficients determined in this case?

Problem 6.16

Develop a finite difference representation for multigroup photon diffusion, including emission and absorption terms for one dimensional geometry. Discuss the behavior of the coupling coefficient in the strong and in the weak coupling limits.

The transport approximation (6.100) gives the rate of change in F_ν; summing (6.100) over $\Delta\varepsilon_{j+1/2}$ gives the change in the neutrino energy density (see (6.45)), while summation over $(\Delta\varepsilon/\varepsilon)_{j+1/2}$ gives the change in the neutrino number density (see (6.46)). Thus, (6.100) transports neutrino energy and number together in a consistent manner. Furthermore, (6.100) has been differenced in conservative form, so that the net change in either neutrino number or energy throughout the grid is due (to within numerical roundoff) to boundary effects, and to emission and absorption by the material.

Problem 6.17

Assuming boundary conditions of the form (6.111) and (6.112), show that the finite difference representation (6.100) with $K_{eff} \equiv 0$ conserves neutrino energy and number. Do not assume uniform spatial zoning.

Material Coupling

The material coupling and composition terms (6.94) and (6.95) are solved once $F^{n+1}{}_{k+1/2}$ has been obtained. This is done energy group by energy group so that the diffusion and coupling coefficients, and the emission terms, for the group $j+1/2$ use the temperature T' and the charge Z' reflecting the change in $F_{j-1/2}$ due to coupling from the previous group. Thus, having solved (6.104) for group $j-1/2$, we define

$$\Delta F_{c,\,j-1/2} \equiv K_{eff,\,j-1/2}\,(A_{j-1/2} - F^{n+1}_{j-1/2})\,\Delta t^{\,n+1/2}, \tag{6.113}$$

and find a new (partial) temperature and change:

$$T'_{j-1/2} = T_{j-1/2} - \Delta F_{c,\,j-1/2}\,\frac{\Delta\varepsilon_{j-1/2}}{\rho\, C_V} \tag{6.114}$$

$$Z'_{j-1/2} = Z_{j-1/2} - \Delta F_{c,\,j-1/2}\,\frac{\Delta\varepsilon_{j-1/2}}{n_b\,\varepsilon_{j-1/2}} \tag{6.115}$$

(common spatial index $k+1/2$ omitted). In applying (6.113)-(6.115) the initial and final partial temperature and charge satisfy $T_{3/2} = T^n$,

$Z_{3/2} = Z^n$ and $T_{J+1/2} = T^{n+1}$, $Z_{J+1/2} = Z^{n+1}$. It should be noted that ΔF_C gives the change in T and Z due to emission and absorption; spatial diffusion does not directly change either of these variables. Finally, the swing in T' and Z' from group to group should be monitered and if they are too large either the time step should be lowered, or an iterative scheme [such as the one discussed below (6.68b)] should be used to find the final values of the temperature and charge.

Energy Group Structure

Ideally one would construct a uniform energy group structure, but if the energy span is very large, as for example in stellar core collapse problems where the neutrino energies may range from a few MeV to several hundred MeV, this approach is impractical. The neutrino energy groups may be structured in several ways so as to minimize the actual number of groups needed to resolve the spectrum thoughout a collapse calculation. One choice that works well with about twenty groups is the geometric energy zoning in which the group boundaries are given by

$$\varepsilon_{j+1} = (\varepsilon_2/\varepsilon_1)\,\varepsilon_j \; . \tag{6.116}$$

With this choice of energy group zoning each group spans the same interval in $\Delta(\log\varepsilon)$. The group centered energies can then be chosen to be (see Problem 6.26)

$$\varepsilon_{j+1/2} = (\varepsilon_2/\varepsilon_1)\,\varepsilon_{j-1/2} \; . \tag{6.117}$$

Finally we choose $\Delta\varepsilon_{j+1/2} / \varepsilon_{j+1/2}$ to be a constant, which leads to the relation between group boundary and center values

$$\varepsilon_{j+1/2} = (\varepsilon_2/\varepsilon_1)^{1/2}\,\varepsilon_j \; . \tag{6.118}$$

It is also possible to dynamically modify ε_I and ε_J as well as the group structure during a calculation. This is formally analogous to spatial rezoning, and should be constructed in such a way as to guarantee that energy conservation and number conservation are retained.

6.5 ENERGY EXCHANGE MODEL

In many astrophysical situations electromagnetic radiation will reach thermal equilibrium on time scales that are short compared to the characteristic dynamical time scales of the system. In these instances the spectrum will be essentially Planckian. The radiation energy density is then given by (6.38) where T_R is the radiation temperature. The material may exist in thermodynamic equilibrium at a material temperature $T \neq T_R$. The exchange of energy between the radiation field and matter occurs primarily through electron-photon interactions. If the radiation-electron coupling rates based on these interactions are long compared with electron and ion collision times, then energy exchange between the radiation field and the material's thermal energy will result. In this case, radiation transport will be described by (6.39)-(6.42), and the coupling to matter can be described by

$$\rho C_V \frac{\partial T}{\partial t} = -\rho a c \left[\kappa_p(T) T^4 - \kappa_p(T, T_R) T_R^4 \right] . \tag{6.119}$$

The evolution equation for the radiation field (6.39) gives the rate of change of the total radiation energy density in terms of the diffusion of the energy density (first term on the right) and the negative of the coupling terms in (6.119). When solving these equations it is often convenient to be able to express the right side of (6.119) as a difference between T^4 and $T_R{}^4$. If $\kappa_p(T) = \kappa_p(T,T_R)$, then the emission-absorption term would be proportional to $(T^4 - T_R{}^4)$. Although this is not true in general, an effective coupling coefficient can be found which reduces the emission and absorption term to a quantity proportional to $T^4 - T_R{}^4$ (see Problem 6.18).

Problem 6.18

Use the definition of the generalized Planckian opacity (6.42) to show that the quantity in brackets on the right of (6.119) can be written in the form

$$\kappa_p(T)\, f(T, T_R)\, (T^4 - T_R^4) = K_{ea}(T^4 - T_R^4)$$

(6.120)

where $f(T,T_R)$ depends only on T and T_R, and is well behaved. Estimate the variation in $f(T,T_R)$ assuming that $\kappa_a \approx \varepsilon^{-3} T^{-1/2}$.

Inelastic scattering terms in the transfer equation and the multigroup diffusion approximation were deferred to Chapter 7, because they require special treatment. The effects of inelastic non-relativistic Compton scattering may, however, be included in the energy diffusion approximation. The change in radiation energy density due to these processes can be written in the form

$$\left(\frac{\partial a T_R^4}{\partial t}\right)_c = 4\, \sigma_T n_e \frac{a\, c\, k_B}{m_e c^2}\, T_R^4\, (T - T_R)\,,$$

(6.121)

where σ_T is the Thomson cross section, k_B is Boltzmann's constant, and n_e is the free electron number density. This term is to be added to the right side of (6.39) and subtracted form the right side of (6.119). If the evolution equation is being solved in the form (6.39), then it is convenient rewrite (6.121) as a difference between T^4 and T_R^4. To accomplish this we can multiply (6.121) by $(T^4 - T_R^4)/(T^4 - T_R^4)$ and factor $(T - T_R)$ out of the denominator, in which case (6.121) may be expressed in the equivalent form

$$\left(\frac{\partial aT_R^4}{\partial t}\right)_c = \rho a c K_c (T^4 - T_R^4) . \tag{6.122}$$

It is then possible to express the rate of change of $aT_R{}^4$ due to transport, emission and absorption and inelastic Compton scattering in the form

$$\frac{\partial aT_R^4}{\partial t} = \nabla \cdot D_R \nabla aT_R^4 + a K_{er} (T^4 - T_R^4) , \tag{6.123}$$

where the effective electron-radiation coupling coefficient is

$$K_{er} \equiv (K_{ea} + K_c) \rho c . \tag{6.124}$$

The material coupling, including inelastic Compton scattering, is then given by

$$\rho C_V \frac{\partial T}{\partial t} = - a K_{er} (T^4 - T_R^4) . \tag{6.125}$$

It should be noted that the effective coupling coefficient appearing in (6.125) can be highly nonlinear in temperature; thus instabililities may arise even when using the implicit methods developed above if the temperature changes by a large factor in a single time step. Iterative methods (or a small time step) may be required in solving the equations in this case.

The energy transport equation (6.123) bears a close resemblance to the multigroup transport equation (6.97), and may be solved by similar means. First, a finite difference representation may be obtained by integrating (6.123) over the volume $\Delta V_{k+1/2}$ as in (6.99). The result will be developed for two dimensional, axially symmetric geometry. A modification of these results has been applied to neutrinos in two dimensional spherical coordinates.

Electron-Radiation Coupling

The last term on the right of (6.123) describes the coupling between radiation and matter due to electron-photon interactions. A finite difference representation of this term, and the coupling equation (6.125) may be constructed by methods similar to those used to treat emission and absorption in the multigroup diffusion approximation. Ignoring the diffusion term for the moment, and defining the total radiation energy density

$$u_R = a T_R^4 ,$$

(6.126)

(6.123) and (6.125) may be written in the form

$$\frac{\partial u_R}{\partial t} = \left(\frac{\partial u_R}{\partial t}\right)_c ,$$

(6.127)

where the rate of change includes only the material coupling terms:

$$\left(\frac{\partial u_R}{\partial t}\right)_c \equiv -\rho C_V \frac{\partial T}{\partial t} = \alpha \; K_{er}(aT^4 - u_R) .$$

(6.128)

The constant α will be specified below. In solving (6.128), the material temperature in the coupling term will be taken to be T^{n+1}. Expanding $(T^{n+1})^4$ in (6.128) about T^n and keeping only first order terms leads to an expression for $(\partial u_R / \partial t)_c$ depending on T^n and u_R. If this is substituted into (6.126), we find the finite difference representation (common indices $k+1/2, j+1/2$ omitted):

$$\frac{u_R^{n+1} - u_R^{n}}{\Delta t^{\,n+1/2}} = K_{er,\,eff} \left[a(T^n)^4 - u_R^{n+1}\right] ,$$

(6.129)

$$K_{er,\,eff} \equiv \frac{K_{er}\,\alpha}{1+\frac{4\,a\,T^{3}}{\rho C_V}K_{er}\;\alpha\;\Delta t^{\,n+1/2}}\;.$$

(6.130)

As in the multigroup diffusion approximation for photons, the effective radiation electron coupling coefficient satisfies $K_{er,eff} \leq K_{er}$. Since the electron-radiation coupling is linear in u_R, as the spatial diffusion term, (6.123) may be solved implicitly guaranteeing stability for any $\Delta t^{n+1/2}$ (this is strictly true only for the linearized equations).

Problem 6.19

In the strong coupling limit the electron-radiation coupling satisfies $K_{er}\Delta t \rightarrow \infty$. Find $u_R{}^{n+1}$ from (6.129)-(6.130) under the assumption: (1) $u_R{}^n << u_m{}^n = \rho C_V T^n$, where the matter is hot but very little radiation energy density exists; (2) $u_R{}^n >> u_m{}^n$, cold matter in a hot radiation field. In each case assume for simplicity that $\alpha = 1$.

Spatial Diffusion

In two dimensional axially symmetric systems the spatial diffusion term in (6.123) may be expressed, using cylindrical coordinates, in the form

$$\frac{1}{\Delta V}\int\frac{\partial u_R}{\partial t}dV = \left(\frac{\partial u_R}{\partial t}\right)_{k+1/2,\,j+1/2} = \frac{1}{\Delta V}\int \nabla\cdot D_R \nabla u_R\; dV$$

$$= \frac{1}{\Delta V}\int D_R \nabla u_R \cdot dS$$

$$= \frac{1}{\Delta V}\{2\pi \int [r D_R \frac{\partial u_R}{\partial r}]_j^{j+1} dz + 2\pi \int [D_R \frac{\partial u_R}{\partial z}]_k^{k+1} r\, dr\}$$

(6.131)

where ΔV is the zonal volume centered at $k+1/2$, $j+1/2$, and the surface integral is over the zone. The first term on the right side of (6.131) may be approximated by (common axial index $k+1/2$ omitted):

$$\Delta z_{k+1/2} \frac{2\pi}{\Delta V}\{r_{j+1} D_{k,j+1} \frac{(u^{n+1}_{R,j+3/2} - u^{n+1}_{R,j+1/2})}{\Delta r_{j+1}} - r_j D_{R,j} \frac{(u^{n+1}_{R,j+1/2} - u^{n+1}_{R,j-1/2})}{\Delta r_j}\} .$$

(6.132)

The radiation energy density is evaluated at the new time. A similar expression may be obtained for the axial gradient term in (6.131):

$$\frac{2\pi r_{j+1/2} \Delta r_{j+1/2}}{\Delta V}\{D_{R,k+1} \frac{(u^{n+1}_{R,k+3/2} - u^{n+1}_{R,k+1/2})}{\Delta z_{k+1}} - D_{R,k} \frac{(u^{n+1}_{R,k+1/2} - u^{n+1}_{R,k-1/2})}{\Delta z_k}\}$$

(common radial subscript $j+1/2$ omitted). Equation (6.132) couples $u_R{}^{n+1}$ at three points in the radial direction, but involves only one value of the axial index, namely $k+1/2$. Furthermore (6.129) for the material coupling depends only on variables at $k+1/2, j+1/2$. An operator split approach to (6.123) is to express each spatial diffusion term is finite difference form and include with each the source coupling (6.129)-(6.130) with $\alpha = 1/2$. The net result is equivalent to (6.123). Selecting the axial direction gives the finite difference representation (common radial index $j+1/2$ omitted):

$$\frac{u^{n+1}_{R,k+1/2} - u^{n}_{R,k+1/2}}{\Delta t^{n+1/2}} = \frac{2\pi r \Delta r}{\Delta V_{k+1/2}}\{D_{R,k+1} \frac{(u^{n+1}_{R,k+3/2} - u^{n+1}_{R,k+1/2})}{\Delta z_{k+1}}$$

$$- D_{R,\,k} \frac{(u^{n+1}_{R,\,k+1/2} - u^{n+1}_{R,\,k-1/2})}{\Delta z_k} \Big\} + K_{er,\,eff} \left[a\,(T^{n}_{k+1/2})^4 - u^{n+1}_{R,\,k+1/2} \right] .$$

(6.133)

Problem 6.20

Derive a finite difference representation analogous to (6.133) for the change in radiation energy density (6.123) due to diffusion in the radial direction which includes the coupling between radiation and matter. Compare your results with (6.133).

The diffusion coefficient D_R in (6.133) is evaluated at the zone edge. However, according to (6.40) D_R involves the Rosseland mean opacity, which is naturally defined at the zone center in terms of the density and temperature there. Zone edge centered opacities have been encountered in (6.74) and (6.101) for the transfer equation. Under many astrophysical conditions the frequency dependence of the opacity is approximately of the form (6.75), and the equivalent Rosseland mean opacity (6.41) is approximately proportional to $T^{-3.5}$.

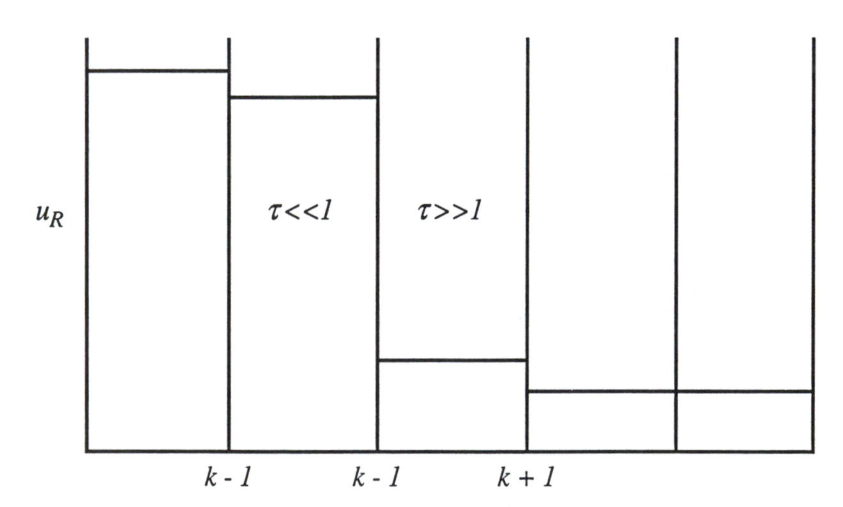

Figure 6.9. A diffusion front at k, assuming a Rosseland mean opacity $\kappa_R \sim T^{-3.5}$. The optical depth $\tau \sim \Delta x/\lambda_R \sim \rho \kappa_R \Delta x$. The material to the right (left) of the zone boundary k is assumed to be cold (hot).

Consider the propagation of a radiation front into cold matter, and suppose that the optical depth average (6.74) were used for $D_{R,k}$. Figure 6.9 shows the finite difference representation of a diffusion front located near zone k. If the zone edge centered opacity is represented by the optical depth average (6.74), then $D_{R,k}$ will be determined essentially by the opacity ahead of the front, which is large, and the flow of radiation will be impeded (recall that in the transfer or multigroup approximations the front is less steep, since $\lambda_\nu \approx \nu^3$; high energy photons can, therefore, move ahead of the bulk of the wave and heat the material, which usually reduces the opacity for lower energy photons comming from behind). A better approximation, particularly in two dimensional simulations where zoning resolution is limited, is the inverse average give by

$$\frac{1}{\tau_{R,\,k}} = \frac{1}{2}\left(\frac{1}{\tau_{R,\,k+1/2}} + \frac{1}{\tau_{R,\,k-1/2}}\right) \tag{6.134}$$

(common index $j+1/2$ omitted). When (6.134) is used to describe the front in Figure 6.9, $\tau_{R,k} \approx \tau_{R,k-1/2}$ which is small; this facilitates radiation flow into optically thick regions. The averaging procedure (6.134) works best when steep radiation gradients are present. For regions where the gradients are small, an average similar to (6.101) is most accurate.

Problem 6.21

Show that when steady state radiation flow occurs

$$\tau_k = \frac{1}{2}\left(\tau_{k+1/2} + \tau_{k-1/2}\right)$$

represents a better average optical depth.

As in the multigroup diffusion approximation, a radiation front may propagate through optically thin matter with a speed exceeding the speed of light. The use of a flux limited diffusion coefficient (see (6.102) and the accompanying discussion) circumvents this difficulty. The following form has been found to be useful in practice:

$$D_{R,\,k} = \frac{\Delta z_k \, c/3}{\rho_k \, \kappa_{R,\,k} \, \Delta z_k + C_L \dfrac{|u_{R,\,k+1/2} - u_{R,\,k-1/2}|}{\frac{1}{2}(u_{R,\,k+1/2} + u_{R,\,k-1/2})}} \tag{6.135}$$

(common index $j+1/2$ omitted). A similar expression is readily derived for $D_{R,\,j,\,k+1/2}$ which appears in the equation for diffusion in the radial direction. Typically the same coefficient $C_L = 1/3$ may be used for radial or axial diffusion.

Solution Method for Energy Transport

The implicit transport equation (6.133) may be expressed in the general form (6.104), and may be solved assuming that $u^{n+1}{}_{R,k-1/2}$ satisfies an expression of the form (6.108), subject to suitable boundary condition (see (6.111) and the accompanying discussion). This may be done in two dimensional axially symmetric geometry in the following order. First, the radiation energy density u'_R due to diffusion in one direction plus one half of the material coupling is computed. Then one half of the material coupling given by (6.128) is computed using u'_R and T^n to obtain the new (partial) temperature T: in finite difference form this is given by ($\alpha = 1/2$)

$$\frac{T' - T^n}{\Delta t^{n+1/2}} = - K_{er,\,eff} \frac{[a(T^n)^4 - u'_R]}{\rho \, C_V} \tag{6.136}$$

(common subscripts $k+1/2, j+1/2$ omitted). Next, the final radiation energy density $u^{n+1}{}_R$ due to diffusion in the remaining direction plus one half of the coupling is obtained. This step uses T' in the coupling term, but does not recompute the opacity. The last half of the material coupling is computed using (6.136) with T and u'_R replaced by T' and $u^{n+1}{}_R$, respectively. The result is T^{n+1}, the final temperature. Note that neither ρ nor C_V are changed in the process (see Chapter 3). The order of solution is summarized in Table 6.1, where each step is expressed in terms of a formal partial differential equation.

Table 6.1

Order if solution of the diffusion plus emission and absorption terms for an axially symmetric two dimensional problem. Finite difference equations are written for each partial differential equation below, and are solved in the order of appearance (from top to bottom).

$$\frac{\partial u_R}{\partial t} = \frac{\partial}{\partial z}\left(D_R \frac{\partial u_R}{\partial z}\right) + K_{er,\,eff}\,(T^4 - u_R)$$

$$\rho C_V \frac{\partial T}{\partial t} = -\,K_{er,\,eff}\,(T^4 - u_R)$$

$$\frac{\partial u_R}{\partial t} = \frac{1}{r}\frac{\partial}{\partial r}\left(r D_R \frac{\partial u_R}{\partial r}\right) + K_{er,\,eff}\,(T^4 - u_R)$$

$$\rho C_V \frac{\partial T}{\partial t} = -\,K_{er,\,eff}\,(T^4 - u_R)$$

Stability and Accuracy

Because the implicit finite difference approximations described above are unconditionally stable (see Chapter 2), material coupling and diffusion do not place constraints on the computational time step. Nevertheless, if Δt is too large, the radiation field (u_R in the energy exchange formulation, or F_V for multigroup diffusion) may oscillate excessively (for some problems T will oscillate for all Δt, but at an acceptable level) during alternate direction sweeps and yield inaccurate results. In such cases it is necessary to reduce the time step in order to obtain more accurate results. This is a consequence of calculations involving implicit methods; while time step limits are not necessary to insure stability, they may be necessary to guarantee accuracy.

The use of operator splitting to solve separately for the change in u_R associated with radial and axial diffusion in (6.131) does not always yield physically reasonable results. The difficulty arises because more radiation can sometimes be transported along one direction when operator splitting is used than is permitted if diffusion in both directions were computed simultaneously. A better approach in such instances is to retain all of the terms on the right side of (6.131) at the finite difference level. The resulting scheme, which must be solved implicitly in practice, is a five point scheme in which the temperature in four surrounding zones is coupled to the temperature in a given zone. The system of equations is formally identical to approximations to Poisson's equation, and can be expressed in band matrix form. Methods for solving Poisson's equation will be discussed in Chapter 9.

The use of quantities at the advanced time to represent linearized implicit diffusion equations is unconditionally stable, but is only first order accurate in time. Using this approach for problems involving only radiation diffusion without an accuracy time step constraint may produce poor results. The method usually works well when employed in conjunction with other physical processes, such as hydrodynamics, because the latter generally control the time step. Therefore, some form of time centering should be considered for problems involving only radiation processes (a discussion of time averaging methods may be found in Richtmyer and Morton, 1967).

6.6 EQUILIBRIUM DIFFUSION

The equilibrium diffusion equation (6.50a) was discussed at the end of Section 6.2. The major complication in solving (6.50a) is the nonlinear temperature dependence of the time derivative. In systems where the material energy density greatly exceeds the radiation energy density, $aT^3/\rho C_V << 1$, the rate of change in aT^4 appearing on the left side of (6.50a) may be neglected, and it may be written in the following Lagrangian form:

$$\rho C_V \frac{dT}{dt} = \nabla \cdot \left(\frac{4acT^3}{3\kappa\rho} \nabla T\right) \equiv \nabla \cdot K_{th} \nabla T \, . \tag{6.137}$$

The thermal diffusion equation (which has the same structure as the other radiation diffusion equations discussed above) is usually solved implicitly. If the thermal diffusion coefficient K_{th} does not change rapidly in time or in space, this equation may be solved by writing it in an implicit form analogous to (6.100), and by using Gaussian elimination to solve for the new temperature (Problem 6.22).

Problem 6.22

Construct a linear implicit solution to (6.137) for a one dimensional spherically symmetric system. Discuss the spatial centering of the thermal diffusion coefficient. Write out the back substitution coefficients. What boundary conditions would be appropriate for a spherically symmetric star?

Thermal conduction occurs when electrons (and, to a lesser extent, ions) transport energy from regions of higher to lower temperature (Reif 1965). The rate of change in material energy due to thermal conduction is described by an equation of the form (6.137)

$$\rho C_V \frac{dT}{dt} = \nabla \cdot K_c \nabla T, \tag{6.138}$$

where K_c is the thermal conductivity. This equation may be solved by essentially the same methods as (6.137).

Problem 6.23

Construct a single partial differential equation describing thermal conduction and thermal diffusion for a one dimensional spherically symmetric system. From it, find a finite difference representation, including boundary conditions, which could be applied to the cooling of a white dwarf.

When $\rho C_V / aT^3 \geq 1$ the radiation energy density appearing in (6.50a) can not be ignored. However, in this regime (typical of the hot cores of stars less massive than the sun) it may be possible to linearize the temperature dependence (evaluate the thermal diffusion coefficient and the effective heat capacity (6.50b) using the temperature at the beginning of a cycle) and solve the resulting equation implicity. Alternately, the fully implicit methods described for Lagrangian hydrodynamics in Section 4.7 may be adopted. One advantage of the latter approach is that very long time steps may often be used for stages of stellar evolution which are long compared with hydrodynamic time scales. Another advantage is that the iterative approach includes nonlinear effects which may be important if the temperature changes rapidly (as, for example, when a strong radiation front moves into cooler matter). The method described below may be implemented in conjunction with implicit hydrodynamics using, for example, operator splitting (Oliphant and Witte 1987).

For a one dimensional spherically symmetric system (6.50a) may be written in the following Lagrangian form:

$$\rho C_V \frac{dT}{dt} + \frac{d aT^4}{dt} = \rho \frac{\partial}{\partial m}\left(\tilde{D} \frac{\partial aT^4}{\partial m}\right), \tag{6.139}$$

where

$$\tilde{D} \equiv (4\pi r^2)^2 \rho D_R \, . \tag{6.140}$$

The density and radii to be used in solving a finite difference representation of (6.139) are the latest values obtained from the hydrodynamics, and thus represent their values at time t^{n+1}. It may be recalled that the operator split form of the diffusion equation does not change the material velocity or density, and that radiation acceleration of the material is to be included in the momentum equation. The material heat capacity is the value at time t^n. Operator splitting described in Chapter 3 may be used following the solution of (6.139) to obtain final values of the material heat capacity and temperature (see Problem 6.25). The hydrodynamics may have changed the material internal energy; denoting the value of the temperature corresponding to the internal energy at the end of the hydrodynamics by T', a finite difference representation of (6.139) is

$$\frac{1}{a}\rho^{n+1}_{k+1/2}C^{n}_{V,k+1/2}\left(\frac{T^{n+1}_{k+1/2}-T'_{k+1/2}}{\Delta t^{n+1/2}}\right)+\frac{(T^{n+1}_{k+1/2})^4-(T'_{k+1/2})^4}{\Delta t^{n+1/2}}=$$

$$\frac{\rho^{n+1}_{k+1/2}}{\Delta m_{k+1/2}}\left\{\tilde{D}_{k+1}\frac{(T^{n+1}_{k+3/2})^4-(T^{n+1}_{k+1/2})^4}{\Delta m_{k+1}}-\tilde{D}_{k}\frac{(T^{n+1}_{k+1/2})^4-(T^{n+1}_{k-1/2})^4}{\Delta m_k}\right\}.$$

(6.141)

The opacity appearing in the diffusion coefficient D_R depends on the density and temperature. The temperature dependence is another source of nonlinearity in (6.141). It is usually sufficient to use T' in evaluating the opacity. While this approximation is not necessary in the iterative approach discussed below, its use can represent a significant time saving. Therefore, we shall assume that

$$D_R = D_R(\rho^{n+1}, T').$$

(6.142)

Whichever value of the temperature is used in (6.142), inspection of (6.141) shows that it can be written as a set of functions of the form

$$f_{k+1/2}(T^{n+1}_{k+3/2}, T^{n+1}_{k+1/2}, T^{n+1}_{k-1/2}) = 0.$$

(6.143)

A similar set of function of zone radiii (4.123) served as the starting point for the solution of the implicit hydrodynamic equations in one dimensional Lagrangian form (Section 4.7). That method may be used to solve (6.143). In particular, the partial derivatives of $f_{k+1/2}$ with respect to each argument are constructed (analytically or numerically) and the results used to construct the set of tridiagonal equations

$$\left(\frac{\partial f_{k+1/2}}{\partial T^n_{k+3/2}}\right) \delta T^n_{k+3/2} + \left(\frac{\partial f_{k+1/2}}{\partial T^n_{k+1/2}}\right) \delta T^n_{k+1/2} + \left(\frac{\partial f_{k+1/2}}{\partial T^n_{k-1/2}}\right) \delta T^n_{k-1/2}$$

$$= f_{k+1/2}(T^n_{k+3/2}, T^{n+1}_{k+1/2}, T^{n+1}_{k-1/2}).$$

(6.144)

The increments in temperature, $\delta T^{n+1}{}_{k+1/2}$, may be found using Gaussian elimination. Once a convergent solution has been obtained, the temperature is given by

$$T^{n+1}_{k+1/2} = T'_{k+1/2} + \delta T^{n+1}_{k+1/2} \,.$$

(6.145)

The method can also be applied to (6.141) when the opacity is evaluated implicitly $[D_R = D_R(\rho^{n+1}, T^{n+1})]$. The calculation of the opacity for each interation leading to (6.145) may be time consuming, particularly if composition changes occur, but the results represent a solution to the fully nonlinear problem. Finally, it should be evident that the implementational procedure described in Section 4.7 can be applied to (6.141) with negligible changes in coding.

Problem 6.24

A simplified stellar opacity is given by

$$\kappa(\rho, T) = \kappa_o \, \rho \, T^{-\alpha}$$

where $\alpha > 0$, and κ_o is a constant (which may depend on composition). Using this approximation, find the coefficients of the tridiagonal equation (6.144) corresponding to the thermal diffusion equation (6.141).

Problem 6.25

The thermal diffusion equation may be expressed in conservative form as

$$\frac{d(\rho C_V T)}{dt} + \frac{d(aT^4)}{dt} = A, \tag{6.146}$$

where A represents the spatial diffusion term. This may be rewritten in an equivalent, operator split form consisting of (6.50a) and the additional equation

$$\frac{d(\rho C_V T)}{dt} + \frac{d(aT^4)}{dt} = 0. \tag{6.147}$$

Describe the solution of this equation following the solution of (6.146) to obtain the final temperature and heat capacity of the material (see Section 3.2).

Problem 6.26

(a). The Planck function (6.18a) satisfies the analytic relationship (6.18b). Consider a group structure defined by $\varepsilon_{j+1} = (\varepsilon_2/\varepsilon_1)\varepsilon_j$ consisting of twelve energy groups with $\varepsilon_1 = 1\ MeV$. Replace the integral

$$\int_0^\infty \frac{\varepsilon^3 d\varepsilon}{e^{\varepsilon/kT} - 1} = \frac{\pi^4}{15}$$

by a finite sum on the energy grid, and compare the sum to the exact value of the integral. Assume that the group centered energies are given by $\varepsilon_{j+1/2} = (\varepsilon_1/\varepsilon_2)^{1/4}\varepsilon_j$. Take the material temperature $T = 5\ MeV$.

(b). Repeat the exercise above for an energy group structure defined by $\varepsilon_{j+1} = \eta\varepsilon_j$, and $\varepsilon_{j+1/2} = (\eta+1)\varepsilon_j/2$.

Problem 6.27

Develop the antineutrino emission and absorption terms and material heating terms analogous to (6.94)-(9.97). In particular show what form these equation take when the material temperature, the neutrino and antineutrino spectra and Z are to be solved simultaneously. Assume that the change in antineutrino spectum is given by

$$\frac{\partial \bar{F}_\nu}{\partial t} = \frac{1}{r^2}\frac{\partial}{\partial r}\left(r^2 \bar{D}_\nu \frac{\partial \bar{F}_\nu}{\partial r}\right)$$

and the change in charge is given

$$n_b \frac{\partial Z}{\partial t} = -\left(\frac{\partial F_\nu}{\partial t} - \frac{\partial \bar{F}_\nu}{\partial t}\right)\frac{\Delta\varepsilon}{\varepsilon} .$$

The rate of change in the neutrino spectum is given by (6.93), which includes the scattering term. Omit scattering to simplify the analysis. Do your results conserve baryon and lepton number?

Problem 6.28

The usual form of the one-temperature diffusion equation can be derived from (6.29) by assuming that the rate of change of the specific intensity is small relative to its spatial gradient. Ignoring scattering, show that to lowest order

$$I(\Omega,\varepsilon) \cong B(T,\varepsilon) - \frac{1}{\kappa_a(T,\varepsilon)\,\rho}\,\Omega\cdot\nabla B(T,\varepsilon) .$$

Use this result to evaluate the radiation flux (6.5), the radiation energy density (6.3). Then construct the one-temperature form of the radiation diffusion equation, and show that the Rosseland mean opacity is given by (6.41) with $T_R=T$.

References

M. L. Alme and J. R. Wilson, *Astrophys. J.*, **194**, 147, 1974.

R. L. Bowers and T. Deeming, *Astrophysics* (Boston: Jones and Bartlett, 1984).

R. L. Bowers and J. R. Wilson, *Ap. J. Suppl.* **50**, 115, 1982.

S. I. Braginskii, "Transport Processes in a Plasma," in *Reviews of Plasma Physics*, Vol. 1, 1965.

C. D. Levermore and G. C. Pomraning, *Astrophys. J.*, **248**, 321, 1981.

C. Lund, "Radiation Transport in Numerical Astrophysics," in *Numerical Astrophysics*, edited by J. M. Centrella, J. M. LeBlanc and R. L. Bowers (Boston: Jones and Bartlett, 1985).

D. Mihalas, *Stellar Atmospheres*, Second Edition (San Francisco: Freeman, 1978).

D. Mihalas and B. W. Mihalas, *Foundations of Radiation Hydrodynamics* (New York: Oxford University Press, 1984).

G. C. Pomraning, The Equations of Radiation Hydrodynamics (Oxford: Pergamon Press, 1973).

F. Reif, Fundamentals of Statistical and Thermal Physics (New York: McGraw-Hill, 1965).

R. D. Richtmeyer and K. W. Morton, *Difference Methods for Initial—Value Problems*, Second Edition (New York: Interscience, 1967).

G. B. Rybicki, and A. P. Lightman, *Radiative Processes in Astrophysics,* (New York: John Wiley and Sons, 1979).

S. L. Shapiro and S. A. Teukolsky, *Black Holes, White Dwarfs and Neutron Stars* (New York: Wiley-Interscience, 1983).

L. Spitzer, *The Physics of Fully Ionized Gases* (New York: Interscience, 1962).

D. L. Tubbs and D. N. Schramm, "Neutrino Opacities at High Temperatures and Densities," *Ap. J.* **201**, 467, 1975.

J. R. Wilson, "Neutrino Flow and Stellar Core Collapse," *Ann. N. Y. Acad. Sci.* **336**, 358, 1980.

Ya. B. Zel'dovich, and Yu. P. Raizer, *Physics of Shock Waves and High-Temperature Hydrodynamic Phenomena* (New York: Academic Press, 1966), Vol. I.

7

RADIATION-MATTER COUPLING

The spatial transport of radiation and its coupling to matter via radiation emission and absorption were discussed in the previous chapter under the assumption that the matter was at rest. In this chapter the effects of material motion on radiation transport and the coupling of radiation to hydrodynamics are introduced. The resulting radiation hydrodynamic equations are accurate through terms of order v/c, and are developed in a form appropriate for a one dimension Lagrangian description or for a moving grid Eulerian description. The centering of variables has already been established in Section 6.2, and will be used here.

7.1 RADIATION HYDRODYNAMICS

The derivation of the transfer equation (and of the diffusion limit obtained from it) in Section 6.1 is strictly valid for material at rest. When the frame of reference used to describe the radiation field is noninertial (as in Lagrangian descriptions), or when it is inertial but fluid motion occurs (Eulerian descriptions), the finite speed of light leads to additional terms in the transfer equation, and terms which couple the radiation to the hydrodynamic equations. A full treatment of these effects requires the use of special relativity; a comprehensive review of the physics of radiation hydrodynamics including references may be found, for example, in Mihalas and Mihalas (1984). The results developed here are approximate because they ignore all quantities of order $(v/c)^2$ and higher in the material velocity. Including these higher order terms would require a relativistic treatment of hydrodynamics, which we defer until Chapter 10. There it will be found that relativistic corrections to the nonradiative hydrodynamic equations are of order $(v/c)^2$ or Φ/c^2 and higher, where Φ/c^2 is the Newtonian gravitational potential.

The relativistic terms which are retained below are of order v/c, and are associated with the classical Doppler shift. These terms arise in the Eulerian formulation of radiation hydrodynamics because the radiation field is usually described in the laboratory frame of reference (or with respect to an externally imposed, moving grid), but the absorption coefficients and scattering cross sections which are generally functions of material density and temperature are most naturally defined in the material or comoving frame of reference. Consequently, the source and the observer are in relative motion. In the Lagrangian description, v/c corrections arise because the radiation field is transformed to the material rest frame. Finally, since terms of order $(v/c)^2$ are assumed to be small, and the effects of strong gravitational fields (general relativity) will be ignored, the results developed here may be combined into a consistent framework with the Newtonian equations of hydrodynamics developed in Chapter 4 and Chapter 5. It should be noted that for most applications envisioned here, $v/c << 1$; nevertheless these terms cannot be ignored. The reasons for this will become evident below.

Comoving and Laboratory Frames of Reference

In the material or comoving reference frame (the Lagrangian frame) the absorption coefficient and scattering cross sections will be assumed to be isotropic. If the material is moving relative to an observer, these quantities will no longer appear isotropic to that observer, and angle dependent terms will arise. In keeping with the notation of Chapter 6 we shall use the photon energy $\varepsilon = h\nu$ rather than the frequency as an independent variable. In the comoving frame the absorption coefficient will be denoted by $\kappa_o(\varepsilon_o)$, where ε_o is the energy in that frame (subscript zero will be used throughout to denote quantities measured in this frame). In the laboratory frame the absorption coefficient is denoted by $\kappa(\varepsilon,\mu)$, the angle dependence

μ arising because absorption no longer appears to be isotropic. In general, Lorentz transformations must be used to relate quantities observed in the laboratory (inertial) frame with quantities defined in another inertial frame whose velocity coincides with the instantaneous velocity $\mathbf{v}$ of the material.

Consider the specific intensity, which was defined in (6.2) for matter at rest, and which will be denoted here by $I_o(\varepsilon_o, \mu_o)$. Using the fact that the number of quanta passing through a unit area per unit time, per unit energy, per unit solid angle must be the same for observers in different inertial frames, it can be shown that the quantity $I(\varepsilon, \mu) / \varepsilon^3$ is a Lorentz invariant, and that

$$I(\varepsilon, \mu) = \left(\frac{\varepsilon}{\varepsilon_o}\right)^3 I_o(\varepsilon_o, \mu_o) , \tag{7.1}$$

where the photon energy transforms according to

$$\varepsilon_o = \varepsilon \frac{(1 - \Omega \cdot \mathbf{v} / c)}{[1 - (v/c)^2]^{1/2}} . \tag{7.2}$$

In (7.2) Ω is the unit vector specifying the direction of the radiation in the laboratory frame. In the following discussion all terms of order $(v/c)^2$ and higher will be ignored; consequently the factor $[1 - (v/c)^2]^{1/2}$ in the denominator of (7.2) may be ignored.

The number of quanta emitted per unit energy, per unit time, per unit solid angle from a unit volume of matter must also be invariant with respect to Lorentz transformations. Formally this constraint on the source function is expressed by

$$j(\varepsilon, \mu) = \left(\frac{\varepsilon}{\varepsilon_o}\right)^3 j_o(\varepsilon_o) . \tag{7.3}$$

Similar arguments may be made about the invariance of the number of quanta absorbed by matter from the radiation field, a result which may be expressed formally as

$$\kappa(\varepsilon, \mu) = \left(\frac{\varepsilon}{\varepsilon_o}\right) \kappa_o(\varepsilon_o).$$

(7.4)

Note that the source function (7.3) and the absorption coefficient (7.4) are independent of μ_o in the comoving frame. This is clear motivation for expressing all material or dependent variables in the comoving frame.

Comoving Transfer Equation

In an inertial frame of reference the transfer equation is, in one dimensional spherical coordinates,

$$\left\{\frac{1}{c}\frac{\partial}{\partial t} + \mu\frac{\partial}{\partial r} + \frac{(1-\mu^2)}{r}\frac{\partial}{\partial \mu}\right\} I(\varepsilon, \mu) = j(\varepsilon, \mu) - k(\varepsilon, \mu)\, I(\varepsilon, \mu)$$

(7.5)

where $k(\varepsilon,\mu)$ is the absorption coefficient of the material as observed in the laboratory frame of reference. Similarly, $j(\varepsilon,\mu)$ is the source function as observed in the laboratory frame. In (7.5) the independent variables are r, t, ε and μ, and $k(\varepsilon,\mu) \equiv \kappa_a(\varepsilon,\mu)\rho$. In particular the quantities ε and μ are independent of t. Equations (7.1) through (7.4) may be used to transform the dependent variables appearing in (7.5) from the inertial frame to the comoving frame. When this is done it must be remembered that because the comoving frame is noninertial ε_o and μ_o will vary with time. Taking the first term in (7.5) as an example, and substituting (7.1)

$$\frac{1}{c}\left[\frac{\partial I(\varepsilon, \mu)}{\partial t}\right]_{\varepsilon, \mu, r} = \frac{1}{c}\frac{\partial}{\partial t}\left[\left(\frac{\varepsilon}{\varepsilon_o}\right)^3 I_o(\varepsilon_o, \mu_o)\right]$$

$$= \left(\frac{\varepsilon}{\varepsilon_o}\right)^3 \left\{ \frac{1}{c} \left[\frac{\partial I_o(\varepsilon_o, \mu_o)}{\partial t} \right]_{\varepsilon, \mu, r} - \frac{I_o(\varepsilon_o, \mu_o)}{\varepsilon_o} \frac{\partial \varepsilon_o}{\partial t} \right\} .$$

(7.6)

The chain rule for partial differentiation may be used to express the partial derivative of I_o in the equivalent form

$$\left(\frac{\partial I_o}{\partial t}\right)_{\varepsilon, \mu, r} = \left(\frac{\partial I_o}{\partial t}\right)_{\varepsilon_o, \mu_o, r}$$

$$+ \left(\frac{\partial I_o}{\partial \varepsilon_o}\right)_{\mu_o, r, t} \left(\frac{\partial \varepsilon_o}{\partial t}\right)_{\varepsilon, \mu, r} + \left(\frac{\partial I_o}{\partial \mu_o}\right)_{\varepsilon_o, r, t} \left(\frac{\partial \mu_o}{\partial t}\right)_{\varepsilon, \mu, r} .$$

(7.7)

Similar expressions may be obtained for the terms $(\partial I_o / \partial r)_{\varepsilon, \mu, t}$ and $(\partial I_o / \partial \mu)_{\varepsilon, r, t}$ which appear in (7.5). Next, the transformation of the energy ε and the angle μ to the comoving frame are used to evaluate terms such as $(\partial \varepsilon_o / \partial t)_{\varepsilon, \mu, r}$. Through order (v/c), these are easily found to be

$$\varepsilon_o = \varepsilon\,(1 - \mu v/c)$$

(7.8)

$$\mu_o = \mu - \frac{v}{c}(1 - \mu^2) ,$$

(7.9)

and we find, for example, that

$$\frac{\partial \varepsilon_o}{\partial t} = \frac{\varepsilon \mu}{c} \frac{\partial v}{\partial t} \approx \frac{\varepsilon_o \mu_o}{c} \frac{\partial v}{\partial t} .$$

Proceeding in this manner each term in (7.5) may ultimately be reexpressed in terms of r, t, μ_o, ε_o and variables defined in the comoving frame. The result is the radiative transfer equation in the comoving frame

$$\frac{1}{c}\left(\frac{\partial I_o}{\partial t} + v\frac{\partial I_o}{\partial r}\right) + \frac{\mu_o}{r^2}\frac{\partial (r^2 I_o)}{\partial r} + \frac{1}{r}\frac{\partial}{\partial \mu_o}\{(1-\mu_o^2)[1+\frac{r\mu_o}{c}(\frac{v}{r}-\frac{\partial v}{\partial r})]\, I_o\}$$

$$-\frac{\partial}{\partial \varepsilon_o}\{\varepsilon_o[(1-\mu_o^2)\frac{v}{rc} + \frac{\mu_o^2}{c}\frac{\partial v}{\partial r}]\, I_o\} + \frac{v}{c}\frac{(3-\mu_o^2)}{r}I_o$$

$$+\frac{(1+\mu_o^2)}{c}\frac{\partial v}{\partial r}I_o = j_o(\varepsilon_o) - k_o(\varepsilon_o)I_o(\varepsilon_o,\mu_o)\ . \tag{7.10}$$

In the derivation the terms that are proportional to the fluid acceleration, $(\partial v / \partial t)$, have been ignored since they can be shown to be of order $(v / c)^2$. Finally we note that $I_o = I_o(\varepsilon_o,\mu_o)$. Equation (7.10) represents the transfer equation in the comoving frame through terms of order (v / c). When $v = 0$, (7.10) reduces to the static transfer equation given by (6.10), (6.17) and (6.19) when scattering is omitted. The first two terms on the left side of (7.10) are just the Lagrangian time derivative of the specific intensity:

$$\frac{\partial I_o}{\partial t} + v\frac{\partial I_o}{\partial r} = \frac{dI_o}{dt}\ . \tag{7.11}$$

Finally, the photon energy gradient terms, which vanish for matter at rest, represent the shift in the shape of the spectrum due, in part, to Doppler effects. These terms, and the angular gradient terms are expressed in conservative form in (7.10).

Problem 7.1 (Straightforward, but tedious)

Use the transformations (7.1), (7.3)-(7.4) and (7.8)-(7.9) to obtain the transfer equation to order (v / c) in the comoving frame. Evaluate (7.6)-(7.7), and then evaluate in similar fashion the derivatives $(\partial I_o / \partial r)_{\varepsilon,\mu,t}$ and $(\partial I_o / \partial \mu)_{\varepsilon,r,t}$. Finally, collect terms so that the radial and angular derivatives are expressed in conservative form.

The transfer equation (7.10) may be cast into Lagrangian form using (7.11) and the following identities (which follow from the mass continuity equation)

$$\frac{\mu_o}{r^2}\frac{\partial(r^2 I_o)}{\partial r} = 4\pi\mu_o\rho\frac{\partial(r^2 I_o)}{\partial m},$$

$$\frac{\partial v}{\partial r} = -\frac{2v}{r} - \frac{1}{\rho}\frac{d\rho}{dt}. \tag{7.12}$$

Multigroup Equations (Diffusion Regime)

For many applications it is not necessary to follow the evolution of the radiation field for each angular group, and the angle averaged multigroup approximation is sufficient. We shall start from (7.10), but shall omit for notational convenience the subscript zero from the specific intensity and all other variables. The multigroup approximation is obtained by integrating (7.10) over solid angle for spherical geometry and using the diffusion limit

$$P(\varepsilon) = \frac{1}{3}E(\varepsilon) \qquad \lambda \ll 1. \tag{7.13}$$

This leads to the energy equation (zeroth moment of the transfer equation)

$$\frac{dE}{dt} + \frac{1}{r^2}\frac{\partial(r^2F)}{\partial r} + \frac{4}{3}E\frac{1}{r^2}\frac{\partial(r^2v)}{\partial r}$$

$$- \frac{1}{3} \frac{1}{r^2} \frac{\partial(r^2 v)}{\partial r} \frac{\partial(\varepsilon E)}{\partial \varepsilon} = 4\pi \eta(\varepsilon) - c k(\varepsilon) E. \tag{7.14}$$

The absorption mean free path $\lambda_a(\varepsilon) = k(\varepsilon)^{-1}$, and l is the characteristic gradient scale length for physical variables.

Problem 7.2

Derive the multigroup radiation energy equation (7.14) in the diffusion limit $\lambda(\varepsilon) << l$. Integrate the result over the radiation energy to obtain the total radiation energy equation

$$\frac{dE_R}{dt} + \frac{1}{r^2}\frac{\partial(r^2 F_R)}{\partial r} + \frac{4}{3}\frac{E_R}{r^2}\frac{1}{}\frac{\partial(r^2 v)}{\partial r} = 4\pi\int \eta(\varepsilon)\, d\varepsilon - c\int \kappa(\varepsilon)\rho\, E\, d\varepsilon, \tag{7.15}$$

where E_R is the total radiation energy density and F_R is the total radiation flux. Explain physically why the term proportional to $\partial(\varepsilon E)$ / $\partial\varepsilon$ drops out.

Equation (7.15) is a simple energy balance equation for the radiation field, as can be seen by integrating it over a Lagrangian volume element δV, and using the identity

$$\frac{4}{3}\int E_R\, \nabla\cdot\mathbf{v}\, \delta V = \frac{4}{3}\int E_R \frac{d\,\delta V}{dt} = \int E_R \frac{d\,\delta V}{dt} + \int P_R \frac{d\,\delta V}{dt},$$

which is obtained with the use of the equation below (4.1) for the time rate of change of δV, and the relation between the radiation energy density and radiation pressure [which follows from (7.13)] $P_R = E_R/3$. Using this in (7.15), and collecting terms leads to

$$\frac{d}{dt}\int E_R\,\delta V + \int_S \mathbf{F}_R \cdot dS + \int P_R \frac{d\,\delta V}{dt} = \rho\int_0^\infty d\varepsilon \int \delta V\,\kappa(\varepsilon)\,[4\pi B(T,\varepsilon) - cE].$$

The first term on the left is the rate of change of the radiation energy in δV; the second term is the radiation energy flux integrated over the surface of the volume element; and the third term gives the mechanical work done on the volume element by radiation pressure. The sum of these three terms just equals the difference between the amount of radiation energy emitted by the matter in δV (first term on the right) and the amount which it absorbs (second term on the right).

The first moment of the transfer equation is obtained by averaging (7.10) over $\Omega\,d\omega$. In the diffusion regime this leads to the multigroup radiation momentum equation, which reduces to [recall (6.32b)]

$$F = -\frac{\lambda_{tr}(\varepsilon)\,c}{3}\frac{\partial E}{\partial r}. \tag{7.16}$$

This is identical to (6.34) for one dimensional spherical geometry, with $\sigma(\varepsilon) = 0$. Notice that the flux contains the transport mean free path as discussed in Section 6.1.

If the material is in thermal equilibrium the emission term $\eta(\varepsilon) = k(\varepsilon)\,B(T,\varepsilon)$ (Kirchoff's law, see (6.17)). Using (7.16) to eliminate the spectral radiation flux, the multigroup transport equation (7.14) becomes

$$\frac{dE}{dt} - \frac{1}{r^2}\frac{\partial}{\partial r}\left\{r^2\frac{\lambda_{tr}c}{3}\frac{\partial E}{\partial r}\right\} +$$

$$\left[\frac{4E}{3} - \frac{1}{3}\frac{\partial(\varepsilon E)}{\partial \varepsilon}\right]\frac{1}{r^2}\frac{\partial(r^2 v)}{\partial r} = k\,\left[4\pi B(T,\varepsilon) - cE\right]. \tag{7.17}$$

This should be compared with (6.35) for a medium at rest, to which it reduces if $v = 0$. The two additional terms appearing in (7.17) are a result of fluid motion. The term

$$E \ \nabla \cdot \mathbf{v} = 3P \frac{1}{r^2} \frac{\partial (r^2 v)}{\partial r}$$

corresponds to the work done by the radiation field due to compression of the material, and is analogous to the mechanical work term appearing in the nonradiative fluid equations. The term proportional to $\partial (\varepsilon E)/\partial \varepsilon$ gives the change in the spectrum due to the Doppler shift.

The continuity equation (7.12), and the identity $dm = 4\pi r^2 \rho \, dr$ may be used to rewrite (7.17) in Lagrangian coordinates.

Problem 7.3

Show that $dE/dt = -(4/3)E \nabla \cdot \mathbf{v}$ corresponds to adiabatic compression or expansion of the radiation energy density in the frequency range ε to $\varepsilon + d\varepsilon$, and explain physically why such a term should appear in the multigroup transport equation (7.17).

Energy Diffusion Equation

Adopting the definition (6.38) for the radiation temperature T_R, and noting that the material and radiation temperatures need not be equal, we may integrate (7.17) over energy to construct the energy diffusion equation in Lagrangian form:

$$\frac{d \, aT_R^4}{dt} = \nabla \cdot D_R \nabla aT_R^4 + \rho a c \left[\kappa_P(T) T^4 - \kappa_P(T, T_R) T_R^4 \right]$$

$$-\frac{4}{3} aT_R^4 \, \nabla \cdot \mathbf{v} \, .$$

(7.18)

The Rosseland mean free path is defined by (6.40)-(6.41), and may at this stage include the scattering cross section. The Planckian opacities are given by (6.42). The comoving (Lagrangian) energy diffusion equation differs from (6.39) for a static medium to order v/c in two ways. First, the Lagrangian time derivative appears and second a term proportional to $\nabla \cdot \mathbf{v}$ (volume contraction) appears. It is instructive to rewrite these terms in the following equivalent form:

$$\frac{d\,aT_R^4}{dt} + \frac{4}{3} aT_R^4 \nabla\cdot\mathbf{v} = \frac{\partial aT_R^4}{\partial t} + \mathbf{v}\cdot\nabla aT_R^4 + aT_R^4\,\nabla\cdot\mathbf{v} + \frac{1}{3} aT_R^4\,\nabla\cdot\mathbf{v}$$

$$= \frac{\partial\,aT_R^4}{\partial t} + \nabla\cdot(aT_R^4\mathbf{v}) + P_R\,\nabla\cdot\mathbf{v} \tag{7.19}$$

where the radiation pressure is

$$P_R \equiv \frac{1}{3} aT_R^4 . \tag{7.20}$$

The time derivative in (7.19) is now with respect to the laboratory frame, and is the usual Eulerian derivative. Finally we note that because the derivation of (7.19) retains terms of order v/c, a Galilean transformation may be performed without affecting the validity of the equations. It is therefore permissible to transform to an arbitrary moving grid with grid velocity $\mathbf{v}_g$ at each point. This implies that the Lagrangian time derivative

$$\frac{d}{dt} \to \frac{\partial}{\partial t} + (\mathbf{v} - \mathbf{v}_g)\cdot\nabla$$

and the right side of (7.19) becomes

$$\frac{\partial\,aT_R^4}{\partial t} + \nabla\cdot[aT_R^4(\mathbf{v} - \mathbf{v}_g)] + P_R\,\nabla\cdot\mathbf{v} - aT_R^4\,\nabla\cdot\mathbf{v}_g . \tag{7.21}$$

This has the same form as the material energy equation (5.10) with $dq/dt = 0$, and includes a radiation work term $P_R \nabla \cdot \mathbf{v}$, and radiation energy advection in conservative form.

As expected, the Doppler shift terms which describe the redistribution of the radiation spectrum due to material motion drop out in (7.18).

Hydrodynamic Equations

Scattering and absorption result in momentum exchange between radiation and matter which leads to additional terms in the fluid momentum equations. Similarly, energy exchange will occur which couples the radiation field to the fluid energy.

The radiation force can be obtained directly from the specific intensity. Choose the z axis to coincide with the direction of the material velocity $\mathbf{v}$ (see Figure 6.1). Then the component of momentum along $\mathbf{v}$ associated with the radiation field is $\mu I(\mu)/c$, and the average of this quantity over solid angle (which is $1/c$ times the flux $\mathbf{F}$) is the net radiation momentum associated with radiation of energy ε in the direction of $\mathbf{v}$. The change in momentum of a fluid element of volume ΔV due to momentum transfer from the field is given by [recall (6.32b) and the discussion of the transport opacity]

$$\frac{d}{dt}\int_{\Delta V} \rho\, \mathbf{v}\, \delta V = \frac{1}{c}\int_0^\infty d\varepsilon \int_{\Delta A} [\kappa_a + \kappa_s(1-\tilde{\mu})]\, \rho\, \mathbf{F}\, dA\ ds \tag{7.22}$$

where dA is the area bounding the Lagrangian zone, and is chosen to be normal to $\mathbf{v}$. Note that $k ds$ represents the fractional reduction in beam flux along the path ds. Since $\rho\,\delta V$ is a constant and, by construction $dAds = \delta V$, we obtain the radiation acceleration

$$\rho\left(\frac{d\mathbf{v}}{dt}\right)_{rad} = \frac{1}{c}\int_0^\infty [\kappa_a + \kappa_s(1-\tilde{\mu})]\, \rho\ \mathbf{F}\, d\varepsilon\,, \tag{7.23}$$

which is to be added to the fluid momentum equation.

In the diffusion regime the flux is given by (7.16) and the radiation acceleration becomes

$$\rho\left(\frac{d\mathbf{v}}{dt}\right)_{rad} = -\frac{1}{3}\int_0^\infty \nabla\cdot E \; d\varepsilon \,. \tag{7.24a}$$

Assuming that $P = E/3$, we find that (7.24a) is equivalent to

$$\rho\left(\frac{d\mathbf{v}}{dt}\right)_{rad} = -\nabla P_R \tag{7.25}$$

where

$$P_R = \int P \; d\varepsilon \,.$$

As expected, radiation pressure in the diffusion regime simply adds to the material pressure in the momentum equation.

When flux limited diffusion is used the radiation acceleration must vanish smoothly as $\lambda_{tr} \rightarrow \infty$. For such applications the radiation acceleration can be expressed in the form

$$\rho\left(\frac{d\mathbf{v}}{dt}\right)_{rad} = -\frac{1}{c}\int_0^\infty \frac{D}{\lambda_{tr}} \nabla E \; d\varepsilon \,. \tag{7.24b}$$

The corresponding change in the radiation spectral energy is no longer unique. A prescription for flux limited neutrino diffusion will be considered in Section 7.5 which may be applied to photons as well. Note that as $\lambda_{tr}(\varepsilon) \rightarrow \infty$ the ratio $D(\varepsilon)/\lambda_{tr}(\varepsilon) \rightarrow 0$ smoothly, and the radiation acceleration vanishes.

In the absence of inelastic scattering processes the change in material energy is due to emission-absorption, and is described by (6.43) or (6.44). No additional velocity dependent terms arise through order v/c.

Since the emission and absorption terms and the specific intensity appearing in the transfer equation are defined in the fluid's rest frame, the material coupling equation (6.43) or (6.44) is formally unchanged if the time derivative is taken to be the Lagrangian derivative:

$$\rho C_V \frac{dT}{dt} = -\int d\varepsilon \int d\omega\, k\, \left[B(T,\varepsilon) - I(\Omega)\right]. \tag{7.26}$$

Recalling (3.14), the left side of (7.26) may be written in the equivalent form $\rho\, d\varepsilon/dt$ (in keeping with the notation of the hydrodynamics ε here denotes the material specific energy, and should not be confused with the photon energy). This will contribute to the term $T ds/dt$ appearing in (4.7). In an Eulerian formuation the right side of (7.26) will contribute to the term $dq\,/\,dt$ in (5.10).

Problem 7.4

Show that the Eulerian material energy equation for a radiating fluid is given by (5.10) with dq/dt given by the right side of (7.26).

In summary, when radiation transfer occurs in moving material the hydrodynamic equations of Chapter 4 and Chapter 5 must be modified to include the radiation acceleration using (7.23) in the transport regime, or (7.24) in the diffusion regime. The material coupling associated with emission-absorption is modified to include the comoving time derivative as in (7.26). The radiation transfer equation (7.10) contains additional terms proportional to v/c, and the time derivative is defined with respect to the comoving fluid frame. In the diffusion regime the multigroup transport equation (7.17) also contains a time derivative defined with respect to the comoving frame, and a term proportional to $\nabla \cdot \boldsymbol{v}$ which describes the change in spectral shape associated with changes in volume of fluid elements. Finally, the energy diffusion equation (7.18), which is also expressed in the comoving frame, contains a term identified with compressional work done on the radiation field.

7.2 RADIATION TRANSFER IN A MOVING MEDIUM

A finite difference representation of the radiative transfer equation for moving material (7.10) can be developed as an extension of the results developed in Section 6.3 for a stationary medium. It will be recalled that the subscript denoting the comoving frame has been omitted, since all quantities appearing in (7.10) and in the formulae of Section 6.3 are defined there. Furthermore, only the contributions to $c^{-1}dI/dt$ which are proportional to v/c need be considered here, since the finite difference equations developed previously are unchanged. The change in specific intensity associated with the velocity dependent terms in (7.10) may be written in the form

$$\frac{1}{c}\frac{dI}{dt} = -\frac{1}{c}\left(\frac{v}{r}-\frac{\partial v}{\partial r}\right)\frac{\partial}{\partial \mu}\left[\mu(1-\mu^2)\, I(\mu)\right]$$

$$+\left[\frac{v}{rc}-\frac{\mu^2}{c}\left(\frac{v}{r}-\frac{\partial v}{\partial \rho}\right)\right]\varepsilon\frac{\partial I(\mu)}{\partial \varepsilon}-\frac{I}{c}\,\nabla\cdot\mathbf{v}\,. \tag{7.27}$$

This form is convenient for several reasons, as will be seen below. The last term on the right side gives the work done by the radiation on compression or expansion of the fluid. The second term on the right side can be shown to conserve photon number.

Mechanical Work

Compression or expansion of a fluid element changes the volume occupied by photons and thus does work on the radiation field. This is described by the last term in (7.27):

$$\left(\frac{dI}{dt}\right)_{work} = -I\ \nabla\cdot\mathbf{v} = -\frac{I}{\Delta V}\frac{d\,\Delta V}{dt}\,. \tag{7.28}$$

This is analogous to the change in material energy (4.38), except that the specific intensity is a function only of the independent variables. In the Lagrangian framework, (7.28) may be rewritten as

$$\frac{d(I\ \Delta V)}{dt} = 0 \tag{7.29}$$

and evaluated for each angle and energy by the finite difference representation

$$I^{n+1} = I^{n}(\Delta V^{n} / \Delta V^{n+1}) = I^{n}(\rho^{n+1} / \rho^{n}) . \tag{7.30}$$

Since (7.29) involves only the time, the spatial, angular and frequency indices $k+1/2$, $j+1/2$ and $i+1/2$ have not been exhibited. The volume at t^{n+1} is given by (4.70), and represents the change due to the hydrodynamics. Although (7.30) changes the radiation spectrum, it does so in such a way that the total photon energy and photon number in a Lagrangian mass element of volume ΔV remains constant.

Spectral Doppler Shift

The second term on the right side of (7.27) represents the frequency or energy shift in the photon spectrum due to material motion, and has been written in a form that conserves photon number (photon number is an invariant under Lorentz transformations, and cannot be changed by Doppler effects). To show this, recall that $I/c\varepsilon$ is proportional to the photon number density, and consider its time rate of change due to Doppler effects: from (7.27)

$$\frac{d}{dt}\int_{0}^{\infty} \frac{1}{c} I \frac{d\varepsilon}{\varepsilon} \approx \int_{0}^{\infty} \frac{\partial I}{\partial \varepsilon} d\varepsilon \approx I \Big|_{0}^{\infty} = 0 \tag{7.31}$$

where all frequency independent factors have been dropped. Using the mass continuity equation, the change in I from Doppler effects may be written in the form

$$\frac{1}{c}\left(\frac{dI}{dt}\right)_{Doppler} = \left[\frac{1-3\mu^2}{c}\frac{v}{r} - \frac{\mu^2}{c}\frac{1}{\rho}\frac{d\rho}{dt}\right]\varepsilon\frac{\partial I}{\partial \varepsilon}$$

$$\equiv \frac{1}{c} f(\mu)\, \varepsilon \frac{\partial I}{\partial \varepsilon} . \tag{7.32}$$

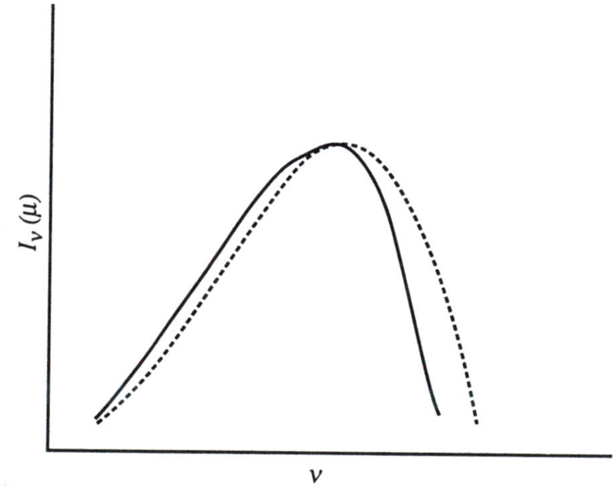

Figure 7.1. Spectral Doppler shift shown schematically, in which radiation is effectively shifted from lower to higher energy groups.

All of the quantities appearing in brackets in (7.32) are naturally zone centered except for the ratio v/r, which may be differenced according to (4.97). Straightforward differencing of the term $\varepsilon\partial I/\partial\varepsilon$ can lead to difficulties, such as nonconservation of photon number, or negative values for I. A finite difference representation which shares the properties of the differential equation can be obtained from physical considerations. First note that the quantity $f(\mu)$ may be positive or negative depending on the angle and velocity. Thus, if $f(\mu) < 0$ [$f(\mu) > 0$] then the right side of (7.32) implies that the photons are upshifted (down shifted) in energy. This is shown schematically in

Figure 7.1 for $f(\mu) < 0$, where radiation is effectively shifted from lower energy groups to higher ones. Thus, for $f(\mu) < 0$, (7.32) may be approximated by

$$I_{j+1/2}^{n+1} - I_{j+1/2}^{n} = \frac{1}{c} f(\mu)\, \Delta t^{n+1/2} \frac{(I_{j+1/2}^{n} - I_{j-1/2}^{n})}{\Delta\varepsilon_{j+1/2}} \varepsilon_{j+1/2}$$

(7.33a)

for $2 \le j \le J - 1$. This represents upwind differencing in energy space, and is analogous to donor cell advection discussed for Eulerian hydrodynamics in Chapter 5. In practice a higher order accurate advection scheme should be used. If photon energy is assumed not be advected out of the energy group structure (the spectrum corresponds to all of the radiation energy), then the lowest energy group $j = 1$ may be described by

$$I_{3/2}^{n+1} = I_{3/2}^{n} \left\{1 + \frac{1}{c} \frac{\varepsilon_{3/2}}{\Delta\varepsilon_{3/2}} f(\mu)\, \Delta t^{n+1/2}\right\} ,$$

(7.33b)

while for the uppermost energy group $j = J$ may be described by

$$I_{J+1/2}^{n+1} = I_{j+1/2}^{n} - \frac{1}{c} f(\mu)\, I_{J-1/2}^{n} .$$

(7.33c)

The first term on the right side of (7.33a) describes the energy shifted from energy group $j+1/2$ into $j+3/2$; the second term describes energy shifted into $j+1/2$ from group $j-1/2$.

For $f(\mu) > 0$, the spectrum is downshifted, and (7.32) may be approximated by the finite difference representation for $2 \le j \le J - 1$

$$I_{j+1/2}^{n+1} - I_{j+1/2}^{n} = \frac{1}{c} f(\mu)\, \Delta t^{n+1/2} \frac{(I_{j+3/2}^{n} - I_{j+1/2}^{n})}{\Delta\varepsilon_{j+1/2}} \varepsilon_{j+1/2} .$$

(7.34a)

For $j = J$,

$$I^{n+1}_{J+1/2} = I^{n}_{J+1/2} \left\{ 1 - \frac{1}{c} f(\mu) \frac{\varepsilon_{J+1/2}}{\Delta\varepsilon_{J+1/2}} \Delta t^{n+1/2} \right\} ,$$

(7.34b)

and for $j = 1$,

$$I^{n+1}_{3/2} = I^{n}_{3/2} + \frac{1}{c} f(\mu) \frac{\varepsilon_{3/2}}{\Delta\varepsilon_{3/2}} \Delta t^{n+1/2} I^{n}_{5/2} .$$

(7.34c)

The spatial and angular indices have not been shown in (7.33)-(7.34). It may be readily verified that either scheme conserves photon number exactly, as does the differential equation (see (7.31)). Using (7.33) or (7.34) we find

$$\sum_{j=1}^{J} \frac{(I^{n+1}_{j+1/2} - I^{n}_{j+1/2})}{\Delta t^{n+1/2}} \frac{\Delta\varepsilon_{j+1/2}}{\varepsilon_{j+1/2}} \equiv 0$$

(7.35)

which guarantees conservation of photon number.

The finite difference equations (7.33)-(7.34) are not coupled in space or angle. Therefore, a practical approach is to solve them zone by zone. Starting with $\mu = -1$, either (7.33) or (7.34) is solved for all j depending on the value of $f(\mu, \nu)$. The angle is then incremented, and the process repeated. The finite difference representation above is explicit, and can result in negative values of $I(\mu)$ for some energy group structures if the time step becomes too large. For many astrophysical applications the energy groups may be related geometrically,

$$\varepsilon_{j+1/2} / \varepsilon_{j-1/2} = const.,$$

which implies that $\varepsilon / \Delta\varepsilon$ is bounded. A reasonable limit on Δt in (7.33)-(7.34) would then be

$$|f(\mu, \nu)| \frac{\varepsilon}{\Delta\varepsilon} \Delta t^{n+1/2} \leq 1 .$$

Alternately, the terms in (7.33)-(7.34) which correspond to the removal of energy from a given group could be limited to the amount actually present. While this approach may not be exact, it has the practical advantage of eliminating an additional time step control.

The Doppler shift described by (7.32) conserves photon number, but will in general change the total energy in the spectrum.

Problem 7.5

In a spherically symmetric stellar core undergoing homologous contraction the velocity satisfies $v = \alpha_o r$, where α_o is a constant. Evaluate the rate of change of the radiation energy density, and the total radiation energy of a Lagrangian mass element using (7.27). Discuss each case physically.

Angular Advection

The first term on the right side of (7.27) represents a velocity dependent contribution to the change in $I(\mu)$ due to transport in angle. As in the $v = 0$ limit, this term redistributes radiation in angle without changing the photon number or total energy in the spectrum. The mass continuity equation may be used to rewrite this contribution as

$$\frac{dI}{dt} = -\left(\frac{3v}{r} + \frac{1}{\rho}\frac{d\rho}{dt}\right)\frac{\partial}{\partial\mu}\left[\mu(1-\mu^2)\, I(\mu)\right]. \tag{7.36}$$

The $v = 0$ change in $I(\mu)$ due to advection in angle was derived in Section 6.1 [see (6.9)], and accounts for the fact that as radiation propagates, its angle relative to the outward radial direction decreases. When the material is moving, (7.36) expresses the fact that an additional change in angle occurs due to that motion. A finite difference representation of this effect may be constructed following arguments used to evaluate (6.82). Thus, to (6.83) we add the term (common subscript $j+1/2$ omitted)

$$\left(\frac{dI}{dt}\right)_{i+1/2,\,k+1/2} = -\left(\frac{3v}{r} + \frac{1}{\rho}\frac{d\rho}{dt}\right)_{k+1/2} \{(1-\mu_{i+1}^2)\,\mu_{i+1}\,I_{i+1,\,k+1/2}^{n+1}$$

$$- (1-\mu_i^2)\,\mu_i\,I_{i,\,k+1/2}^{n+1}\}\,\frac{1}{\Delta\mu_{i+1/2}}\,.$$

(7.37)

As in (7.33) this is an upwind differencing in angle, and is analogous to first order advection in Eulerian hydrodynamics (Chpater 5). The edge centered intensity is defined by (6.84), and may be evaluated at the new times. The result may be written in the form

$$I_{i+1/2}^{n+1} = I_{i+1/2}^{n} + A\,I_{i+1/2}^{n+1} + B\,I_{i-1/2}^{n+1}$$

(7.38)

which may be solved for $I^{n+1}{}_{i+1/2}$ in terms of $I^{n+1}{}_{i-1/2}$. The approximations used to center $f(\mu)$ in (7.32) should be used in evaluating the velocity and density dependent coefficient of (7.36). Note that for homologous motion this term vanishes identically (see (7.27)), but that the sign of the contribution will in general depend on the velocity field, as well as on the angle.

7.3 Radiation Force

The radiation force acting on a Lagrangian fluid element (7.23) may be differenced for the volume element ΔV as follows:

$$\int_{\Delta V_k} \rho\left(\frac{dv}{dt}\right)\delta V = \Delta m_k \left(\frac{dv}{dt}\right)_k$$

$$= \int_{\Delta V_k} \delta V \int_0^\infty d\varepsilon\ \rho \int_{-1}^{1} 2\pi\,\mu\,\kappa(\varepsilon,\mu)\,I(\mu)\,d\mu$$

$$= 2\pi \int \delta V \int_0^{\infty} d\varepsilon \, \rho \, [\int_{-1}^{0} \kappa \mu \, I(\mu) \, d\mu + \int_0^1 \kappa \mu \, I(\mu) \, d\mu] . \tag{7.39}$$

In these expressions the mass density may be taken outside the angular and energy integrations. Consider the second term on the right, which corresponds to outward moving radiation $\mu > 0$. The change in v_k associated with these photons arises from $I_{k-1/2}$, and the absorption coefficient $(\kappa\rho)_{k-1/2}$ evaluated in the zone behind v_k. Therefore, for each energy group and each angular group $\mu > 0$, a change in material velocity will result given by (for $\mu > 0$)

$$\frac{v_k' - v_k}{\Delta t^n} = 2\pi \, \Delta V_k \, \kappa_{j+1/2,\, k-1/2} \, \rho_{k-1/2} \, I_{i+1/2,\, j+1/2,\, k-1/2} \, \frac{\mu_{i+1/2} \Delta\mu_{i+1/2} \Delta\varepsilon_{j+1/2}}{\Delta m_k} . \tag{7.40}$$

Similarly, for inward moving photons ($\mu < 0$) a change in v_k will be associated with $I_{k+1/2}$ in the zone ahead, as well as the absorption coefficient in that zone:

$$\frac{v_k' - v_k}{\Delta t^n} = 2\pi \, \Delta V_k \, \kappa_{j+1/2,\, k+1/2} \, \rho_{k+1/2} \, I_{i+1/2,\, j+1/2,\, k+1/2} \, \frac{\mu_{i+1/2} \Delta\mu_{i+1/2} \Delta\varepsilon_{j+1/2}}{\Delta m_k} . \tag{7.41}$$

Equation (7.40) or (7.41) are to be applied successively for each angle and energy group to produce a partial change in the fluid velocity. The value of v_k obtained at the end of this process may be taken to be $v^{n+1}{}_k$. This process is similar in spirit to the partial temparature approach used to evaluate the emission-absorption terms in Section 6.3.

Solution of the Moving Medium Transfer Equation

We now summarize a proceedure for solving the radiative transfer equation and the radiation acceleration term for a one dimensional system. A solution of the $v = 0$ terms was discussed in Chapter 6, and

---

Table 7.1

Summary of computational loops for solution of the transfer equation in a moving medium. The DO loop indices correspond to the energy and angular groups and radial zones used in the text.

DO $j = 1, J$ — *Main loop including all terms* ($v = 0$ *and* $v \neq 0$)

  DO $i = 1, I_o$ — *(ingoing radiation)*

    DO $k = K, 1, -1$

  DO $i = I_o, I$ — *(outgoing radiation)*

    DO $k = 1, K$

This block above solves the $v = 0$ *transfer equation, the emission and absorption terms and material heating terms (see Figure 6.8);*

  DO $i = 1, I_o$ — *(ingoing radiation)*

    DO $k = 1, K$ — *(solve (7.4.1) using new value of* I*)*

  DO $i = I_o, I$ — *(outgoing radiation)*

    DO $k = 1, K$ — *(solve (7.40) using new* I*)*

This block above solves the radiation acceleration using the new intensity I *from the* $v = 0$ *solution to get the new material velocity;*

  DO $i = 1, I$

    DO $k = 1, K$ — *(using latest* v *and* I*, evaluate* $f(\mu, v)$ *to choose between (7.33) and (7.34); also solve (7.38) and (7.30))*

The block above solves the Doppler shift, radiation compression and angular advection for $v \neq 0$.

This ends the main loop.

---

is summarized in Figure 6.8. The $\nu = 0$ terms are solved first, giving new values for $I_{i+1/2,\ j+1/2,\ k+1/2}$, which are used as initial values for the radiation hydrodynamics. The radiation hydrodynamics and velocity dependent terms in the transfer equation are solved within the energy group loop after the $\nu = 0$ terms have been solved. This may be accomplished as in Table 7.1.

7.4 SCATTERING

The general form of the scattering term appearing in the transfer equation (6.22)-(6.23) for matter at rest is nonlinear in $I(\Omega)$, and couples all energy and angular groups together. When scattering is strong (implying large changes in specific intensity per computational time step) the scattering terms should be represented by an implicit finite difference reperesentation. The solution of such a system of equations would require iterative matrix inversion techniques which will not be considered further.

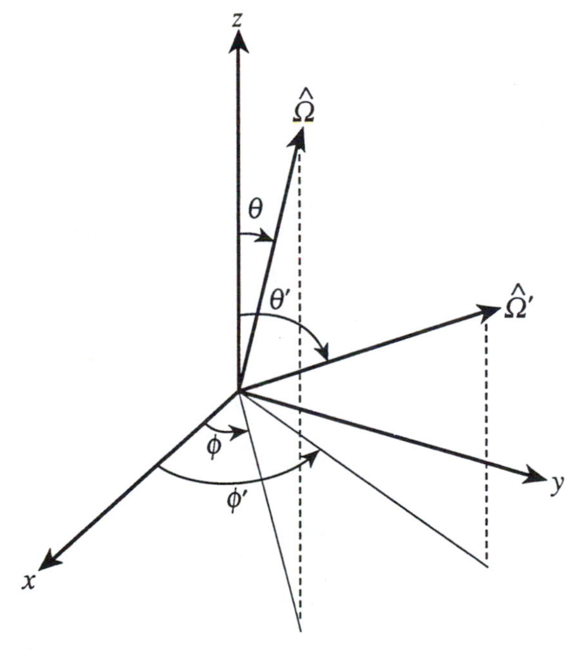

Figure 7.2. Angles defining Ω and Ω' with respect to an outward direction along the z-axis. Note that $\cos\gamma = \Omega{\cdot}\Omega'$.

The discussion of scattering for the transfer equation will be confined to the simpler case of Thomson scattering with $I << \varepsilon^3/h^2c^2$ (where nonlinear effects may be ignored). The scattering terms appearing in the transfer equation are then the last two terms in (6.29):

$$\left(\frac{dI}{dt}\right)_S = -\sigma I(\Omega) + \frac{3}{16\pi}\sigma \int d\omega' \left[1 + (\Omega \cdot \Omega')^2\right] I(\Omega') \,. \tag{7.42}$$

The effects of material motion may be incorporated using (7.1)-(7.4) and the methods described in Section 7.1, but lead to no additional complications and will be ignored. Figure 7.2 shows the relation between Ω, Ω' with the outward direction assumed to lie along the z-axis. Using the definitions $\mu = \cos\theta, \mu' = \cos\theta'$, and the identity

$$\begin{aligned} \Omega \cdot \Omega' &= \gamma \\ &= \cos\theta \cos\theta' + \sin\theta \sin\theta' \cos(\phi - \phi') \,, \end{aligned} \tag{7.43}$$

(7.42) is readily shown to reduce to

$$\left(\frac{dI(\mu)}{dt}\right)_S = -\sigma I(\mu) + \frac{3\sigma}{8}\int_{-1}^{1} d\mu' f(\mu', \mu) I(\mu') \,, \tag{7.44}$$

where

$$f(\mu, \mu') \equiv \frac{1}{2}(3 - \mu'^2) + \frac{1}{2}\mu^2(3\mu'^2 - 1) \tag{7.45a}$$

and

$$\int_{-1}^{1} f(\mu, \mu')\, d\mu' = \frac{8}{3}. \tag{7.45b}$$

The integro-differential equation defined by (7.44) and (7.45a) has two important properties which should be preserved by any finite difference representation. First, because Thomson scattering as described by (6.27) is conservative, the net effect of (7.42) is to redistribute photons among the angular groups without changing the total energy in the spectrum. In other words, (7.42) satisfies the relation

$$\int_{-1}^{1} \left(\frac{dI(\mu)}{dt}\right)_S d\mu = \frac{d}{dt}\int_{-1}^{1} I(\mu)\, d\mu \equiv 0. \qquad (7.46)$$

Second, if the radiation is isotropic ($I(\mu) = I$, independent of angle) then the left side of (7.44) must vanish identically:

$$\left(\frac{dI}{dt}\right) = 0, \qquad I(\mu) = I. \qquad (7.47)$$

This is, in fact, the steady state solution to the scattering term (7.42).

Problem 7.6

Use the identity (7.43) to obtain (7.4.3)-(7.4.4) from (7.4.1). Verify (7.4.5) and (7.4.6).

A finite difference representation of (7.42) may be found by replacing the integral over the element of solid angle $d\omega' = 2\pi\, d\mu'$ by a summation over angular groups (common subscripts $k+1/2, j+1/2$ omitted):

$$I^{n+1}_{i+1/2} = I^{n}_{i+1/2} - \sigma \Delta t^{n+1/2} I_{i+1/2}$$

$$+ \frac{3}{8}\sigma \Delta t^{n+1/2} \sum_{i'=1}^{I} f(\mu_i, \mu_{i'}) I_{i'+1/2}\, \Delta\mu_{i'+1/2} \qquad (7.48)$$

where $f(\mu_i, \mu_{i'})$ is defined as in (7.45a). The time centering of the intensity on the right side has not been shown, but will be considered below. It is also necessary to impose a constraint to guarantee that the conditions (7.46) and (7.47) will be satisfied by the solution to (7.48). The flexibility to accomplish this arises because the centering of the variable $\mu_{i+1/2}$ has not been specified (recall that the centering of the angular group boundaries μ_i fixes $\Delta\mu_{i+1/2}$ but not $\mu_{i+1/2}$). Thus will the angular group structure is constructed to satisfy the finite difference equivalent of (7.45b), or

$$\sum_{i=1}^{I} f(\mu_i, \mu_i)\, \Delta\mu_i = \frac{8}{3} \; . \tag{7.49}$$

Straightforward evaluation of this sum using (7.44a) leads to the requirement that

$$\sum_{i=1}^{I} \mu_{i+1/2}^2 \, \Delta\mu_{i+1/2} = \frac{2}{3} \tag{7.50}$$

as well as (6.58)

$$\sum_{\mu_i} \Delta\mu_{i+1/2} = 2$$

which guarantees that

$$\int d\omega = 4\pi \, .$$

Any angular group structure satisfying (7.50) and (6.58) will guarantee that the finite difference representation satisfies (7.46) and (7.47). Notice that while these constraints limit the number of group structures, they do not in general determine them uniquely.

Problem 7.7

Construct an angular group structure for (7.48) which satisfies the constraints (7.46) and (7.47) assuming six angular groups ($I = 4$).

Note that the choice is not unique. Will the simple structure $\Delta\mu_i =$ *const.* be acceptable?

In the weak scattering regime ($\sigma\, \Delta t << 1$), each specific intensity undergoes small changes per time step, and (7.48) may be evaluated explicitly by setting $I = I^n$ on the right side of the finite difference equation. In the strong scattering regime ($\sigma\, \Delta t >> 1$), it is necessary to replace (7.48) by some implicit scheme. An obvious choice is to set $I = I^{n+1}$ on the right side of the finite difference equation, thereby obtaining a fully implicit scheme (which will be unconditionally stable). In this case the new (unknown) values of the intensity are completely coupled in angle, and (7.48) leads to a matrix equation which must be inverted during each computational cycle.

It may be inconvenient to readjust the angular group structure to satisfy (7.4.4b). A practical alternative is to replace $f(\mu,\ \mu')$ in (7.4.3) by

$$\tilde{f}(\mu, \mu') = \frac{1}{2}(\alpha - \mu'^{\,2}) + \frac{1}{2}\mu^2(\alpha\, \mu'^{\,2} - 1)\,, \tag{7.51}$$

and then determine the constant α such that

$$\sum_i (\alpha\, \mu_i^2 - 1)\, \Delta\mu_i = 0\,. \tag{7.52}$$

This is readily done for any angular group structure once it has been specified. We then replace the factor $3/8$ in (7.44) by the normalization constant N given by

$$N^{-1} = \sum_i \frac{1}{2}(\alpha - \mu_{i+1/2}^2)\, \Delta\mu_{i+1/2}. \tag{7.53}$$

Together, (7.51)-(7.53) leads to a conservative scattering model which satisfies (7.46) and (7.47) for an arbitrary angular group structure. Generally, the error in α (relative to its exact value of three) decreases as the number of groups increases.

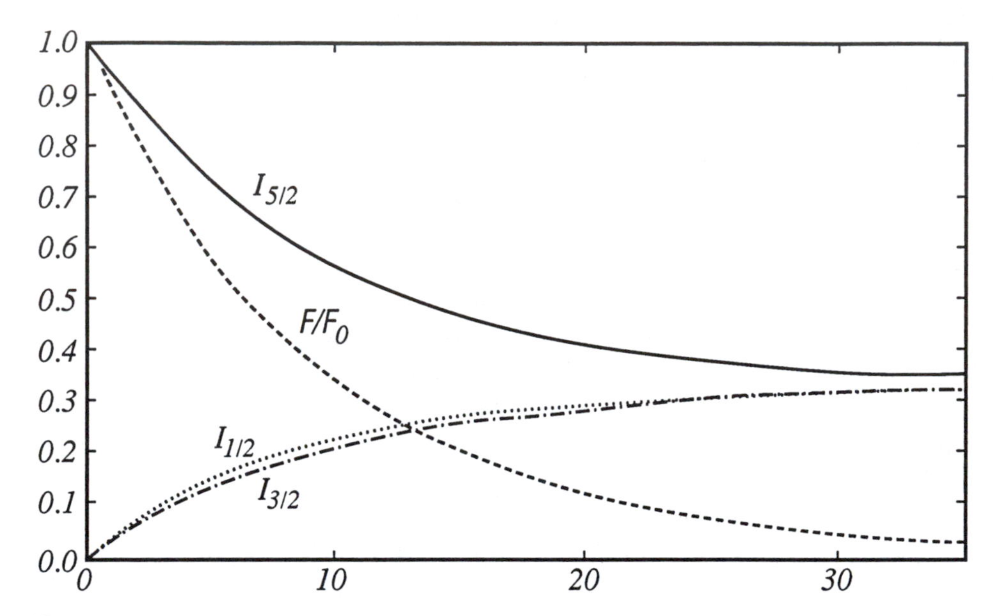

Figure 7.3. Specific intensity of photons versus time showing the effect of scattering on a spectrum described by three angular groups.

Figure 7.3 shows $I_{i+1/2}(t)$ obtained from (7.48) assuming three angular groups and the initial conditions $I_{5/2}(0) = 1.0, I_{3/2}(0) = I_{1/2}(0) = 0$, and a constant time step satisfying $\sigma(\varepsilon)\Delta t = 0.1$. Also shown by x is the net flux normalized to its initial value. The angular group structure is defined by $\Delta\mu_{i+1/2} = 2/3$, $\mu_{1/2} = -1/2^{1/2}$, $\mu_{3/2} = 0$ and $\mu_{5/2} = 1/2^{1/2}$.

7.5 Multigroup Diffusion in a Moving Medium

When the angular distribution of the radiation spectrum is not required, and multigroup diffusion is used in a moving medium, the results of Section 6.4 must be extended to include the velocity dependent term in (7.17). As in Section 6.4, we will consider neutrinos, whose energy spectrum is denoted by F_ν, but the approach can easily be modified to apply to photons. One dimensional spherical symmetry will also be assumed.

Spectral Doppler Shift

In the absence of a flux limiter the change in $F_\nu(\varepsilon)$ due to material motion is given by [see (7.17)]

$$\left(\frac{dF_\nu}{dt}\right)_{Doppler} = -\left[\frac{4}{3}F_\nu - \frac{1}{3}\frac{\partial(\varepsilon F_\nu)}{\partial\varepsilon}\right] \nabla\cdot\mathbf{v}$$

$$= -\left(F_\nu - \frac{\varepsilon}{3}\frac{\partial F_\nu}{\partial\varepsilon}\right) \nabla\cdot\mathbf{v}. \tag{7.54}$$

This equation must be modified if flux limited diffusion is desired (see the discussion of the radiation force below). The first term on the right can be differenced as in (7.28)-(7.30);

$$F^{n+1}_{j+1/2} = F^{n}_{j+1/2}\left(\rho^{n+1}/\rho^{n}\right). \tag{7.55}$$

This scheme conserves the neutrino energy and the neutrino number in the spectrum, as does the partial differential equation (7.54). The second term on the right of (7.54) also conserves the total neutrino number in the spectrum as long as $F_\nu(\varepsilon)$ vanishes as $\varepsilon \to 0$ and $\varepsilon \to \infty$. It does not conserve the total energy in the spectrum since work must obviously be done to red shift or blue shift it (see below). A finite difference representation of the last term can be found by noting that for a zone which is undergoing compression (expansion) $\nabla\cdot\mathbf{v} < 0$ ($\nabla\cdot\mathbf{v} > 0$) and the neutrinos throughout the spectrum are shifted to higher (lower) energies. If the neutrinos are constrained to shift one energy group at at time, then the change in energy group $j+1/2$ for a zone being compressed should include a term proportional to $\nabla\cdot\mathbf{v}\, F_{j+1/2}$ (representing the energy shifted out of the group into the next higher one) and a term proportional to $-\nabla\cdot\mathbf{v} F_{j-1/2}$ (representing the energy shifted into the group from the lower one). Guided by these observations we adopt the following finite difference representation for (7.56) when $\nabla\cdot\mathbf{v} < 0$:

$$\frac{F^{n+1}_{j+1/2} - F^{n}_{j+1/2}}{\Delta t^{n+1/2}} =$$

$$\frac{1}{3} K \frac{\varepsilon_{j+1/2}}{\Delta\varepsilon_{j+1/2}} \left(F_{j+1/2} - F_{j-1/2} \right), \quad \nabla\cdot\mathbf{v} < 0$$

(7.56a)

where the coefficient K will be discussed below. Notice that (7.56a) conserves the total neutrino number in the spectrum as long as no neutrinos are produced outside the bounds of the energy group structure; this constraint is incorporated by the boundary conditions

$$F_{-1/2} = F_{J+1/2} \equiv 0.$$

(7.56b)

In zones which are expanding, the change in energy group $j+1/2$ should include a term proportional to $\nabla\cdot\mathbf{v}\, F_{j+3/2}$ (representing the energy downshifted into the group from the next higher one) and a term proportional to $-\nabla\cdot\mathbf{v}\, F_{j+1/2}$ (representing the energy shifted out of the group into the lower one). These arguments suggest the following finite difference representation for (7.54) when $\nabla\cdot\mathbf{v} > 0$:

$$\frac{F^{n+1}_{j+1/2} - F^{n}_{j+1/2}}{\Delta t^{n+1/2}} =$$

$$\frac{1}{3} K \frac{\varepsilon_{j+1/2}}{\Delta\varepsilon_{j+1/2}} \left(F_{j+3/2} - F_{j+1/2} \right), \quad \nabla\cdot\mathbf{v} > 0$$

(7.56c)

In optically thick matter (to which the multigroup diffusion approximation strictly applies), the coefficient $K = \nabla\cdot\mathbf{v}$. We will return to the finite difference form of this coefficient later. If the finite difference equations above are used with an explicit Lagrangian hydrodynamic scheme such as the one discussed in

Chapter 4, the hydrodynamics will limit the time step such that the coefficient $K \Delta t^{n+1/2} = \nabla \cdot \mathbf{v} \, \Delta t^{n+1/2}$ is less than or equal to unity, and (7.56) may be evaluated explicitly (the quantities $F_{j+1/2}$ on the right side of the equations represent values at time t^n). Finally, the new spectrum is assumed to satisfy the boundary conditions (7.56b). The second term on the right side of (7.56) is an advection term so a higher order accurate advection scheme similar to those discussed in Chapter 5 for mass advection should be used when higher accuracy is necessary.

We may also need the neutrino distribution function in phase space, $f_\nu(\varepsilon)$ which is related to $F_\nu(\varepsilon)$ by

$$F_\nu = \frac{1}{2\pi^2}\left(\frac{2\pi k}{c}\right)^3 \varepsilon^3 f_\nu \equiv \alpha \varepsilon^3 f_\nu \, , \tag{7.57a}$$

and which satisfies the constraint

$$0 \le f_\nu \le 1 \tag{7.57b}$$

characteristic of fermions. The distribution function is zone centered, as is F_ν, and it is convenient to set (in one dimension)

$$F_{k+1/2,\, j+1/2} = \alpha \varepsilon_{j+1/2}^3 \, f_{k+1/2,\, j+1/2} \, . \tag{7.58}$$

In degenerate matter f_ν is very nearly unity for many of the lower energy states. It is important that the value of f_ν derived from (7.57a) never exceed unity. The form of (7.56) does not guarantee that (7.57b) will always be satisfied, so the results should be monitored to insure that overpopulation of degenerate states does not result.

Problem 7.8

Use (7.57a) to show that the Doppler shift (7.54) can be expressed in the form

$$\frac{df_{\nu}}{dt} = \frac{\varepsilon}{3} \frac{\partial f_{\nu}}{\partial \varepsilon} \nabla \cdot \mathbf{v} . \tag{7.59}$$

Try to construct a finite difference representation of this equation which conserves neutrino number in the spectrum and which guarantees that the distribution function will always satisfy (7.57b). Test your results on sample spectra for a zone which is undergoing compression and for one undergoing expansion.

Radiation Force

The radiation force experienced by matter is given by (7.23). For neutrino radiation the force in the multigroup diffusion approximation is given by (7.24a) [or by (7.24b) for flux limited diffusion], with $E_{\nu}(\varepsilon)$ replaced by $F_{\nu}(\varepsilon)$. For flux limited diffusion

$$\rho \left(\frac{d\mathbf{v}}{dt}\right)_{rad} = -\frac{1}{c}\int_{0}^{\infty} \frac{D_{\nu}(\varepsilon)}{\lambda_{\nu}(\varepsilon)} \nabla F_{\nu}(\varepsilon)\, d\varepsilon. \tag{7.60}$$

In the diffusion regime the term $D_{\nu} / \lambda_{\nu} = c / 3$ (which uses the transport mean free path as discussed in Chapter 6) is appropriate in the deep stellar core during core collapse. As they stand, these equations cannot be used correctly outside the diffusion regime, where neutrinos stream freely, since there should be essentially no coupling between the radiation and matter. By retaining the general form D_{ν} / λ_{ν} where the diffusion coefficient D_{ν} includes a flux limiter, the radiation acceleration will smoothly vanish during the transition from optically thick to optically thin matter. A finite difference representation of (7.60) which includes the flux limited diffusion coefficient is given by

$$\frac{v_k^{n+1/2} - v_k^{n-1/2}}{\Delta t^n} = -\frac{\Delta V_k}{\Delta m_k} \sum_{j=1}^{J} \left(\frac{D_\nu}{\lambda_\nu c}\right)_{k,\, j+1/2} \frac{F^D_{k+1/2,\, j+1/2} - F^D_{k-1/2,\, j+1/2}}{\Delta r_k} \Delta\varepsilon_{j+1/2}$$

(7.61)

where F^D includes the changes associated with emission, absorption and spatial diffusion, and D_ν represents the diffusion coefficient appearing in (6.100).

The change in neutrino energy to be associated with the flux limited acceleration (7.61) is no longer given by (7.54). For flux limited diffusion the choice is not unique. We give below a method that conserves neutrino number and kinetic energy. Let

$$\frac{dF_\nu}{dt} = \varepsilon \frac{\partial}{\partial \varepsilon} \left[F_\nu \nabla\cdot\left(\frac{D_\nu \mathbf{v}}{\lambda_{tr} c}\right)\right].$$

(7.62a)

It follows immediately that

$$\frac{dn_\nu}{dt} = \int \frac{dF_\nu}{dt} \frac{d\varepsilon}{\varepsilon} = 0 .$$

The energy conservation equation

$$\frac{1}{\Delta V} \int dV \int \frac{dF_\nu}{dt} d\varepsilon = -\rho\, \mathbf{v}\cdot \frac{d\mathbf{v}}{dt}$$

(7.62b)

can be derived by use of equation (7.60) and (7.62a) together with repeated integrations by parts. Equation (7.62a) has the form of an advection equation with an effective advection velocity $\mathbf{v}_a = \nabla\cdot(D_\nu \mathbf{v} / \lambda_{tr} c)$. A simple upwind difference scheme such as the one discussed for the spectral Doppler shift can be used to represent (7.62a). If higher accuracy is desired a second order scheme similar to the ones discussed in Chapter 5 should be used. However, since (7.62a) itself is somewhat ad hoc the advantage in using a higher order scheme may be questionable.

Problem 7.9

Integrate (7.62a) over energy to obtain the change in neutrino energy associated with the neutrino radiation force. Then using integration by parts and (7.60) verify (7.62b).

7.6 Elastic Scattering of Electrons and Radiation

When scattering occurs between charged particles and photons (Compton scattering) energy and momentum are transferred from the faster particle to the slower one. The most important electromagnetic process is scattering between electrons or positrons and photons

$$e^{\pm} + \gamma \rightarrow e^{\pm} + \gamma . \tag{7.63}$$

Compton scattering of radiation and ions will also occur, but the cross section for this process is smaller than for (7.63) by several orders of magnitude, and can usually be ignored. Another important process which occurs is the scattering between electrons and neutrinos (electon, mu or tau neutrinos or their antiparticles), for example

$$e^{\pm} + \nu_e \rightarrow e^{\pm} + \nu_e . \tag{7.64}$$

Kinetic energy and momentum are also transferred from the faster to the slower particle by the weak interaction process (7.64). We shall refer to photon or neutrino electron scattering processes generically as Compton scattering. The scattering processes (7.63) represents the primary mechanism by which radiation thermalizes in a hot nondegenerate plasma. The neutrino scattering process (7.64) represents an important means of neutrino thermalization in stellar core collapse during a supernova explosion. In these circumstances scattering will occur between an arbitrary distribution of radiation quanta and a Maxwellian distribution of electrons which is characterized by a temperature T_e. Compton scattering may be formulated within the context of the transfer equation (see, for example, Pomraning 1973 for a discussion of electron-photon scattering).

Consider a dense plasma in local thermodynamic equilibrium (in this case the material temperature is T_e so the subscript may be dropped) at density ρ and material temperature T containing free electrons and an arbitrary distribution of radiation (photons or neutrinos). The radiation need not be in equilibrium, but will be assumed to be characterized by an arbitrary spectral distribution. If the mean energy of the radiation is greater than the thermal energy of the matter, Compton scattering will tend to thermalize the radiation field at an effective radiation temparature T_R, and after a sufficient time will tend to equilibrate the radiation and material temperatures. Expressions for the time rate of change of the radiation spectrum may be obtained from quantum scattering theory (Cooper 1971), or from the Boltzmann transport equation (Kompaneets 1957). The analysis, which is outlined in Appendix B, establishes the following expression (in which the energy is taken as the independent variable) for the time rate of change of the photon distribution function $f_B(\varepsilon)$, assuming that the electrons are nonrelativistic and that the scattering cross section can be approximated by the Thomson cross section σ_T:

$$\frac{\partial f_B}{\partial t} = \frac{n_e \sigma_T}{m_e c} \frac{1}{\varepsilon^2} \frac{\partial}{\partial \varepsilon} \left\{ \varepsilon^4 \left[kT \frac{\partial f_B}{\partial \varepsilon} + f_B (1 + f_B) \right] \right\}. \tag{7.65}$$

The right hand side has the general form of a diffusion equation in energy space but, because of the last term, it is nonlinear in f_B. The factor $n_e \sigma_T \varepsilon / m_e c$ may be thought of as a diffusion coefficient [see (7.71b)]. This expression, known as the Fokker-Planck equation, is valid in the nonrelativistic regime as long as the energy exchange per collision, $\Delta\varepsilon$, satisfies the constraint $\Delta\varepsilon << kT$, where T is the material temperature, and for photon energies satisfying $\varepsilon << m_e c^2$. The equation is also applicable to systems containing relativistic electrons if the scattering cross section is constructed using the Klein-Nishina formula and a relativistic Maxwellian distribution (see Pomraning 1973). The number density of free electrons is n_e, and m_e is the mass of the electron. The photon distribution function expresses the occupancy of quantum states for photons having energies in the range ε to $\varepsilon + d\varepsilon$. The total photon number density is given by

$$n_\gamma = \int_0^\infty \alpha \, \varepsilon^2 f_B(\varepsilon) \, d\varepsilon \tag{7.66}$$

where α is a normalization constant.

The Fokker-Planck equation (7.65) exhibits several intrinsic properties which should be preserved by any solution (see Chang and Cooper 1970). The first of these is photon number conservation; multiplying (7.65) by $\alpha\varepsilon^2$, integrating over energy and using (7.66) lead to the result

$$\frac{\partial n_\gamma}{\partial t} = \int_0^\infty \frac{\partial g(\varepsilon, T)}{\partial \varepsilon} \, d\varepsilon = g(\varepsilon, T) \Big|_0^\infty .$$

Assuming that $g(\varepsilon,T)$ vanishes at $\varepsilon = 0$ and $\varepsilon = \infty$, as it will for physically reasonable distribution functions, (7.65) guarantees that solutions to the Fokker-Planck equation will conserve the total number of photons in the system:

$$\frac{\partial n_\gamma}{\partial t} = 0. \tag{7.67}$$

The physical interpretation of this property, as is evident from

(7.63), is that Compton scattering does not change the total photon number.

Since $f(\varepsilon,T)$ represents a probability density [recall (7.66)] it must be positive definite. It can be shown that if at any time t_o, the distribution function satisfies $f(\varepsilon, \,) \geq 0$ for all energies, then the solution of (7.65) is also positive definite:

$$f(\varepsilon, t_o) \geq 0 \quad then \quad f(\varepsilon, t) \geq 0. \tag{7.68}$$

Clearly this is necessary if $f(\varepsilon, t)$ is to represent a probability density.

Finally, the steady state solution to (7.65) is obtained when the quantity in square brackets vanishes. If the photons are in thermodynamic equilibrium at temperature T the distribution function will reduce to the familiar Bose-Einstein result

$$f_B^{eq}(\varepsilon) = \frac{8\pi\varepsilon^2}{(hc)^3} \frac{d\varepsilon}{e^{\varepsilon / kT} - 1}. \tag{7.69}$$

Problem 7.10

Show that

$$f(\varepsilon) = \frac{1}{exp\,[(\varepsilon - \phi)/k_BT] - 1}$$

is a steady state solution to the Fokker-Planck equation (7.65) for photons. Explain the significance of the constant of integration ϕ in the solution. Under what circumstances does the solution reduce to (7.69)?

The spectral energy density in the radiation field is related to the distribution function by (α is a normalization constant)

$$F_B(\varepsilon) \equiv \alpha\, \varepsilon^3 f_B(\varepsilon). \tag{7.70}$$

Using (7.70) the Fokker-Planck equation can be written in terms of the spectral energy density (the subscript B will be omitted for notational convenience):

$$\frac{\partial F}{\partial t} = \varepsilon \frac{\partial}{\partial \varepsilon}\{K[kT(\frac{\partial F}{\partial \varepsilon} - \frac{3F}{\varepsilon}) + F(1 + \frac{F}{\alpha \varepsilon^3})] \tag{7.71a}$$

$$K \equiv \frac{n_e \sigma_T \varepsilon}{m_e c}. \tag{7.71b}$$

The net change in radiation energy density per unit volume is obtained by multiplying (7.65) by the photon energy, ε, and integrating over photon energies

$$(\frac{dE}{dt})_{Compton} = \frac{8\pi}{(hc)^3} \int \frac{\partial f_B}{\partial t} \varepsilon^3 d\varepsilon. \tag{7.72a}$$

The results above apply to an arbitrary radiation spectrum, describing in particular its evolution towards a Planckian distribution.

The results above may be employed to describe Compton scattering in the energy diffusion regime (Chapter 6). Denoting the material temperature by T and approximating the radiation spectrum by a Planckian spectrum at the effective temperature T_R, the radiation energy density is $E = aT_R{}^4$ and (7.72a) reduces to the simple form

$$(\frac{d\, aT_R^4}{dt})_{Compton} = \frac{4\, \sigma_T n_e ack}{m_e c^2} T_R^4 (T - T_R). \tag{7.72b}$$

As expected physically, the rate is proportional to the product of the photon energy density aT_R^4 and the electron flux $n_e c$, to the scattering cross section and to the separation in material and radiation temperatures. The change in radiation energy density is, for example, positive if $T > T_R$, since collisions tend to transfer energy from the electron's thermal bath to the photon spectrum.

Problem 7.11

Find the time scale for the equilibration process (7.72b), and apply it to (a) the solar interior ($T = 1.2 \times 10^7$ K, $n_e = 5 \times 10^{25}$ cm^{-3}); (b) the solar atmosphere ($T = 6000$ K, $n_e = 3 \times 10^{11}$ cm^{-3}); (c) an HII region ($T = 10^4$ K, $n_e = 10^4$ cm^{-3}).

Problem 7.12

Carry out the analysis leading to (7.72b) assuming that the photon distribution function is given by (7.69). First use integration by parts to show that

$$\left(\frac{dE}{dt}\right)_{Compton} = -\frac{8\pi n_e \sigma_T c}{m_e c^2 (hc)^3} \int \varepsilon^4 \left[kT \frac{\partial f}{\partial \varepsilon} + f(1+f) \right] d\varepsilon$$

and then show that (7.69) implies that

$$\frac{\partial f}{\partial \varepsilon} = -\frac{f(1+f)}{kT_R}.$$

Compton Scattering (Energy Diffusion Regime)

Compton scattering between electrons and photons may be rather easily incorporated in the energy diffusion regime where the radiation distribution is characterized by the effective temperature T_R (see Section 6.1). It will be recalled that the emission-absorption terms in this regime could be expressed as an effective coupling coefficient times $T^4 - T_R{}^4$ [see (6.120)]. The results of Problem 6.18 may be used to write (7.72b) in the form

$$\left(\frac{d\,aT_R^4}{dt}\right)_{Compton} = \kappa_C (T^4 - T_R^4), \tag{7.73}$$

where κ_C is an effective Compton coupling coefficient which is easily shown to be

$$\kappa_C = \frac{4\,\sigma_T n_e c\,k\,aT_R^4}{m_e c^2}\,\frac{1}{T^3 + T^2 T_R + T T_R^2 + T_R^3}. \tag{7.74}$$

This proves the assertion made in Chapter 6 that the radiative diffusion equation including emission-absorption and Compton scattering can be written in the form (6.123) with the effective radiation-electron coupling coefficient given by (6.124) and (7.74).

Multigroup Compton Scattering

Electron-neutrino scattering may also be described by a Fokker-Planck type equation which differs from the electron-photon equation primarily in the form of the coefficient K, and the sign of the blocking factor (see Appendix B). The result for scattering between electrons and electron neutrinos (7.64), expressed in terms of the neutrino spectral energy density, F_ν, is

$$\frac{\partial F_\nu}{\partial t} = \varepsilon \frac{\partial}{\partial \varepsilon}\{K_\nu [kT(\frac{\partial F_\nu}{\partial \varepsilon} - \frac{3F_\nu}{\varepsilon}) + F_\nu(1 - \frac{F_\nu}{\alpha\varepsilon^3})]\} \tag{7.75a}$$

$$K_\nu = [(\frac{\Delta\varepsilon}{\varepsilon})^2_{coll}]^{1/2}\,\tau_c^{-1}. \tag{7.75b}$$

The diffusion coefficient for neutrinos has been expressed as a function of the mean energy exchange per collision and the neutrino-electron relaxation time τ_c. Both are functions of the density and temperature of the material which must be supplied during computations. Analytic expressions for cross sections, and for $|\Delta\varepsilon/\varepsilon|_{coll}$ are given in the literature (see, for example, Tubbs and Schramm 1975). Similar expressions may be found for electron scattering off other neutrino species (see, for example, Bowers and Wilson 1982). In thermal equilibrium the neutrino distribution function has the usual form for Fermi-Dirac particle:

$$f_{eq}(\varepsilon) = \frac{1}{exp[(\varepsilon - \mu_\nu)/kT] + 1} \tag{7.76}$$

where μ_ν, the neutrino chemical potential, is determined by the total electron neutrino number density. For an arbitrary electron neutrino spectrum, the neutrino number density is given by

$$n_{\nu_e} \equiv \int_0^\infty F_\nu \frac{d\varepsilon}{\varepsilon}.$$

The effects of Compton scattering on electron neutrinos may be included in the multigroup diffusion equation by adding (7.75a) to the right hand side of (6.47):

$$\frac{\partial F_\nu}{\partial t} = \nabla \cdot D_\nu \nabla F_\nu + K_\nu (A_\nu - F_\nu) + \left(\frac{dF_\nu}{dt}\right)_{Compton} \tag{7.77}$$

Since Compton scattering will also alter the material's thermal energy, (6.48) becomes

$$\rho C_V \frac{\partial T}{\partial t} = -\int \kappa_\nu (A_\nu - F_\nu)\, d\nu - \int \left(\frac{dF_\nu}{dt}\right)_{Compton} d\nu. \tag{7.78}$$

Fokker-Planck Finite Difference Equation

When Compton scattering is included with radiative diffusion and material heating the resulting equations will introduce spatial gradients as well as gradients in energy space. One approach to solving (7.77)-(7.78) is to operator split the Compton scattering and associated material heating terms from the radiative diffusion and emission-absorption terms, solving the latter as described in Chapter 6. It should be remarked that there is no direct coupling between Compton scattering as described by the Fokker-Planck equation and the fluid motion. Using operator splitting the Compton terms may then be solved in energy space one spatial zone at a time (note that there is no coupling between spatial zones in (7.75a) or in the heating term).

Electron-Photon Scattering

We shall illustrate one approach (Chang and Cooper 1970) to the Fokker-Planck equation by first considering the linearized form of (7.65) where, for notational convenience the distribution function f_B will be denoted by f:

$$\frac{\partial f}{\partial t} = \frac{n_e \sigma_T}{m_e c} \frac{1}{\varepsilon^2} \frac{\partial}{\partial \varepsilon}\{ \varepsilon^4 [kT \frac{\partial f}{\partial \varepsilon} + f]\}. \tag{7.79}$$

The arguments used to construct a finite difference representation of this simplified equation may be extended in a straightforward manner to (7.71) or (7.75). Many finite difference representations to (7.78) may by constructed, all of which reduce to the correct partial differential equation in the limit $\Delta t \to 0$ and $\Delta\varepsilon \to 0$. The intrinsic properties of the Fokker-Planck equation discussed earlier in this section may be used to construct a finite difference equation whose solution preserves these properties. The derivative on the right side of (7.79) may be carried out explicitly to form a second order partial differential equation. Straightforward finite difference approximations to $\partial f / \partial \varepsilon$ and $\partial^2 f / \partial \varepsilon^2$ may then be used to obtain an equation for $f^{n+1}{}_{j+1/2}$. Unfortunately it can be shown that this approach does not satisfy (7.67). In order to satisfy number conservation to machine roundoff, (7.79) will be differenced as it stands since the right side can be interpreted as the gradient of a generalized number flux in energy space:

$$\frac{f^{n+1}_{k+1/2,\,j+1/2} - f^{n}_{k+1/2,\,j+1/2}}{\Delta t^{n+1/2}} = K_{o,\,k+1/2} \frac{(F_{k+1/2,\,j+1} - F_{k+1/2,\,j})}{\varepsilon^2_{j+1/2}\Delta\varepsilon_{j+1/2}}. \tag{7.80}$$

Here the generalized number flux $F_{k+1/2,j}$ is defined by a finite difference representation of

$$F = \varepsilon^4 \left(k_B T \frac{\partial f}{\partial \varepsilon} + f\right) \tag{7.81}$$

and

$$K_o \equiv \frac{n_e \sigma_T}{m_e c}.$$

The factor K_o could have been incorporated into the definition of F; it has been factored out here for notational convenience. All spatial dependent variables are evaluated at the zone center. Finally, since there is no spatial coupling in (7.79) the spatial subscript (in one dimension) $k+1/2$ will be omitted below for notational convenience.

The finite difference form of the generalized number flux F_ν, denoted by F_j, will be discussed below. First, however, note that the form (7.80) automatically guarantees number conservation if the generalized number flux satisfies

$$F_{J+1} = F_0 = 0. \tag{7.82}$$

One way to enforce this condition is to set the coupling coefficient K_o to zero for the boundary zones. These two constraints, which represent boundary conditions in energy space, state that there is no number flux into or out of the finite energy group structure.

Problem 7.13

Prove that the finite difference form (7.80) with the boundary conditions (7.82) guarantee the finite difference equivalent of

$$\left(\frac{d n_\gamma}{d t}\right)_{Compton} = 0.$$

A finite difference representation of the generalized number flux (7.81) must be defined to complete (7.80). The first term on the right side is naturally centered, so we may take

$$F_{k+1/2,\,j} \equiv \varepsilon_j^4\left[kT_{k+1/2}\,\frac{f_{j+1/2}^{n+1}-f_{j-1/2}^{n+1}}{\Delta\varepsilon_j} + f_j^{n+1}\right].$$

(7.83)

When this is substituted into (7.80) the result resembles the finite difference representation of the thermal diffusion equation. Since that equation was most conveniently evaluated implicitly, we anticipate that the same advantages will obtain for the Fokker-Planck equation, and have specified the temporal indices in (7.83) accordingly.

Finally a prescription must be established for the distribution function at each energy group boundary in terms of the group centered values. An obvious average is

$$f_j^{n+1} \equiv \frac{f_{j+1/2}^{n+1}+f_{j-1/2}^{n+1}}{2}.$$

(7.84)

This choice is unacceptable, as can be shown by the following arguments (Chang and Cooper 1970). In a steady state the generalized number flux must vanish, in which case (7.84) and (7.83) can be solved for the ratio

$$\frac{f_{j+1/2}^{n+1}}{f_{j-1/2}^{n+1}} = \frac{2\,kT - \Delta\varepsilon_j}{2\,kT + \Delta\varepsilon_j}. \tag{7.85}$$

Since $f_{j+1/2}$ cannot change sign, and in fact must be positive, [recall (7.68)], the group structure must satisfy the constraint $kT > \Delta\varepsilon_j / 2$ if f_j is to be defined by (7.84). This will generally not be feasible. In more complicated situations [nonlinear terms included as in (7.65), for example] a physically reasonable constraint may occur, but it will generally impose such a severe limitation on the energy group structure as to be impractical.

The requirement that in steady state the generalized number flux (7.81) vanish leads to the partial differential equation

$$kT\frac{\partial f}{\partial \varepsilon} + f = 0, \tag{7.86}$$

whose exact solution is

$$f(\varepsilon) = A\, e^{-\varepsilon / kT}. \tag{7.87}$$

The finite difference form of this solution may be used to show that in steady state the ratio

$$\frac{f_{j+1/2}^{eq}}{f_{j-1/2}^{eq}} = \frac{exp(-\varepsilon_{j+1/2} / kT)}{exp(-\varepsilon_{j-1/2} / kT)} = exp(-\Delta\varepsilon_j / kT) \tag{7.88}$$

which would guarantee that (7.68) is satisfied in equilibrium for any energy group structure. The problem with the interpolation (7.84) is that it causes the spectrum to fall off too fast, becoming negative for some values of j, whereas the equilibrium spectrum (7.88) does not fall off as fast.

The problem with (7.84) is that it mixes too much of $f_{j-1/2}$ into the average for f_j. The solution to this problem is to define a constant δ_j such that

$$f_j^{n+1} = (1-\delta_j) f_{j+1/2}^{n+1} + \delta_j \, f_{j-1/2}^{n+1} \tag{7.89}$$

with $0 \le \delta_j \le 1/2$. The constant δ_j may then be chosen so that the equilibrium solution (7.6.26) is satisfied exactly. In this way we enforce the second and third intrinsic properties of the differential form of the Fokker-Planck equation which were discussed above. In equilibrium $F_j = 0$; equations (7.83) and (7.88) can then be solved for the ratio $f_{j+1/2}/f_{j-1/2}$, which in combination with (7.89) yields

$$\delta_j \equiv \frac{1}{x_j} - \frac{1}{exp(x_j) - 1}$$

$$x_j = \Delta\varepsilon_j / kT . \tag{7.90}$$

Equations (7.80), (7.83) and (7.88)-(7.90) may be taken as a finite difference representation of the Fokker-Planck equation which conserves photon number, and guarantees that the spectrum is nonnegative and that the correct thermal equilibrium solution results. Collecting terms, the resulting finite difference equation can be written in the tridiagonal form $(0 \le j \le J)$

$$A_{j+1/2} f_{j+1/2}^{n+1} - B_{j+1/2} f_{j+1/2}^{n+1} + C_{j+1/2} f_{j-1/2}^{n+1} + D_{j+1/2} = 0 \tag{7.91}$$

where the coefficients are given by

$$A_{j+1/2} = \frac{K_o \, \Delta t^{n+1/2}}{\varepsilon_{j+1/2}^2 \, \Delta\varepsilon_{j+1/2}} \varepsilon_{j+1}^4 \left(\frac{kT}{\Delta\varepsilon_{j+1}} + 1 - \delta_{j+1}\right)$$

$$B_{j+1/2} = 1 + \frac{K_o \, \Delta t^{n+1/2}}{\varepsilon_{j+1/2}^2 \, \Delta\varepsilon_{j+1/2}} \left\{\varepsilon_{j+1}^4 \left(\frac{kT}{\Delta\varepsilon_{j+1}} - \delta_{j+1}\right)\right.$$

$$\left. + \varepsilon_j^4 \left(\frac{kT}{\Delta\varepsilon_j} - \delta_j + 1\right)\right\}$$

$$C_{j+1/2} = \frac{K_o \, \Delta t^{n+1/2}}{\varepsilon_{j+1/2}^2 \, \Delta\varepsilon_{j+1/2}} \varepsilon_j^4 \left(\frac{kT}{\Delta\varepsilon_j} - \delta_j\right)$$

$$D_{j+1/2} = f_{j+1/2}^n \, . \tag{7.92}$$

The temperature and the coupling coefficient K_o are evaluated at the spatial zone center $k+1/2$. The back substitution algorithm used to solve the thermal diffusion equation [see (6.104)-(6.110) and the accompanying discussion] may be used to solve (7.91) subject to appropriate boundary conditions. It follows from (7.90) that all of the coefficeints in (7.91) are nonnegative.

The boundary conditions which must be imposed on the finite difference equation follow from (7.6.29) for $j = 0$ and $j = J$. Since $f_{J+1/2}$ and $f_{-1/2}$ lie outside the energy group structure, the coefficients of these terms must vanish identically:

$$A_{J+1/2} = 0,$$

$$C_{1/2} = 0. \tag{7.93}$$

The solution to (7.91)-(7.93) may easily be constructed in analogy with (6.104).

In order to solve the full nonlinear Fokker-Planck equation a practical method is to replace the term $f^{n+1}{}_j$ in (7.83) by $f^{n+1}{}_j(1+f^n{}_j)$ where $f^n{}_j$ is defined in the same way as $f^{n+1}{}_j$ as regards energy averaging. The resulting difference equation is stable for an arbitrary time step, but since the stimulated emission term is evaluated at the old time, a short enough time step must be used so that the factor $1+f_j$ only changes modestly per time step. This constraint is very important for neutrino processes where the nonlinear term is $f_j(1-f_j)$, and the use of a large time step can result in $f_j > 1$.

Electron-Neutrino Scattering

A modification of Chang and Cooper's approach to the Fokker-Planck equation has been adopted for Compton scattering of electrons and various neutrino species which occurs during stellar core collapse (see Tubbs *et. al.* 1980 and Bowers and Wilson 1982 for a discussion of electron neutrino scattering and the Fokker-Planck equation). Applications to stellar core collapse have treated Compton scattering in terms of the spectral energy distribution by (7.75). The equations could also be formulated in terms of the neutrino distribution function f_ν. It is desirable to use a difference scheme which preserves the positivity of the energy spectrum, approaches the correct equilibrium distribution and also conserves neutrino number. Recalling the discussion above (7.80) we proceed to associate the quantity in braces in (7.75a), which is a generalized neutrino energy flux in energy space, by F_ν , and write the Fokker-Planck equation in the following finite difference form:

$$F^{n+1}_{k+1/2,j+1/2} - F^{n}_{k+1/2,j+1/2} = \frac{\varepsilon_{j+1/2}}{\Delta\varepsilon_{j+1/2}}(F_{k+1/2,j+1/2} - F_{k+1/2,j-1/2})\,\Delta t^{n+1/2} . \tag{7.94}$$

Even though this equation expresses the rate of change in neutrino energy, the form above will conserve neutrino number exactly [as can be seen from the definition below (7.76)] if the generalized energy flux satisfies boundary conditions similar to (7.82). This can always be accomplished because the definition of F_j is arbitrary [recall the discussion preceeding (7.88)].

The generalized neutrino energy flux in energy space due to Compton scattering is given by

$$\mathcal{F}_\nu \equiv K_\nu \Big[F_\nu (1 - F_\nu / \alpha\varepsilon^3) + kT\big(\frac{\partial F_\nu}{\partial \varepsilon} - \frac{3F_\nu}{\varepsilon}\big)\Big]. \tag{7.95}$$

The electron-neutrino coupling coefficient K_ν is given by (7.75b), and the first two terms in square brackets above result from the Fermi-Dirac nature of the neutrinos. The coupling coefficient (7.75b) is a function of the collisional energy exchange and the collision time. These quantities will in general be functions of neutrino energy, and the material density, temperature and composition, which are spatially zone centered variables. The blocking factor $1 - F_\nu/\alpha\varepsilon^3$ arises because no two fermions can occupy the same quantum state. A fully implicit treatment of these terms would result in a nonlinear finite difference equation, which would be difficult to solve. One can evaluate the blocking terms using $F_\nu{}^n$ and then evaluating the rest of (7.95) using $F_\nu{}^{n+1}$. In this way a difference equation can be derived by the same methods used for photons. The generalized energy flux (7.95) is then linear in the new value of the spectral energy density F_ν.

Other finite difference approximations to (7.95) are possible; an approach which has been found to work well for stellar core collapse calculations using an exponential energy group structure is

$$\mathcal{F}_j \equiv K_j \Big[kT \frac{F^{n+1}_{j+1/2} - F^{n+1}_{j-1/2}}{\Delta\varepsilon_j} - \frac{3kT}{2}\big(\frac{F^{n+1}_{j+1/2}}{\varepsilon_{j+1/2}} + \frac{F^{n+1}_{j-1/2}}{\varepsilon_{j-1/2}}\big)$$

$$+ F_j^{n+1}\Big(1 - \frac{F^n_{j+1/2}}{2\alpha\varepsilon^3_{j+1/2}} - \frac{F^n_{j-1/2}}{2\alpha\varepsilon^3_{j-1/2}}\Big)\Big] \tag{7.96}$$

where the energy group structure has zone centered energies

$$\varepsilon_{j+1/2} = (\varepsilon_{3/2} / \varepsilon_{1/2})\, \varepsilon_{j-1/2} \, . \tag{7.97a}$$

The energy at the group boundary is related to the zone centered value by

$$\varepsilon_{j} = (\varepsilon_{3/2} / \varepsilon_{1/2})^{1/2} \varepsilon_{j-1/2} \, . \tag{7.97b}$$

The energy group width is

$$\Delta\varepsilon_{j+1/2} = \varepsilon_{j+1} - \varepsilon_{j-1/2} \, . \tag{7.97c}$$

For simplicity, all spatial indices have been omitted. The first term in (7.96) is naturally centered in energy space. The quantity multiplying the last term is edge centered. A simple arithmetic average has been used as a finite difference approximation to $3F_{\nu} / \varepsilon$ in (7.96). The resulting spectrum will not change sign as long as the energy group structure above is constructed with $\varepsilon_{3/2} / \varepsilon_{1/2} < 3$. Finally, since the blocking factor multiplying $F^{n+1}{}_{j}$ contains the old value of the spectral energy, we have simply averaged it to obtain the boundary value.

The edge centered quantity $F^{n+1}{}_{j}$ can be defined as in (7.89), and the constant δ_j determined by requiring that the equilibrium spectrum satisfy (7.82). In practice it has been found sufficient to impose the less restrictive equilibrium constraint obtained by setting the blocking factor equal to unity in (7.96);

$$F + kT\left(\frac{\partial F}{\partial \varepsilon} - \frac{3F}{\varepsilon}\right) = 0. \tag{7.98}$$

The difference equation corresponding to (7.98) follows from the definition of the generalized energy flux, and is

$$0 = kT\frac{F^{n+1}_{j+1/2} - F^{n+1}_{j-1/2}}{\Delta\varepsilon_j} - \frac{3kT}{2}\left(\frac{F^{n+1}_{j+1/2}}{\varepsilon_{j+1/2}} + \frac{F^{n+1}_{j-1/2}}{\varepsilon_{j-1/2}}\right) + F^{n+1}_j \tag{7.99}$$

with

$$F^{n+1}_j = (1-\delta_j)F^{n+1}_{j+1/2} + \delta_j F^{n+1}_{j-1/2} . \tag{7.100}$$

The finite difference equations (7.99)-(7.100) may be solved analytically for δ_j. The results above may be used to express the Fokker-Planck equation for the neutrino spectral energy density in tridiagonal form.

An application of neutrino-electron scattering under conditions expected in stellar core collapse using the finite difference methods above has been compared with thermal relaxation and Monte Carlo methods (Tubbs *et. al.* 1980).

References

M. L. Alme and J. R. Wilson *Ap. J.*, **186**, 1015, 1973.

-- *Ap. J.*, **194**, 147, 1974.

-- *Ap. J.*, **210**, 233, 1976.

R. L. Bowers and J. R. Wilson *Ap. J. Suppl.*, **50**, 115, 1982.

J. S. Chang and G. E. Cooper *J. Comp. Phys.* **6**, 1, 1970.

A. S. Kompaneets *J. E. P. T.*, **4**, 730, 1957.

D. Mihalas and B. W. Mihalas, *Foundations of Radiation Hydrodynamics*, (New York, Oxford Univeristy Press, 1984).

G. C. Pomraning, *The Equations of Radiation Hydrodynamics*, (Oxford: Pergamon Press, 1973).

D. L, Tubbs and D. N. Schramm *Ap. J.*, **201**, 467, 1975.

D. L, Tubbs, T. A. Weaver, R. L. Bowers, J. R. Wilson and D. N. Schramm *Ap. J.*, **239**, 271, 1980.

J. R. Wilson, R. Couch, S. Cochran, J. LeBlanc and Z. Barkat *Ann. NY Acad. Sci.*, **262**, 54, 1975.

MAGNETOHYDRODYNAMICS

Magnetic fields are observed in many astrophysical systems, including solar and stellar atmospheres, the interstellar medium, pulsars and radio galaxies, and are believed to play an important role in such processes as accretion onto neutron stars and white dwarfs, star formation and galactic evolution. In this chapter finite difference methods for resistive magnetohydrodynamics in one and two dimensions will be considered. We assume that the hydrodynamic, gravitational and radiative behavior of the system can be treated separately by means of operator splitting using the methods described elsewhere in the text.

It will be assumed here that all magnetohydrodynamic time scales are long compared with electron and ion collision time scales τ_{ee}, τ_{ei}, and τ_{ii} (see Chapter 6 and Appendix E), and that the material under consideration is a completely ionized plasma. The first assumption implies that the plasma is relatively dense, and guarantees that fluid equations of state exist, as does an electrical resistivity.

In the following discussion Gaussian units will be employed for the electromagnetic fields and magnetohydrodynamic variables. In Gaussian units $\varepsilon_o = 1$ and $\mu_o = 1$.

8.1 PLASMA ELECTRODYNAMICS

The electrodynamic fields in a plasma may be externally imposed, or may arise directly from the plasma motion itself. In this section the electric field and magnetic intensity vectors in the local rest frame of the plasma will be denoted by $\boldsymbol{E}'$ and $\boldsymbol{B}'$ respectively. The electrical current density and electrical resistivity in the plasma frame (denoted by primes) are $\boldsymbol{J}'$ and η' respectively. Assuming that the plasma is spatially isotropic, the resistivity will be a scalar and Ohm's law may be used to relate the local electric field to the current density

$$\boldsymbol{E}' = \eta' \boldsymbol{J}' . \tag{8.1}$$

The electrical resistivity represents a constitutive relation for the plasma, and is usually given as a function of the material density, temperature and composition (the dependence on the latter will usually not be exhibited explicitly):

$$\eta = \eta(\rho, T). \tag{8.2}$$

In principle the resistivity may be expressed as a function of density and specific energy. A relation of the functional form (8.2) will be assumed below, although the details of its form are not important for the following discussions.

Two approximations will be considered. The first approximation relates the electromagnetic fields in both the comoving and the laboratory frames of reference. The second approximation will allow us to ignore the electric field in the fluid frame, and leads to the standard magnetohydrodynamic approximation. The nonrelativistic approximation has been adopted for simplicity, and is independent of the magnetohydrodynamic approximation. First we consider the relation between the fields as seen in the fluid, or comoving frame, and as seen in the laboratory (Eulerian) frame.

The components of the electromagnetic field in the laboratory frame (relative to which the plasma moves) will be denoted by $\boldsymbol{E}$ and $\boldsymbol{B}$. These components are related to the components in the comoving (plasma) frame by means of Lorentz transformations (see, for example, Jackson 1975). The discussions of this chapter will be limited to the case $v/c \ll 1$, where the single fluid plasma velocity relative to the laboratory is $\boldsymbol{v}$. Then the fields in the laboratory frame and in the plasma frame are related by

$$E' = E + \frac{\mathbf{v} \times B}{c} \tag{8.3a}$$

$$B' = B - \frac{\mathbf{v} \times E}{c} \, . \tag{8.3b}$$

Even though the ratio $\mathbf{v}/c$ is often quite small, terms of order $\mathbf{v}/c$ usually cannot be ignored (recall the discussion of $\mathbf{v}/c$ terms in radiation theory discussed in Chapter 6). The current density J and the local charge density q_e in the laboratory frame and those in the plasma frame are related by

$$J' = J - q_e \mathbf{v} \tag{8.3c}$$

$$q_e' = q_e - \frac{\mathbf{v} \cdot J}{c^2} \, . \tag{8.3d}$$

Terms of order $\mathbf{v}/c$ in these expressions will be identified below. Finally, the electric displacement D and the magnetic field H are related to the electric field and the magnetic displacement in the plasma by

$$D = \varepsilon E \quad \text{and} \quad H = \mu^{-1} B. \tag{8.3e}$$

The plasma shall be assumed to be nonmagnetic, so that $\mu = 1$, and we need not distinguish between the magnetic field and the magnetic intensity.

A fundamental approximation to be used below is that the plasma has zero space charge density (local charge neutrality). Combining this assumption, which is equivalent to setting $q_e' = 0$, (8.3d) implies that $q_e = \mathbf{v} \cdot J / c^2$. When this is substituted into (8.3c) we find:

$$J' = J - q_e v \sim J - \left(\frac{v}{c}\right)^2 J$$

$$\sim J + O(v^2/c^2). \tag{8.4}$$

Thus, to order v/c the current in the comoving frame and the current in the laboratory frame are the same. Next, consider the transformation of the magnetic fields in (8.3b). In a highly ionized plasma the local electric field is usually small, and so to order of magnitude $E \sim vB/c$. It then follows that

$$B' = B + O(v^2/c^2). \tag{8.5}$$

Thus, retaining terms of order v/c the magnetic field in the comoving frame and the magnetic field in the laboratory frame are also the same. In view of (8.5) and (8.3a) the electric field is not the same in the two frames of reference. Using Ohm's law (8.1) to eliminate the local electric field

$$E = \eta' J' - \frac{v \times B}{c}$$

$$= \eta' J - \frac{v \times B}{c} + O(v^2/c^2). \tag{8.6}$$

The last form of the equation above contains the resistivity expressed in the fluid frame, where it is most naturally defined, but involves the magnetic field and the current density as measured in the laboratory frame.

Finally, the net electromagnetic force acting on the fluid in its rest frame is given by $q_e' E'$. When this is transformed to the laboratory frame using (8.3a) one finds:

$$F' = q_e' E' = q_e E + q_e \frac{v \times B}{c}$$

$$= q_e E + \frac{J \times B}{c}. \tag{8.7}$$

But $q_e E c \sim (vJ / c^2)(vB)$, which is of order $(v/c)^2$ times the magnetic force $J \times B$. Therefore the electromagnetic force as seen in the laboratory frame is simply

$$F = \frac{J \times B}{c} + O(v^2 / c^2). \tag{8.8}$$

To lowest order in v/c we may consider both the magnetic field and current density to be frame independent, but must retain the lowest order correction to the electric field (8.6).

Magnetohydrodynamic Approximation

We now consider within the nonrelativistic approximation the equations governing the behavior of a plasma in the reduced set of Maxwell's equations which apply in the magnetohydrodynamic regime. The electrodynamic behavior of a plasma may be incorporated with the fluid dynamic description in a relatively straightforward way when the plasma exhibits approximate local charge neutrality, and when electric displacement currents can be neglected. These two requirements represent the underlying assumptions of magnetohydrodynamics. The implications of these assumptions may be appreciated by considering the following examples. For simplicity we shall assume that the plasma is fully ionized.

Approximate local charge neutrality means that the local net charge is always small relative to either the ion or the electron charge density. It may be expressed in the form

$$|q_e'| = |Ze\, n_i - en_e| \ll en_e. \tag{8.9}$$

For the local charge density q_e', Gauss's law is $\nabla \cdot \boldsymbol{E}' = (4\pi / \varepsilon)\, q_e'$. If l represents a typical scale length over which changes in the local electric field $\boldsymbol{E}'$ occur, then we obtain the following order of magnitude relation: $\varepsilon E' \sim 4\pi q_e' l$.

Next, consider a nearly static plasma which is permeated by electromagnetic fields. In the plasma frame of reference, the net force acting on it locally is given by

$$\rho \frac{d\boldsymbol{v}}{dt} = -\nabla P + q_e \boldsymbol{E}' \sim 0$$

or $\nabla P \sim q_e \boldsymbol{E}'$. To order of magnitude, the pressure gradient for an ideal gas may be written as $\nabla P \sim n_e kT/l$. Setting the local charge density $q_e' \sim e\, n_e$, the quasi-static condition becomes

$$\frac{n_e kT}{l} \sim e n_e E' \sim (4\pi n_e e\, l)\, q_e << 4\pi n_e^2 e^2 l$$

which may be rewritten in the form

$$\lambda_D^2 \equiv \frac{kT}{4\pi e^2 n_e} << l^2. \tag{8.10}$$

The quantity λ_D is called the Debye length. Approximate local charge neutrality implies that the scale length over which fluid variables, such as the pressure, change, is large compared with the Debye length. Thus, as long as $l >> \lambda_D$ we may ignore q_e'. Note, however, that the net local charge density is not assumed to vanish identically.

Local charge conservation requires that the local charge density and the current density satisfy the relation

$$\frac{\partial q_e'}{\partial t} + \nabla \cdot \boldsymbol{J}' = 0.$$

Under some conditions the time rate of change of the local charge density in a plasma is very small. Then it is, to order of magnitude,

$$\frac{\partial q_e'}{\partial t} \sim \frac{q_e'}{\tau} \sim \omega_P q_e' \sim \left(\frac{4\pi n_e e^2}{m_e}\right)^{1/2} q_e'$$

where ω_P is the plasma frequency, which characterizes the frequency of space charge oscillations. Next, consider the divergence of the current density; it is, to order of magnitude,

$$\nabla \cdot \boldsymbol{J}' \sim J'/l \sim e n_e v/l\,.$$

The ratio of the rate of change in local charge density to the divergence of the current density as given by the two expressions above can be shown to be of order v_A/c, where

$$v_A \equiv \left(\frac{B^2}{4\pi\rho}\right)^{1/2} \tag{8.11}$$

is called the Alfven velocity of the plasma. For fluid densities and magnetic field strengths such that $v_A << c$, the ratio is $<< 1$, and the time rate of change in the local charge density can be neglected compared with the divergence of the current density. We then can set

$$\nabla \cdot \boldsymbol{J}' = \nabla \cdot \boldsymbol{J} = 0, \tag{8.12}$$

where (8.4) has been used. This condition is equivalent to neglecting electric displacement currents in Maxwell's equations.

The magnetohydrodynamic approximation in the nonrelativistic limit $\boldsymbol{v} << c$ allows us to ignore the local charge density and displacement currents, and write Maxwell's equations in the reduced form:

$$\nabla \times \boldsymbol{E} = -\frac{1}{c}\frac{\partial \boldsymbol{B}}{\partial t} \tag{8.13a}$$

$$\nabla \times \boldsymbol{B} = \frac{4\pi}{c}\boldsymbol{J}$$

(8.13b)

$$\nabla \cdot B = 0.$$

(8.13c)

The first two equations, (8.13a) and (8.13b) may be combined with Ohm's law (8.6) to obtain dynamic equations for the magnetic field. Equation (8.13c) represents a constraint equation which must be satisfied by the solutions to the dynamic equations.

Finally, we note that Coulomb's law,

$$\nabla \cdot \boldsymbol{E} = \frac{4\pi}{\varepsilon} q_e ,$$

(8.13d)

is not used directly in magnetohydrodynamics as long as the assumption of approximate local charge neutrality holds. It may be recalled that in electrodynamics the local charge density is usually given, from which the electric fields are found. In magnetohydrodynamics, the local electric field, if needed, is to be obtained from Ohm's law (8.6). Gauss's law is used only to find the local charge density. Equation (8.9) is then used to verify that the charge density is sufficiently small to justify using the magnetohydrodynamic approximation.

For some applications it is convenient to employ the magnetic vector potential, $\boldsymbol{A}$, from which the magnetic field may be calculated:

$$\boldsymbol{B} = \nabla \times \boldsymbol{A} .$$

(8.14a)

In the transverse gauge (see, for example, Jackson 1975) the vector potential satisfies the gauge conditions:

$$\nabla \cdot \boldsymbol{A} = 0 \qquad \boldsymbol{E} = -\frac{1}{c}\frac{\partial A}{\partial t}.$$

(8.14b)

8.2 MAGNETOHYDRODYNAMICS

The magnetohydrodynamic equations consist of an evolution equation and a constraint equation for the magnetic field, and the momentum, energy and mass continuity equations for the plasma. In this section a set of differential equations will be reviewed in Eulerian and Lagrangian form. Subsequent sections will develop Lagrangian one dimensional representations and a two dimensional Eulerian representation of these equations. Although the discussion below will not include radiation transport or gravitation, the methods developed for these elsewhere in the text may be incorporated using operator splitting.

Evolution Equation

The electric field may be eliminated from the reduced set of Maxwell's equations using (8.6) to obtain an Eulerian form of the evolution equation for the magnetic field. First, (8.6) is substituted into (8.13a) and then (8.13b) is used to eliminate the current density. Hereafter the resistivity will always be expressed in the fluid frame and the prime may be deleted. The result expresses the rate of change of $\boldsymbol{B}$ in the form

$$\frac{\partial \boldsymbol{B}}{\partial t} = - \nabla\times\left(\frac{\eta c^2}{4\pi} \nabla\times\boldsymbol{B}\right) + \nabla\times(\boldsymbol{v}\times\boldsymbol{B}) . \tag{8.15}$$

The first term on the right represents current diffusion through the plasma, with an effective diffusion coefficient given by

$$D_B = \frac{\eta c^2}{4\pi}. \tag{8.16}$$

The second term in (8.15) represents advection of the magnetic field associated with motion of the plasma. For example, if the resistivity is small (corresponding to a large conductivity) then the diffusion term in (8.15) may be ignored and the change in magnetic field is due entirely to material motion.

Problem 8.1

Evaluate the curl of (8.15) and show that for a constant diffusion coefficient the first term on the right side is formally identical to the thermal diffusion equation. Show that to order of magnitude the current diffusion time scale τ_B is given by

$$\tau_B \sim \frac{4\pi l^2}{\eta c^2}, \tag{8.17}$$

where l is a characteristic scale length for the current (and thus for the magnetic field) distribution in the plasma.

The relative importance of the two terms on the right side of (8.15) is measured by the ratio of the magnitude of the transport term to the magnitude of the diffusion term. The ratio is called the magnetic Reynolds number, and is

$$R_B \equiv \frac{v l_B}{\eta c^2}, \tag{8.18}$$

where l_B is the magnetic field scale length. When $R_B \gg 1$ the magnetic field behaves as if it were frozen into the plasma, and its motion is determined entirely by the plasma velocity. In the opposite extreme, $R_B \ll 1$, the magnetic field moves rapidly by diffusion through the plasma. The finite difference methods discussed below allow for either extreme, as well as for intermediate regions where both magnetic field diffusion and transport by the plasma occur.

An alternate approach, which is sometimes useful for numerical models, is to construct an evolution equation for the magnetic vector potential. Starting with (8.14b), we use (8.6) to eliminate the electric field, and then replace the magnetic field with (8.14a) to obtain

$$\frac{\partial \boldsymbol{A}}{\partial t} = -D_B \nabla \times (\nabla \times \boldsymbol{A}) + \boldsymbol{v} \times (\nabla \times \boldsymbol{A}). \tag{8.19}$$

The first term on the right represents diffusion of the vector potential, as can be shown by expanding the cross products and using the first gauge condition in (8.14a). The diffusion coefficient is the same as appears in (8.15). The second term on the right corresponds to transport of the vector potential associated with material motion. Under some circumstances it may be easier to evolve some components of $\boldsymbol{A}$, and then obtain the magnetic field from it using (8.14a) when needed. The relative advantages and disadvantages of this approach will be noted below.

The results above represent a convenient starting point for Eulerian magnetohydrodynamics. Equation (8.15) may also be recast into a form suitable for Lagrangian methods. Combining the transport term with the time derivative, and expanding the cross product we obtain the following identity:

$$\frac{\partial \boldsymbol{B}}{\partial t} - \nabla \times (\boldsymbol{v} \times \boldsymbol{B}) = \frac{\partial \boldsymbol{B}}{\partial t} + \boldsymbol{v} \cdot \nabla \boldsymbol{B} + \boldsymbol{B}(\nabla \cdot \boldsymbol{v}) - \boldsymbol{B} \cdot \nabla \boldsymbol{v} .$$

The first two terms on the right constitute the Lagrangian derivative of the magnetic field vector. The velocity divergence in the third term may be eliminated using the mass continuity equation (4.3). When this is done (8.15) becomes:

$$\frac{d\boldsymbol{B}}{dt} - \frac{\boldsymbol{B}}{\rho}\frac{d\rho}{dt} - \boldsymbol{B} \cdot \nabla \boldsymbol{v} =$$

$$\rho \frac{d}{dt}\left(\frac{\boldsymbol{B}}{\rho}\right) - \boldsymbol{B} \cdot \nabla \boldsymbol{v} = -\nabla \times (D_B \nabla \times \boldsymbol{B}). \tag{8.20}$$

This form of the evolution equation is particularly useful for one dimensional Lagrangian simulations in planar and cylindrical coordinates.

Problem 8.2

Can a spherically symmetric magnetic field exist which satisfies Maxwell's equations? If so, describe how it enters into the fluid dynamic equations.

Constraint Equation

A solution to the evolution equation must also satisfy the constraint equation (8.13c). If the initial value of $\boldsymbol{B}$ satisfies (8.13c) everywhere, subsequent values obtained as solutions of the evolution equation (8.15) will also satisfy the constraint equation. Thus, when solving the evolution equation in partial differential form the constraint equation is essentially part of an initial value condition. When (8.15) is solved only approximately (as is always the case when numerical methods are employed), the approximate solution will generally not satisfy the constraint equation, and the error can be cumulative. Thus after many computational cycles the value of $\boldsymbol{B}$ may lead to unphysical results. Therefore, it is usually necessary to guarantee that the constraint equation is also satisfied along with the evolution equation for each computational cycle.

Momentum Equation

Electromagnetic fields couple primarily to free electrons and to ions in a plasma (the magnetohydrodynamics to be discussed below is applicable primarily to an ionized plasma). The equations of motion for the free electrons and ions, called two-fluid magnetohydrodynamic equations, are discussed in Appendix E. Numerical methods can be developed which separately follow the dynamics of each individual species in a plasma. For many astrophysical purposes it is sufficient to consider the single fluid approximation, in which the individual change species are replaced by an electrical current density which couples to the magnetic field. The Lagrangian form of the momentum equation (including gravitation) in this case is given by

$$\rho \frac{d\mathbf{v}}{dt} = -\nabla P + \rho g + \frac{\boldsymbol{J} \times \boldsymbol{B}}{c}. \tag{8.21}$$

The magnetic field simply enters as a body force acting on the Lagrangian fluid element.

Problem 8.3

The Lorentz force in (8.21) may be converted into a magnetic stress. Show that the Cartesian components of the Lorentz force are

$$\left(\frac{\boldsymbol{J} \times \boldsymbol{B}}{c}\right)_k = -\frac{\partial \Sigma_{ik}^B}{\partial x^i}$$

where the magnetic stress is given by

$$\Sigma_{ik}^B \equiv \frac{B^2}{2}\delta_{ik} - B_i B_k.$$

Internal Energy

Electron-ion scattering, which is the principle source of electrical resistivity in a dense plasma, converts electrical energy into thermal energy of the electrons [see Appendix E, in particular (E.23)]. Electron-ion scattering also transfers energy from the electronic component of the plasma to the ionic component. If the electron-ion relaxation time scale, τ_{ei}, is comparable to the dynamic time scale, τ_D, it may be necessary to follow the change in the electron energy and ion energy separately. One way of accomplishing this is to assign separate temperatures to the electrons and the ions. It is then necessary to solve the single fluid internal energy equations (E.27) and (E.30) for each species. The methods developed below may be extended to this case. In this Chapter, however, it will be assumed that $\tau_D >> \tau_{ei}$ so that the electrons and the ions are completely coupled. The advantage of this assumption is that the electron and ion internal energy equations may be added, giving the single fluid equation

$$\rho \frac{d\varepsilon}{dt} = -P\, \nabla\cdot\boldsymbol{v} + \eta J^2 - \nabla\cdot\boldsymbol{q} \tag{8.22}$$

where P is the total fluid pressure, and $\boldsymbol{q} = \boldsymbol{q}_e + \boldsymbol{q}_i$ represents the total energy flux associated with dissipative processes, such as thermal diffusion. We note that the electron-ion coupling term cancels, and that $\varepsilon = \varepsilon_e + \varepsilon_i$ is just the total fluid specific internal energy.

The mechanical work and the dissipative terms can be solved using operator splitting and the methods developed in previous chapters. If this is done, it is often convenient to use the effective heat capacity (3.14), and to express the change in internal energy associated with Joule heating in the form

$$\frac{d(\rho C_V T)}{dt} = \eta J^2. \tag{8.23}$$

In general the resistivity will be given as a function of temperature (8.2), and the effects of Joule heating are most easily calculated in temperature mode. For problems where the production and transport of radiation is unimportant the magnetohydrodynamic equations may be formulated in terms of the specific energy (energy mode description) as long as the equations of state (including the electrical resistivity) are available as functions of energy and density. The finite difference representations obtained below are essentially independent of whether the hydrodynamics is expressed in temperature or energy mode.

Mass Continuity Equation

The form of the mass continuity equation is unchanged by the presence of magnetic fields, and is given by (4.3).

Hydrodynamic and Magnetic Time Scales

Astrophysical systems whose mass distributions change dynamically are most efficiently described using explicit hydrodynamic methods. If the resistive diffusion time scale τ_B is large relative to the dynamic time scale τ_D, magnetic diffusion may also be modeled using explicit methods. This represents the simplest numerical approach to magnetohydrodynamics. When resistive and dynamic time scales are comparable, it is often sufficient to use linearized implicit methods to describe magnetic field diffusion. Because these algorithms are not much more complicated than explicit ones they should probably be used even for regions where $\tau_B >> \tau_D$.

Implicit hydrodynamics schemes are only necessary for systems whose mass distributions are changing quasistatically. The discussion of magnetic diffusion above also applies to this case as long as $\tau_D \leq \tau_B$. The primary increase in complexity in this case involves the hydrodynamics.

For quasistatic systems in which $\tau_B << \tau_D$, a fully implicit treatment of the magnetohydrodynamics may be required. This approach is sufficiently more complex, and requires several times more computation per cycle, than explicit or linearized implicit methods; it should be used only when clearly necessary.

Section 8.3 will discuss several approaches to Lagrangian magnetohydrodynamics for one dimensional systems based on the observations above. Two dimensional Eulerian magnetohydrodynamic methods will be discussed in Section 8.4. Because of the complexity of implicit methods in two or more spatial dimensions, we will only consider linearized implicit methods for magnetic diffusion.

8.3 Lagrangian Magnetohydrodynamics

In this section we shall develop a one dimensional magnetohydrodynamic scheme for cylindrically symmetric systems. The results may be used in conjunction with the operator split equations discussed in Chapter 4 and in Chapter 6. The most common nontrivial magnetic configurations in one dimension are systems exhibiting planar or cylindrical symmetry. The equations and a finite difference representation corresponding to cylindrical symmetry will be considered first.

The only nonzero components of $\boldsymbol{B}$ in a one dimensional cylindrically symmetric system are the torroidal component B_θ, and the axial component B_z. Denoting the only nonzero component (radial) of the velocity in this symmetry by v, it is readily shown that

$$\boldsymbol{B}\cdot\nabla\, \boldsymbol{v} = \frac{B_\theta v}{r}\boldsymbol{e}_r\ ,$$

where $\boldsymbol{e}_r$ is a unit vector in the radial direction, and thus the radial component of the evolution equation (8.20) is

$$r\rho\frac{d}{dt}\left(\frac{H}{r^2\rho}\right) = \frac{\partial}{\partial r}\left(\frac{D_B}{r}\frac{\partial H}{\partial r}\right). \tag{8.24}$$

The variable H , defined by

$$H \equiv r B_\theta \tag{8.25}$$

appears frequently in the equations below, and is a particularly convenient variable for numerical purposes. The form of the evolution equation has several properties which will be exploited below. First, we note that

$$\frac{H}{r^2\rho} = \frac{2\pi\,\Delta\Phi_R}{\Delta m}, \tag{8.26}$$

where Δm is the Lagrangian mass of an infinitesimal fluid element, and $\Delta\Phi_R$ is the torroidal magnetic flux throught the fluid element. In a perfectly conducting plasma ($D_B = 0$), the right side of (8.24) vanishes and the left side shows that the torroidal flux is transported with the mass (the magnetic field is frozen into the plasma). Thus, magnetic flux conservation can be easily maintained to the same accuracy as mass conservation (machine roundoff). Furthermore, if the magnetohydrodynamics is to be operator split from the other physics packages, then the solution of (8.24) has the form of a simple diffusion equation.

The axial component of the evolution equation is easily shown to be

$$r\rho \frac{d}{dt}\left(\frac{B_z}{\rho}\right) = \frac{\partial}{\partial r}\left(r D_B \frac{\partial B_z}{\partial r}\right) \tag{8.27}$$

where

$$\frac{B_z}{\rho} = L\frac{\Delta\Phi_z}{\Delta m} \tag{8.28}$$

is the ratio of the axial magnetic flux through a Lagrangian volume element of mass Δm, and L is the axial length of the system. The parameter L may be eliminated if a mass per unit axial length is used for the Lagrangian coordinate.

Problem 8.4

Derive the two components of the evolution equation (8.20) for a one dimensional cylindrically symmetric system. Why is there no radial magnetic field component for this symmetry? Show that (8.26) and (8.28) represent the ratio of torroidal and axial flux to mass as claimed.

Problem 8.5

Find the steady state solution for the torroidal magnetic field component using (8.24), and assuming a constant resistivity for a cylindrical plasma of radius R and length L through which a net current I_o is flowing.

The Lorentz force density in the momentum equation (8.21) for a one dimensional cylindrically symmetric system has on a radial component which is given by

$$\rho\frac{dv}{dt} = -\frac{\partial P}{\partial r} + \rho g + \frac{1}{c}(J_\theta B_z - J_z B_\theta),$$

(8.29)

where the torroidal current density is

$$J_\theta = -\frac{c}{4\pi}\frac{\partial B_z}{\partial r}$$

(8.30a)

and the axial current density is

$$J_z = \frac{c}{4\pi r}\frac{\partial(rB_\theta)}{\partial r} = \frac{c}{4\pi r}\frac{\partial H}{\partial r}.$$

(8.30b)

Combining (8.29) and (8.30), the radial momentum equation can be written in the form

$$\rho\frac{dv}{dt} = -\frac{\partial}{\partial r}\left(P + \frac{B_z^2}{8\pi}\right) - \frac{1}{8\pi r^2}\frac{\partial H^2}{\partial r} + \rho g.$$

(8.31)

Finally, the internal energy equation in cylindrical coordinates becomes

$$\rho\frac{d\varepsilon}{dt} = -P\frac{1}{r}\frac{\partial(rv)}{\partial r} + \eta(J_\theta^2 + J_z^2).$$

(8.32)

Only the adiabatic fluid terms and gravitation have been retained in (8.29)-(8.32) for simplicity. In practice the dissipative terms corresponding to shock acceleration, shock heating and radiative transport may also be included using operator splitting. In the discussion below only the torroidal terms will be considered (finite difference representations of the axial terms will be left to the exercises).

Solution of the Lagrangian Equations

As a first step toward solving the Lagrangian equations we rewrite them in Lagrangian coordinates. A convenient choice in one dimensional cylindrical symmetry is the mass per unit axial length, which is given by

$$dm = 2\pi \rho r \, dr . \tag{8.33}$$

The momentum equation (8.31) is then (the axial component of the field has been omitted for simplicity)

$$\frac{dv}{dt} = -2\pi r \frac{\partial P}{\partial m} - \frac{1}{4r}\frac{\partial H^2}{\partial m} + g , \tag{8.34}$$

where H is defined by (8.25). The internal energy equation is

$$\frac{d\varepsilon}{dt} = -2\pi P \frac{\partial (rv)}{\partial m} + \frac{\eta J_z^2}{\rho}, \tag{8.35}$$

where the axial current density is

$$J_z = -\frac{1}{2}\rho c \frac{\partial H}{\partial m}. \tag{8.36}$$

The evolution equation for the torroidal component of the magnetic field is

$$\frac{d}{dt}\left(\frac{H}{r^2 \rho}\right) = 4\pi^2 \frac{\partial}{\partial m}\left(\rho D_B \frac{\partial H}{\partial m}\right). \tag{8.37}$$

Finally, it should be noted that in one dimensional cylindrical symmetry the constraint equation (8.13c) is automatically satisfied since $B_r = 0$ and B_θ is a function only of the radius (the axial field component, if included, is also only a function of radius).

The coupled system of equations above can be solved in several steps using operator splitting. The first step solves the momentum equation (8.34), including the gravitational force and the Lorentz force. Next, we solve the energy equation (8.35) (ignoring the Joule heating term), which then takes the form

$$\frac{d\varepsilon}{dt} = -2\pi P \frac{\partial (rV)}{\partial m} . \tag{8.38}$$

The evolution equation (ignoring diffusion) takes the simple form

$$\frac{d}{dt}\left(\frac{H}{r^2 \rho}\right) = 0 \tag{8.39}$$

and may be solved directly. This equation describes the motion of the fluid as if it were frozen into the plasma. Thus the first step accounts for all changes in the plasma and field due to plasma motion.

The next step solves for the change in the internal energy due to Joule heating, which was omitted from (8.38) in the first step. Since η is usually a function of temperature, it is often convenient to solve (8.23) and to do it in the operator split form

$$\rho C_V \frac{dT}{dt} = \eta J_z^2 \tag{8.40}$$

$$\frac{d(C_V T)}{dt} = 0. \tag{8.41}$$

In the last equation the density has been omitted since it does not change during this stage of the calculation. If radiation diffusion is to be included it may be done at this stage by adding, for example, the thermal diffusion term to the right side of the last equation above.

The final step in the process consists of solving for the field diffusion, which is given by

$$\frac{dH}{dt} = 4\pi r^2 \rho \frac{\partial}{\partial m}\left(\rho D_B \frac{\partial H}{\partial m}\right). \tag{8.42}$$

In writing this equation specific use has been made of the fact that the density and radius do not change in time during this step. In particular, the values of these variables obtained from the first step are used as initial values. The solution of (8.42) accounts for torroidal field diffusion through the plasma.

Space-Time Centering

A finite difference representation of the magnetohydrodynamic equations requires, in addition to the usual hydrodynamic variables, a representation of the magnetic field components and of the resistivity. The torroidal field B_θ is most naturally treated as a zone centered quantity, as is the resistivity which depends on the temperature and composition of the plasma (at low energies it may also depend on the plasma density). Therefore we associate (Figure 8.1)

$$B_\theta \rightarrow B_{k+1/2} \quad \text{or} \quad H \rightarrow H_{k+1/2} = r_{k+1/2} B_{k+1/2} \tag{8.43}$$

$$\eta \rightarrow \eta_{k+1/2}. \tag{8.44}$$

The Lagrangian mass per unit axial length (8.33) will be represented by the zone centered quantity

$$dm \rightarrow \Delta m_{k+1/2}. \tag{8.45}$$

It should be noted that the variable Δm has been used previously for

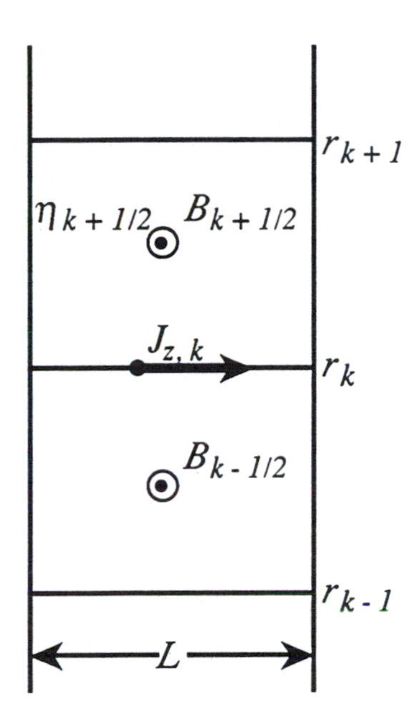

Figure 8.1. Spatial centering of the Lagrangian magnetohydrodynamic variables in a one dimensional cylindrically symmetric system.

the zonal mass; in this section it represents the mass per unit length. All remaining variables will be centered as for the one dimensional Lagrangian scheme discussed in Section 4.3. It follows from (8.44) that the torroidal flux per unit length passing through a zone of radial thickness $\Delta r_{k+1/2}$ is

$$2\pi\, r_{k+1/2}\, \Delta r_{k+1/2}\, B_{k+1/2} \;=\; 2\pi\, H_{k+1/2}\, \Delta r_{k+1/2}\,.$$

The way in which these equations are to be solved will depend on the nature of the problem under consideration, as noted in the discussion concluding Section 8.2. We first consider a system in which the resistive diffusion time scale is comparable to or larger than the dynamic time scale. In this case the magnetohydrodynamics may be differenced explicitly in time.

The finite difference equations to be solved in the first step correspond to (8.34), (8.38) and (8.39). An explicit representation of the momentum equation in spherical coordinates was discussed in Section 4.3. A slight modification of those results which is applicable to cylindrical Lagrangian coordinates gives the finite difference equation:

$$\frac{V_k^{n+1/2} - V_k^{n-1/2}}{\Delta t^n} = -\,2\pi(r_k^n)\,\frac{P^n_{k+1/2} - P^n_{k-1/2}}{\Delta m_k}$$

$$-\,\frac{1}{4\,r_k^n}\,\frac{H^2_{k+1/2} - H^2_{k-1/2}}{\Delta m_k} + g_k^n\,. \tag{8.46}$$

The pressure is assumed to be a function of the density and specific energy. New zonal radii are found using the new velocities from (8.46) and an equation of the form (4.48).

The internal energy equation in cylindrical coordinates which is to be solved in the first step is given by the finite difference equation

$$\frac{\varepsilon'_{k+1/2} - \varepsilon^n_{k+1/2}}{\Delta t^{n+1/2}} = -\,2\pi P^{n+1/2}_{k+1/2}\,\frac{r^n_{k+1}\,V^{n+1/2}_{k+1} - r^n_k\,V^{n+1/2}_k}{\Delta m_{k+1/2}}. \tag{8.47}$$

Note that the new value of the specific energy is denoted by a prime and that the time centered pressure is evaluated from ε^n, ρ^n and ε' and ρ^{n+1}. Except for the dependence on cylindrical coordinates, this has the same form as the energy equation in (4.54). The torroidal field transport term (8.39) may be expressed in the form

$$\frac{H'_{k+1/2}}{\rho^{n+1}_{k+1/2}\,(r^{n+1}_{k+1/2})^2} = \frac{H^n_{k+1/2}}{\rho^n_{k+1/2}(r^n_{k+1/2})^2}. \tag{8.48}$$

Note that the new values of H are denoted by a prime. Finally the mass continuity equation gives the density per unit axial length and is

$$\rho^n_{k+1/2} = \frac{1}{\pi}\frac{\Delta m_{k+1/2}}{[(r^n_{k+1})^2 - (r^n_k)^2]} . \tag{8.49}$$

The next step consists of solving the Joule heating term (8.40). The second equation, (8.41), which completes this step will not be discussed here, since it may be solved as outlined in Section 3.2. The right side of (8.40) involves the axial current density and the resistivity. Using the magnetic field from the first step above the axial current density is given by

$$J'_{z,k} = -\frac{c}{2}\rho^{n+1}_k \frac{H'_{k+1/2} - H'_{k-1/2}}{\Delta m_k}. \tag{8.50}$$

The face centered mass density is the same as was used in the hydrodynamics. It will be noticed that the current density as defined by (8.50) is an edge centered variable (Figure 8.2). The Joule heating term, which changes the zone's internal energy, therefore involves two quantities that are centered at different spatial points on the grid. Since the internal energy (or equivalently, the temperature) is zone centered it will be necessary to find a zone centered representation of $\eta J_z{}^2$, which is considered next.

One representation of the Joule heating term may be constructed using the following physical arguments. The current $J_{z,k}$ may be considered to generate thermal energy at a rate $\eta_k J_{z,k}{}^2$ where η_k is an appropriately defined edge centered resistivity. In fact, we shall define the latter so as to guarantee that the correct heating rate is obtained. To do so, suppose that the current associated $J_{z,k}$ is made up of two equivalent parallel currents. The first current, associated with the current density $J_{z,k-1/4}$ flows through a zone containing material from the upper half of $\Delta m_{k-1/2}$, whose naturally centered resistivity is $\eta_{k-1/2}$. The other current, associated with $J_{z,k+1/4}$ flows through a zone containing material in the lower half of $\Delta m_{k+1/2}$, whose naturally centered resistivity is $\eta_{k+1/2}$. The equivalent resistance of the zone through which $J_{z,k}$ flows (shown shaded in Figure 8.2a) may

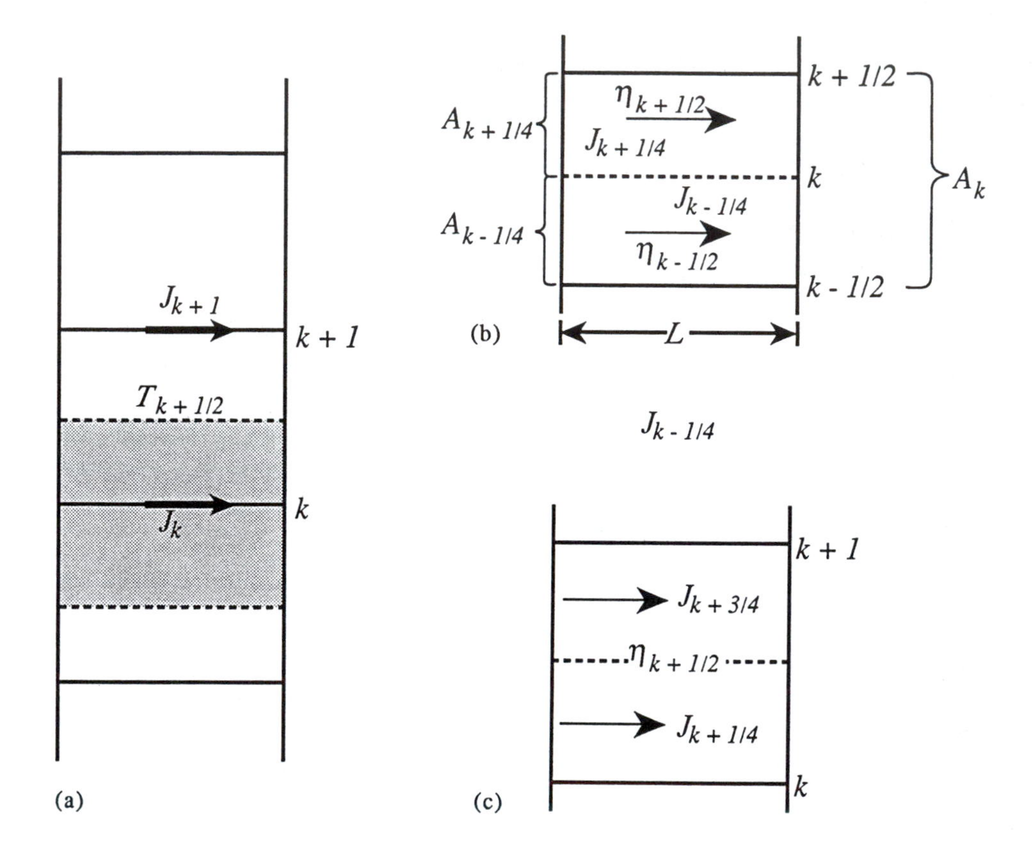

Figure 8.2. Spatial centering of the current density and resistivity used to calculate Joule heating associated with B_θ fields in one dimensional cylindrically symmetric systems. (a) Natural centering of the current density, and the zone through which it flows (shown shaded). (b) Parallel currents equivalent to J_k. (c) The equivalent current densities and the resistivity used to Joule heat zone $k+1/2$.

be expressed in terms of the local resistivity of the shaded zone, the area through which the current flows A_k, and the (axial) length L of the "wire" by:

$$R \equiv \frac{\eta_k L}{A_k}. \tag{8.51}$$

It is equally correct to consider the equivalent resistance to be due to the resistance of the two parallel current paths (the upper half of zone $k-1/2$, and the lower half of zone $k+1/2$, which is shown in Figure 8.2b). Each of these resistances is given by an expression analogous to (8.51)

$$R_{\pm} \equiv \frac{\eta_{k\pm 1/2}}{A_{k\pm 1/4}} L. \tag{8.52}$$

Adding these two resistances in parallel,

$$\frac{1}{R} = \frac{1}{R_{+}} + \frac{1}{R_{-}}$$

and using (8.51) and (8.52) in the result leads to the following relation between the edge centered resistivity and the zone centered resistivities:

$$\frac{r_{k+1/2}^2 - r_{k-1/2}^2}{\eta_k} = \frac{r_{k+1/2}^2 - r_k^2}{\eta_{k+1/2}} + \frac{r_k^2 - r_{k-1/2}^2}{\eta_{k-1/2}} . \tag{8.53}$$

This equation, in fact, may be used to define η_k. Notice that η_k depends on the resistivity above and below the zone edge, as well as on the three zonal radii r_{k-1}, r_k, and r_{k+1}.

The principle of equivalent circuits may also be used to construct representations of the current densities $J_{z,\,k+1/4}$ and $J_{z,\,k-1/4}$. This may be done by noting that the potential drop across the resistance R, which is just $I_k R_k = J_k \eta_k$, is the same as the potential drop across either of the parallel resistances R_+ or R_-. From this observation it follows immediately that:

$$J_{k\pm 1/4} = \frac{\eta_k J_{z,k}}{\eta_{k\pm 1/2}} . \tag{8.54}$$

Using (8.53) the right side of (8.54) can be expressed in terms of variables that are naturally centered on the grid.

A finite difference representation of the change in internal energy of a zone associated with Joule heating may now be constructed by integrating (8.44) over the zonal volume $\Delta V_{k+1/2}$:

$$\int \rho\, C_V \frac{dT}{dt}\, \delta V = \int \eta\, J_z^2\, \delta V. \tag{8.55}$$

Each quantity on the left side is naturally centered. Furthermore the internal energy is a constant throughout the zone, as is the resistivity. Using the results above the current appearing on the right side may be considered to consist of the two equivalent components (Figure 8.2c) which are given by (8.54). Thus if $J_{k+1/4}$ ($J_{k+3/4}$) is used for the lower (upper) half of the zone, the integral on the right of (8.55) is

$$\rho^{n+1}_{k+1/2}\, C^n_{V,\,k+1/2} \frac{T''_{k+1/2} - T'_{k+1/2}}{\Delta t^{n+1/2}}\, \Delta V_{k+1/2} = \eta_{k+1/2} \left[J^2_{z,\,k-1/4} \frac{\Delta V_{k+1/2}}{2} + J^2_{k+1/4} \frac{\Delta V_{k+1/2}}{2} \right].$$

In this expression T' is the temperature corresponding to the specific energy obtained from the solution of (8.47). Finally, cancelling common factors, and using (8.54) to express the result in terms of the naturally defined current density (8.50), the change in internal energy of a zone associated with Joule heating becomes:

$$\rho^{n+1}_{k+1/2}\, C^n_{V,\,k+1/2} \frac{T''_{k+1/2} - T'_{k+1/2}}{\Delta t^{n+1/2}} = \frac{\eta_k^2\, (J'_{z,\,k})^2 + \eta^2_{k+1}\, (J'_{k+1})^2}{2\,\eta_{k+1/2}}\,. \tag{8.56}$$

All quantities needed to calculate Joule heating are now available in terms of variables naturally specified on the computational grid.

Problem 8.6
Use the definitions (8.50) and (8.51) to establish (8.54)

Problem 8.7
Develop a one dimensional finite difference representation of thermal diffusion in cylindrical coordinates which may be used with the torroidal magnetohydrodynamic equations discussed above.

The final step in solving the one dimensional magnetohydrodynamic equations is to obtain a representation of the evolution equation (8.37) assuming no material motion [the effect of material motion on the field has been taken into account in the first step by (8.48)]. For systems in which $\tau_B >> \tau_D$ an explicit representation of the diffusion equation may be used. We shall not consider this case further, since it can be obtained simply from the results below. The explicit finite difference equations are subject to a stability time step constraint. For the remainder of this section we consider systems in which $\tau_B \approx \tau_D$. The explicit magnetohydrodynamics discussed above may be used in this case, but a linearized implicit form of the magnetic diffusion equation should be employed. The latter may be constructed by integrating (8.37) over the Lagrangian mass $\Delta m_{k+1/2}$:

$$\frac{1}{\Delta m_{k+1/2}} \int \frac{dH}{dt} \delta m = \frac{H^{n+1}_{k+1/2} - H'_{k+1/2}}{\Delta t} = \frac{4\pi r^2_{k+1/2} \rho_{k+1/2}}{\Delta m_{k+1/2}} \left(\rho D_B \frac{\partial H}{\partial m}\right)^{k+1}_{k} .$$

Expanding the right hand side results in a diffusion equation of the form:

$$\frac{H^{n+1}_{k+1/2} - H'_{k+1/2}}{\Delta t^{n+1/2}} = \frac{4\pi^2 \rho^{n+1}_{k+1/2} (r^{n+1}_{k+1/2})^2}{\Delta m_{k+1/2}} \left\{ \rho^{n+1}_{k+1} D_{B,k+1} \frac{(H^{n+1}_{k+3/2} - H^{n+1}_{k+1/2})}{\Delta m_{k+1}} - \rho^{n+1}_{k} D_{B,k} \frac{(H^{n+1}_{k+1/2} - H^{n+1}_{k-1/2})}{\Delta m_k} \right\} . \tag{8.57}$$

The magnetic diffusion coefficient corresponding to (8.16) is

$$D_{B,k} = \frac{\eta_k c^2}{4\pi}. \tag{8.58}$$

The latest values of the zonal radii and density are used on the right side of (8.57). The resistivity for most materials will be a function of the plasma density and temperature, and may also depend on the local magnetic field (Braginskii 1965). Although the linearized implicit diffusion equation is unconditionally stable, accuracy constraints should be observed, particularly whenever the resistivity is changing rapidly. If the values ρ^{n+1}, T'' and H' are used to evaluate (8.58), the diffusion equation is linear in H^{n+1}. It may then be expressed in the standard form (6.103), and may be solved using Gaussian elimination [see, for example, (6.104)-(6.110)].

The edge centered resistivity appearing in $D_{B,k}$ should be represented by (8.53), that is by the same value as was used to evaluate the Joule heating term. This is to be expected physically since the change in flux through a zone is associated with the radial derivative of the local electric field ηJ_z, and the Joule heating resulting from the change in flux is also proportional to the electric field.

The resistivity in a low density, high temperature plasma is proportional to $T^{-3/2}$, and the magnetic Reynolds number $R_B \sim v l_B T^{3/2}$. In many astrophysical regimes l_B is large, $R_B >> 1$, and the magnetic field is frozen into the plasma. In this regime magnetic diffusion and Joule heating may be neglected. In this case the second and third steps discussed below (8.37) can be skipped, and the only terms containing the magnetic field which need to be solved are the Lorentz acceleration and the transport term (8.39).

For quasistatic systems where $\tau_B << \tau_D$ an implicit treatment of the magentohydrodynamic equations (such as the one discussed in Section 4.7) is necessary. This can be done as follows. First, all factors on the right side of (8.46)-(8.47), except for the magnetic field, are to be evaluated at the advanced time. Then the set of equations (8.46)-(8.49), (4.116) and an equation of state represent six equation in six unknowns. Thus (8.48) may be used to eliminate

$H'_{k+1/2}$ from the right side of (8.46), and then the substitutional procedure outlined in Section 4.7 leads to a tridiagonal equation of the form (4.123) which may be solved by Newton-Raphson interation. In this way the first step solves for the values of the unknowns $r^{n+1}{}_k$, $v^{n+1/2}{}_k$, $\rho^{n+1}{}_{k+1/2}$, $\varepsilon^{n+1}{}_{k+1/2}$,, and $H'_{k+1/2}$. The zonal radii, velocities and density obtained in this way represent the final values of the variables for the computational cycle. The specific energy will require a further update (because of Joule heating), as will the torroidal magnetic field (because of resistive diffusion) so the values of these variables obtained above have been denoted by a prime.

Joule heating is calculated next using (8.50)-(8.56). The resistivity appearing in (8.56) is usually a function of temperature. If it is to be evaluated fully implicitly as a function of T'' then The Newton-Raphson method developed in Section 4.7 can be used to solve the resulting nonlinear equation for T''.

Finally, the diffusion equation is solved. For a fully implicit treatment (8.57) is used with all quantities on the right side evaluated at the new time. This includes the nonlinear dependence of the resistivity on density and temperature. The resulting equations may be solved using Gaussian elimination and Newton-Raphson iterations.

Problem 8.8

Another way to obtain (8.57) is to start from Farraday's law in the comoving frame, which is

$$\frac{1}{c}\frac{\partial \mathbf{B}}{\partial t} = -\nabla \times \mathbf{E},$$

where the local electric field $\mathbf{E} = \eta \mathbf{J}$. Integrate this over the cross sectional area $L\Delta r$ of a cylindrical zone through which the field $\mathbf{B}$ passes. Show that the change in magnetic flux Φ through the zone is represented by

$$\Delta\Phi = (\eta_{k+1} J_{z,k+1} - \eta_k J_{z,k})\, c\Delta t\, \Delta z. \tag{8.59}$$

Finally, show that this leads to the Lagrangian finite difference representation (8.57).

Problem 8.9

Develop a linearized implicit magnetohydrodynamic scheme assuming that the only nonzero component of the magnetic field is the axial component B_z. Use a three step splitting method similar to the one discussed above for the torroidal field component.

Consistent Diffusion and Heating

Finite difference representations of the magnetic field diffusion and resistive heating terms discussed above required values of the electrical resistivity at a zone boundary. Many possible forms for an edge centered finite difference representation of the resistivity can be constructed. However, it is important to consider the internal consistency of such choices.

The electrical resistivity is responsible for the spatial diffusion of magnetic field in a medium, and field diffusion is accompanied by energy dissipation in the form of Joule heating. For example, in a stationary medium the diffusion equation (8.15) may be rewritten in terms of the local electric field (using (8.13b), and omitting the prime) and (8.1) in the form

$$\frac{\partial \boldsymbol{B}}{\partial t} = - \nabla \times (c\eta \boldsymbol{J}) = - \nabla \times (c\boldsymbol{E}) . \tag{8.60}$$

Taking the dot product of this equation with the magnetic field $\boldsymbol{B}$, and using the vector identity

$$\begin{aligned} \boldsymbol{B} \cdot \nabla \times (\eta \boldsymbol{J}) &= \nabla \cdot (\eta \boldsymbol{J} \times \boldsymbol{B}) + \eta \boldsymbol{J} \cdot \nabla \times \boldsymbol{B} \\ &= \nabla \cdot (\eta \boldsymbol{J} \times \boldsymbol{B}) + \frac{4\pi}{c} \eta J^2 , \end{aligned}$$

the diffusion equation (8.60) becomes an energy equation of the form

$$\frac{\partial (B^2/8\pi)}{\partial t} + \eta J^2 + \nabla\cdot(\frac{c}{4\pi}\eta \boldsymbol{J}\times\boldsymbol{B}) = 0. \tag{8.61}$$

Integrating over a volume element δV and recalling that $\eta \boldsymbol{J} = \boldsymbol{E}$ is the local electric field, gives

$$\int\frac{\partial}{\partial t}(\frac{B^2}{8\pi})\,\delta V + \int \eta J^2 \delta V + \int \nabla\cdot(\frac{c}{4\pi}\boldsymbol{E}\times\boldsymbol{B})\,\delta V = 0. \tag{8.62}$$

The first term on the left is the rate of change in the energy of the magnetic field in δV, and the second term is the Joule heating rate of the material in the volume. The last term on the left contains the divergence of the local Poynting flux.

Simple physical arguments were used earlier as motivation for the centering of the resistivity appearing in the finite difference representations of the magnetic field diffusion and in the Joule heating terms. Since the two processes are physically related, as shown in (8.62), it is important that they be represented consistently at the finite difference level. In fact, it can be shown that once a choice of a finite difference representation has been made for the diffusion terms, the same choice should be used for the Joule heating term. The argument presented below assumes for simplicity a one dimensional planar geometry (with spatial coordinate z). We will assume that the current J_z flows in the z direction, and that the field B points in the y direction. However, it can be readily extended to two or more spatial dimensions. Let us start with the diffusion equation (again restricting attention to a stationary medium) in the finite difference form

$$\frac{B'_{k+1/2} - B_{k+1/2}}{c\,\Delta t} = -\frac{\eta_{k+1}J_{z,k+1} - \eta_k J_{z,k}}{\Delta z_{k+1/2}}\,, \tag{8.63}$$

where the prime denotes the new time value for the magnetic field (all other time superscripts will be omitted for notational convenience). Finally note that only the axial current density J_z appears. Multiplying (8.63) by the time centered magnetic field

$$\tilde{B}_{k+1/2} = \frac{1}{2}(B'_{k+1/2} + B_{k+1/2})$$

and retaining terms on the right side of first order in the time step

$$\frac{(B'^{2}_{k+1/2} - B^{2}_{k+1/2})}{2c\,\Delta t} = -\frac{(\eta_{k+1} J_{z,k+1} - \eta_k J_{z,k})}{\Delta z_{k+1/2}} \frac{(B'_{k+1/2} + B_{k+1/2})}{2}$$
$$\approx -\frac{(\eta_{k+1} J_{z,k+1} B_{k+1/2} - \eta_k J_{z,k} B_{k+1/2})}{\Delta z_{k+1/2}}. \tag{8.64}$$

The right side may be rewritten by adding and substracting like terms to obtain

$$-\eta_{k+1} J_{z,k+1} \frac{(B_{k+1/2} - B_{k+3/2})}{\Delta z_{k+1}} \frac{\Delta z_{k+1}}{\Delta z_{k+1/2}} - \frac{\eta_{k+1} J_{k+1} B_{k+3/2}}{\Delta z_{k+1/2}}$$
$$+ \eta_k J_{z,k} \frac{(B_{k+1/2} - B_{k-1/2})}{\Delta z_k} \frac{\Delta z_k}{\Delta z_{k+1/2}} + \frac{\eta_k J_{z,k} B_{k-1/2}}{\Delta z_{k+1/2}}.$$

It will be seen that the first and third terms above are finite difference representations of the radial current density, and that the right side of (8.64) may be expressed in the form

$$-4\pi \eta_{k+1} J^2_{k+1} \frac{\Delta z_{k+1}}{\Delta z_{k+1/2}} - 4\pi \eta_k J^2_k \frac{\Delta z_k}{\Delta z_{k+1/2}}$$
$$+ \frac{\eta_k J_{z,k} B_{k-1/2} - \eta_{k+1} J_{z,k+1} B_{k+3/2}}{\Delta z_{k+1/2}}.$$

Proceding in this manner, additional terms may be added and subtracted to obtain a finite difference representation of (8.61), which when summed over the entire grid has the form

$$\sum_{k=1}^{K} \left\{ \frac{B'^{2}_{k+1/2} - B^{2}_{k+1/2}}{8\pi\, \Delta t} + \frac{\eta_{k+1} J^{2}_{z,k+1} \Delta z_{k+1} + \eta_{k} J^{2}_{z,k} \Delta z_{k}}{2\, \Delta z_{k+1/2}} \right.$$

$$\left. + \frac{c}{4\pi} \frac{\eta_{k+1} J_{z,k+1} B_{k+1} - \eta_{k} J_{z,k} B_{k}}{\Delta z_{k+1/2}} \right\} \Delta z_{k+1/2} = 0. \tag{8.65}$$

Notice that $\Delta z_{k+1/2}$ is proportional to the volume element in one dimensional planar geometry, and the edge centered magnetic field appearing in the definition of the Poynting flux above is just

$$B_k \equiv (1/2)(B_{k+1/2} + B_{k-1/2}).$$

Problem 8.10

Supply the steps leading to the finite difference representation (8.65) of the energy equation (8.61).

The last term on the left side of (8.65) may be rewritten in terms of the local electric field, and is just the Poynting flux. Notice that it vanishes termwise, except at the grid boundaries $k = 1$ and $k = K$. Equation (8.65) may now be used to show that the definition of the resistivity in the diffusion coefficient is determined by the choice used to represent Joule heating. The second term in brackets in (8.65) represents Joule heating; if the Poynting flux into the grid is zero, then a reduction in the magnetic energy density is associated with an increase in material temperature through Joule heating. Equating the heating term in (8.65) with the right side of (8.56) summed over the entire grid we obtain

$$\sum_{k=1}^{K}\left\{\frac{\eta_k J_{z,k}^2 \Delta z_k + \eta_{k+1} J_{z,k+1}^2 \Delta z_{k+1}}{\Delta z_{k+1/2}} - \frac{\eta_k^2 J_{z,k}^2}{\eta_{k+1/2}} - \frac{\eta_{k+1}^2 J_{z,k+1}^2}{\eta_{k+1/2}}\right\} \Delta z_{k+1/2} = 0.$$

Noting that $\Delta z_{k+1/2}$ is arbitrary, we may rearrange terms in the summation and show that the resistivity at a zone edge is related to the zone centered resistivity by

$$\frac{\Delta z_k}{\eta_k} = \frac{1}{2}\left(\frac{\Delta z_{k+1/2}}{\eta_{k+1/2}} + \frac{\Delta z_{k-1/2}}{\eta_{k-1/2}}\right). \tag{8.66}$$

This expression is the planar equivalent of (8.53) which was derived by physical arguments for cylindrical geometry. As demonstrated by the derivation of (8.56) (see Problem 8.10) the same expression for the edge centered resistivity appears in the magnetic diffusion equation and in the Joule heating equation. As a bonus, we also obtain a definition of the Poynting flux at a zone boundary which is consistent with the diffusion and heating terms above.

8.4 EULERIAN MAGNETOHYDRODYNAMICS

A two dimensional Eulerian magnetohydrodynamic scheme will be developed below for axially symmetric systems in cylindrical coordinates. It may be combined with the treatment of hydrodynamics discussed in Chapter 5 using the method of operator splitting. The general approach may be extended to other coordinate systems, or to three dimensional systems in a straightforward manner. The partial differential equation in cylindrical coordinates will be developed first, and then finite difference representations will be constructed for the torroidal and for the poloidal components of the magnetic field.

For axially symmetric systems in two dimensional cylindrical coordinates r and z, all spatial partial derivatives with respect to the angle θ vanish identically. All three components of the magnetic field can be nonzero in such a system while still preserving two dimensionality (this is analogous to having axial rotation in a two dimensional system as discussed in Chapter 5). In general the magnetic field may be described as consisting of a torroidal (B_θ) component, associated with axial and radial currents, and a poloidal component (B_r and B_z) associated with angular currents.

The evolution equation (8.15) under the assumptions above can easily be reduced to the following component forms: the torroidal component described by

$$\frac{\partial B_\theta}{\partial t} = \frac{\partial}{\partial r}\left(\frac{D_B}{r}\frac{\partial rB_\theta}{\partial r}\right) + \frac{\partial}{\partial z}\left(\frac{D_B}{r}\frac{\partial rB_\theta}{\partial z}\right)$$

$$- \frac{\partial}{\partial r}(vB_\theta) - \frac{\partial}{\partial z}(uB_\theta)$$

$$+ \frac{\partial}{\partial r}(v_\theta B_r) + \frac{\partial}{\partial z}(v_\theta B_z); \tag{8.67}$$

and the poloidal component described by the two equations

$$\frac{\partial B_r}{\partial t} = \frac{\partial}{\partial z}\left(D_B\frac{\partial B_r}{\partial z}\right) - \frac{\partial}{\partial z}\left(D_B\frac{\partial B_z}{\partial r}\right)$$

$$-\frac{\partial}{\partial z}(uB_r - vB_z) \tag{8.68}$$

and

$$\frac{\partial B_z}{\partial t} = -\frac{1}{r}\frac{\partial}{\partial r}\left(rD_B\frac{\partial B_z}{\partial r}\right) + \frac{1}{r}\frac{\partial}{\partial r}\left(rD_B\frac{\partial B_r}{\partial z}\right)$$

$$+ \frac{1}{r}\frac{\partial}{\partial r}\left[r(uB_r - vB_z)\right]. \tag{8.69}$$

The rotational velocity $v_\theta = v_\theta(r,z)$ has been included in the evolution equations. It may be noted that it appears only in the torroidal field equation, and can mix poloidal and torroidal fields. In particular, it follows from (8.67)-(8.69) that if initially the poloidal fields vanish, then they will remain zero unless angular currents are introduced. Furthermore, in a nonrotating system there is no mixing between torroidal and poloidal components.

In addition to the evolution equations above the constraint equation (8.13c) must also be satisfied. In cylindrical geometry this becomes an issue only if poloidal fields are nonzero. It may then be advantageous to replace the evolution equations for the poloidal fields by an equivalent equation for the vector potential A_θ, since the two components B_r and B_z are given by

$$B_r = -\frac{\partial A_\theta}{\partial z} = -\frac{1}{r}\frac{\partial (rA_\theta)}{\partial z} \tag{8.70}$$

$$B_z = \frac{1}{r}\frac{\partial (rA_\theta)}{\partial r}. \tag{8.71}$$

Thus it is possible to use as primary variables either (B_r, B_θ, B_z) or (A_θ, B_θ). The relative advantages of these two formulations will be considered when we discuss finite difference representations for the poloidal field below.

The evolution equation for A_θ follows from (8.19), and is easily shown to be

$$\frac{\partial A_\theta}{\partial t} = D_B\left[\frac{\partial}{\partial r}\left(\frac{1}{r}\frac{\partial (rA_\theta)}{\partial r}\right) + \frac{\partial}{\partial z}\left(\frac{1}{r}\frac{\partial (rA_\theta)}{\partial z}\right)\right]$$
$$-\frac{u}{r}\frac{\partial (rA_\theta)}{\partial z} - \frac{v}{r}\frac{\partial (rA_\theta)}{\partial r}. \tag{8.72}$$

The simple equation (8.72) replaces (8.68)-(8.69), with the poloidal field components given by differentiation according to (8.70) and (8.71).

The first two terms on the right side of (8.67) represent the diffusion of the torroidal field in a resistive medium with a diffusion coefficient given by (8.16). The third and fourth terms on the right side of (8.67) are advection terms, and can be shown to describe the transport of the torroidal magetic flux, which may be defined by

$$\Phi_\theta \equiv \int \boldsymbol{B} \cdot d\boldsymbol{S}_\theta = \int B_\theta \, dr \, dz \tag{8.73}$$

where $\boldsymbol{B}$ is the vector magnetic field, and $d\boldsymbol{S}_\theta$ is an element of area lying within the r-z plane (see Problem 8.11). The last two terms on the right side of (8.67) arise, for example, because poloidal fields in differentially rotating systems are stretched in the angular direction and give rise to torroidal components. The first two terms on the right side of (8.68) and (8.69) correspond to diffusion of the poloidal field components, and the last two terms in each equation can be related to magnetic flux transport due to fluid motion.

In nonrotating axially symmetric systems the torroidal and poloidal fields are independent and may often be treated separately. We shall assume this to be the case initially, and shall use operator splitting to treat them separately.

Problem 8.11

Consider an infinitesimal area $dS_\theta = dr\,dz$ normal to B_θ, and show that the third and fourth terms on the right of (8.67) describe the net change in torroidal magnetic flux due to transport with the fluid.

The momentum equation (8.21) in axially symmetric cylindrical coordinates can be solved as described in Chapter 5 with the Lorentz force terms treated by means of operator splitting. In keeping with this approach the change in velocity due to the Lorentz force occurs at the end of the magnetic diffusion step. The magnetic acceleration terms are then

$$\rho \frac{\partial v}{\partial t} = \frac{1}{c}(J_\theta B_z - J_z B_\theta), \tag{8.74}$$

$$\rho\frac{\partial u}{\partial t} = \frac{1}{c}(-J_\theta B_r + J_r B_\theta).$$

(8.75)

If the system's rotation is as described in Section 5.7, then the rate of change in angular momentum L associated with magnetic forces is described by

$$\frac{dL}{dt} = \frac{r}{c}(J_z B_r - J_r B_z).$$

(8.76)

If the right side of (8.76) is nonzero, the system will start to rotate. The current density can be obtained from the magnetic field using (8.13b);

$$J_r = -\frac{c}{4\pi}\frac{1}{r}\frac{\partial(rB_\theta)}{\partial z},$$

(8.77a)

$$J_\theta = \frac{c}{4\pi}\left(\frac{\partial B_r}{\partial z} - \frac{\partial B_z}{\partial r}\right),$$

(8.77b)

$$J_z = \frac{c}{4\pi}\frac{1}{r}\frac{\partial(rB_\theta)}{\partial r},$$

(8.77c)

Finally, the change in internal energy due to Joule heating is described by

$$\rho\frac{\partial \varepsilon}{\partial t} = \eta(J_r^2 + J_z^2 + J_\theta^2).$$

(8.78)

The changes associated with mechanical work and artificial viscous heating if present, are to be included with the solution to the fluid hydrodynamic step as described in Chapter 5.

The finite difference representation developed below may be used along with the hydrodynamic representations discussed in

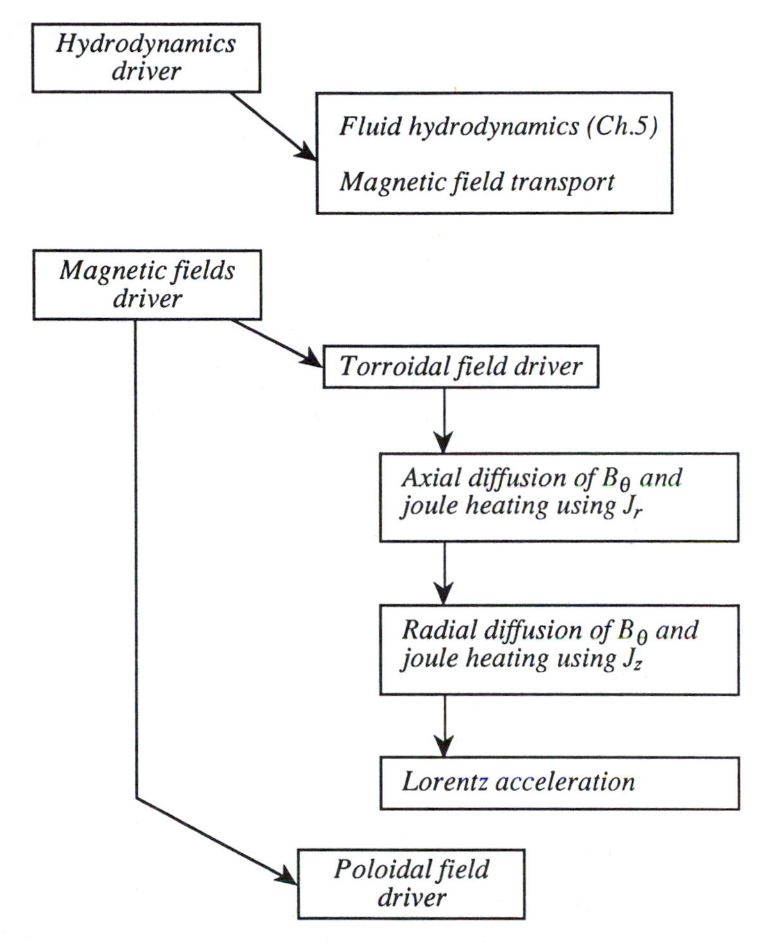

Figure 8.3. Example of the calling sequence used to solve the magnetic diffusion, Joule heating and Lorentz force equation along with the Eulerian fluid dynamic equations of Chapter 5.

Chapter 5 as shown schematically in Figure 8.3. The magnetic field advection terms are usually solved along with the advection of the mass density and other scalar quantities. Magnetic field diffusion and the associated Joule heating can be incorporated in a separate physics package with its own driver. The Lorentz acceleration may be done at the end of the field diffusion and Joule heating in preparation for the next computational cycle. In this approach the magnetic driver and its associated coding does not change the material density, but it does change the specific energy and the velocity of the fluid. The torroidal and poloidal components may also be operator split for many problems, particularly if there is very little rotation present. In the discussions below a finite difference representation for the torroidal components will be developed first, and then the poloidal components will be discussed.

Problem 8.12

The quantity $F = rA_\theta$ is often called the flux function; using (8.14a) find an expression for the magnetic flux obtained from A_θ through an area $2\pi\, rdr$ and $2\pi r\, dz$. Interpret the terms on the right side of (8.72) in terms of their affect on the torroidal flux.

Problem 8.13

One advantage of using A_θ and B_θ primary variables for two dimensional axially symmetric systems is that they automatically satisfy the constraint equation (8.13c). Show that A_θ and B_θ satisfy (8.13c) exactly, where the poloidal field components are given by (8.70)-(8.71).

Space-Time Centering

The magnetohydrodynamic variables will be centered on the space-time grid as defined in Chapter 5. If a hydrodynamic scheme based on vertex centered velocity components is in use, then a suitable average of the velocities at the zone faces may be assumed. As we shall see, face centered components are more natural for the advection of torroidal flux. However, vertex centering is more natural for the vector potential or the flux function. In either case, some averaging will be necessary. The new variables describing the magnetic fields and associated quantities can be centered as shown in Figure 8.4. In a highly conducting plasma the torroidal field moves with the fluid mass (see (8.26) and Problem 8.4). Furthermore, the diffusion of torroidal magnetic field can be associated with the flux of electrical current across a zone boundary. Consequently, it is natural to consider the torroidal field B_θ to be a zone centered variable:

$$B_\theta \to B^n_{\theta,k+1/2,j+1/2}. \tag{8.79a}$$

From (8.25) it follows that

$$H \to H^n_{k+1/2,j+1/2} = r_{j+1/2} B^n_{\theta,k+1/2,j+1/2}. \tag{8.79b}$$

Having chosen B_θ to be a zone centered variable, it is natural to define the poloidal flux in terms of the surface bounding the volume element $\Delta V_{k+1/2,\ j+1/2}$. The radial and axial magnetic field components may then be chosen to be face centered, as shown in Figure 8.4:

$$B_r \to B^n_{r,k+1/2,j} \tag{8.80}$$

$$B_z \to B^n_{z,k,j+1/2}. \tag{8.81}$$

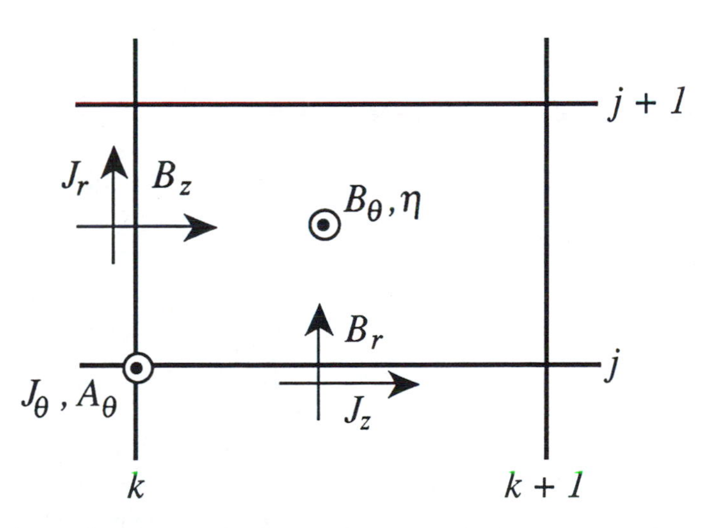

Figure 8.4. Spatial centering of the Eulerian magnetohydrodynamic variables for a two dimensional cylindrically symmetric system.

The centering of the components of the current density follow immediately from (8.77). If the vector potential A_θ is to be used in place of B_r and B_z as a primary variable then A_θ is naturally corner centered. Because the evolution equations and the advection terms for vector potential all involve the flux function, it will be used in the finite difference equations, and will be centered at the zone vertex:

$$F \rightarrow F^{n}_{k,j} \, . \tag{8.82}$$

It follows, then, from (8.71) that the polodial field components are centered as in (8.80)-(8.81).

8.5 TORROIDAL FIELD COMPONENT

The solution of the torroidal field equations may be found by the following steps. First, the transport of B_θ may be done along with the advection of scalar quantities in the hydrodynamics step. The remaining processes are incorporated under the torroidal field driver (Figure 8.3). Finally, the axial and radial steps will be solved using operator splitting.

Torroidal Field Transport

The transport of the torroidal component is described by the third and fourth terms in (8.67). First, consider the axial transport which is given by:

$$\frac{\partial B_\theta}{\partial t} = - \frac{\partial (uB_\theta)}{\partial z} . \tag{8.83}$$

This equation actually describes the advection of torroidal flux, Φ_θ, due to material motion. In fact, in the limit of a perfectly conducting medium, the magnetic flux is frozen into the material. Integrating this equation over the zonal area $\Delta r \Delta z$ which is normal to B_θ gives

$$\int \frac{\partial B_\theta}{\partial t} dS_\theta = - \int \frac{\partial (uB_\theta)}{\partial z} dS_\theta = - \int \frac{\partial (uB_\theta)}{\partial z} dr\, dz$$

$$= - \int dr \left(uB_\theta \right)_k^{k+1} .$$

A possible finite difference representation of this equation is (common subscript $j+1/2$ omitted)

$$(B'_{\theta,k+1/2} - B^n_{\theta,k+1/2}) \Delta r \, \Delta z_{k+1/2} = - (u^{n+1/2}_{k+1} \tilde{B}_{\theta,k+1} - u^{n+1/2}_k \tilde{B}_{\theta,k}) \Delta t^{n+1/2}, \tag{8.84}$$

where the fields at the zone boundaries $B_{\theta,k}$ will be defined below. Note that the new value of the magnetic field, $B'_{\theta,k+1/2}$, will serve as the initial value for the magnetic driver to be described below. It will be recognized that the left side of (8.84) is simply the change in the torroidal flux through zone $\Delta V_{k+1/2,j+1/2}$ during the time step $\Delta t^{n+1/2}$. Physically this change must be due in part to the difference between the flux carried into and out of the zone by mass transport (see Figure 8.5).

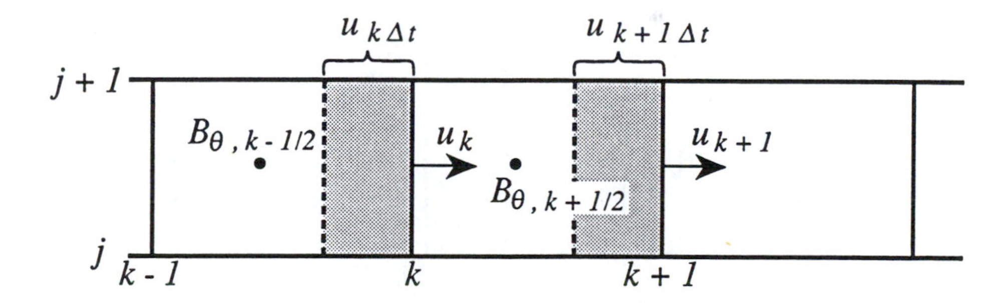

Figure 8.5. Advection of torroidal magnetic flux due to axial fluid motion. The fraction of the flux transported is shown by the shaded region.

A scheme that is second order in space results if for the advected field $B_{\theta,k}$ appearing in (8.84) one uses the value interpolated between $B_{\theta,k+1/2}$ and $B_{\theta,k-1/2}$, given by:

$$B_{\theta,k} = \frac{1}{2} B^{n}_{\theta,\,k-1/2} \frac{(\Delta z_{k+1/2} + u_k^{n+1/2} \Delta t^{n+1/2})}{\Delta z_k} + \frac{1}{2} B^{n}_{\theta,k+1/2} \frac{(\Delta z_{k+1/2} - u_k^{n+1/2} \Delta t^{n+1/2})}{\Delta z_k}. \tag{8.85}$$

Figure 8.6 shows how the interpolated values are constructed assuming that both u_k and u_{k+1} are positive. The interpolation (8.86b) holds for nonuniform zoning, and for either sign of the material velocity. Taken together, equations (8.84) and (8.85) are second order accurate in spatial derivatives, and conserve magnetic flux to machine roundoff. Finally we note that (8.85) may be modified along the lines discussed for the mass density.

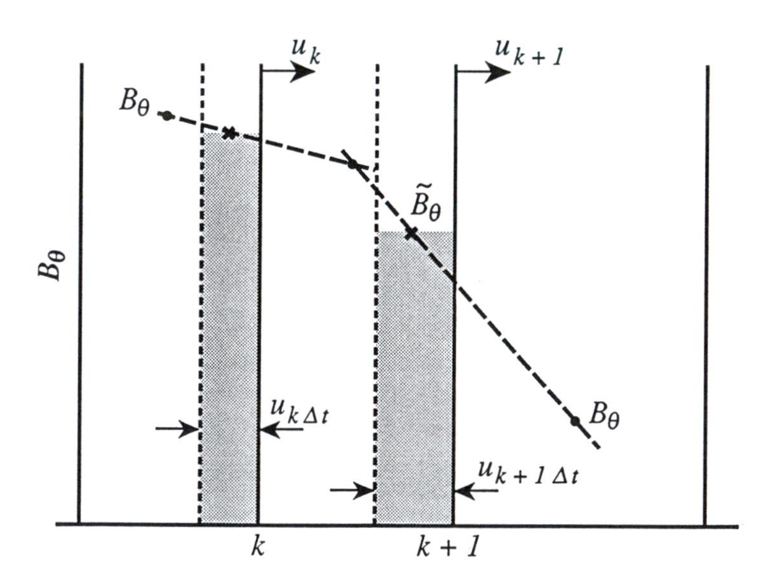

Figure 8.6. Magnetic field interpolation used to construct a second order accurate advection scheme for axial advection of torroidal flux. The fluid flow is to the right.

A finite difference representation for the radial transport of B_θ, corresponding to radial advection of torroidal magnetic flux, can be constructed in a similar way. The result, which is second order accurate in space is given by (common subscript $k+1/2$ omitted):

$$B'_{\theta,j+1/2} - B^n_{\theta,j+1/2} = -\frac{v_{j+1}^{n+1/2} B_{\theta,j+1} - v_j^{n+1/2} B_{\theta,j}}{\Delta r_{j+1/2}} \Delta t^{n+1/2}, \tag{8.86}$$

where

$$B_{\theta,j} = \frac{1}{2} B^n_{\theta,j-1/2} \frac{(\Delta r_{j+1/2} + v_j^{n+1/2} \Delta t^{n+1/2})}{\Delta r_j} + \frac{1}{2} B^n_{\theta,j+1/2} \frac{(\Delta r_{j+1/2} - v_j^{n+1/2} \Delta t^{n+1/2})}{\Delta r_j}. \tag{8.87}$$

The second order scheme (8.87) may also be modified to include a monotonicity constraint as discussed in Chapter 5.

Problem 8.14

Show that the radial transport equation

$$\frac{\partial B_\theta}{\partial t} = -\frac{\partial (vB_\theta)}{\partial r} \tag{8.88}$$

corresponds to the radial advection of torroidal flux. Construct a finite difference representation for the flux advection and from it obtain (8.86).

Torroidal Field Diffusion

The first two terms on the right of (8.67) describe the diffusion of B_θ in a resistive medium. The quantities in parentheses are proportional to the axial and radial components of the current density respectively. It is easy to show, for example, that the gradient of the radial current density results in the axial diffusion of torroidal flux, and that the net change in torroidal flux in the grid associated with diffusion is due entirely to boundary contributions (Problem 8.15). In situations where the change in B_θ per cycle in a zone is small, it is reasonable to operator split the axial and radial diffusion terms in (8.67). We then calculate, for example, the change in $B^n{}_\theta$ throughout the grid first in the axial direction to obtain temporary values B'_θ, and then use the results to calculate the change due to radial gradients. Following the calculation of axial changes, the Joule heating associated with radial currents is obtained and the new temperatures (or energies) are used to evaluate the resistivity for the radial calculation. The Joule heating associated with axial currents is then done and the diffusion step is complete. It is possible to solve the full diffusion equation without resort to operator splitting using methods to be discussed for Poisson's equation in Chapter 9 (see also the discussion of the vector potential for poloidal fields in Section 8.6).

The change in B_θ, due to resistive diffusion is best represented by implicit methods similar to those used for the thermal diffusion equation (see Chapter 6). The equations below treat the diffusion equation in linearized form, that is, all coefficients use initial values of the temperature or energy and density. A finite difference representation of radiation diffusion was obtained by integrating an equation of the form (6.39), for example, over a zonal volume to find the rate of change in energy. Since (8.67) represents the diffusion of magnetic flux rather than energy, we construct a finite difference representation by integrating over the zonal area normal to the r-z plane. Considering only the diffusion terms in (8.67) the integral over the zonal surface area $\Delta r\, \Delta z$ yields

$$\int \frac{\partial B_\theta}{\partial t} dr\, dz = \int \frac{\partial}{\partial r}\left(\frac{D_B}{r}\frac{\partial rB_\theta}{\partial r}\right) dr\, dz$$

$$+ \int \frac{\partial}{\partial z}\left(\frac{D_B}{r}\frac{\partial rB_\theta}{\partial z}\right) dr\, dz. \tag{8.89}$$

The right side is just the line integral of the current along the closed boundary of the zonal area $\Delta r\, \Delta z$. Consider first the contribution to the change in torroidal field due to axial diffusion [first term on the right of (8.89)]; the change in flux may be represented by the implicit form (common subscript $j+1/2$ omitted):

$$\frac{B'_{\theta,\,k+1/2} - B^n_{\theta,\,k+1/2}}{\Delta t^{n+1/2}} \Delta r\, \Delta z_{k+1/2} =$$

$$= \left\{ D^z_{B,k+1}\left(\frac{\partial B_\theta}{\partial z}\right)^{n+1}_{k+1} - D^z_{B,k}\left(\frac{\partial B_\theta}{\partial z}\right)^{n+1}_{k} \right\} \Delta r\,, \tag{8.90}$$

which reduces to (common subscript $j+1/2$ omitted):

$$B'_{\theta,\,k+1/2} = B^n_{\theta,\,k+1/2} + \frac{D^z_{B,k+1}\,\Delta t^{n+1/2}}{\Delta z_{k+1/2}} \frac{B'_{\theta,\,k+3/2} - B'_{\theta,\,k+1/2}}{\Delta z_{k+1}}$$

$$- \frac{D^z_{B,k}\,\Delta t^{n+1/2}}{\Delta z_{k+1/2}} \frac{B'_{\theta,\,k+1/2} - B'_{\theta,\,k-1/2}}{\Delta z_k}. \tag{8.91}$$

This will be recognized as a tridiagonal form in B'_θ, which can be solved by means of Gaussian elimination as described in detail in Chapter 6. The new values of the field, B'_θ, are used to construct the radial currents for Joule heating, and as the initial field values for the radial diffusion calculation.

The diffusion equation requires the resistivity at each zone boundary. These may be constructed in a manner consistent with the Joule heating equations. In fact, the discussion leading to (8.53) for the edge centered resistivity appearing for the one dimensional Lagrangian diffusion coefficient can be extended to the Eulerian equations to give a relation between $D^z{}_{B,k,j+1/2}$ and the zone centered resistivities $\eta_{k+1/2,j+1/2}$ and $\eta_{k-1/2,j+1/2}$. The same physical constraints apply in both cases, and the derivation is straightforward, leading to the relation (common subscript $j+1/2$ omitted):

$$\frac{\Delta z_k}{\eta_k} = \frac{1}{2}\left(\frac{\Delta z_{k+1/2}}{\eta_{k+1/2}} + \frac{\Delta z_{k-1/2}}{\eta_{k-1/2}}\right). \tag{8.92}$$

As in the Lagrangian model, this value of the edge centered resistivity will be used to construct the Joule heating associated with the axial currents.

The change in the torroidal magnetic field due to diffusion in the radial direction is described by the last term on the right of (8.90) and leads to the following implicit finite difference representation (common subscript $k+1/2$ omitted):

$$B^{n+1}_{\theta,j+1/2} = B'_{\theta,j+1/2} + \frac{D^r_{B,j+1}\Delta t^{n+1/2}}{\Delta r_{j+1/2}} \frac{r_{j+3/2}B^{n+1}_{\theta,j+3/2} - r_{j+1/2}B^{n+1}_{\theta,j+1/2}}{r_{j+1}\Delta r_{j+1}} - \frac{D^r_{B,j}\Delta t^{n+1/2}}{\Delta r_{j+1/2}} \frac{r_{j+1/2}B^{n+1}_{\theta,j+1/2} - r_{j-1/2}B^{n+1}_{\theta,j-1/2}}{r_j\Delta r_j}. \tag{8.93}$$

This equation is tridiagonal in $B^{n+1}{}_\theta$, and can be solved by means of Gaussian elimination as described in detail in Chapter 6.

The radial diffusion coefficent must also be defined in terms of the zone centered resistivity; using arguments similar to those above (8.53) leads to the result

$$\frac{r_{j+1/2}^2 - r_{j-1/2}^2}{\eta_j} = \frac{r_{j+1/2}^2 - r_j^2}{\eta_{j+1/2}} + \frac{r_j^2 - r_{j-1/2}^2}{\eta_{j-1/2}} . \tag{8.94}$$

This value of the edge centered resistivity will also be used to construct the Joule heating associated with the radial currents. It should be noted that if the resistivities of two adjacent zones are equal, the edge centered resistivity will be the same, regardless of the zoning.

The finite difference equations for the diffusion of torroidal magnetic fields are complete, subject to suitable boundary conditions. We note that in cases where the material acts as a perfect conductor, the electrical resistivity becomes small, and the effects of magnetic diffusion will generally be small compared to those associated with mass motion. The equations above will then describe the field motion as though it were frozen into the fluid.

Torroidal Boundary Conditions

The specification of consistent boundary conditions is more complex in magnetohydrodynamics than it is for hydrodynamics or radiation transport. This is because the fields depend not only on the state of the material at the boundary, but on the instantaneous distribution of currents throughout the grid. In cylindrical symmetric systems in two dimensions the specification of torroidal boundary conditions is relatively simple, but the specification of poloidal boundary conditions can lead to a significant complication.

The tridiagonal solution of the torroidal magnetic diffusion equations (8.91) and (8.93) each require two boundary conditions. The radial boundary condition on the axis $r = 0$ follows from Ampere's law, (8.13b), which can be used to relate the net current $I_z(r)$ passing through the surface area πr^2 to $B_\theta(r)$:

$$B_\theta(r) = \frac{2\, I_z(r)}{r\, c}. \tag{8.95}$$

In the limit $r \to 0, B_\theta(r) \to r$, which leads to the boundary condition

$$B_\theta(r,z) = 0, \qquad r = 0. \tag{8.96a}$$

At the outer boundary of the grid, $r = R$, (8.95) can be used to specify the torroidal magnetic field in terms of the total current I in the grid:

$$B_\theta(R) = \frac{2I}{Rc}, \qquad r = R. \tag{8.96b}$$

Alternately, the value of the torroidal magnetic field at the outer boundary may be set to a prescribed value:

$$B_\theta(r,z) = f(z,t), \qquad r = R. \tag{8.97}$$

Boundary conditions along $k = 1$ and $k = K$ (left and right boundaries, respectively) will depend on the problem being solved, but must also be specified to complete the solution.

Torroidal Joule Heating

The change in the material internal energy associated with Joule heating has two contributions from the torroidal field which are given by the first two terms on the right side of (8.78). These contributions may be found in either energy or temperature base:

$$\rho \frac{\partial \varepsilon}{\partial t} = \rho \frac{\partial (C_V T)}{\partial t} = \eta (J_r^2 + J_r^2). \tag{8.98}$$

The mass density has been taken outside the partial derivative because in operator split form it remains constant during the solution of the magnetic diffusion and Joule heating. The radial and axial current density components are constructed using the latest values of the torroidal field from the diffusion step. We shall assume for definiteness that η can be expressed as a function of internal energy and solve the energy based form of (8.98). The temperature based equation may be solved in essentially the same manner.

At the end of the diffusion in the axial direction, (8.91), the currents $J'_{r,k,j+1/2}$ are found in the form (common subscript $j+1/2$ omitted):

$$J'_{r,k} = -\frac{c}{4\pi}\frac{B'_{\theta,k+1/2} - B'_{\theta,k-1/2}}{\Delta z_k}. \tag{8.99}$$

The centering of the currents which contribute to the change in the specific energy of a zone is shown in Figure 8.7. The specific energy,

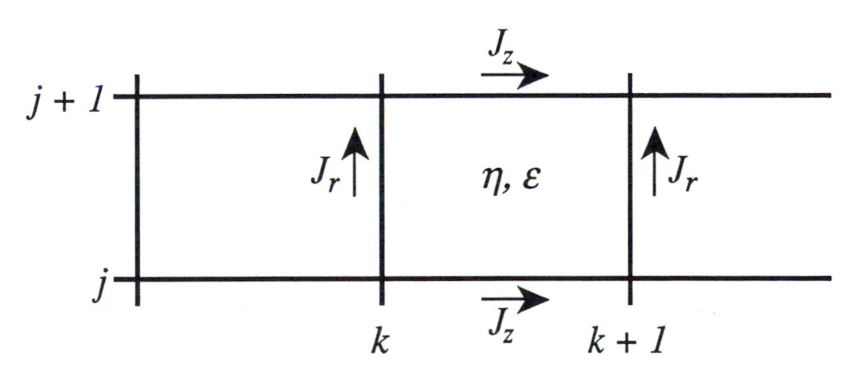

Figure 8.7. Spatial centering of the current density, resistivity and internal energy required to calculate the Joule heating of matter in zone $k+1/2,j+1/2$.

density and resistivity are zone centered, but all four current components are face centered. The discussion of one dimensional Lagrangian Joule heating leading to (8.56) is immediately applicable with only slight modifications to the two dimensional Eulerian case in either energy or temperature base. In particular, the axial current component $J_{r,k,j+1/2}$ contributes both to $\Delta\varepsilon_{k+1/2,j+1/2}$ and $\Delta\varepsilon_{k-1/2,j+1/2}$.

Extending the arguments preceding (8.56) it is easily shown that the heating associated with radial currents can be represented by (common subscript $j+1/2$ omitted):

$$\rho_{k+1/2}\frac{\varepsilon'_{k+1/2}-\varepsilon_{k+1/2}}{\Delta t^{n+1/2}} = \frac{\eta_k^2(J'_{r,k})^2+\eta_{k+1}^2(J'_{r,k+1})^2}{2\eta_{k+1/2}}, \tag{8.100}$$

where η_k, given by (8.92), is the same edge centered resistivity used for the axial diffusion, and (common subscript $j+1/2$ omitted):

$$\eta_{k+1/2} = \eta(\varepsilon_{k+1/2}, \rho_{k+1/2}). \tag{8.101}$$

If the radial diffusion is constructed following the axial diffusion, as is assumed here, then the energy ε' obtained from (8.100) can be used to construct the resistivity appearing in (8.94). The Joule heating associated with the axial currents (common subscript $k+1/2$ omitted)

$$J_{z,j}^{n+1} = \frac{c}{4\pi}\,\frac{H_{j+1/2}^{n+1}-H_{j-1/2}^{n+1}}{r_{j+1/2}\Delta r_{j+1/2}}, \tag{8.102}$$

shown in Figure 8.7, can be represented by (common subscript $k+1/2$ omitted)

$$\rho_{j+1/2}\frac{\varepsilon_{j+1/2}^{n+1}-\varepsilon'_{j+1/2}}{\Delta t^{n+1/2}} = \frac{\eta_j^2(J_{z,j}^{n+1})^2+\eta_{j+1}^2(J_{z,j+1}^{n+1})^2}{2\eta_{j+1/2}}. \tag{8.103}$$

Note that since H^{n+1} is given by (8.93) all quantities except $\varepsilon^{n+1}{}_{k+1/2}$ in (8.103) are known.

Problem 8.15

Carry out a derivation of the contributions to the Joule heating associated with currents given by (8.100). Discuss the modifications needed if the equations are to be solved in temperature base.

Torroidal Lorentz Force

Following the torroidal field diffusion and the Joule heating discussed above, the change in velocity associated with the last term in (8.74) and in (8.75) may be obtained. In keeping with the operator split approach, the mass density will be assumed to remain constant during this step. Consider first the axial acceleration integrated over the momentum zone $\Delta V_{k,j+1/2}$; denoting the axial velocity at the beginning of the magnetic step by $u_{k,j+1/2}$, and the value after the axial Lorentz acceleration by $u'_{k,j+1/2}$, we find

$$\frac{1}{\Delta V_k}\int \rho \frac{\partial u}{\partial t} dV = \rho_k \frac{u'_k - u_k}{\Delta t^{n+1/2}} = \frac{1}{c\,\Delta V_k}\int J_r B_\theta \, dV. \tag{8.104}$$

The spatial centering is shown in Figure 8.8. The edge centered mass density is defined as in (5.34). The radial current in the momentum zone is just $J^{n+1}{}_{r,k,j+1/2}$, and the right side of (8.104) reduces to (common subscript $j+1/2$ omitted)

$$\frac{2\pi r\,\Delta r}{c\,\Delta V_k} J^{n+1}_{r,k} \int_{k-1/2}^{k+1/2} B^{n+1}_\theta dz = \frac{J^{n+1}_{r,k}}{c\,\Delta z_k}\left(B^{n+1}_{\theta,\,\kappa+1/2}\frac{\Delta z_{k+1/2}}{2} + B^{n+1}_{\theta,\,k-1/2}\frac{\Delta z_{k-1/2}}{2}\right), \tag{8.105}$$

where $J^{n+1}{}_{r,k,j+1/2}$ is constructed as in (8.99). Substituting (8.105) into the right side of (8.104) yields a finited difference representation of the axial acceleration:

$$\rho_k \frac{u'_k - u_k}{\Delta t^{n+1/2}} = \frac{J^{n+1}_{r,k}}{2c\,\Delta z_k}\left(B^{n+1}_{\theta,k+1/2}\Delta z_{k+1/2} + B^{n+1}_{\theta,k-1/2}\Delta z_{k-1/2}\right). \tag{8.106}$$

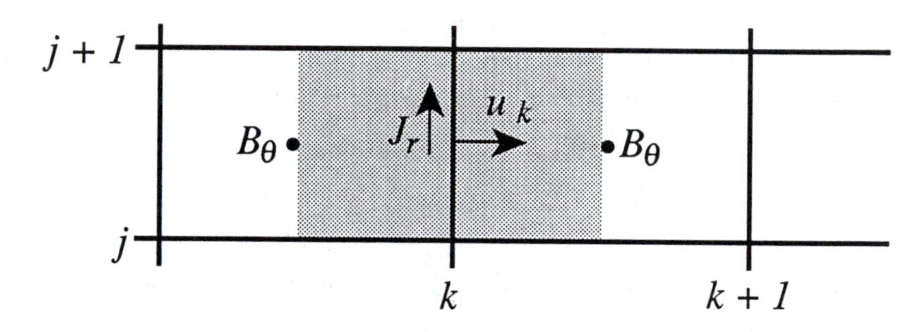

Figure 8.8. The momentum zone (shaded) used to construct the axial component of the Lorentz acceleration associated with torroidal fields.

It should be noted that (8.105) vanishes identically whenever the radial current density is zero, as it should. For uniform zoning (8.105) reduces to the simple gradient of the magnetohydrostatic pressure $B^2{}_\theta/8\pi$.

The radial Lorentz acceleration associated with torroidal fields may be constructed by integrating over the momentum zone $\Delta V_{k+1/2,j}$, which is shown in Figure 8.9:

$$\frac{1}{\Delta V_j}\int \rho \frac{\partial v}{\partial t} dV = \rho_j \frac{v'_j - v_j}{\Delta t^{n+1/2}} = \frac{-1}{c\,\Delta V_j}\int J_z\, B_\theta\, dV. \tag{8.107}$$

The axial current density $J^{n+1}{}_{z,k+1/2,j}$ may be removed from the integral, and the right side written in the form (common subscript $k+1/2$ omitted)

$$\frac{2\pi\,\Delta z}{c\,\Delta V_j} J^{n+1}_{z,j} \int_{j-1/2}^{j+1/2} B^{n+1}_{\theta}\, r\, dr = \frac{J^{n+1}_{z,j}}{c\, r_j\, \Delta r_j} \Big(B^{n+1}_{\theta, j+1/2} \frac{(r^2_{j+1/2} - r^2_j)}{2} + B^{n+1}_{\theta, j-1/2} \frac{(r^2_j - r^2_{j-1/2})}{2} \Big), \tag{8.108}$$

where $J^{n+1}{}_{z,k+1/2,j}$ is constructed as in (8.102). Substituting (8.108) into the right side of (8.107) gives (common subscript $k+1/2$ omitted)

$$\rho_j \frac{v'_j - v_j}{\Delta t^{n+1/2}} = -\frac{J^{n+1}_{z,j}}{c\, r_j\, \Delta r_j} \Big(B^{n+1}_{\theta, j+1/2} \frac{(r^2_{j+1/2} - r^2_j)}{2} + B^{n+1}_{\theta, j-1/2} \frac{(r^2_j - r^2_{j-1/2})}{2} \Big). \tag{8.109}$$

As expected, the acceleration vanishes identically if the axial current density is zero.

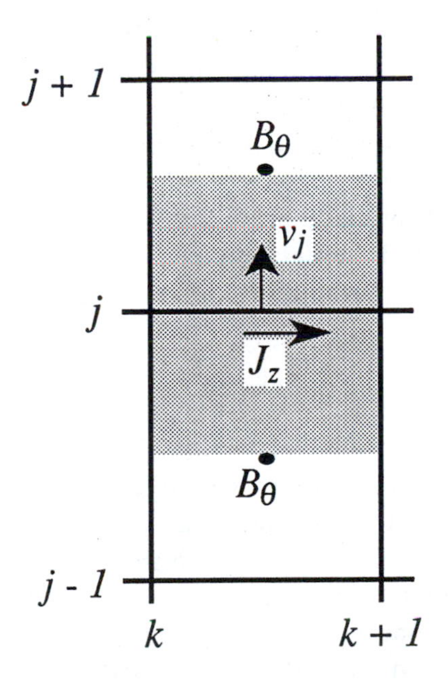

Figure 8.9. The momentum zone used to represent the radial component of the Lorentz acceleration for torroidal fields.

Whenever the magnetic acceleration is comparable to either the pressure gradient acceleration or to the gravitational acceleration (8.109) should be combined with them at the start of the hydrodynamic cycle. In general, operator splitting of these three forces should not be employed.

Problem 8.16
Write the magnetic stress components for an axially symmetric system using the results of Problem 8.3, which gives the Lorentz acceleration directly in terms of the magnetic field. Construct a finite difference representation of the radial and axial acceleration starting from the magnetic stress. Compare your results with (8.106) and (8.109).

8.6 POLOIDAL FIELD COMPONENT

The magnetohydrodynamic equations for the poloidal field components B_r and B_z in an axially symmetric system may be solved either in terms of the flux function $F = rA_\theta$ using the evolution equation (8.72), from which the fields may be derived, or by solving directly for the radial and axial fields themselves using (8.68)-(8.69). Each approach has its advantages. Perhaps the strongest argument for using the flux function is that it requires less memory (one less full grid array compared with schemes based on the two field components B_r and B_z). Furthermore, in an axially symmetric system in cylindrical coordinates use of the flux function $F = rA_\theta$ and B_θ automatically guarantees that the constraint equation (8.13c) is satisfied at all times (Problem 8.13). Finally, there is a savings in the amount of computation required since only the one evolution equation (8.72) needs to be solved, whereas the two equations (8.68) and (8.69) must be solved if the poloidal field components are taken as primary variables. The major disadvantage of using the vector potential is that second derivatives are required to compute the current density which appears in the Lorentz acceleration and in the Joule heating terms. Consequently small variations in F may lead to serious inaccuracies in J_θ.

A finite difference representation based on the poloidal fields as primary variables for transport will be developed below. The development of a finite difference representation of poloidal field transport based on the vector potential is less complex, and will only be sketched, with details left to the problems. The space-time centering of the respective variables is given by (8.80)-(8.82). The poloidal field transport, diffusion, Joule heating and Lorentz acceleration will be solved using operator splitting as shown schematically in Figure 8.3. We consider first the poloidal evolution equation (8.68)-(8.69) and the constraint equation (8.13c), which reduces to the form

$$\frac{1}{r}\frac{\partial (rB_r)}{\partial r} + \frac{\partial B_z}{\partial z} = 0 \tag{8.110}$$

in axially symmetric systems. Equation (8.110) could be reduced to finite difference form and solved along with the evolution equation. Instead we shall impose (8.110) as a constraint on the form of the finite difference equations used to represent the evolution equations (Evans and Hawley 1988).

The constraint equation implies that the total polodial flux out of a zone $\Delta V_{k+1/2,\ j+1/2}$ is zero at all times:

$$\Phi_{poloid,\, k+1/2\, j+1/2} = 0. \tag{8.111}$$

Thus, if the poloidal flux is initially zero (initial conditions) then a finite difference representation of the evolution equations which guarantees (8.111) will automatically satisfy (8.110). We recall that the term $(1/r)\partial B_\theta/\partial\theta$ appearing in the constraint equation vanishes by symmetry. Consequently the total flux through a zone remains a constant in time, which is equivalent to the constraint $\nabla \cdot \boldsymbol{B} = 0$.

Poloidal Field Transport

The transport terms in (8.68) and (8.69) arise from the last term on the right of (8.15). Integrating (8.15) over an arbitrary part of a zone's surface S gives the following expression for the rate of change of the flux through that surface:

$$\frac{d\Phi}{dt} = \int_S \frac{\partial \boldsymbol{B}}{\partial t} \cdot d\boldsymbol{S} = \int_S \nabla \times (\boldsymbol{v} \times \boldsymbol{B}) \cdot d\boldsymbol{S} = \int_C (\boldsymbol{v} \times \boldsymbol{B}) \cdot d\boldsymbol{l}. \tag{8.112}$$

Stokes' theorem has been used in the last step to relate the surface integral to a line integral bounding the surface S. We shall use this expression to construct a finite difference representation for the rate of change of Φ through the radial and axial surface elements of a zone, and from them extract expressions which reduce to the transport term in (8.68) and (8.69) in the differential limit.

Consider first the change in axial flux through the zone surface centered at $(k+1/2,\ j+1/2)$. For the moment we assume that axial and radial velocity components are defined at each zone node (k,j) since this is the most natural centering for poloidal flux advection (if a hydrodynamic scheme based on node centered velocities is used this becomes the natural velocity centering). Figure 8.10a shows the field and velocity centering, and the path chosen for the line integeral. From (8.112) we find

$$\begin{aligned}
\frac{\Delta \Phi^z_{k,j+1/2}}{\Delta t} &= -2\pi \left[r\,(\boldsymbol{v} \times \boldsymbol{B})_\theta \right]^{k,j+1}_{k,j} = -2\pi \left[r\,(uB_r - vB_z) \right]^{k,j+1}_{k,j} \\
&= -2\pi r_{j+1} u_{k,j+1} \tilde{B}_{r,k,j+1} + 2\pi r_j u_{k,j} \tilde{B}_{r,k,j} \\
&\quad + 2\pi r_{j+1} v_{k,j+1} \tilde{B}_{z,kj+1} - 2\pi r_j v_{k,j} \tilde{B}_{z,k,j}\,.
\end{aligned} \tag{8.113}$$

The magnetic field components appearing in the last step are not naturally centered, but must be related to the components $B_{r,k+1/2,j}$ and $B_{z,k,j+1/2}$. This may be done by noting that the radial velocity $v_{k,j}$ either carries axial flux into or out of the surface $S^z{}_{k,j+1/2}$. In fact (see Figure 8.10) the flux carried into or out of $S^z{}_{k,j+1/2}$ is approximately

$$B_{z,k,j\pm 1/2}\, v_{k,j}\, \Delta t\, 2\pi r_j \tag{8.114}$$

where the plus (minus) sign holds for $v_{k,j} < 0$ ($v_{k,j} > 0$). Equation (8.114) simply represents donor cell advection of axial magnetic flux resulting from radial fluid motion. Consequently, we may define

$$\tilde{B}_{z,k,j} = \begin{cases} B^n_{z,k,j+1/2}, & v_{k,j} < 0 \\ B^n_{z,k,j-1/2}, & v_{k,j} > 0 \end{cases} \tag{8.115}$$

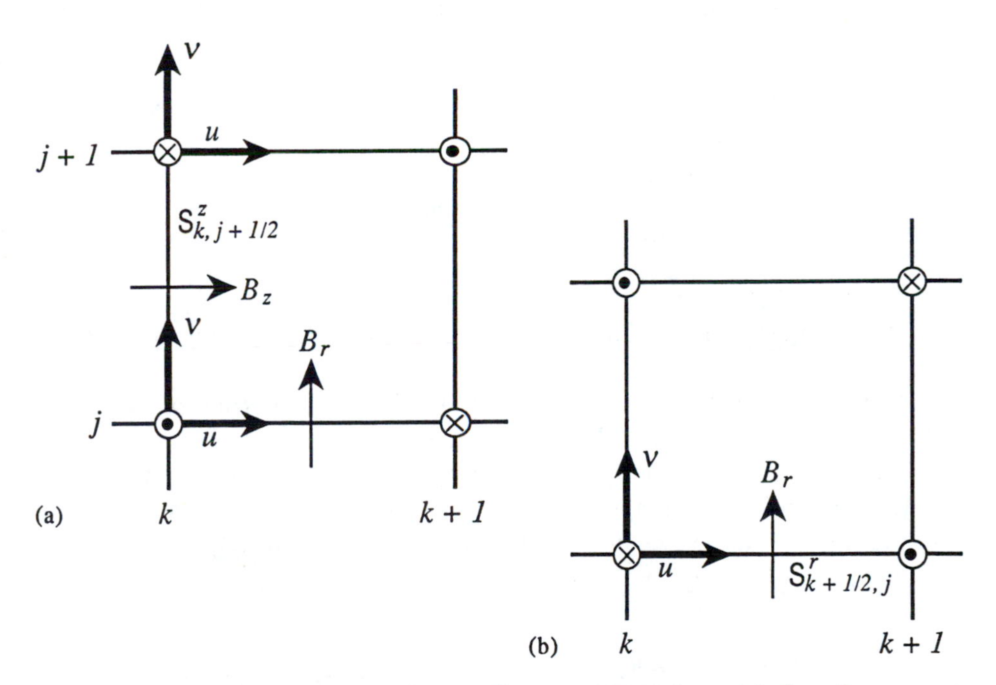

Figure 8.10. (a) The surface $S^z{}_k$ used to construct the axial flux for a zone. The curve enclosing $S^z{}_k$ consists of two circles of radius r_{j+1} (directed out of the figure) and r_j (directed into the figure). (b) The surface $S^r{}_j$ used to construct the radial flux for a zone.

The choice of time centering will be discussed below. The centering of the radial field may be considered by evaluating the change in flux through the surface $S^r{}_{k+1/2,j}$ and applying the constraint (8.111).

The change in radial flux passing through the surface $S^r{}_{k+1/2,j}$ is constructed from (8.112) and is easily shown to be

$$\frac{\Delta \Phi^z_{k+1/2,j}}{\Delta t} = 2\pi \left[r (\mathbf{v} \times \mathbf{B})_\theta \right]^{k+1,j}_{k,j}$$

$$= 2\pi\, r_j (u_{k+1,j} \tilde{B}_{r,k+1,j} - u_{k,j} \tilde{B}_{r,k,j})$$

$$- 2\pi\, r_j (v_{k+1,j} \tilde{B}_{z,k+1,j} - v_{k,j} \tilde{B}_{z,k,j}) \, .$$

(8.116)

The first two terms on the right represent radial flux carried into or out of the surface $S^r{}_{k+1/2,j}$ resulting from axial fluid motion, and we may define:

$$\tilde{B}_{r,k,j} = \left[\begin{array}{ll} B^n_{r,k+1/2,j}, & u_{k,j} < 0 \\ & \\ B^n_{r,k-1/2,j}, & u_{k,j} > 0 \end{array} \right. \, .$$

(8.117)

The centering of the axial field variables in (8.116) has yet to be specified.

Now consider the constraint (8.111). If the centering of the axial field in the last term on the right side of (8.116) is centered as in (8.115), then it will exactly cancel the last term on the right side of (8.113) when the changes in radial and axial flux are added. Continuing in this way, the change in flux through $S^z{}_{k+1,j}$ is easily shown to be given by

$$\frac{\Delta \Phi^z_{k+1,j+1/2}}{\Delta t} = 2\pi \left[r (\mathbf{v} \times \mathbf{B})_\theta \right]^{k+1,j+1}_{k+1,j}$$

$$= 2\pi (r_{j+1} u_{k+1,j+1} \tilde{B}_{r,k+1,j+1} - r_j u_{k+1,j} \tilde{B}_{r,k+1,j})$$

$$-2\pi\,(r_{j+1}\,v_{k+1,j+1}\,\tilde{B}_{z\,k+1,j+1} - r_j\,v_{k+1,j}\,\tilde{B}_{z\,k+1,j})$$

(8.118)

where the field components are centered as in (8.115) and (8.117). The second term on the right of (8.118) now appears with the opposite sign as the first term on the right of (8.116). Finally, the change in flux through $S^r{}_{k,j+1}$ is

$$\frac{\Delta\,\Phi^r_{k+1/2,j+1}}{\Delta t} = 2\pi\,r_{j+1}\,[\,(\mathbf{v}\times\mathbf{B})_\theta]^{k+1,j+1}_{k,j+1}$$

$$= -2\pi\,r_{j+1}(u_{k+1,j+1}\,\tilde{B}_{r,k+1,j+1} - u_{k,j+1}\,\tilde{B}_{r,k,j+1})$$

$$+\;2\pi\,r_{j+1}(v_{k+1,j+1}\,\tilde{B}_{z\,k+1,j+1} - v_{k,j+1}\,\tilde{B}_{z\,k,j+1})$$

(8.119)

with the field components centered as in (8.115) and (8.117). The net change in poloidal flux passing through $\Delta V_{k+1/2,j+1/2}$ is obtained by adding (8.113), (8.116), (8.118) and (8.119). The result

$$\Delta\Phi^z_{k,j+1/2} + \Delta\Phi^r_{k+1/2,j} + \Delta\Phi^z_{k+1,j+1/2} + \Delta\Phi^r_{k+1/2,j+1} \equiv 0$$

(8.120)

vanishes identically because for each term in any one of these expressions, there is a corresponding term of opposite sign in another.

Finite difference representations for the evolution equation (8.68) and (8.69) may now be extracted from the expressions for the change in flux. Dividing (8.113) by the surface area

$$S^z_{k,j+1/2} = 2\pi\,r_{j+1/2}\,\Delta r_{j+1/2}$$

(8.121)

we find for the rate of change of the axial field component

$$\frac{B'_{z,k,j+1/2} - B^n_{z,k,j+1/2}}{\Delta t^{n+1/2}} = -\frac{r_{j+1} v_{k,j+1} \tilde{B}_{z,k,j+1} - r_j v_{k,j} \tilde{B}_{z,k,j}}{r_{j+1/2} \Delta r_{j+1/2}}$$

$$+ \frac{r_{j+1} u_{k,j+1} \tilde{B}_{r,k,j+1} - r_j u_{k,j} \tilde{B}_{r,k,j}}{r_{j+1/2} \Delta r_{j+1/2}}. \tag{8.122}$$

Dividing (8.116) by the surface area

$$S^r_{k+1/2,j} = 2\pi r_j \Delta z_{k+1/2} \tag{8.123}$$

we find for the rate of change of the radial field component

$$\frac{B'_{r,k+1/2,j} - B^n_{r,k+1/2,j}}{\Delta t^{n+1/2}} = -\frac{u_{k+1,j} \tilde{B}_{r,k+1,j} - u_{k,j} \tilde{B}_{r,k,j}}{\Delta z_{k+1/2}}$$

$$+ \frac{v_{k+1,j} \tilde{B}_{z,k+1,j} - v_{k,j} \tilde{B}_{z,k,j}}{\Delta z_{k+1/2}}. \tag{8.124}$$

There is an overall change in sign in (8.122) and (8.124) because the field component and the surface normal are oppositely directed.

Since the first term on the right side of (8.122) and (8.124) are formally identical to the mass continuity equation, it may be differenced explicitly in time, as has already been assumed in (8.115) and (8.117). Equation (8.122) is a first order finite difference representation of the last term on the right side of (8.69), while (8.124) is a first order in space finite difference representation of the last term on the right side of (8.69). These equations are, in fact, a rather obvious representation of the differential equations if we recall that the transport terms describe the advection of radial and axial magnetic flux into a zone.

Problem 8.17

A better approximation than (8.114) to the axial flux advected in the radial direction results if the area

$$\frac{1}{2} v_{k,j} \Delta t \, (r_j \pm v_{k,j} \Delta t) \tag{8.125}$$

is used. How should (8.122) and (8.124) be modified to accomplish this? Does the result still satisfy the constraint equation?

Problem 8.18

What are the stability requirements associated with the explicit field advection equations?

Problem 8.19

Develop a second order monotonicity scheme to be used with the advection equations (8.122) and (8.124) which preserves the constraint condition.

The solution of these equations represents the advection of poloidal flux, and formally maintains the constraint equation $\nabla \cdot \boldsymbol{B} = 0$. They may be modified to give a second order monotonic representation of the spatial derivatives (Problem 8.19). When the advection scheme above is implemented, the cancellation of terms will not be exact because of machine roundoff, and as a result $\nabla \cdot \boldsymbol{B}$ will begin to deviate from zero as the calculation proceeds. The error tends to grow monotonically with cycle, so periodic testing should be employed to determine if the error becomes significant. An estimate of the error is given by evaluating

$$max \{(\nabla \cdot \boldsymbol{B}) \, \Delta x / B\} \tag{8.126}$$

over each zone, with the maximum taken over the grid. Here Δx is the maximum of Δr and Δz, and $\boldsymbol{B}$ is the magnitude of the poloidal field in the zone.

Poloidal Transport by the Vector Potential

An alternate description of the transport of poloidal field components in an axisymmetric system uses the vector potential, A_θ, and corresponds to the last two terms in (8.72). A finite difference representation of these terms will be sketched below, but the details will be left to the exercises. If the vector potential is used as the primary variable for the poloidal field for axisymmetric calculations, the constraint equation will automatically be satisfied. Consider first the axial term in (8.72), written in terms of the flux function $F \equiv rA_\theta$.; in the absence of grid motion

$$\frac{\partial F}{\partial t} = - u \frac{\partial F}{\partial z} . \tag{8.127}$$

Equation (8.127) is a remap equation for the flux function. A finite difference representation of the radial magnetic flux is given by (common subscript j omitted)

$$\frac{F'_k - F_k}{\Delta t} = - u_k \left(\frac{\partial F}{\partial z} \right)_k , \tag{8.128}$$

where the term on the right side denotes a second order approximation to the derivative centered a distance $u_k \Delta t$ upstream from the zone boundary. It may be constructed using the interpolation formulae from Appendix C (see Problem 5.). The radial transport term in (8.72) corresponds to the radial advection of axial magnetic flux. When written in terms of the flux function it also reduces to a remap equation

$$\frac{\partial F}{\partial t} = - v \frac{\partial F}{\partial r} , \tag{8.129}$$

and may be represented by the upwind difference (common subscript k omitted)

$$\frac{F'_j - F_j}{\Delta t} = - v_j \left(\frac{\partial F}{\partial r}\right)_j . \tag{8.130}$$

The derivative on the right is again to be interpolated a distance $v_j \Delta t$ upstream from the zone boundary.

The finite difference representation (8.128) and (8.130) are explicit in time, and are stable subject to the usual Courant time step restriction. It is also straightforward to develop monotonic corrections to these equations which are second order in the magnetic flux. Values of the flux function denoted by a prime are to be used as initial values for the calculation of the diffusion terms in (8.72), which will be discussed below. The advection of the flux may be calculated during the hydrodynamic step along with the advection of the density and other scalar variables and B_θ.

Problem 8.20

Derive the second order finite difference representation for the axial transport of the flux function (8.129). First show that the time rate of change in the radial flux through the bottom of a zone $\Delta V_{k+1/2,\, j+1/2}$ is given by (common subscript j omitted)

$$\frac{\partial \Phi_r}{\partial t} = 2\pi \left(\frac{\partial F_{k+1}}{\partial t} - \frac{\partial F_k}{\partial t}\right). \tag{8.131}$$

Next, construct $\partial \Phi_r / \partial t$ directly from the advection of the radial magnetic flux due to $u_{k,j}$ and $u_{k+1,j}$. Then use (8.70) to express the result in terms of F. From your result show that a finite difference representation of (8.127) is given by (8.128).

Poloidal Field Diffusion

The diffusion of the poloidal magnetic field components described by the first two terms on the right side of (8.68) and (8.69) can be represented by finite difference equations, but their solution is more complicated than may be warranted in most problems of astrophysical interest. A far simpler (and more economical) approach is to solve for the evolution of the axial flux as described by the first two terms in (8.72):

$$\frac{\partial A_\theta}{\partial t} = D_B\left[\frac{\partial}{\partial r}\left(\frac{1}{r}\frac{\partial(rA_\theta)}{\partial r}\right) + \frac{\partial}{\partial z}\left(\frac{1}{r}\frac{\partial(rA_\theta)}{\partial z}\right)\right] . \tag{8.132}$$

We will consider two approaches to this equation below. Because (8.132) has the general form of a diffusion equation, an implicit finite difference representation is desirable. The first approach employs operator splitting in the spatial gradients and solves for the change in A_θ in two stages. Either of the spatial gradients may be taken first; for example, the change in vector potential due to the axial gradient may be written as

$$\frac{\partial A_\theta}{\partial t} = D_B\left[\frac{\partial}{\partial z}\left(\frac{1}{r}\frac{\partial(rA_\theta)}{\partial z}\right)\right] \tag{8.133}$$

and may be solved to obtain $A'_{\theta,k,j}$. Then the change associated with the radial gradient is described by

$$\frac{\partial A_\theta}{\partial t} = D_B\left[\frac{\partial}{\partial r}\left(\frac{1}{r}\frac{\partial(rA_\theta)}{\partial r}\right)\right] \tag{8.134}$$

which is solved to give the final value of the vector potential resulting from spatial diffusion, $A^{n+1}{}_{\theta,k,j}$. Finally, note that (8.133)-(8.134) have been written in terms of the quantity

$$F \equiv r A_{\theta}, \tag{8.135a}$$

which is just proportional to the flux function $2\pi\, rA_{\theta}$. This quantity has the natural finite difference representation

$$F^{n}_{k,j} = r_j A^{n}_{\theta,\, k,j}\, . \tag{8.135b}$$

It follows from (8.70) and (8.71) that the quantities in parentheses in (8.133) and (8.134) are directly proportional to the components of the poloidal magnetic field B_r and B_z. The differencing of these quantities should be consistent with their use in constructing the Lorentz acceleration and Joule heating. With this in mind (8.133) may be represented by the following (fully implicit) finite difference representation (see Figure 8.11):

$$\frac{F'_{k,j} - F^{n}_{k,j}}{\Delta t^{n+1/2}} = \frac{D_{B,k,j}}{r_j \Delta z_k}\left(\frac{F'_{k+1,j} - F'_{k,j}}{\Delta z_{k+1/2}} - \frac{F'_{k,j} - F'_{k-1,j}}{\Delta z_{k-1/2}}\right). \tag{8.136}$$

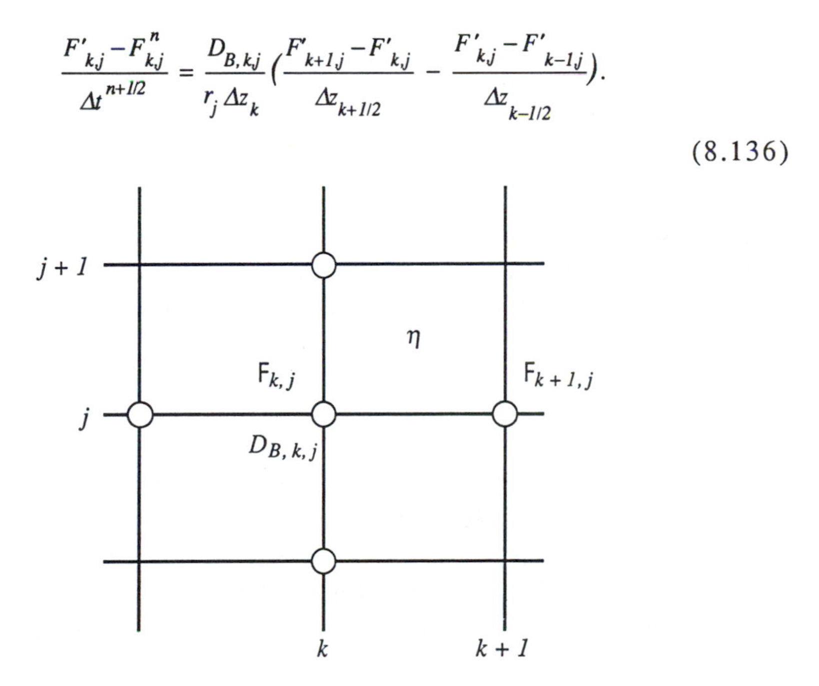

Figure 8.11. Spatial centering of the flux function, resistivity and diffusion coefficient for poloidal fields.

The diffusion coefficient $D_{B,\,k,j}$ which contains the resistivity from the four zones surrounding the node (k,j) will be discussed below. For now we simply assume that it is known at the node.

The change in the vector potential described by (8.134) may be represented as

$$\frac{F^{n+1}_{k,j} - F'_{k,j}}{\Delta t^{n+1/2}} = \frac{D_{B,\,k,j}}{\Delta r_j}\left(\frac{F^{n+1}_{k,j+1} - F^{n+1}_{k,j}}{r_{j+1/2}\,\Delta r_{j+1/2}} - \frac{F^{n+1}_{k,j} - F^{n+1}_{k,j-1}}{r_{j-1/2}\,\Delta r_{j-1/2}}\right). \tag{8.137}$$

It follows from comments above that the poloidal fields are to be constructed from the vector potential as follows:

$$B_{r,\,k+1/2,j} = -\frac{F_{k+1,j} - F_{k,j}}{r_j\,\Delta z_{k+1/2}}, \tag{8.138}$$

$$B_{z,\,k,j+1/2} = \frac{F_{k,j+1} - F_{k,j}}{r_{j+1/2}\,\Delta r_{j+1/2}}. \tag{8.139}$$

Equations (8.136) and (8.137) resemble the thermal diffusion equation discussed in Chapter 6, and can be easily expressed in tri-diagonal form and solved by Gaussian elimination.

Poloidal Boundary Conditions

Boundary values of the vector potential or the flux function must be specified in order to solve (8.136) and (8.137). If the lower radius r_2 of the computational grid corresponds to $r = 0$, then the vector potential must vanish there since the radial component of the magnetic field vanishes as $r \to 0$. This may be expressed in terms of the flux function

$$F_{k,2} = 0 \qquad (r_2 = 0) . \tag{8.140a}$$

In order for (8.143a) to be satisfied for all times, $(\partial F/\partial t)_{j=2}$ must also vanish. In combination with (8.143a) this leads to the additional boundary condition

$$F_{k,3} = - F_{k,1} \qquad (r_2 = 0) . \tag{8.140b}$$

The remaining boundary conditions will depend on the nature of the problem under consideration.

Problem 8.21

Consider the magnetic flux through a circular area normal to the z axis, and show that

$$\lim_{r \to 0} A_\theta = \frac{1}{2} r B_z . \tag{8.141}$$

From this result justify the boundary conditions (8.140).

Problem 8.22

What is an appropriate bondary condition for the vector potential for an axisymmetric system with reflecting boundary conditions at $z = 0$?

The use of operator splitting to solve the evolution equation for the vector potential in the two steps (8.133) and (8.134) may not always yield reasonable results. It is then necessary to express the entire equation (8.132) in implicit form, which results in a five point scheme (Figure 8.11) where the unknowns are $F_{k,j}$, $F_{k+1,j}$, $F_{k-1,j}$, $F_{k,j+1}$, and $F_{k,j-1}$. The finite difference equations for this system of variables may be written in the form of a block diagonal matrix, whose solution may be found by standard means (these will be reviewed in Chapter 9).

Poloidal Joule Heating

The diffusion of a poloidal magnetic field is accompanied by Joule heating of the material, as described by the last term on the right side of (8.78). The material specific energy is zone centered, but the theta component of the current current J_θ is node centered, so obvious centering problems arise. It may be recalled that a similar problem was encountered for torroidal Joule heating in the previous section, and that a consistent solution was obtained which could be understood in terms of parallel equivalent resistances of neighboring zones. Similar reasoning may be used to arrive at a finite difference representation of the term $\eta J_\theta{}^2$ (see Problem 8.23). This yields not only the Joule heating contribution to (8.73), but defined the node centered resistivity $\eta_{k,j}$ which is to be used in the diffusion equation (8.132). It is given by

$$\frac{4\,\Delta r_j\,\Delta z_k}{r_j\,\eta_{k,j}} = \frac{\Delta r_{j+1/2}}{r_{j+1/4}}\left(\frac{\Delta z_{k+1/2}}{\eta_{k+1/2,j+1/2}} + \frac{\Delta z_{k-1/2}}{\eta_{k-1/2,j+1/2}}\right) + \frac{\Delta r_{j-1/2}}{r_{j-1/4}}\left(\frac{\Delta z_{k+1/2}}{\eta_{k+1/2,j-1/2}} + \frac{\Delta z_{k-1/2}}{\eta_{k-1/2,j-1/2}}\right) \tag{8.142}$$

which in fact defines $\eta_{k,j}$. The radii centered at $j\pm 1/4$ are defined by

$$r_{j\pm 1/4} \equiv \frac{1}{2}(r_j + r_{j\pm 1/2}) \,. \tag{8.143}$$

The diffusion coefficient appearing in (8.136) and (8.137) is given by

$$D_{B,k,j} = \frac{\eta_{k,j}}{4\pi} \tag{8.144}$$

with the node centered resistivity defined as in (8.142).

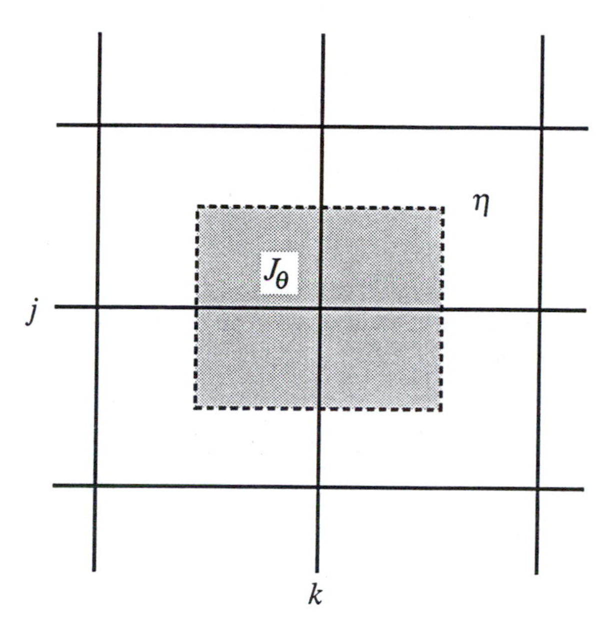

Figure 8.12. Node centered zone associated with the axial current density. Joule heating of the shaded portion contributes to the change in material energy of the top right zone.

The heating due to the theta component of the current can be found formally by integrating (8.78) over the zone volume $\Delta V_{k+1/2,j+1/2}$. When this is done the right side may be written as

$$\int \eta J_\theta^2 \, dV = \eta_{k+1/2,j+1/2} \int J_\theta^2 \, dV. \tag{8.145}$$

The resistivity has been taken outside the integral since it is assumed to be a constant across the zone. The volume integral of the current J_θ can be evaluated by returning to the idea of parallel resistances. First, represent $\Delta V_{k+1/2,\ j+1/2}$ by four subzones through which an equivalent current flows [one of these is shown shaded in Figure (8.12)]. We now wish to represent the equivalent currents in terms of the node centered ones. Noting that the electric potential drop in the angular direction through the current zone (dashed in Figure 8.12) and each subzone must be the same, it is straightforward to show that

$$J^{\theta}_{k+1/4,j+1/4} \equiv \frac{J_{\theta,k,j}\,\eta_{k,j}\,r_j}{\eta_{k+1/2,j+1/2}\,r_{j+1/4}}$$

(8.146)

with similar definitions for the three other current densities $J^{\theta}_{k+3/4,j+1/4}$, $J^{\theta}_{k+1/4,\,j+3/4}$, and $J^{\theta}_{k+3/4,j+3/4}$. The right side of (8.145) will contain four contributions, and the Joule heating contribution to (8.148) can be expressed in the form:

$$[\rho\,(\varepsilon^{n+1}-\varepsilon^{n})]_{k+1/2,j+1/2} = \frac{\Delta t^{n+1/2}}{4\,\eta_{k+1/2,j+1/2}\,r_{j+1/2}}\Big[\frac{r_j^2}{r_{j+1/4}}(\eta^2_{k,j}\,J^2_{\theta,k,j} + \eta^2_{k+1,j}\,J^2_{\theta,k+1,j})$$

$$+ \frac{r^2_{j+1}}{r_{j+3/4}}(\eta^2_{k,j+1}\,J^2_{\theta,k,j+1} + \eta^2_{k+1,j+1}\,J^2_{\theta,k+1,j+1})\Big]\,.$$

(8.147)

Problem 8.23

Derive the node centered resistivity (8.142) using the rule of parallel resistances. First find the resistance of the shaded zone in Figure 8.12, whose area is $(\Delta r_{j+1/2}\Delta z_{k+1/2})/4$. The resistance of the current zone (dashed) is defined to be

$$R_{k,j} \equiv \frac{\eta_{k,j}\,2\pi r_j}{\Delta r_j\,\Delta z_k}.$$

Problem 8.24

Carry out the derivation of (8.146), and write down expressions for the remaining three effective theta components of the currents flowing in $\Delta V_{k+1/2,\,j+1/2}$.

Poloidal Joule heating may be computed using the vector potential obtained from the diffusion of the poloidal fields (vector potential) to obtain the material energy (and temperature). The final step in the solution of the magnetohydrodynamic equations for a nonrotating system is the calculation of the Lorentz acceleration associated with the theta component of the current.

Poloidal Lorentz Force

The Lorentz acceleration of a fluid element associated with the poloidal field components are described by the first term on the right side of (8.74) and (8.75) (the acceleration terms proportional to B_θ were discussed in the previous section). In keeping with the results of Problem 8.20 the components of the fluid velocity will be assumed to be node centered. This simplifies the derivation of the Lorentz acceleration associated with poloidal currents.

The change in the axial momentum of a fluid element of volume $\Delta V_{k,j}$ follows from the first term on the right side of (8.75); integrating over the volume of the element

$$\int \rho \frac{\partial u}{\partial t} dV = -\frac{1}{c} \int J_\theta B_r \, dV = -\frac{1}{c} J_{\theta,k,j} \int B_r \, dV \tag{8.148}$$

where the current density, which is assumed to be constant throughout $\Delta V_{k,j}$ has been removed from the integral. The left hand side may be evaluated over the volume element much as in Chapter 5 where face centered velocity components were considered. The result is easily shown to be

$$\int \rho \frac{\partial u}{\partial t} dV = \rho_{k,j} \frac{u'_{k,j} - u_{k,j}}{\Delta t^n} \Delta V_{k,j} \tag{8.149a}$$

$$\rho_{k,j} \Delta V_{k,j} \equiv \frac{\pi}{2} \{ r_{j-1/4} \Delta r_{j-1/2} (\rho_{k-1/2,j-1/2} \Delta z_{k-1/2} + \rho_{k+1/2,j-1/2} \Delta z_{k+1/2}$$

$$+ r_{j+1/4} \Delta r_{j+1/2} (\rho_{k-1/2,j+1/2} \Delta z_{k-1/2} + \rho_{k+1/2,j+1/2} \Delta z_{k+1/2}) \} \tag{8.149b}$$

where $u'_{k,j}$ denotes the axial component of the velocity after the Lorentz acceleration. The axial component $u_{k,j}$ is the final value of the velocity after the hydrodynamic cycle. If the magnetohydrodynamic stage represents the last process in a computational cycle, then $u^{n+1}{}_{k,j} = u'_{k,j}$. Finally we note that $\rho_{k,j}$ is, in fact, the node centered density which appears in the finite difference representations of the momentum equation based on node centered velocities.

Next consider the right side of (8.148); assuming that B_r is a constant along each zone face, we find that

$$\int B_r \, dV = \pi (B_{r,k-1/2,j} \Delta z_{k-1/2} + B_{r,k+1/2,j} \Delta z_{k+1/2}) r_j \Delta r_j \, . \tag{8.150}$$

Combining (8.149) and (8.150), and cancelling common factors

$$\rho_{k,j} \frac{u'_{k,j} - u_{k,j}}{\Delta t^n} = -\frac{1}{c} J_{\theta,k,j} \frac{(B_{r,k-1/2,j} \Delta z_{k-1/2} + B_{r,k+1/2,j} \Delta z_{k+1/2})}{2 \Delta z_k} \, . \tag{8.151}$$

The finite difference representation of the change in axial momentum will vanish identically if the theta component of the current is zero. Note that the radial field component appearing on the right side of the acceleration equation is just the volume average of the components from adjacent zones.

The change in radial momentum associated with the Lorentz acceleration can be obtained in a similar manner, and is

$$\rho_{k,j} \frac{v'_{k,j} - v_{k,j}}{\Delta t^n} = \frac{1}{c} J_{\theta,k,j} \frac{(r_{j+1/4} \Delta r_{j+1/2} B_{z,k,j+1/2} + r_{j-1/4} \Delta r_{j-1/2} B_{z,k,j-1/2})}{2 r_j \Delta r_j} \tag{8.152}$$

which involves the same value of the current density and mass density as does (8.151).

Stability and Accuracy

The calculation of the components of the Lorentz acceleration completes the magnetohydrodynamic phase of the computation. The algorithms developed for two dimensional configurations must be numerically stable to be of practical use. A set of stability criteria will be found below using heuristic arguments, and analogy with previous analyses. Formal arguments for the linearized equations may be interesting, but will not be given here (see, for example, Richtmyer and Morton 1967). The magnetohydrodynamic equations for cylindrical symmetric systems considered in this and the previous section can be thought of as involving three principle parts: (1) the acceleration terms in the Lagrangian step of the calculation, (2) the transport, or advection-like terms, and (3) the diffusion terms. We consider each of these processes below.

First, consider the transport terms for the torroidal field. These resemble the planar advection of a scalar quantity [recall, for example, (5.50) with $u_g = 0$]. The corresponding stability condition (5.51) is independent of the quantity being advected, and thus we might expect that it would apply here as well. In practice this is often found to be an adequate stability constraint on the transport of the torroidal field. Physically, the constraint simply guarantees that the torroidal magnetic flux will be advected less than one zone per computational cycle. The transport equation for the poloidal components are also similar to the advection equation (with each field component considered to be a scalar). Finally, the equation for the transport of the vector potential (8.127) resembles a remap equation (2.48) whose stability constraint was given by (2.54).

The finite difference representation of the magnetic diffusion equations (8.91)-(8.92) and (8.136)-(8.137) are fully implicit equations which resemble the thermal diffusion equation, and thus are expected to be stable for arbitrary time step. The linearized equations are, in fact, unconditionally stable for all Δt. In practice (8.91)-(8.92) or (8.136)-(8.137) can usually be employed without time step constraints even in the nonlinear regime. We recall that although large time steps may not result in instabilities, they may result in inaccurate solutions to the partial differential equations.

The Lorentz acceleration terms in the momentum equation do lead to a time step constraint, as can be seen by the following heuristic arguments. First, consider the acceleration associated with a torroidal field. Expanding the axial and radial current appearing on the right side of (8.74) and (8.75) (we ignore the poloidal field components here) in terms of the magnetic field using (8.77a) and (8.77c) we find that the momentum equation can be expressed as

$$\rho \frac{\partial \boldsymbol{v}}{\partial t} = -\nabla \left(\frac{B_\theta^2}{8\pi} \right) - \frac{B_\theta^2}{4\pi r} \boldsymbol{e}_r , \tag{8.153}$$

where $\boldsymbol{e}_r$ is a unit vector in the radial direction. An equation including the vector acceleration associated with poloidal field components can be written in an analogous form. Apart from the second term on the right side of (8.153), the momentum equation resembles the usual fluid equation with a pressure gradient involving the magnetic pressure

$$P_B \equiv \frac{B_\theta^2}{8\pi} . \tag{8.154}$$

In nonplanar coordinates the pressure is replaced by a stress which complicates the analysis. In analogy with the stability analysis of the fluid equations, it might be supposed that a Courant-like time step constraint of the form

$$\Delta t < \frac{\Delta x}{(B_\theta^2 / 8\pi\rho)^{1/2}}$$

would apply to the magnetohydrodynamic equations. The fluid pressure acceleration must be included and the stability condition which results is

$$\Delta t = C_B \frac{\Delta x}{(c_s^2 + v_A^2)^{1/2}} , \tag{8.155}$$

where Δx represents either Δz or Δr, and c_s is the adiabatic sound speed. The smallest value of the time step (multiplied by a factor C_B

$= 0.5$) is used. Physically, $v_A \equiv (B_\theta^2/4\pi\rho)^{1/2}$ is the Alfen velocity, and (8.155) simply expresses the requirement that magnetosonic disturbances travel less than a zone per time step. Detailed stability analyses in fact show that the time step constraint actually depends on the relative direction of the magnetic field and the disturbance. For disturbances that travel perpendicular to the field, the time step must satisfy (8.155), while for disturbances travelling along the field lines the constraint reduces to

$$\Delta t = C_B \frac{\Delta x}{v_A} . \tag{8.156}$$

The final time step is then taken as the minimum of (8.156) and any other time step limits (such as the Courant or transport time step).

If the magnetic diffusion equation is differenced explicitly then it will impose additional stability constraints on the time step. If a linearized implicit treatment is used, then there will be no additional constraints on the time step as long as the coefficients in the diffusion equation (the resistivity) change only slightly per cycle. As with all diffusion equations, if the right side involves quantities defined at the advanced time only [see, for example, (8.91)], then the results may be inaccurate if the time step is too large.

Problem 8.25

Using an analysis similar to the one preceding (8.155), develop a time step constraint for the magnetohydrodynamic equations which include the effects of poloidal and torroidal field components.

Problem 8.26

Construct the finite difference representation of the change in radial momentum (8.152) assuming that the fluid velocity components are node centered.

Problem 8.27

Construct a finite difference representation for the advection of torroidal magnetic flux assuming that the fluid velocity components $u_{k,j}$ and $v_{k,j}$ are node centered.

8.7 THERMOELECTRIC EFFECTS IN A PLASMA

Whenever physical quantities vary on time scales that are comparable to or smaller than the electron-ion collisional time scale τ_{ei}, the plasma is no longer characterized by a single temperature. In the simplest case it is necessary to follow separately the thermal energy of the electrons and the ions (see Appendix E), although it is still possible to use a single velocity to characterize the fluid dynamics. It may also be necessary to use a generalized form of Ohm's law (a simple example of which is derived in Appendix E). A simple example will be considered here which illustrates some of the computational issues arising when these more complex plasma conditions are modeled. We shall retain the single fluid velocity approximation which has been used in the earlier sections of this chapter. The momentum equation is then given by (8.21), but the total pressure must be expressed as the sum of an electron pressure (E.2) and an ion pressure (E.3):

$$P = P_e(\rho, T_e) + P_i(\rho, T_i) . \tag{8.157}$$

The mass continuity equation is unchanged, and is given by (4.3).

When τ_{ei} is comparable to the other time scales, the energies of the electrons and the ions will change in different ways (as was discussed in Section 6.1), and thus the rate of change of each must be followed. In particular the ion specific energy ε_i will satisfy (E.30), where

$$Q_{ie} = -\frac{T_i - T_e}{\tau_{ei}} , \tag{8.158}$$

and a simple model of the flux of thermal energy carried by ions (ion heat flux) is

$$q_i = -K_i \nabla T_i . \tag{8.159}$$

Here K_i is the (scalar) ion thermal conductivity. In more nearly realistic models of a plasma the thermal conductivities are tensors and (8.159) can be expressed in terms of the temperature gradient parallel to and normal to the magnetic field. We shall not consider these complications here (see, for example, Braginskii 1965). If shocks occur then an artificial viscous stress is to be added to the momentum equation in the usual way (Chapter 4), and the artificial viscous heating can be deposited in the ions. Electron-ion coupling will then tend to equilibrate the electron and ion temperatures.

A simple model for the rate of change of the electron specific energy ε_e is given by (E.27), with

$$Q_{ei} = \eta J^2 - Q_{ie} + \frac{J \cdot f_T}{e n_e} - [\frac{d}{dt}(\frac{B^2}{8\pi})]_{TE} \tag{8.160}$$

$$f_T = n_e \beta_T \nabla k_B T_e \ . \tag{8.161}$$

The first term in (8.160) represents the usual Joule heating of the plasma, which in the two temperature model is deposited in the electrons. The second term is the energy exchange between electrons and ions. The third term describes thermal energy transport associated with electric currents in the presence of a gradient in the electron temperature. The final term is the change in the magnetic field energy associated with the thermoelectric force (8.161).

Besides the energy change Q_{ei}, a flow of electron thermal energy down the temperature gradient will occur which can be expressed by a simple electron heat flux model:

$$q_e = - K_e \nabla T_e \ , \tag{8.162}$$

where K_e is the (scalar) electron thermal conductivity.

Finally, a generalized form of Ohm's law (such as E.45) is used to express the electric field in the laboratory frame in terms of the magnetic field and the plasma variables. The expressions above constitute a relatively simple model of a high temperature, low density plasma. A detailed discussion of these effects (including tensor thermal conductivities and stresses) may be found in Braginskii (1965).

The electron and ion specific energies and temperatures may be related by a simple generalization of (3.14):

$$C_{V,e} = \varepsilon_{\varepsilon} / T_e \tag{8.163a}$$

$$C_{V,i} = \varepsilon_i / T_i \; . \tag{8.163b}$$

Here $C_{V,e}$ and $C_{V,i}$ are the electron and ion pseudo specific heats. These quantities are to be obtained from the equations of state by a straightforward generalization of the discussion in Section 3.2.

Problem 8.28

(a) Consider a one dimensional Lagrangian model of the plasma described above which ignores all dissipative effects except for an artificial viscous stress and artificial viscous heating of the ions. Develop an explicit finite difference representation of the equations analogous to those discussed in Chapter 4.

(b) Discuss the modifications needed to extend part (a) above to a one dimensional Eulerian regime. Describe in particular how the advection stage should be modified to include a second order scheme which transports the thermal energies ε_e and ε_i with the mass.

Thermoelectric Field Generation

When the simple form of Ohm's law (8.1) is replaced by an expression such as (E.45), the evolution equation for the magnetic field (8.6) includes additional terms. Straightforward substitution of (E.45) for the electric field in (8.13a) results in an evolution equation of the form

$$\frac{\partial \boldsymbol{B}}{\partial t} = \left(\frac{\partial \boldsymbol{B}}{\partial t}\right)_o - \frac{c}{e} \nabla \times \left\{ \frac{1}{n_e} \nabla \left(P_e + \frac{B^2}{8\pi}\right) - \frac{(\boldsymbol{B} \cdot \nabla)\boldsymbol{B}}{4\pi \; n_e} \right\} , \tag{8.164}$$

where the first term on the right is just (8.15). The second term on the right represents thermoelectric generation of magnetic fields. For example, if $\boldsymbol{B} = 0$ initially, but a region develops in which ∇P_e and ∇n_e do not point in the same direction, a magnetic field may develop spontaneously (Chase, LeBlanc and Wilson 1973). The rate of change of magnetic energy associated with this last term corresponds to the term on the right in (8.160).

Problem 8.29

Carry out the derivation of (8.164). Then use (8.13c) to show that the last term on the right can be expressed in the form

$$\left(\frac{\partial \boldsymbol{B}}{\partial t}\right)_{TE} = - \frac{c}{e} \; \nabla \times \left(\frac{\boldsymbol{F}}{n_e}\right) , \tag{8.165}$$

where the Cartesian components of the vector $\boldsymbol{F}$ are given by

$$F_k = \frac{\partial \Sigma_{ik}}{\partial x^i} ,$$

and Σ_{ik} is a stress tensor whose Cartesian components are

$$\Sigma_{ik} = \left(P_e + \frac{B^2}{8\pi}\right) \delta_{ik} - \frac{B_i B_k}{4\pi} .$$

The introduction of thermoelectric effects requires modification of the hydrodynamic equations which are straightforward (see Problem 8.28), as well as changes associated with the dissipative effects discussed above. Electron and ion thermal conduction (8.159) and (8.162) in the energy equations reduce to diffusion equations which may be solved by standard linearized implicit algorithms, such as those discussed in Chapter 6. The extension to tensor thermal conductivities complicate the finite difference representations somewhat, and will not be discussed here. These issues are left as exercises for the reader. The remainder of this section will address several aspects of the thermoelectric effects represented by the last term in (8.164), and the transport of thermal energy by electric currents [the third term on the right side of (8.162)].

When the magnetic fields change slowly on dynamic time scales, it is possible to operator split the thermoelectric terms so that the equations are solved in two stages. In the first, the magnetohydrodynamics, excluding thermoelectric effects, is solved by the methods discussed in the first part of this chapter. The primary modifications required for this part of the process are to deposit the thermal energy associated with Joule heating in the electrons (in the single temperature model Joule heating changed the total material energy), and to use the electron temperature when evaluating the electrical resistivity. The second step solves the thermoelectric terms

$$\left(\frac{\partial \boldsymbol{B}}{\partial t}\right)_{TE} = -\frac{c}{e}\nabla\times\left(\frac{\boldsymbol{F}}{n_e}\right) \tag{8.166}$$

and

$$\rho\left(\frac{\partial \varepsilon_e}{\partial t}\right)_{TE} = \frac{\boldsymbol{J}\cdot \boldsymbol{f}_T}{e\, n_e} = \frac{1}{e}\beta_T k_B \nabla\cdot(\boldsymbol{J}\, T_e)\,. \tag{8.167}$$

The density is held constant during this step of the process. Once the change in the magnetic field has been found, the last term in (8.161) may be used to find the corresponding change in the electron energy due to thermal forces.

As an example of thermoelectric forces we shall consider the special case of spontaneous field generation for an axisymmetric system in which initially $\boldsymbol{B} = 0$. In this instance it is assumed that the gradient of the electron pressure and the gradient of the electron number density point in different directions. Then defining

$$F = \nabla \Pi \equiv \nabla(P_e + B^2/8\pi) , \tag{8.168}$$

(8.166) becomes

$$\frac{\partial B_\theta}{\partial t} = \frac{c}{e n_e^2} (\nabla n_e \times \nabla \Pi)_\theta$$

$$= \frac{c}{e n_e^2} (\frac{\partial n_e}{\partial z}\frac{\partial \Pi}{\partial r} - \frac{\partial n_e}{\partial r}\frac{\partial \Pi}{\partial z}) . \tag{8.169}$$

It is important to retain the magnetic pressure $B^2/8\pi$ in (8.168). Even though this term is assumed to be small compared with the material pressure, it can play an important role in determining the rate of change of spontaneously generated fields. Finally, we note that the change in the poloidal component of the field (if initially zero) in this geometry is zero.

We now wish to find a finite difference representation of (8.169) which is zone centered, and which vanishes whenever the gradients ∇n_e and $\nabla \Pi$ point in the same direction [as required by the partial differential equation (8.169)]. This is a particularly important constraint since we wish to avoid the generation of fields associated with noise in the computation. Standard definitions of the derivatives in the two terms on the right side of (8.169) are badly centered, and simple spatial averages will not guarantee that the change in flux vanishes when it should. One way to construct zone centered derivatives is from a quadratic fit through three neighboring points (see Appendix C). If the same procedure is used to construct each component of ∇n_e and $\nabla \Pi$, then if $\nabla n \times \nabla \Pi$ vanishes, so will the finite difference representation for it. For example, an interpolated value for $\partial n_e/\partial z$ at the zone center is given by (common subscript $j+1/2$ omitted)

$$(\frac{\partial n_e}{\partial z})_{k+1/2} = \frac{(\partial n_e/\partial z)_k \Delta z_{k+1} + (\partial n_e/\partial z)_{k+1} \Delta z_k}{2\Delta} , \tag{8.170}$$

where Δ is defined by (C.10). Similar expressions may be found for the remaining derivatives of Π and n_e. When these are combined in (8.169) a finite difference representation is obtained for the spontaneous generation of B_θ due to thermoelectric effects.

Problem 8.30

Derive expressions for the derivatives of Π and n_e which appear on the right side of (8.169). Show that as long as the gradients of ∇n_e and $\nabla \Pi$ point in the same direction, the finite difference representation for right side of (8.169) vanishes.

Approximations for the change in ε_e associated with thermal energy carried by electric currents (8.167) also leads to spatial averages which are badly centered. Integrating (8.169) over the zonal volume $\Delta V_{k+1/2,j+1/2}$ gives

$$\frac{[\rho(\varepsilon' - \varepsilon)]}{\Delta t} = \frac{1}{e}\beta_T k_B \frac{1}{\Delta V}\int \nabla\cdot(J\,T_e)\,dV$$

$$= \frac{1}{e}\beta_T k_B \frac{1}{\Delta V}\left\{\int \frac{1}{r}\frac{\partial}{\partial r}(r\,J_r T_e)\,dV + \int \frac{\partial}{\partial z}(J_z T_e)\,dV\right\}. \tag{8.171}$$

Consider the axial derivative in the last expression above. The integral reduces to (common subscript $j+1/2$ omitted)

$$\frac{2\pi}{\Delta V}\int (J_z T_e)_k^{k+1}\, r\,dr = \frac{J_{z,k+1}\tilde{T}_{e,k+1} - J_{z,k}\tilde{T}_{e,k}}{\Delta z_{k+1/2}}. \tag{8.172}$$

It is immediately obvious that none of the variables are properly centered. The simplest approach is to use (C.7) to find the axial components at k,j and at $k,j+1$, and then to average these to obtain

$J_{z,k,j+1/2}$. For example (common subscript $j+1/2$ omitted),

$$J_{z,k} = \frac{J_{z,k+1/2}\,\Delta z_{k-1/2} + J_{z,k-1/2}\,\Delta z_{k+1/2}}{\Delta z_k}, \tag{8.173a}$$

and finally,

$$J_{z,k,j+1/2} = \frac{1}{2}\left[J_{z,k,j} + J_{z,k,j+1}\right]. \tag{8.173b}$$

A similar choice of centering may be used to define the radial components of the current at the zone faces.

The electron temperature at the zone face may be found in several ways. One choice would be to interpolate as in (8.173a), using the values on either side of the zone face. Another approach, which is more in keeping with the concept of thermal energy transport by electric currents, is to use an upstream value (with respect to the direction of current flow) of the temperature, such as (common subscript $j+1/2$ omitted)

$$T_{e,k} = T_{e,k-1/2}\left(\frac{\Delta z_{k+1/2}}{2} \pm v_k \Delta t\right) + T_{e,k+1/2}\left(\frac{\Delta z_{k-1/2}}{2} \pm v_k \Delta t\right). \tag{8.174}$$

The plus sign (minus sign) is used if $J_{z,k} < 0$ ($J_{z,k} > 0$). This second order interpolation may also be modified to include monotonicity constraints if appropriate. Similar second order interpolation may be used in the radial direction.

Combining the approximations above, the resulting change in electron energy associated with thermal transport by axial currents can be expressed in the form (common subscript $j+1/2$ omitted)

$$(\rho C_{V,e})_{k+1/2} \frac{T'_{e\,k+1/2} - T_{e,k+1/2}}{\Delta t} = \frac{1}{e}\beta_T k_B \left\{ \frac{J_{z,k+1}\tilde{T}_{e,k+1} - J_{z,k}\tilde{T}_{e,k}}{\Delta z_{k+1/2}} \right\}. \tag{8.175}$$

This finite difference representation resembles an advection equation and as such may fail if too large a time step is used in calculations.

Problem 8.31

Find an expression for the change in electron energy associated with radial components of the electric current analogous to (8.175). Include expressions for the face centered values of J_r and T_e.

Problem 8.32

Assume that the electric current distribution changes slowly with T_e, and estimate the maximum time step for which the explicit representation (8.175) should yield reasonable answers.

Poloidal Fields

Thermoelectric effects on the poloidal component of the magnetic field in two dimensional cylindrically symmetric systems can be described using a vector potential, or in terms of the field components themselves. In the discussion below the angular component A_θ, from which B_r and B_z may be obtained, will be used. If Ohm's law is taken in the form (E.45) then the evolution equation for the vector potential becomes:

$$\frac{\partial A}{\partial t} = -E_o - \frac{1}{en_e}(J \times B - \nabla P_e) \tag{8.176}$$

where the electric field E_o gives the diffusion and transport terms appearing on the right side of (8.19) in the absence of thermoelectric effects. We will assume that (8.19) has been solved, and shall use operator splitting to solve for the thermoelectric change in the angular component of the vector potential. In a cylindrically symmetric system the pressure gradient term appearing in Ohm's law contributes only to the torroidal component of the field, and thus the only term in (8.176) which we consider here is

$$\left(\frac{\partial A_\theta}{\partial t}\right)_{TE} = -\frac{1}{e n_e} (\boldsymbol{J} \times \boldsymbol{B})_\theta .$$

(8.177)

Notice that if B_r and B_z both vanish, then poloidal fields are not generated by the thermoelectric effect in this symmetry. The evolution equation can be rewritten in the form

$$\frac{\partial F}{\partial t} = \frac{1}{e n_e}\left(J_r \frac{\partial F}{\partial r} + J_z \frac{\partial F}{\partial z}\right)$$

(8.178)

where the components of the current density depend on B_θ through (8.77a)-(877c.). Since $\boldsymbol{J} / en_e$ is the current drift velocity, (8.178) is formally a remap equation. For situations where the radial and axial components of the current density change slowly during a typical computational time step, it is reasonable to use operator splitting to solve for the change in the vector potential associated with (8.177) and (8.19) separately. When this approach is valid (8.178) may be solved as a remap, using current values for J_r and J_z. For example, the change associated with the axial gradient can be represented by

$$\frac{F'_{k,j} - F_{k,j}}{\Delta t} = \frac{J_{z,kj}}{e\,(n_e)_{k,j}} \left(\frac{\partial F}{\partial z}\right)_{k,j},$$

(8.179)

where the derivative on the right side is interpolated to the vertex using a second order scheme similar to (8.170). The electron number density at the vertex may be defined in terms of the mass density at the vertex by

$$(n_e)_{k,j} = (Z/m_i)\,\rho_{k,j} .$$

(8.180)

The radial gradient in (8.178) may be approximated in the same way, using (8.180) for the electron number density. Finally, the centering of the components of the current density needs to be specified. Since J / en_e is the current drift velocity, a reasonable choice is to use the upstream value of J_z in (8.179), and the upstream value of J_r to multiply the radial gradient.

The finite difference representations above are fully explicit in the flux function and in the components of the current density, and thus stability time step constraints may be required (see Problem 8.32).

The change in the electron internal energy associated with thermoelectric effects on the poloidal field does not contribute to the third term on the right side of (8.160) in a cylindrically symmetric system. Thus (8.175) remains valid in this case. However, the last term in (8.160) must include losses associated with any thermoelectric change in the poloidal field. If the latter are obtained from an explicit model, such as the one discussed here, care must be exercised because the last term could produce a negative electron specific energy. In some problems it may be necessary to solve the torroidal and poloidal thermoelectric equations implicitly, possibly without recourse to operator splitting. While techniques exist to accomplish this, they are much more difficult to implement, and should only be used when necessary.

References

S. I. Braginskii, "Transport Processes in a Plasma," in *Reviews of Plasma Physics,* Vol. 1 (New York: Consultants Bureau, 1965).

J. B. Chase, J. M. LeBlanc, and J. R. Wilson, Phys. Fluids. **16** (1973), pg.1142.

T. G. Cowling, 1976, *Magnetohydrodynamics* (New York: Adam Hilger, Publisher, 1976), distributed by Crane, Russak and Company, Inc., New York, New York.

C. R. Evans and J. F. Hawley, 1988 preprint.

W. D. Jackson, 1975, *Classical Electrodynamics,* Second Edition (New York: Wiley, 1975).

L. E. Kalikhman, 1967, *Elements of Magnetohydrodynamics,* edited by A. G. W. Cameron (London: W. B. Saunders Company, 1967).

R. D. Richtmyer and K. W. Morton, *Difference Methods for Initial-Value Problems*, Second Edition (New York: Interscience Publishers, 1967).

9

GRAVITATION

In descriptions of some astrophysical systems gravitation does little more than establish a background mass distribution within which other processes occur. Often, however, it plays an integral role in determining how the system evolves. When the velocities in a mass distribution are small compared with the speed of light c, or when its effective scalar gravitational potential $\phi << c^2$, the effects of gravity can be accurately described in terms of Newtonian forces through the use of Poisson's equation (4.109), and the momentum equation (4.108). Newtonian gravitation within the framework of a one dimensional system was discussed in Section 4.6. In this case the gravitational potential can usually be obtained directly, and the Newtonian force terms can be easily added to the momentum equation. Thus Poisson's equation need not be solved directly during the computation of such systems. In two or three spatial dimensions this simple approach is no longer possible, and the fluid equations must be augmented by Poisson's equation. Newtonian gravitation affects fluid motion only through the momentum equation (4.108). Furthermore, since there is no time scale defined by Poisson's equation, changes in the Newtonian gravitational potential occur instantaneously throughout the system, and the system's response occurs on hydrodynamic timescales. Poisson's equation is an elliptic differential equation whose solution is constrained by boundary conditions as described in Chapter 2.

For compact objects, or when fluid velocities are comparable to the speed of light, a relativistic theory, such as Einstein's General

Theory of Relativity, should be used to describe the effects of gravitation. Relativistic theories of gravitation are much more complicated than Newtonian theory in part because changes in the fields (or equivalently in the structure of space-time) occur at a finite rate (determined by the speed of light), and are accompanied by energy transport in the form of gravitational radiation. The development of numerical methods to describe relativistic gravitation and fluid dynamics have been presented in the literature. In this chapter we shall discuss one way of incorporating hydrodynamics within the frame work of General Relativity. For systems which can be described by a single spatial dimension, a generalization of the Lagrangian approach may be considered. However, for problems requiring more than one spatial dimension (and these will include all problems in which gravitational radiation is possible), the Eulerian approach may be preferred. Therefore, in the remainder of this chapter attention will be restricted to a one dimensional Eulerian description. While this approach is not the most practical for one dimensional problems, it illustrates many of the features which are important for problems in higher dimensions. Our approach represents a relatively straightforward extension of the basic methods discussed in Chapter 5, with which the reader is assumed to be familiar. Relativistic models involving a single spatial dimension are considerably simplier than models which require two or three spatial dimensions. This is similar to the simplification which results in one spatial dimension in Newtonian theory. Finally, no attempt will be made here to develop the underlying relativistic theory since general discussions may be found in standard texts (see, for example, Misner, Thorne and Wheeler 1973; Weinberg 1972; Narliker 1983).

9.1 Newtonian Gravitation

Newtonian gravitation may be incorporated with the finite difference representations of the fluid dynamic equations developed in Chapters 4 and 5. Newtonian gravity for spherically symmetric systems described by explicit fluid dynamic equations in Section 4.6, and was described by fully implicit equations in Section 4.7 (Problem 4.25). In one dimensional systems gravitation can be included by a simple addition of the Newtonian force to the momentum equation.

For systems involving motion in two or three spatial dimensions this simple approach is not generally possible. Instead, Poisson's equation

$$\nabla^2\Phi = 4\pi G \rho \tag{9.1}$$

must be solved along with the hydrodynamic equations. Here Φ is the Newtonian gravitational potential, ρ is the mass density and G is the gravitational constant. In this section a finite difference representation of (9.1) applicable to two dimensional systems will be developed. The approach can be extended to apply to three dimensional problems. The resulting equations may be solved by different methods, one of which will be outlined below. In the remainder of this section it will be assumed that the fluid motion is two dimensional, and is to be described by the axially symmetric Eulerian hydrodynamic equations of Chapter 5, and the finite difference representations discussed there. Application to other coordinate systems may be in a straightforward manner.

Poisson's equation for a two dimensional axially symmetric system can be written in the form

$$\frac{1}{r}\frac{\partial}{\partial r}\left(r\frac{\partial \Phi}{\partial r}\right) + \frac{\partial^2 \Phi}{\partial z^2} = 4\pi G \rho . \tag{9.2}$$

The only modification required of the hydrodynamic equations to account for (9.2) is the addition of the Newtonian gravitational force term to the right side of the momentum equation, whose components (in the absence of gravitation) are given by (5.12) and (5.13). Omitting the terms representing advection and the acceleration associated with the pressure gradient, the change in the radial component of the momentum equation associated with Newtonian gravitation can be written as

$$\frac{\partial T}{\partial t} = -\rho \frac{\partial \Phi}{\partial r} \tag{9.3a}$$

while the change in the axial component can be written as

$$\frac{\partial S}{\partial t} = -\rho \frac{\partial \Phi}{\partial z} . \tag{9.3b}$$

The momenta S and T are defined in terms of the density and velocity components as in (5.11). It may be noted that grid velocities do not appear in the gravitational acceleration terms, or in Poisson's equation. Thus the use of grid motion does not effect how the Newtonian gravitational equations are to be solved. The first step in formulating a finite difference representation of Poisson's equation and the acceleration (9.3) is to settle on a centering for the gravitational potential.

Space-Time Centering

The Newtonian gravitational potential is found using the local mass density through (9.1). Since we wish to incorporate the results into a general hydrodynamic scheme, such as the one discussed in Chapter 5, we shall adopt the centering of fluid variables shown in Figure 5.3. It is then natural to center the potential Φ on the space time grid along with the mass density; thus

$$\Phi(r,z) \rightarrow \Phi^{n}{}_{k+1/2,\, j+1/2} . \tag{9.4}$$

The gravitational potential depends implicitly on the time through the mass density. An obvious advantage of the centering (9.4) is that the axial and radial gradients of Φ which appear in (9.3) are zone-face centered, and this coincides with the spatial centering of the momenta S and T. Furthermore, the Laplacian of the potential (9.2) will also be a zone centered quantity so that both sides of (9.1) are naturally centered.

The method of operator splitting may be used to solve (9.2) along with other physics packages (such as hydrodynamics). It is possible, for example, to solve (9.2) at the end of the hydrodynamics step. The new value of the potential may then be used in the subsequent computational cycle along with the material pressure gradient (and any other acceleration terms) to obtain changes in the velocity field and the material density. Finite difference representations of the gravitational acceleration will be considered later. First, we shall consider the problem of solving an elliptic partial differential equation such as Poisson's equation.

Poisson's Equation

Using the space-time centering adopted in (9.4) we may obtain a finite difference representation of Poisson's equation by integrating (9.2) over a zonal volume $\Delta V_{k+1/2,j+1/2}$. The computational grid will be assumed to be nonuniform. The result is easily shown to be

$$\left\{ r_{j+1} \frac{\Phi_{j+3/2} - \Phi_{j+1/2}}{\Delta r_{j+1}} - r_j \frac{\Phi_{j+1/2} - \Phi_{j-1/2}}{\Delta r_j} \right\}_{k+1/2} \Delta z_{k+1/2} +$$

$$\left\{ \frac{\Phi_{k+3/2} - \Phi_{k+1/2}}{\Delta z_{k+1}} - \frac{\Phi_{k+1/2} - \Phi_{k-1/2}}{\Delta z_k} \right\}_{j+1/2} r_{j+1/2} \Delta r_{j+1/2} -$$

$$- 2G\, \Delta m_{k+1/2,j+1/2} = 0. \tag{9.5}$$

The mass $\Delta m_{k+1/2,j+1/2}$ is known from the density, so the only unknowns appearing in (9.5) are the potentials $\Phi_{k+1/2,j+1/2}$. Thus, equation (9.5) is a five-point scheme as shown in Figure 9.1, and is formally similar the the finite difference representation of the radiation diffusion equation discussed in Chapter 6. If (9.5) is divided by $r_{j+1/2}$ we obtain an equation for the potential whose coefficients are of order $\Delta r/\Delta z$ or $\Delta z/\Delta r$. In the special case of Poisson's equation in Cartesian coordinates on a uniform spatial grid

($\Delta x = \Delta y = constant$), the coefficient of $\Phi_{k+1/2,j+1/2}$ is -4 and the coefficients of the remaining potential terms are $+1$.

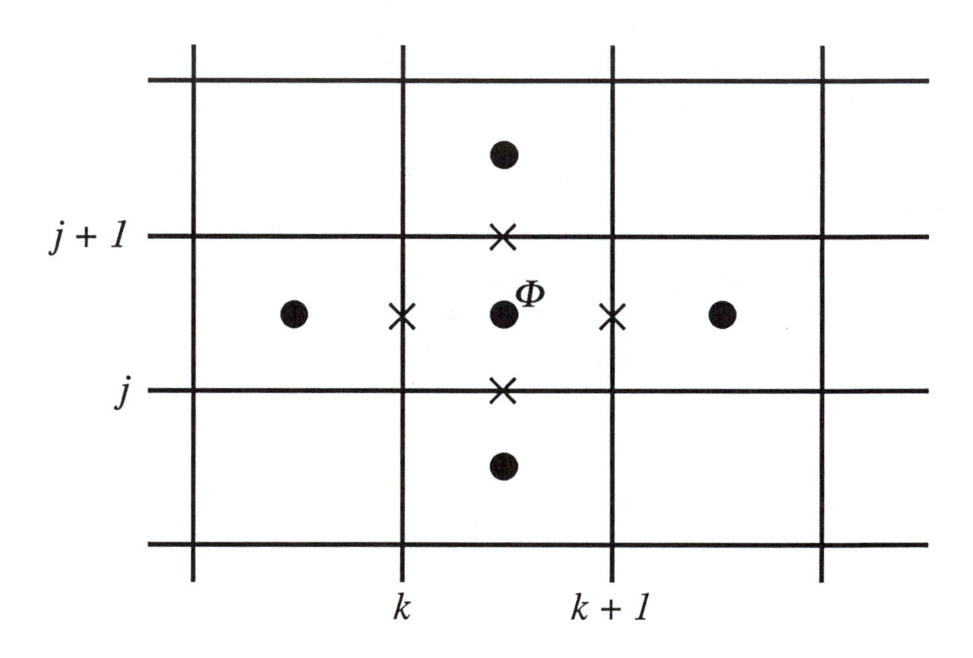

Figure 9.1. Five point scheme for a finite difference representation of Poisson's equation for the Newtonian gravitational potential.

Problem 9.1

Starting with Poisson's equation for a two dimensional system expressed in spherical coordinates (r,θ), obtain a finite difference representation analogous to (9.5).

Poisson's equation is an elliptic partial differential equation; the boundary conditions for such equations, as discussed in Section 2.4, require the specification of the function along a boundary enclosing the region of interest. An equation of the form (9.5) is defined for each zone on the grid, and the finite difference representation consists of, in effect, a large set of coupled algebraic equations which must be solved simultaneously. One method for solving coupled algebraic equations is to represent them in matrix form, and then invert the matrix.

The coupled set of finite difference equations represented by (9.5) can be rewritten in matrix form as follows. Define the vector index

$$l = (j-1)n_k + k,$$

where n_k is the number of zones in the k direction. Then the elements of the two dimensional array $\Phi_{k+1/2,j+1/2}$ can be used to form an equivalent vector whose elements can be denoted by Φ_l. Similarly, a vector ρ_l can be formed from the elements of the density array $\rho_{k+1/2,j+1/2}$. Using these two vectors, the set of finite difference equations (9.5) can be expressed in the equivalent matrix form

$$M_{lm}\Phi_m = 4\pi G \rho_m$$

where M is a matrix of coefficients which can be identified from (9.5). It has a special structure that is shown schematically in Figure 9.2. The only nonzero matrix elements lie along the diagonal, and immediately above and below it (as in a tridiagonal system), and in two additional bands displaced n_k elements horizontally to the left and to the right of the diagonal. All other elements are identically zero. This type of matrix is called a banded matrix. For a modestly resolved problem in two spatial dimensions on a grid containing $50x100$ zones, (9.5) corresponds to 5000 unknowns and the equivalent matrix of coefficients (which in principle must be inverted to solve for the potential at each point on the grid) would contain $(5x10^3)^2$ or 25 million elements. Direct inversion of matricies of this size is clearly impractical on any conceivable machine.

Problem 9.2

Assume that a computer can perform 10 million floating point operations per second, and estimate the time that it would take to invert a matrix using Cramer's rule. What is the maximum size matrix that could be inverted in one hour? How long would it take to invert an nxn matrix where $n = 25$?

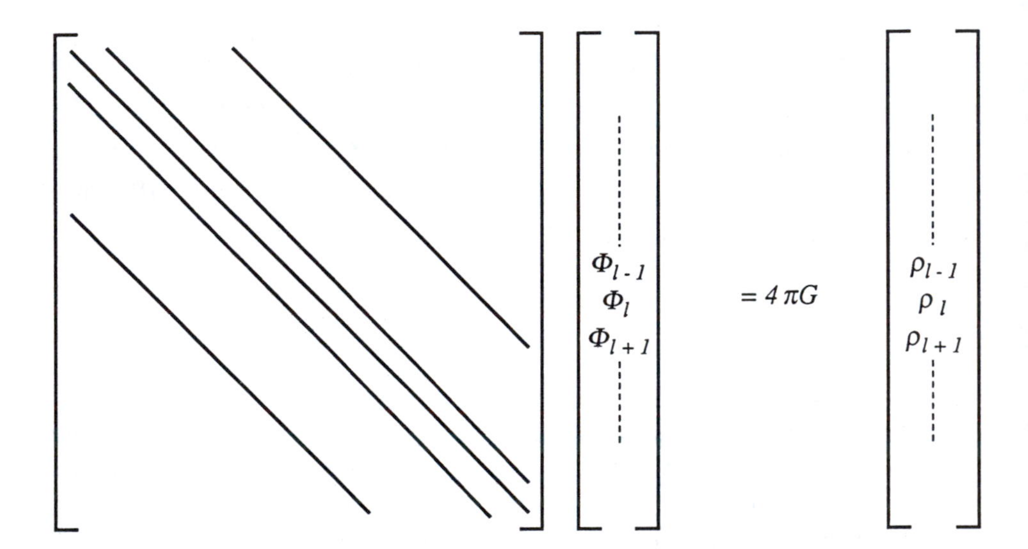

Figure 9.2. Schematic representation of the matrix of coefficients arising in the solution of Poisson's equation for the Newtonian gravitational potential.

Solving elliptic equations reduces essentially to finding practical ways of inverting gigantic matrices (see Problem 9.2). An extensive body of work exists in the literature which describes the mathematics of large scale matrix inversion suitible for use on large scale computers. An introduction to the subject, (which includes references and FORTRAN programs for several methods) can be found in Press, et. al. (Press, Flannery, Teukolsky and Vetterling 1986). In the remainder of this section we shall discuss a very simple finite difference scheme to solve Poission's equation which is both easy to implement and is relatively foolproof. Although the method is simple, it is not very fast. If speed is important, then the more advanced methods (noted below) should be considered. Many of these methods require extensive coding and logic (and often memory) to achieve both speed and accuracy.

A relatively simple algorithm which can be used to solve Poisson's equation starts with the parabolic partial differential equation

$$\frac{\partial \Phi}{\partial \tau} = \nabla^2 \Phi - 4\pi G \rho, \tag{9.6a}$$

where $\Phi = \Phi(\boldsymbol{r}, \tau)$ is a function of the spatial coordinates and a parameter, τ. If τ were the physical time, then (9.6a) would be formally identical to the thermal diffusion equation (with the addition of a sink term proportional to the mass density). Although the parameter τ does not represent physical time, it is convenient to exploit the formal similarity between (9.6a) and the thermal diffusion equation discussed in Chapter 6. Therefore, we will refer to τ as a pseudotime, and can think of (9.6a) as a pseudoevolution equation. The density appearing on the right side of (9.6a) is a function of the spatial coordinates and the physical time t. The pseudotime τ is used to evolve Φ to steady state. By anology with a time evolution equation the steady state condition corresponding to (9.6a) occurs when

$$\partial \Phi / \partial \tau = 0. \tag{9.6b}$$

In effect, we shall replace the steps necessary to invert a large matrix by a series of iterations in which the pseudoevolution of the function is carried out to steady state. In steady state (9.6a) reduces to Poisson's equation (9.1). Thus, the steady state solution of (9.6) will also be a solution of (9.1).

We may incorporate (9.6) into a computational scheme in the following way. Each computational cycle solves the hydrodynamic equations, obtaining the new density $\rho(\boldsymbol{r},t)$. This new density is then used in a finite difference representation of (9.6a) which is evolved to the steady state defined by (9.6b). The steady state solution for the potential is then used along with $\rho(\boldsymbol{r},t)$ and the other variables to start the next computational cycle. This means that a number of interations are required each cycle to solve (9.6a), and a convergence criterion is needed to determine when the solution has converged, that is satisfies (9.6b). In practice the number of iterations needed to satisfy a reasonable convergence value is not excessively large since the density field in a hydrodynamic calculation cannot change much per computational cycle.

The formal similarity between (9.6a) and the thermal diffusion equation suggests that we consider a finite difference representation similar to those discussed in Chapter 6. Thus, integrating the left side over a zonal volume gives

$$\int \frac{\partial \Phi}{\partial \tau} dV \rightarrow \frac{\Phi' - \Phi}{\Delta \tau} \Delta V, \tag{9.7}$$

where $\Delta\tau$ is a pseudotime step to be discussed below, and $\Phi' \equiv \Phi(\tau+\Delta\tau)$. Each iteration will consist of finding the change in Φ in a pseudotime $\Delta\tau$. A finite difference representation of (9.6a) in axial coordinates can be found by setting the left side equal to (9.7):

$$\left\{ r_{j+1} \frac{\Phi_{j+3/2} - \Phi_{j+1/2}}{\Delta r_{j+1}} - r_j \frac{\Phi_{j+1/2} - \Phi_{j-1/2}}{\Delta r_j} \right\}_{k+1/2} \Delta z_{k+1/2} +$$

$$\left\{ \frac{\Phi_{k+3/2} - \Phi_{k+1/2}}{\Delta z_{k+1}} - \frac{\Phi_{k+1/2} - \Phi_{k-1/2}}{\Delta z_k} \right\}_{j+1/2} r_{j+1/2} \Delta r_{j+1/2} -$$

$$- 2G\, \Delta m_{k+1/2,\,j+1/2} = \frac{\Phi'_{k+1/2,\,j+1/2} - \Phi_{k+1/2,\,j+1/2}}{\Delta \tau} \Delta V_{k+1/2,\,j+1/2} \tag{9.8}$$

The pseudotime centering of the potentials on the left side will be specified below. This centering actually represents the stage in the process of interation to steady state, and its similarity to time centering is purely formal.

Solution of Poisson's Equation

The algorithm for solving elliptic equations such as (9.8) which will be considered here is called successive overrelaxation. The approach does not require detailed knowledge of the physics described by the elliptic partial differential equation, nor does it depend on any special averaging procedures. In the interest of notational simplicity (and to focus attention on the procedure itself) we shall use as our example a finite difference representation of (9.6) in a two dimensional Cartesian coordinate system whose zones are defined by Δx and Δy, both of which are constants. The approach can be extended in a straightforward manner to other coordinate systems, or to grids with varying zone widths. The extension essentially involves the appearance of more complicated coefficients in the equations. Finally, the method is independent of the centering of the independent variable, and may be applied to quantities which are zone face or node centered. Formulating (9.6) on a grid defined by Δx and Δy, and rearranging terms gives

$$\Phi'_{k+1/2,j+1/2} - \Phi_{k+1/2,j+1/2} = \frac{\Delta\tau}{\Delta x^2}(\Phi_{k+3/2,j+1/2} - 2\,\Phi_{k+1/2,j+1/2} + \Phi_{k-1/2,j+1/2}) +$$

$$\frac{\Delta\tau}{\Delta y^2}(\Phi_{k+1/2,j+3/2} - 2\,\Phi_{k+1/2,j+1/2} + \Phi_{k+1/2,j-1/2}) - f_{k+1/2,j+1/2}\,\Delta\tau \tag{9.9}$$

where $f = 4\pi G\rho$. In the Cartesian coordinate system used here k is used for the x direction and j is used for the y direction. The size of the pseudotime step need not remain constant during the course of the iterations. Equation (9.9) may be rewritten in terms of the aspect ratio of the grid, $\alpha \equiv \Delta x/\Delta y$

$$\Phi'_{k+1/2,j+1/2} = \Phi_{k+1/2,j+1/2}[1 - \frac{2\,\Delta\tau}{\Delta x^2}(1 + \alpha^2)] +$$

$$+ \frac{\Delta\tau}{\Delta x^2} \{ \Phi_{k+3/2,j+1/2} + \Phi_{k-1/2,j+1/2} + \alpha^2(\Phi_{k+1/2,j+3/2} + \Phi_{k+1/2,j-1/2}) - f_{k+1/2,j+1/2} \Delta x^2 \} .$$

(9.10)

Although τ is not physical time, the formal analogy of (9.10) to an evolution equation suggests that not all solution to it will be numerically stable. Formally (9.10) resembles an explicit finite difference representation to which one might apply the Von Neumann stability analysis (see, for example, Roache 1976). The stability of the explicit diffusion equation in one spatial dimension was discussed in Chapter 2, where it was shown that the time step must satisfy (2.31). Extending the analysis to the pseudoevolution equation (9.10) in two spatial dimensions results in the constraint on the pseudotime step

$$\frac{\Delta\tau}{\Delta x^2}(1 + \alpha^2) = \frac{\Delta\tau}{\Delta x^2} + \frac{\Delta\tau}{\Delta y^2} \leq \frac{1}{2} .$$

(9.11)

[note that the "diffusion coefficient" in (9.10) is unity]. Thus, if the explicit representation of (9.6a) is to produce stable a solution, the pseudotime step must satisfy (9.11). Since the pseudotime step bears no relation to the physical time step used in the hydrodynamics, the maximum value of $\Delta\tau$ should be used to reduce the number of iterations. Then (9.10) reduces to the simple expression

$$\Phi'_{k+1/2,j+1.2} = \frac{1}{2(1 + \alpha^2)} \{ \Phi_{k+3/2,j+1/2} + \Phi_{k-1/2,j+1/2} + \alpha^2(\Phi_{k+1/2,j+3/2} + \Phi_{k+1/2,j-1/2}) - f_{k+1/2,j+1/2} \Delta x^2 \} .$$

(9.12)

This is known as Richardson's formula, and simply states that the value of the potential is determined, in part, by an average of the potential in four surrounding zones. On a uniform grid ($\alpha = 1$) the new value of the potential is a simple average of four surrounding values. In principle, (9.12) represents one interation of the potential to steady state.

Several improvements may be made to this approximation which are relatively easy to implement and which significantly reduce the amount of computing needed to converge (a complete discussion of several methods may be found in Roach 1976, and in Press *et. al.* 1986). If (9.12) is solved by sweeping along the x direction and then incrementing in the y direction it is possible to use new values of the potential for $\Phi_{k-1/2,j+1/2}$ and $\Phi_{k+1/2,j-1/2}$. This may be accomplished simply by replacing in the storage array for the potential its new value whenever it is computed (this procedure also saves computer storage since only one array is needed for the potential). It can be shown that this simple modification cuts in half the number of iterations required to reach steady state. Another improvement can be made as follows. Suppose that the value of the potential at the start of the ith iteration is denoted by Φ^i. Then another gain in efficiency results if at the beginning of the $(i+1)^{th}$ iteration the value of the potential is taken to be a linear combination of Φ^i and Φ':

$$\Phi^{i+1} = \omega\Phi' + (1-\omega)\Phi^i \tag{9.13}$$

where Φ' is given by

$$\Phi'_{k+1/2,j+1.2} = \frac{1}{2(1+\alpha^2)}\{\Phi^i_{k+3/2,j+1/2} + \Phi'_{k-1/2,j+1/2} + \alpha^2(\Phi^i_{k+1/2,j+3/2} + \Phi'_{k+1/2,j-1/2}) - f_{k+1/2,j+1/2}\,\Delta x^2\} \tag{9.14}$$

rather than by (9.12). The primes appearing on the right side of (9.14) denote that only one value of the potential is stored, as discussed above. The parameter ω appearing in (9.13) is called the relaxation parameter, and for (9.13) to converge it can be shown that it must satisfy the relation

$$1 \leq \omega < 2 . \tag{9.15}$$

Furthermore, it can be shown that there is an optimum value for the relaxation parameter which requires the fewest iterations for convergence, and that the rate of convergence is quite sensitive to deviations from the optimal value (see Roache 1976 for examples). Analytic expressions for the optimal value of ω have been found only for special cases. The results depend on the nature of the grid and the type of boundary conditions imposed (Dirichlet or Neumann). On a rectangular Cartesian grid containing $K-1$ uniform zones in the x direction, and $J-1$ uniform zones in the y direction, with either homogeneous Dirichlet or Neumann boundary conditions, the optimal choice for the relaxation parameter is (Press *et. al.* 1986)

$$\omega = \frac{2}{1+\sqrt{1-r_s^2}} \tag{9.16a}$$

where r_s is called the spectral radius. For the conditions stated above it is given by

$$r_s = \frac{\cos(\pi/J) + \alpha^2 \cos(\pi/K)}{1+\alpha^2} . \tag{9.16b}$$

Although the analytic results such as (9.16) apply only under the special conditions, they serve as useful initial estimates for other types of grids or boundary conditions. In practice it is usually simplest to perform a series of calculations on the grid of interest with a known mass distribution whose gravitational potential is known analytically. It is then straightforward to vary the relaxation parameter and determine the rate of convergence for each value. In the process one may also verify the accuracy of the convergence criterion. Typically the number of iterations during a calculation required to obtain a convergence of one part in 10^4, which is usually sufficient for most purposes, seldom exceeds the larger of K or J.

In practice the number of iterations required to converge using SOR depends on how good the initial guess for the potential is. Thus a good initial guess for the potential should always be sought. Nevertheless, SOR is usually efficient because it involves few arithmetic operations, and in many hydrodynamic situations the density, and hence the potential, change by only a small amount per computational cycle. Of course it is necessary to save the potential from one interation to the next in this type of method.

The method above is called successive over-relaxation (SOR), and has been used extensively for elliptical problems in gravitation and in plasma physics. Further simple modifications such as Chebyshev acceleration (Press, et. al.) can be made which further improve the rate of convergence, and are discussed in the literature.

To summarize the SOR method discussed above, a series of iterative calculations using (9.13)-(9.15) and a numerically determined value of the relaxation parameter is performed during each computational cycle, using as the initial value of the potential on the grid $\Phi^n{}_{k+1/2,j+1/2}$ where the superscript denotes the value of the potential at the start of the cycle. Equation (9.13) is used to obtain the next value in the sequence, and the process is repeated until

$$\frac{|\Phi^{i+1} - \Phi^i|}{\Phi^i} \leq \varepsilon \tag{9.17}$$

where ε is a convergence parameter. It should be stressed that ε does not give a direct estimate of the error. If Φ^{i+1} satisfies (9.17), then $\Phi^{n+1} = \Phi^{i+1}$ is taken as the value of the potential for the next computational cycle.

Problem 9.3

Obtain an equation analogous to (9.14) for Poisson's equation in a two dimensional axially symmetric coordinate system starting from (9.8).

Problem 9.4

The hydrostatic structure of gravitating spheres whose matter is described by polytropic equations of state can be found in standard texts on stellar astrophysics. Select a polytropic mass distribution and compute its gravitational potential using the results of Problem 9.3. Select a convergence criterion, and investigate the sensitivity of the solution to the relaxation parameter. Asume that the entire star is contained inside the computational grid, with the surrounding region containing "background" mass whose density is small compared with the stellar mass density near the surface.

Boundary Conditions

It was shown in Section 2.4 that Dirichelet or Neumann boundary conditions are required to complete the specification of a potential problem such as (9.1). It is natural to specify the value of the gravitational potential along the boundary of the grid. Along reflecting boundaries, the normal derivative of the potential is set to zero. Thus, if the plane $z = 0$ is a reflecting boundary in cylindrical coordinates,

$$\left.\frac{\partial \Phi}{\partial z}\right)_{z=0} = 0.$$

Thus, if $k = 2$ represents the left hand boundary of the physical grid, then the reflecting boundary condition is

$$\Phi_{1,j} = \Phi_{2,j} \qquad 1 \le j \le J.$$

In cylindrically symmetric systems the radial derivative of the potential vanishes along the symmetry axis, and the appropriate boundary condition there is

$$\Phi_{k,1} = \Phi_{k,2} \qquad 1 \le k \le K.$$

Specifying boundary conditions along the outer boundaries of the computational grid is more difficult. If the grid extends well beyond the surface of the mass distribution, then it is possible to express the potential along the grid boundary as a multipole expansion. Consider an axially symmetric system, and let x_1 represent the distance to a point outside the mass distribution, and x_2 the distance to an arbitrary point inside it where the density is $\rho(x_2)$. Then the (axisymmetric) gravitational potential at x_1 may be expressed by the multipole expansion

$$\Phi(x_1) = G \sum_{l=0}^{\infty} \frac{A_l}{x_1^{l+1}} P_l(\cos\gamma)$$

$$A_l = \int x_2^l \rho(x_2)\, P_l(\cos\alpha)\, d^3x_2$$

where P_l are the Legendre Polynomials, γ is the angle between x_1 and the symmetry axis, and α is the angle between x_2 and the symmetry axis. Notice that the variables appearing in the axisymmetric multipole expansion lie in the r-z plane with $x_i = (r_i^2+z_i^2)^{1/2}$, and are easily evaluated by integrating over the grid. The boundary values of the potential are obtained by setting x_1 equal to the position vector of each zone on the outer boundary of the grid. The multipole moments A_l are of order MR^l where M is the total mass of the system, and R is a measure of its radius. When $x_1 >> R$ only the first few moments in the series need to be retained. If $x_i \geq R$, then a sufficient number of moments must be retained such that the terms neglected are small.

Problem 9.5

Use the appropriate multipole expansion to set up boundary conditions for the gravitational potential on the outer boundary of a two dimensional grid using spherical coordinates. Write out a finite difference representation of the first two nonzero multipole moments. Assume that the entire system is to be contained within the computational grid.

Other Methods for Elliptical Equations

The mathematical theory of large scale matrix inversion is an ongoing area of mathematical research, and many improvements over the method of successive over-relaxation have been made (for a review of popular alternatives see Roache 1976, or Press, Flannery, Teukolsky and Vetterling 1986). Each of these methods reduces, in effect, the amount of time required to solve an elliptic partial differential equation. The drawback to these methods is that they often require far more complicated programming in order to achieve increased efficiency.

If increased efficiency is needed, it is probably advisable to adopt more sophisticated methods such as the incomplete Choleski conjugate gradient method, often referred to as ICCG. This approach has been discussed in detail in the literature (Meijerink and van der Vorst 1977; Kershaw 1978). Matrix inversion packages based on this method for matricies such as the one shown schematically in Figure 9.2 have been optimized and are usually available from major computation centers. Many of these assume that the matrix to be inverted is symmetric, as is usually the case for finite difference equations arising in physics. If the reader judges that SOR, or simple modifications of it, is not adequate, then existing packages should be sought and implemented. Fourier transform methods are a very efficient way to solve Poisson's equation on a uniform grid (see, for example *et. al.* 1986).

Problem 9.6

Show that the matrix M_{lm} formed for the Poisson equation in cylindrical coordinates (9.5) is symmetric, and identify the coefficients.

Newtonian Gravitational Acceleration

Once Poisson's equation has been solved for the gravitational potential, the acceleration term (9.3) appearing in the momentum equation can be constructed. Consider first the axial acceleration component (9.3b). To derive a finite difference representation of this equation we integrate the partial differential equation over a momentum volume $\Delta V_{k,j+1/2}$, obtaining (common subscript $j+1/2$ omitted)

$$\frac{1}{\Delta V_k} \int \frac{\partial S}{\partial t}\, dV = \frac{S_k^P - S_k^{n-1/2}}{\Delta t^n} = - \int \rho \frac{\partial \Phi}{\partial z} \frac{dV}{\Delta V_k}.$$

The centering of variables is shown in Figure 9.3. The new value of the axial momentum S^P is expressed at the same time as the change due to the pressure gradient [recall the notation in (5.29)]. Since the density associated with the axial component of the momentum is centered at the zone edge, it is natural to used this value (which was

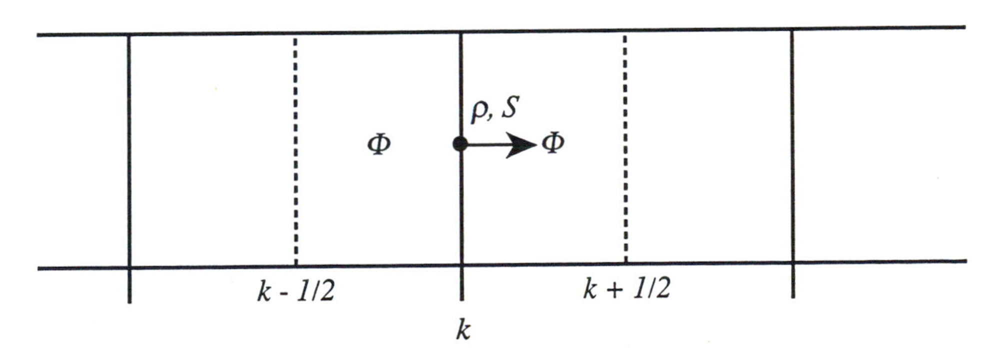

Figure 9.3. Spatial centering of the gravitational potential used to construct the axial gravitational acceleration.

defined in (5.34) in evaluating the integral on the right side. Thus (common subscript $j+1/2$ omitted)

$$\frac{1}{\Delta V_k} \int \rho \frac{\partial \Phi}{\partial z} dV = \frac{2\pi \rho_k}{\Delta V_k} \int r\, dr \left(\frac{\partial \Phi}{\partial z}\right) dz$$

$$= \rho_k \frac{\Phi_{k+1/2} - \Phi_{k-1/2}}{\Delta z_k} .$$

Combining the two equations above gives the finite difference representation for the change in axial momentum associated with Newtonian gravitation (common subscript $k+1/2$ omitted):

$$\frac{S_k^P - S_k^{n-1/2}}{\Delta t^n} = -\rho_k \frac{\Phi_{k+1/2} - \Phi_{k-1/2}}{\Delta z_k} . \tag{9.18}$$

The radial gravitational acceleration can be found in the same way (Problem 9.7), and is (common subscript $k+1/2$ omitted)

$$\frac{T_j^P - T_j^{n-1/2}}{\Delta t^n} = -\rho_j \frac{\Phi_{j+1/2} - \Phi_{j-1/2}}{\Delta r_j} . \tag{9.19}$$

Notice that the densities in (9.18) and (9.19) are centered at different points on the grid.

No additional time step controls are necessary when (9.18) and (9.19) are included with the explicit hydrodynamics developed in Chapter 5. In practice the acceleration associated with pressure gradients and gravitation should be solved in the same step. This is particularly important when the two effects are large and of opposite summation over repeated indicies is implied). The quantity $T_{\mu\nu}$ is the energy-momentum tensor.

The geometry of space-time can be described by the four-metric $g_{\mu\nu}$, and the interval ds between two nearby space-time points x^μ and $x^\mu + dx^\mu$ is given by

$$ds^2 = g_{\mu\nu} dx^\mu dx^\nu . \tag{9.20}$$

Greek indicies will range from 0 to 3, and Roman indicies, which will be used for the space-like components of tensors, will range from 1 to 3. In the remainder of this chapter we choose units such that $c = 1$ and $G = 1/8\pi$. The four velocity is defined by

$$U^\mu = dx^\mu / ds, \tag{9.21a}$$

with the normalization condition

$$U^\mu U_\mu = -1. \tag{9.21b}$$

The gravitational field equations contain ten independent, coupled partial differential equations. Four of these represent constraint equations and the remaining six are evolution equations. It will be assumed here that the *ADM 3+1* formalism is to be used to solve the gravitational field equations (see, for example, Misner, Thorne and Wheeler 1973; York 1985). Projecting out the three space geometry, whose metric is γ_{ik}, the interval between nearby space-time points (9.20) may be conveniently rewritten as

$$ds^2 = -\alpha^2 dt^2 + \gamma_{ik}(dx^i + \beta^i dt)(dx^k + \beta^k dt). \tag{9.22}$$

Here α is the lapse function, and β^k are the three components of the shift vector. The metric coefficients appearing in (9.22) are obtained from Einstein's field equations; assuming maximal slicing,

$$tr\,(K^i_{\ k}) \equiv \sum_k K^k_{\ k} = 0,$$

the field equations reduce to

$$\frac{\partial \gamma_{ik}}{\partial t} = -2\alpha K_{ik} + \beta_{i;k} + \beta_{k;i}$$

where K^i_k is the extrinsic curvature. The last two terms represent the symmetricized covariant derivative of the shift vector, and the covariant derivatives are constructed using the three space defined by the metric $\gamma_{i\,k}$ $(\gamma_{i\,k;\,j} = 0)$.

The primary purpose of this section is to develop a finite difference representation of the relativistic hydrodynamic equations in Eulerian form. The fundamental principles of our approach can be illustrated in one spatial dimension. The extention to describe relativistic hydrodynamics in more than one spatial dimension is then relatively straightforward, while the calculation of the gravitational potentials, $g_{\mu\nu}$, is very complicated. Consequently we assume that the space-time under consideration allows variations in only one spatial dimension. We further assume that it can be described in spherical isotropic coordinates. Other coordinates are also possible (see, for example, Wilson 1979). The interval (9.22) in spherical isotropic coordinates reduces to

$$ds^2 = (-\alpha^2 + A^2\beta^2)dt^2 + 2A^2\beta\, drdt$$

$$+ A^2dr^2 + A^2r^2(d\theta^2 + \sin^2\theta\, d\phi^2)\,. \tag{9.23}$$

where $\beta \equiv \beta^r$ is the radial contravariant component of the shift vector, and $\beta = (\beta,0,0)$. The gravitational field equations in one dimensional spherical isotropic coordinates may be expressed in the following form. Assuming maximal slicing, the lapse function is obtained from the equation

$$\frac{1}{r^2}\frac{\partial}{\partial r}\left(r^2 A \frac{\partial \alpha}{\partial r}\right) = \alpha A^3\left[\rho(1+\varepsilon)(W^2 - 1/2) + P(W^2 + 1/2) + K^k_i K^i_k\right]$$

$$\equiv \frac{\alpha}{2} A^3 \rho_\alpha .$$

(9.24)

The extrinsic curvature tensor is K_{ik}, and $W \equiv \alpha U^t$. For our choice of coordinates the contravariant component of the shift vector is given in terms of the lapse function and the extrinsic curvature by

$$\beta = -\frac{3r}{2}\int_r^\infty \alpha K^r_r \frac{dr}{r} .$$

(9.25)

With the choice (9.25) the three space metric is isotropic and

$$g_{rr} = \frac{1}{r^2} g_{\theta\theta} = \frac{1}{r^2 \sin^2\theta} g_{\phi\phi} .$$

(9.26)

The momentum constraint equation may be written in the form

$$K^r_r = \frac{1}{r^3 A^3}\int_0^r S_r r^3 A^3 dr ,$$

(9.27a)

where the fluid variable S_r is defined below. The remaining components of the extrinsic curvature satisfy

$$K^\theta_\theta = K^\phi_\phi = -\frac{1}{2} K^r_r$$

(9.27b)

$$K^i_k = 0, \quad i \neq k,$$

(9.27c)

$$K^2 \equiv K^i_k K^k_i = \frac{3}{2}(K^r_r)^2. \tag{9.27d}$$

Finally, the Hamiltonian constraint equation is

$$\frac{1}{r^2}\frac{\partial}{\partial r}\left(r^2 \frac{\partial A^{1/2}}{\partial r}\right) = -A^{5/2}\left[\rho(1+\varepsilon)W^2 + P(W^2-1) + \frac{1}{2}K^2\right]$$

$$\equiv -A^{5/2}\rho_H . \tag{9.28}$$

Because three space has been assumed to contain variations in only one spatial dimension, the field equations above do not allow for gravitational radiation, and thus the partial differential equations for the metric coefficients α and A are elliptic in form. While this places a severe constraint on the gravitational field (and results in a set of field equations which are particularly easy to solve numerically), it does allow enough freedom to develop the hydrodynamic equations in a form which can be easily extended to higher dimensions. In particular, the Eulerian finite difference representation of the fluid equations in this simple one dimensional geometry illustrate essentially all of the problems which are encountered in treating relativistic Eulerian systems in two or more spatial dimensions.

Boundary Conditions

Boundary conditions for the field equations must be imposed at the origin of coordinates and at large distances from the mass distribution. In spherical isotropic coordinates the boundary conditions on the lapse function are

$$\lim_{r \to \infty} \alpha = 1,$$

$$\lim_{r \to 0} \left(\frac{\partial \alpha}{\partial r}\right) = 0.$$

At the origin, the Hamiltonian constraint equation satisfies the boundary condition

$$\lim_{r \to 0} \left(\frac{\partial A^{1/2}}{\partial r}\right) = 0.$$

Asymptotically, the gravitational mass of the star is $M = r[A^{1/2}(r) - 1]$. The second boundary condition for A is imposed at the grid boundary, and takes the form

$$\frac{\partial A^{1/2}}{\partial r} + \frac{(A^{1/2} - 1)}{r} = 0.$$

Relativistic Hydrodynamic Equations

The relativistic hydrodynamic equations follow from Einstein's equations, and can be expressed in a number of equivalent forms. We select one which most closely represents the forms used to discuss nonrelativistic hydrodynamics. There are two reasons for this choice. First, the relativistic form must limit to the Newtonian form. Second, by so doing we may use our experience with finite difference representations in the Newtonian regime as a guide in developing the relativistic theory.

The primary fluid variables are the analogue of the density, the specific internal energy, the radial momentum and the radial velocity, and are defined as follows:

$$\begin{aligned}
W &= \alpha U^t \\
D &= \rho W \\
E &= \rho \varepsilon W \\
S_r &= (D + E + PW)U_r \\
V^r &= U^r / U^t
\end{aligned} \tag{9.29}$$

where W is the general relativistic analogue of the special relativistic gamma factor. In the fluid's rest frame the mass density is ρ, the specific internal energy density is ε, and P is the fluid pressure. The total mass-energy density is given by $\rho(1 + \varepsilon)$, and $\rho(1 + \varepsilon) + P$ is the enthalpy. Finally, we note that the adiabatic index Γ is given by $\Gamma = 1 + PW/E$. Henceforth we assume that Γ varies slowly enough that it may be treated as a constant in each zone during the computational cycle. The energy-momentum tensor is assumed to be of the form $T_{\mu\nu} = Pg_{\mu\nu} + [\rho(1 + \varepsilon) + P]U_{\mu}U_{\nu}$. As in the Newtonian regime an equation of state such as (3.1) is assumed. If temperature dependent physics (such as radiation transport) is to be included, then (3.1) may be supplemented by (3.7) and (3.14). We shall not discuss these issues here, but note that the formulation of the hydrodynamics in no way precludes their incorportion.

The fluid equations of motion follow from the covariant derivative of the energy-momentum stress tensor

$$T^{\mu\nu}{}_{;\nu} = 0 .$$

In terms of isotropic spherical coordinates and the fluid variables defined above, we have the mass continuity equation

$$\frac{\partial (Dg)}{\partial t} + \frac{\partial (DgV^r)}{\partial r} = 0; \tag{9.30}$$

the internal energy equation,

$$\frac{\partial (Eg)}{\partial t} + \frac{\partial (EgV^r)}{\partial r} = -P\left[\frac{\partial (\alpha gU^r)}{\partial r} + \frac{\partial (gW)}{\partial t}\right]; \tag{9.31}$$

and the momentum equation

$$\frac{1}{\alpha g}\left[\frac{\partial (S_r g)}{\partial t} + \frac{\partial (S_r gV^r)}{\partial r}\right] = -\frac{\partial P}{\partial r} + a_r \tag{9.32a}$$

where the effective gravitational acceleration is given by

$$a_r = \frac{1}{2}\frac{S_\mu S_\nu}{S^t}\frac{\partial g^{\mu\nu}}{\partial r}.$$

(9.32b)

In isotropic spherical coordinates the acceleration has only a radial component which can be reduced to the form

$$a_r \equiv -\frac{\sigma}{U^t}\Big[\frac{U_r^2}{2}\frac{\partial \gamma^{rr}}{\partial r} - U^t U_r \frac{\partial \beta^r}{\partial r} + WU^t \frac{\partial \alpha}{\partial r}\Big].$$

(9.32c)

In the equations above $g \equiv r^2 A^3$. Finally, the inertial density σ is related to the four velocity and the four momentum by

$$\sigma \equiv S_r/U_r = D + \Gamma E.$$

(9.32d)

The nonzero components of the metric tensor $g_{\mu\nu}$ follow immediately from (9.23), as do the covariant components of γ_{ik}. The nonzero contravariant components of the three space metric γ^{ik} are given by

$$\gamma^{rr} = \frac{1}{A^2} - \frac{\beta^2}{\alpha^2} \qquad \gamma^{\theta\theta} = \frac{1}{A^2 r^2}$$

$$\gamma^{\phi\phi} = \frac{1}{A^2 r^2 \sin^2\theta}.$$

(9.33)

The metric components above and the velocity normalization constraint may be used to express the special relativistic gamma factor in terms of the three space metric and the covariant component of the the four velocity:

$$W = \sqrt{1 + \gamma^{ik} U_i U_k} = \sqrt{1 + \gamma^{rr} U_r^2}\,. \tag{9.34}$$

In similar fashion, the contravariant component of the four velocity may be written as

$$U^r = U_r / A^2 - \beta W / \alpha, \tag{9.35}$$

and the contravariant component of the fluid velocity as

$$V^r = U^r / U^t = \alpha U^r / W. \tag{9.36}$$

It may be recalled that in Newtonian mechanics the velocity characterizing fluid transport, and the specific momentum are the same. In the relativistic regime, however, the specific momentum is given by U^μ and V^μ is the velocity which describes fluid transport. This observation will be useful when momentum advection is discussed below.

In the nonrelativistic approach to hydrodynamics developed in Chapter 5 the internal energy equation was chosen in place of the total energy equation for the fluid. The hydrodynamic equations above give the change in fluid variables (9.29) with coordinate time t, and the spatial derivatives are with respect to the coordinate variable r.

Problem 9.8

Show that the hydrodynamic equations (9.30)-(9.32) reduce in the nonrelativistic limit to the equations in spherical coordinates, where the nonrelativistic limit is taken for A and α.

Problem 9.9

Extend the hydrodynamic equations (9.30)-(9.32) to include grid motion in the radial direction. Use the nonrelativistic equations (as developed in Chapter 5, or Appendix A) as a guide.

The hydrodynamic equations contain, in addition to the fluid variables, the geometric (or metric) quantities $a, A, K^r{}_r$ and β which are specified by the field equations (9.24)-(9.28). In principle the complete, coupled set of equations must be solved. In the numerical approach described below operator splitting will be used to separate the gravitational field equations from the hydrodynamic equations in much the same manner as the solution of Poisson's equation was split off from the Newtonian hydrodynamics (it may be recalled that this step was not necessary for one dimensional Newtonian problems, because there the solution for the gravitational potential was obtained analytically). Thus, we assume that all of the geometric quantities are known at time t from the solution of the gravitational field equations, and proceed to solve for the new values of the fluid variables. This approach is also applicable to problems involving two or more spatial dimensions.

A scheme for solving the nonrelativistic Eulerian hydrodynamic equations has been discussed in Chapter 5, and we wish to adopt as much of that approach as may be applicable in the relativistic regime. One expects the relativistic differential equations to reduce to the Newtonian equations in the limit of small velocities and weak gravitational fields. A natural starting point is to require that in the nonrelativistic limit our relativistic formulation reduce to the results of Chapter 5. This suggests that a staggered space, staggered time approach be used here as well. A true space-time centering for the relativistic equations is more difficult in part because of the factor W appearing in (9.29). Nevertheless, we shall follow as closely as possible the approach of Chapter 5. As a first step equations (9.30)-(9.32) will be rewritten in the following forms.

Mass Continuity

$$\frac{\partial D}{\partial t} = -\frac{1}{r^2 f}\frac{\partial}{\partial r}(r^2 V^r D f) - \frac{D}{f}\frac{\partial f}{\partial t} \tag{9.37a}$$

Internal Energy

$$\frac{\partial E}{\partial t} = - \frac{1}{r^2 f}\frac{\partial}{\partial r}(r^2 V^r E f) - \frac{P}{r^2 f}\frac{\partial}{\partial r}(r^2 \alpha f U^r) - \frac{P}{r^2 f}\frac{\partial}{\partial t}(r^2 f W) - \frac{E}{f}\frac{\partial f}{\partial t}$$

(9.37b)

Momentum Equation

$$\frac{\partial S_r}{\partial t} = - \frac{1}{r^2 f}\frac{\partial}{\partial r}(S_r r^2 V^r f) - \alpha\frac{\partial P}{\partial r} - \alpha a_r - \frac{S_r}{f}\frac{\partial f}{\partial t}$$

(9.37c)

The geometric factor f appearing in these equations is defined by

$$f \equiv A^3.$$

(9.38)

The effective gravitational acceleration in the third term on the right of (9.37c) is given by (9.32c). Most of the terms appearing in the relativistic hydrodynamic equations above have obvious analogues in Newtonian theory. Several terms also appear which do not have a nonrelativistic analogue. For example, the last term on the right side of (9.37a) represents the change in the mass density associated with the change in volume described by the relativistic factor A. Similar terms also appear in the internal energy and momentum equations.

Operator Splitting

Before considering the finite difference representations of the relativistic fluid equations, we shall discuss operator splitting and the order in which the terms in (9.37) are to be solved [the discussion following (5.16) for nonrelativistic Eulerian hydrodynamics should be compared]. For simplicity we shall omit for now the artificial viscous stress. As for nonrelativistic motion the hydrodynamic step involves a Lagrange-like step followed by an advection step. The

Lagrange-like step begins by solving for the change in momentum due to non-viscous accelerations:

$$\frac{\partial S_r}{\partial t} = - \alpha \frac{\partial P}{\partial r} - \alpha a_r. \tag{9.39}$$

This should be compared with (5.22a) for nonrelativistic fluid motion. Equation (9.39) gives the change in the covariant component of the momentum, S_r, which is related to the four velocity by (9.29)

$$S_r = (D + E + PW)U_r = \sigma U_r. \tag{9.40}$$

Here σ is the inertial mass density. The covariant component of the four velocity is obtained in the usual way:

$$U_r = g_{r\mu} U^{\mu}. \tag{9.41}$$

Equations (9.40)-(9.41) and the velocity normalization constraint may be used to solve for W [see (9.34)]. The contravariant component of the transport velocity is then given by

$$V^r = \alpha U^r / W. \tag{9.42}$$

Next, we solve for the change in internal energy associated with the change in W, which is described by the equation

$$\frac{\partial E}{\partial t} + P \frac{\partial W}{\partial t} = 0. \tag{9.43a}$$

We then construct the mechanical work, which is described by

$$\frac{\partial E}{\partial t} = -\frac{P}{r^2 f}\frac{\partial}{\partial r}(r^2 f \alpha U^r) - \frac{E}{f}\frac{\partial f}{\partial t}\ .$$

(9.43b)

The geometrical quantity f is assumed to be known from the previous cycle. It will be assumed that the fluid equation of state can be approximated by the gamma law

$$PW = (\gamma - 1)E$$

which is identical to (3.8). As in Newtonian theory this simplification is convenient, but not essential.

The solution of (9.39)-(9.43) concludes the first step in the solution relativistic hydrodynamics. The next step solves for the change in variables associated with material motion, and is analogous to the advection step discussed in Chapter 5. First, the mass continuity equation

$$\frac{\partial D}{\partial t} = -\frac{1}{r^2 f}\frac{\partial}{\partial r}(r^2 V^r f D)$$

(9.44)

is solved, as is the term describing the advection of specific internal energy:

$$\frac{\partial E}{\partial t} = -\frac{1}{r^2 f}\frac{\partial}{\partial r}(r^2 V^r f E)\ .$$

(9.45)

Finally, the term describing momentum advection

$$\frac{\partial S_r}{\partial t} = -\frac{1}{r^2 f}\frac{\partial}{\partial r}(r^2 V^r S_r f)$$

(9.46)

is solved. The Lagrange-like and advection steps above use the metric components which were obtained from the solution of the gravitational field equations at the end of the previous cycle.

All of the hydrodynamic terms appearing in (9.37a) and (9.37c) have been accounted for except those which are proportional to $f^{-1}\partial f / \partial t$. For example, the total change in the mass density contains a contribution given by

$$\frac{\partial D}{\partial t} = -\frac{D}{f}\frac{\partial f}{\partial t} \, . \tag{9.47}$$

This has the immediate solution

$$f = f_o(D_o/D) \tag{9.48}$$

where D_o represents the density at the end of the hydrodynamic step. Changes in the four momentum are similiarly constructed. In order to carry out this scaling the values of $f_o = A_o^3$ used throughout the hydrodynamic step are saved until after the Hamiltonian constraint equation has been solved to obtain the new value A. The scaling (9.48) may then be completed.

The relativistic hydrodynamic cycle is completed by solving the gravitational field equations for the metric functions α and A. Using the latest fluid variables, the extrinsic curvature (9.27a) is constructed, and then the momentum constraint equation (9.24) is solved. New values of the curvature and lapse are used then to solve for the shift vector (9.25). Finally, the Hamiltonian constraint equation is solved. This completes the computational cycle.

Computational Grid

The relativistic hydrodynamic equations can be solved on a coordinate grid similar to the one used in Chapter 5. In the one spatial dimension corresponding to (9.23) the space-time coordinates will be denoted by r_j (the radial coordinate of the lower edge of a zone) and the coordinate time t^n. As in Chapter 5, we define the radial coordinates of the center of a zone by

$$r_{j+1/2} = (1/2)(r_{j+1} + r_j) \tag{9.49a}$$

and the differences in radial coordinates

$$\Delta r_{j+1/2} = r_{j+1} - r_j \tag{9.49b}$$

$$\Delta r_j = (1/2)(\Delta r_{j+1/2} + \Delta r_{j-1/2})\,. \tag{9.49c}$$

The coordinates r_j may now be used to construct the coordinate volume elements

$$\Delta V_{j+1/2} \equiv (4\pi/3)\,[(r_{j+1})^3 - (r_j)^3]. \tag{9.49d}$$

Finally, we may consider coordinate volume elements ΔV_j associated with zone-edge centered variables

$$\Delta V_j \equiv (1/2)\,(\Delta V_{j+1/2} + \Delta V_{j-1/2}). \tag{9.49e}$$

In the finite difference representations of the nonrelativistic hydrodynamic equations special care was taken to distinguish the two time levels t^n and $t^{n+1/2}$, and the corresponding time steps connecting them. It may be recalled that the appearance of two time levels was required for numerical stability of the finite difference scheme. This requirement is, in fact, satisfied in practice by the order of solution of the equations described by (9.39)-(9.46). In the discussions below we shall make the simplifying assumption of a single time step, denoted by

$$\Delta t^n = t^{n+1} - t^n \tag{9.49f}$$

which will be used to advance all fluid variables. This is not a bad approximation in practice as long as the time step does not undergo large change from one computational cycle to the next.

It is possible to include coordinate grid motion in the relativistic hydrodynamic equations (see Problem 9.9), in which case a set of coordinate grid velocities must also be defined. This may be done as in the nonrelativistic formulation. It should be emphasized that the grid velocities are arbitrary, externally imposed variables whose sole purpose is to move the coordinate grid in a convenient manner.

Finally, it is to be emphasized that the coordinate grid does not represent physical distances or physical time. The association of these quantities with grid variables requires the use of the geometric quantites (the metric tensor) obtained from the gravitational field equations. The spatial centering of the metric variables and the fluid variables will be considered next.

Space-Time Centering

The coordinate grid centering of the dependent variables appearing in the relativistic hydrodynamic equations will be chosen to reduce as closely as possible to the centering used in the nonrelativistic regime. The following choices will guarantee this correspondence:

$$D \rightarrow D^n{}_{j+1/2} \qquad E \rightarrow E^n{}_{j+1/2} \qquad P \rightarrow P^n{}_{j+1/2} \tag{9.50}$$

$$V^r \rightarrow (V^r)^{n+1/2}{}_j \qquad S_r \rightarrow (S_r)^{n+1/2}{}_j . \tag{9.51}$$

The centering of the gamma factor W is more difficult since it involves the four momentum S_r and and the densities D and E. A similar but less troublesome ambiguity arises in nonrelativistic hydrodynamics in that S, ρ, and v involve quantities centered at different space time positions. In fact we shall need both zone centered and face centered values of W. Finally, the centering of the three space component of the four velocity is centered with the velocity

$$U_r \rightarrow (U_r)^{n+1/2}{}_j . \tag{9.52}$$

The choice of centering in (9.50)-(9.51) evidently reduces to the centering of ρ, ε, P, S, and v in the nonrelativistic limit. Since D and E

contain U^t, the latter should also be zone centered:

$$U^t \rightarrow (U^t)_j . \tag{9.53}$$

The special relativistic gamma factor W can be defined in terms of the lapse and U^t through (9.29), or in terms of the spatial components of the four velocity using (9.34). Clearly values of W will be needed at zone centers and at zone edges. We have chosen to use (9.34) as the primary definition of W, so that

$$W \rightarrow (W)_j . \tag{9.54a}$$

Zone centered values are then given by

$$W_{j+1/2} \equiv (1/2)\,(W_{j+1} + W_j) . \tag{9.54b}$$

Figure 9.4 shows the spatial centering of these variables.

Next, consider the metric variables contained in the gravitational field equations. Their centering follows from Einstein's equations and the centering of the fluid variables. The lapse function α is described by an elliptic equation which contains as its source terms the fluid variables ρ, ε, and P. This is analogous to Poisson's equation in Newtonian theory with the lapse function playing the role of a potential. Thus, if we choose

$$\alpha \rightarrow \alpha_{j+1/2} \tag{9.53}$$

then the second derivative of the lapse funtion will be centered with its source terms. Examination of the Hamiltonian constraint equation and similar arguments suggest that A also be centered with density:

$$A \rightarrow A_{j+1/2} . \tag{9.54}$$

The radial component of the four momentum is related to spatial derivatives of the extrinsic curvature tensor. This suggests that the latter be zone centered:

$$K^r{}_r \rightarrow (K^r{}_r)_{j+1/2}\,. \tag{9.55}$$

Finally, because the shift vector is related to spatial derivatives of the extrinsic curvature tensor, it is natural to center it at zone edges:

$$\beta^r \rightarrow (\beta^r)_j\,. \tag{9.56}$$

The spatial centering of the geometric variables is shown in Figure 9.4. The coordinate time centering of D, E, P, S_r, U_r, and V^r should be the same as in Newtonian theory. Coordinate time centering of the

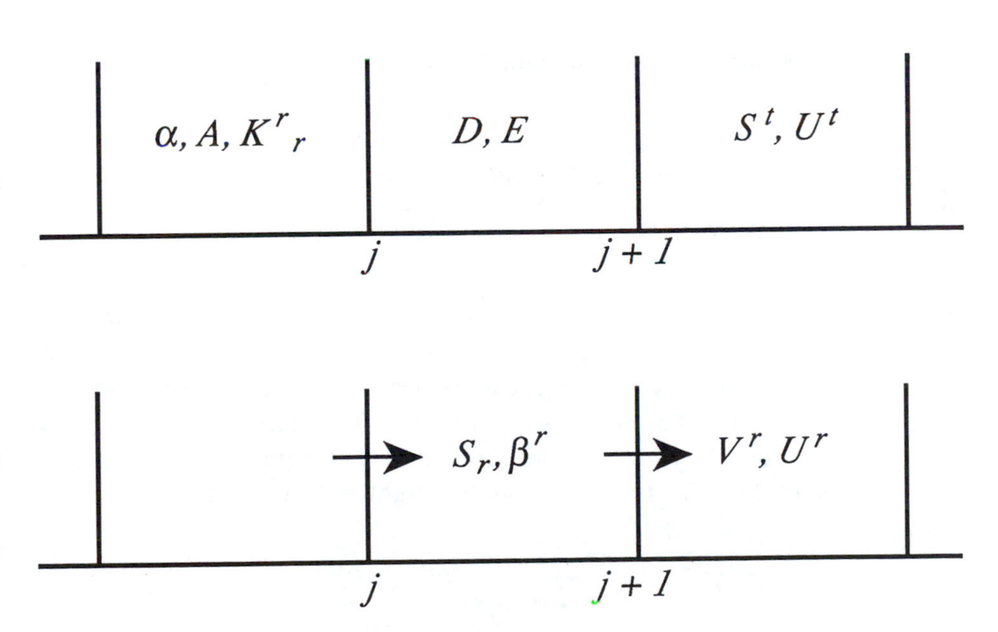

Figure 9.4. Spatial centering of the geometric variables for an Eulerian representation of the one dimensional relativistic hydrodynamic equations.

geometric variables is more difficult. W is not well time centered because it involves the four momentum $(S_r)^{n-1/2}$ and and the densities D^n and E^n. Similar observations show that the lapse function and A have no obvious time centering. In fact, it is difficult to have a true time centered leap-frog relativistic code. A solution to this dilemma would be to adopt a single time centering scheme using, for example, predictor-corrector methods.

Lagrangian Step

The Lagrangian-like step begins with the solution of the acceleration due to the pressure gradient and gravity. Except for the presence of the lapse function α, the acceleration (9.39) may be solved as in Newtonian theory. Integrating (9.39) over the momentum zone ΔV_j, and denoting by S the radial covariant component of the four momentum S_r ,

$$\int \frac{\partial S}{\partial t} dV = -\int \alpha \frac{\partial P}{\partial r} dV - \int \alpha a_r dV. \tag{9.57}$$

The left side is (with $\Delta t = \Delta t^n$ for simplicity)

$$\int \frac{\partial S}{\partial t} dV = \Delta V_j \frac{S_j^P - S_j^{n-1/2}}{\Delta t}. \tag{9.58}$$

The notation used on the right side is the same as used in the discussion below (5.22c). Thus S^P represents the change in the four momentum associated with the pressure gradient. Consider first the acceleration due to the pressure gradient. The lapse function is a zone centered quantity, as is the pressure. Integrating by parts as was done in (5.25), the first term on the right side of (9.57) is

$$\frac{S_j^P - S_j^{n-1/2}}{\Delta t} = -4\pi \alpha_j r_j^2 \frac{(P^n_{j+1/2} - P^n_{j-1/2})}{\Delta V_j}. \tag{9.59}$$

Since α is a zone centered variable, a prescription is needed to define it at the zone edge in (9.59). Noting that the lapse function changes smoothly from zone to zone, a practical choice is

$$\alpha_j \equiv (1/2)(\alpha_{j+1/2} + \alpha_{j-1/2}), \tag{9.60}$$

where $\alpha_{j+1/2}$ is the lapse function at the beginning of the computational cycle. This value will be used throughout the hydrodynamic step.

Problem 9.10

Carry out the derivation of (9.59). How would you relate ΔV_j to $\Delta V_{j+1/2}$?

The gravitational acceleration a_r, which is given by (9.32c), contains zone centered variables as well as edge centered variables. The inertial mass density (9.40) involves E, D, and P, but relates the four momentum to the four velocity which are edge centered. Thus, as in the nonrelativistic theory, a zone centered and edge centered density is needed. The primary variable will be chosen to be the zone centered density

$$\sigma_{j+1/2} \equiv (D + E + PW)_{j+1/2} = (D + \Gamma E)_{j+1/2} \tag{9.61}$$

An edge centered inertial mass density for use in the acceleration can then be constructed in several ways (see Problem 9.11). A simple

$$\sigma_j \equiv (1/2)(\sigma_{j+1/2} + \sigma_{j-1/2}). \tag{9.62}$$

Problem 9.11

Construct an expression for σ_j which reduces in the Newtonian limit to the one dimensional analogue of (5.34). How does your result compare with (9.62)?

The first term in the acceleration (9.32c) can be rewritten using the last expression in (9.29) as

$$\alpha a_1 = -\tfrac{1}{2}\sigma U_r V^t \frac{\partial \gamma^{rr}}{\partial r}.$$

The metric component γ^{rr} given in (9.33) involves A, α and β. If we define a zone centered shift vector by

$$\beta_{j+1/2} \equiv (1/2)(\beta_{j+1} + \beta_j) \tag{9.63}$$

then the first term in the rate of change in momentum associated with gravitation can be expressed as

$$\alpha_j a_{1j} = -\tfrac{1}{2}\alpha_j \sigma_j (U_r)_j V_j^r \frac{\gamma^{rr}_{j+1/2} - \gamma^{rr}_{j-1/2}}{\Delta r_j}, \tag{9.64}$$

where

$$\gamma^{rr}_{j+1/2} \equiv \frac{1}{A^2_{j+1/2}} - \frac{\beta^2_{j+1/2}}{\alpha^2_{j+1/2}}. \tag{9.65}$$

The second term in (9.32c) for the gravitational acceleration is then given by

$$\alpha_j a_{2j} = \alpha_j \sigma_j (U_r)_j \frac{\beta_{j+1/2} - \beta_{j-1/2}}{\Delta r_j}. \tag{9.66}$$

Finally, the last term in the acceleration is given by

$$\alpha_j a_{3\,j} = -\,\alpha_j\,\sigma_j W_j \;\frac{\alpha_{j+1/2} - \alpha_{j-1/2}}{\Delta r_j}.$$

(9.67)

The total change in the four momentum S_r associated with gravitation is then given by the sum of the three terms above:

$$\frac{S_j^G - S_j^P}{\Delta t} = -\,\alpha_j (a_{1j} + a_{2j} + a_{3j}).$$

(9.68)

The formation and propagation of relativistic shock waves can be modeled by introducing an artificial viscous stress in analogy with the Newtonian approach. Artificial viscous acceleration and shock heating would then be included at this point in the Lagrange step. These details will be considered later after completing the discussion of the inviscid equations.

The changes in the four momentum given by (9.59) and (9.68) imply a change in U_r, V^r and W. Since these quantities appear in the mechanical work and advection steps below, new values need to be constructed from the new four momentum. Using (9.40) the specific momentum corresponding to the new four momentum $S_r{}^G$ is given by

$$U_{rj}^G = \frac{S_{rj}^G}{\sigma_j}.$$

(9.69)

This step is analogous to (5.36) in the nonrelativistic, cylindrically symmetric case. In fact, the nonrelativistic limit of (9.69) is just the ratio of the radial momentum to the mass density. The contravariant component of the four velocity will also be needed. It may be constructed most easily from

$$U_r = g_{r\mu} U^\mu = g_{rr} U^r + g_{rt} U^t$$

and is

$$U_j^{r,G} = \frac{U_{r,j}^G}{A_j^2} - \frac{\beta_j W_j}{\alpha_j}.$$

(9.70)

An edge centered expression for the metric factor A_j is needed; since it is a smoothly varying function of position, it may be constructed in analogy with (9.62):

$$A_j = (1/2)\,(A_{j+1/2} + A_{j-1/2})\,.$$

(9.71)

Next, we use (9.69) and (9.65) to find the new values of W associated with S_r. These follow from the four velocity normalization constraint, and yield

$$W_j' = [1 + (\frac{1}{A_j^2} - \frac{\beta_j^2}{\alpha_j^2})\,(U_{r,j}^G)^2\,]^{1/2}.$$

(9.72)

The prime on W_j indicates that (9.72) represents the total change this variable during the current computational cycle. Therefore at this stage (9.54b) is used to construct new zone centered values of W. Finally, (9.70) and (9.72) may be used to construct new values of the transport velocity

$$V_j^{r,G} = \frac{\alpha_j U_j^{r,G}}{W_j'}.$$

(9.73)

The steps leading to (9.70)-(9.73) are, in part, a consequence of our use of curvilinear (in this case spherical) coordinates.

At this point the change in internal energy described by (9.43a) can be found. For the case of a gamma law equation of state, this is trivially accomplished by noting that its solution is then $EW^{\Gamma-1}$, which leads to the finite difference representation

$$E^{W}_{j+1/2} = E^{n}_{j+1/2} (W_{j+1/2} / W'_{j+1/2})^{\Gamma-1}. \tag{9.74}$$

This step can be carried out with (9.72) so that W and W' need not both be saved for over the entire grid (this is not terribly important in one spatial dimension, but can be in higher dimensions).

The mechanical heating described by (9.43b) is relatively easy to solve for the gamma law equation of state (recall the analogy in the Newtonian limit), where it reduces to the expression

$$\frac{1}{E}\frac{\partial E}{\partial t} = -\frac{\Gamma-1}{W}\frac{1}{r^2 A^3}\frac{\partial(\alpha r^2 A^3 U^r)}{\partial r}. \tag{9.75}$$

A finite difference representation of this term may be found by integrating both sides over the proper three volume

$$\Delta V^{(3)}{}_{j+1/2} \equiv A_{j+1/2}{}^3 \Delta V_{j+1/2}$$

assuming that E is a spatial constant over the volume element,

$$A^{3}_{j+1/2}\,\Delta V_{j+1/2}\left(\frac{1}{E}\frac{\partial E}{\partial t}\right)_{j+1/2} = -\frac{\Gamma-1}{W'_{j+1/2}}\int\frac{1}{r^2}\frac{\partial(\alpha r^2 A^3 U^r)}{\partial r}\,dV.$$

Carrying out the coordinate time integration over the interval Δt gives

$$E^{C}_{j+1/2} = E^{n}_{j+1/2}\exp\{-(\Gamma-1)\xi\}$$

$$\xi = \frac{\Delta t}{W'_{j+1/2}}\frac{[(\alpha r^2 A^3 U^{r,G})_{j+1} - (\alpha r^2 A^3 U^{r,G})_j]}{A^{3}_{j+1/2}\,\Delta V_{j+1/2}}. \tag{9.76}$$

This completes the relativistic Lagrangian-like step.

Problem 9.12

The adiabatic change in the internal energy is proportional to the fractional change in volume. Show that ξ is, in fact, proportional to the fraction change in the proper three volume $\Delta V^{(3)}{}_{j+1/2}$.

Relativistic Advection

The relativistic mass continuity equation can be solved using arguments similar to those applied to Newtonian mass advection. A second order monotonic interpolation scheme is needed to describe the (inertial) mass flux crossing zone boundaries. In the relativistic regime where the transport velocities are not necessarily small compared with the speed of light, a simple donor cell approach can give very bad results, and should never be used. The first step in constructing the inertial mass flux is to solve (9.44).

Integrating (9.44) over the proper three volume $\Delta V^{(3)}{}_{j+1/2}$, we note that the change in $D_{j+1/2}\, \Delta V^{(3)}{}_{j+1/2}$ must be due to the net flux of D across the proper zone boundary. Carrying out the integration, the change over a coordinate time interval Δt can then be shown to be

$$\frac{D^{A}_{j+1/2} - D^{n}_{j+1/2}}{\Delta t} = -\frac{[(r^2 A^3 \bar{D}\, V^{r,G})_{j+1} - (r^2 A^3 \bar{D}\, V^{r,G})_j]}{A^3_{j+1/2}\, \Delta V_{j+1/2}}, \tag{9.77}$$

where the density at the zone edge is to be defined by a second order expression analogous with (5.49):

$$\bar{D}_j = \frac{1}{2} D_{j-1/2} \frac{(\Delta r_{j+1/2} + V^r_j\, \Delta t)}{\Delta r_j} + \frac{1}{2} D_{j+1/2} \frac{(\Delta r_{j-1/2} - V^r_j\, \Delta t)}{\Delta r_j}. \tag{9.78}$$

The interpolation may also be made monotonic following the procedure shown in Figure 5.9, and the accompanying discussion. As in the Newtonian theory, (9.78) and the transport velocity may be used to construct the flux

$$F_j^D \equiv V_j^{r,G^-} \bar{D}_j \,. \tag{9.79}$$

This quantity will be used along with an energy flux discussed below to form the inertial mass flux for momentum advection. It may be noted that the relativistic advection, including second order and monotonic interpolations, involves the transport velocity V^r.

Problem 9.13

Integrate (9.44) over the proper three volume and derive (9.77). Justify the definitions (9.78) for the edge centered density.

Problem 9.14

Develop a second order monotonic interpolation scheme for use in the relativistic mass advection equation (9.77). How do the results compare with the nonrelativistic limit?

Problem 9.15

The space-like coordinates of a space-time grid can be allowed to move during a calculation (relativistic grid motion). Define the grid four velocity by

$$U_g{}^\mu \equiv dx_g{}^\mu / ds, \qquad V_g{}^\mu \equiv U_g{}^\mu / U_g{}^t$$

where $x_g{}^\mu$ represents the space-time coordinates of the grid. Use physical arguments to show that the mass continuity equation in spherical isotropic coordinates on a moving grid takes the form (where $g \equiv r^2 A^3$)

$$\frac{\partial D}{\partial t} + \frac{1}{g}\frac{\partial}{\partial r}[g D (V^r - V_g^r)] + \frac{D}{g}\frac{\partial}{\partial r}[\alpha g U_g^r] + \frac{D}{g}\frac{\partial g}{\partial t} = 0.$$

It may help to recall that the nonrelativistic mechanical work is proportional to $\nabla \cdot \mathbf{v}$. How would you modify (9.77) to include simple grid motion?

Since specific internal energy should be transported with mass, a finite difference representation of (9.45) may be constructed using the mass flux. Define the internal energy flux by

$$F_j^E \equiv V_j^{r,G} \bar{E}_j = F_j^D \bar{\varepsilon} \tag{9.80}$$

where the interpolated energy density is defined, for example, as in (9.78). Note that the energy flux in (9.80) has been rewritten as the product of specific energy and the mass flux. Consequently, the specific internal energy is tranported with the mass, as required. In practice, the value of $F^D{}_j$ used to solve the mass continuity equation is stored and used in constructing (9.80). As in the nonrelativistic regime, this reduces diffusion of energy and mass during the advection step. Using the energy flux (9.45) can be integrated over the proper three volume, yielding the approximation

$$\frac{E^A_{j+1/2} - E^C_{j+1/2}}{\Delta t} = -\frac{[(r^2 A^3 F^E)_{j+1} - (r^2 A^3 F^E)_j]}{A^3_{j+1/2}\, \Delta V_{j+1/2}}. \tag{9.81}$$

Here it is understood that the second form of (9.80) is used for the flux. Since (9.78), or an equivalent second order scheme is used to construct the energy flux, (9.81) is also second order accurate in spatial derivatives.

Problem 9.16

Construct a second order, monotonic interpolation scheme for the specific energy to be used in (9.81).

Finally, consider the relativistic advection of momentum, as described by (9.46). Integrating this equation over the momentum proper three volume $\Delta V_j^{(3)} = A^3 \Delta V_j$ gives

$$\frac{S_j^A - S_j^G}{\Delta t} A_j^3 \Delta V_j = - \left[\alpha r^2 A^3 S V^r \right]_j^{j+1} \tag{9.82}$$

where we recall that S denotes the covariant component of the four momentum. Now the right side of (9.82) represents the difference in the flux of momentum carried across the boundaries of the zone ΔV_j. Since four momentum is carried by inertial mass, we use the inertial mass flux σV^r to carry specific four momentum (U^r) into or out of the zone. Thus we use in (9.82)

$$(S V^r)_{j+1/2} = \left(\frac{S}{\sigma}\right)_{j+1/2} (\sigma V^r)_{j+1/2} = (\bar{U}_r)_{j+1/2} (\sigma V^r)_{j+1/2} \tag{9.83}$$

where the four velocity on the right side is a second order monotonic interpolated value, and the inertial mass flux at the zone edge is given by

$$(\sigma V^r)_{j+1/2} \equiv \frac{1}{2} (\sigma_j V_j^r + \sigma_{j+1} V_{j+1}^r) \tag{9.84a}$$

with

$$\sigma_j V_j^r = F_j^D + \Gamma F_j^E = F_j^D (1 + \Gamma \bar{\varepsilon}_j). \tag{9.84b}$$

The last quantity above is the inertial mass flux resulting from the transport of mass and energy. The last form on the right is proportional to the mass flux used in (9.77) and (9.81). In practice the same flux used to solve (9.77) and (9.81) is used here to reduce numerical diffusion of momentum relative to mass. Collecting terms above we have for momentum advection the relativistic finite difference representation

$$\frac{S_j^A - S_j^G}{\Delta t} = - \frac{\left[\alpha r^2 A^3 F^D (1+\Gamma\bar{\varepsilon}) \bar{U}^r\right]_{j-1/2}^{j+1/2}}{A_j^3 \, \Delta V_j} . \tag{9.85}$$

This completes the advection step of the hydrodynamic cycle.

Metric Coefficients

The gravitational field equations are solved now to obtain new values of the metric coefficeints. Since the fluid variables must still be scaled as in (9.47), the values of $A_{j+1/2}$ used in the hydrodynamic stage above must be saved while the new values are constructed from the Hamiltonian constraint equation. This is not a significant problem in one dimensional computations, but can require careful coding in two dimensions.

First, the extrinsic curvature $K^r{}_r$ is constructed from (9.27a) using $S_r A$ as obtained from the final step of the hydrodynamics. The evaluation of the integral is straightforward, except for the centering of A, which can be centered at the zone edge using (9.71), and is

$$(K^r{}_r)_{j+1/2} = \frac{1}{(rA)^3_{j+1/2}} \sum_{n=1}^{n=j} r_n^3 A_n^3 S_n \Delta r_n . \tag{9.86}$$

The outer edge of the physical grid corresponds to r_J, and the origin is $r_2 = 0$.

The density ρ_α appearing as a source term in the equation for the lapse function can now be constructed using $W'_{j+1/2}$, $(K^r{}_r)_{j+1/2}$ and the fluid variables D^A and E^A. Since all of these quantities are zone centered, $\rho_\alpha \rightarrow (\rho_\alpha)_{j+1/2}$, the effective source term is

$$\frac{1}{2}\alpha A^3 \rho_\alpha \rightarrow \frac{1}{2}(\rho_\alpha)_{j+1/2}\, \alpha_{j+1/2} A^3_{j+1/2} \equiv F_{j+1/2} .$$

The equation for the lapse function is an elliptic partial differential equation, and can be represented by

$$r_{j+1}^2 A_{j+1} \frac{(\alpha_{j+3/2} - \alpha_{j+1/2})}{\Delta r_{j+1}} - r_j^2 A_j \frac{(\alpha_{j+1/2} - \alpha_{j-1/2})}{\Delta r_j} = r_{j+1/2}^2 \Delta r_{j+1/2} F_{j+1/2}. \tag{9.87}$$

Using the current value of $A_{j+1/2}$ and (9.71), this reduces to a simple tridiagonal equation for $\alpha_{j+1/2}$, which is to be solved subject to boundary conditions at the origin, and as $r \rightarrow \infty$. The method of back substitution discussed in Chapter 6 can be used to solve (9.87). At the origin, the finite difference representation of the boundary condition is

$$\alpha_{1/2} = \alpha_{3/2} \; . \tag{9.88a}$$

The boundary condition at the outer boundary of the grid is more difficult to implement. First, (9.87) is solved for the temporary solution α_o with the outer boundary condition $\alpha_o(R) = 1$. At large distances from the mass distribution the true lapse function has the asymptotic form $\alpha \approx 1 - k/r$. We take as our solution for the lapse function $\alpha'(r) = \xi\alpha_o(r)$, with the scale factor ξ chosen such that $\alpha(R) = \xi\alpha_o(r)$, and that both $\xi\alpha_o(r)$ and $\alpha(r)$ have the same slope at the edge of the grid $r=R$. These conditions lead to

$$\xi = \frac{1}{1 + R\,(d\alpha_o/dr)_R}. \tag{9.88b}$$

Then $\alpha'_{j+1/2}$ represents the lapse function at the end of the computational cycle, and the corresponding edge centered values can be obtained from (9.71).

The integral for the shift vector (9.25) can now be solved using α' and $(K^r{}_r)'$ above. It may be approximated as follows:

$$\beta'_j = -\frac{3 r_j}{2} \sum_{n=j}^{n=J} \alpha'_{n+1/2} (K^r{}_r)'_{n+1/2} \frac{\Delta r_{n+1/2}}{r_{n+1/2}} - B$$

(9.89)

where B represents the contribution to the integral for $r > R$, and may be found by assuming that

$$K^r{}_r \rightarrow r^{-3}$$

outside the computational grid.

Finally the Hamiltonian constraint equation (9.28) is solved for $A'_{j+1/2}$. Before this step is carried out we rewrite ρ_H to reflect the fact that several quantities appearing in it are conserved when (9.28) is solved (see, for example, Evans 1985). We may also note that the change in the metric coefficient A is analogous to grid motion in the Eulerian equations discussed in Chapter 5 (see the discussion of Grid Compression at the end of Section 5.4). First, define

$$\hat{D} \equiv D\,\phi^6, \quad \hat{E} \equiv E\,\phi^{6\gamma}, \quad \hat{K} \equiv (\Delta K^r{}_r)^2 \phi^{-12}$$

(9.90)

where $\phi \equiv A^{1/2}$. It follows from (9.36) that

$$\int A^3 D\, r^2\, dr = \int \phi^6 D\, r^2 dr = constant$$

where the integral is over the entire grid. An analysis of the specific energy equation (using a gamma law equation of state to replace the pressure with E) also indicates that $E\,\phi^{6\gamma}$ is conserved (Evans 1985). The Hamiltonian constraint equation (9.28) may then be rewritten using (9.90) in the convenient form

$$\frac{1}{r^2} \frac{\partial}{\partial r}\left(r^2 \frac{\partial \phi}{\partial r}\right) = -\,\phi\, \Psi(\phi)$$

(9.91)

where

$$A^{5/2}\rho_H = \phi\,\Psi(\phi) = \phi\{\frac{W\hat{D}}{\phi^2} + \frac{W\hat{E}}{\phi^{6\gamma-4}}[1 + \frac{W(W^2-1)(\gamma-1)}{W^2}] + \frac{\hat{K}^2}{2\phi^8}\}. \tag{9.92}$$

When solving (9.91), all quantities except ϕ should be treated as constants. If Ψ were independent of ϕ, then (9.91) would be a linear equation that could be solved as easily as (9.87). Because of the strong nonlinear nature of this factor, we solve (9.91) iteratively as follows. First form the finite difference representation of (9.91)

$$r^2_{j+1}\frac{(\phi_{j+3/2}-\phi_{j+1/2})}{\Delta r_{j+1}} - r^2_j\frac{(\phi_{j+1/2}-\phi_{j-1/2})}{\Delta r_j} = -\phi_{j+1/2}\,\Psi r^2_{j+1/2}\,\Delta r_{j+1/2}. \tag{9.93}$$

The function Ψ is evaluated using the old values of ϕ and is assumed to be known. Then back substitution is used to solve for new values for $\phi_{\varphi+1/2}$, subject to the boundary conditions given below (9.28). These values are then substituted back into $\Psi(\phi)$ and the procedure repeated until the solution for has converged. The converged solution $(\phi')^2{}_{j+1/2} = A'_{j+1/2}$ is then used for the next computational cycle.

After solving the Hamiltonian constraint equation the density and specific energy are scaled as [recall (9.48)]:

$$D^{n+1}_{j+1/2} = (A_{j+1/2}/A'_{j+1/2})^3 D^A_{j+1/2}$$

$$E^{n+1}_{j+1/2} = (A_{j+1/2}/A'_{j+1/2})^{3\Gamma} E^A_{j+1/2}\,. \tag{9.94}$$

The (9.71) is used to construct A'_j from $A'_{j+1/2}$ and the momentum density is scaled:

$$S^{n+1}_j = (A_j/A'_j)^3 S^A_j. \tag{9.95}$$

The new fluid variables are now used to start the next computational cycle. Before the gravitational acceleration can be calculated new values of U_r and V^r must be obtained. These are given by

$$U'_{r,j} = S_j^{n+1/2} / \sigma'_j , \tag{9.96}$$

where (9.62) is used with

$$\sigma'_{j+1/2} = (D' + \Gamma E')_{j+1/2} \tag{9.97}$$

for a gamma law equation of state. The transport velocity is given by

$$(V_j^r)' = \left[\alpha' \frac{U'_r}{(A')^2} - \beta' W'\right]_j \tag{9.98}$$

with α'_j defined as in (9.50). This concludes the computational cycle.

Problem 9.17

Compare the treatment of grid compression in the nonrelativistic Eulerian case with the solution for the metric coefficient A discussed above. In particular, discuss the equation

$$\frac{\partial A^3}{\partial t} \approx \nabla \cdot \mathbf{v}_g .$$

Relativistic Artificial Viscous Stress

A modification of the treatment given in Section 5.13 may be used to find a representation of the artificial viscous stress for relativistic hydrodynamics. This approach does not produce an invariant tensor quantity, but uses a modification of the nonrelativistic representation discussed in Chapter 5. This is not a serious limitation since the sole purpose for using an artificial viscous stress is to produce a smooth transition in fluid variables across a shock front. The standard von Neumann artificial viscous stress discussed earlier in Chapter 5 does not work well for shock velocities near the speed of light, but the monotonically based viscosity has been found to work quite well. The von Neumann artificial viscous stress is proportional to the jump in fluid velocity. In the relativistic formulation, finite differences of U_μ are taken instead of in V_μ because the coordinate four velocity tends to a constant for highly relativistic flow. For this reason we replace differences ΔV_μ by $\Delta U_\mu / U^t$, which for extreme relativistic flow approaches unity. Thus the radial component of the four velocity U_r is used in (5.180) and in the formulae leading up to it instead of the coordinate velocity V_r. The linear artificial stress is then given by

$$Q_{L,j+1/2} = -C_L\, \sigma_{j+1/2}\, c_{S,j+1/2} \{\Delta U_{j+1/2}\}_{ext} / U^t, \tag{9.99}$$

where the sound speed is just

$$c_S = \left[\frac{\Gamma(\Gamma-1)E}{D+\Gamma E}\right]^{1/2}. \tag{9.100}$$

The quadratic artificial viscous stress is given by

$$Q_{q,j+1/2} = C_q\, \sigma_{j+1/2} \frac{\{\Delta U_{j+1/2}\}^2_{ext}}{(U^t)^2} \tag{9.101}$$

if the zone is undergoing compression, and is zero if it is expanding.

It is also useful to correct for losses in kinetic energy associated with momentum diffusion. In the relativistic regime the kinetic energy diffusion corrector may be expressed as

$$Q_{D,j+1/2} = C_D\, \sigma_{j+1/2} \frac{|U_j + U_{j+1}|}{2} \frac{\{\Delta U_{j+1/2}\}_{ext}}{(U^t)^2} .$$

(9.102)

In the presence of grid motion, the term $|U_j + U_{j+1}|$ appearing in (9.102) should be replaced by the quantity

$$|U_j + U_{j+1} - (V^g{}_j +) V^g{}_{j+1}) / U^t| ,$$

(9.103)

since the stress Q_D is a correction for advection errors associated with fluid motion relative to the grid.

The relativistic analogue of artificial viscous stress acceleration is then given by

$$S'_j = S_j - \alpha_j (Q_{L,j+1/2} + Q_{q,j+1/2} - Q_{L,j-1/2} - Q_{q,j-1/2}) \frac{\Delta t}{\Delta r_j} .$$

(9.104)

The change in momentum density is computed following the change associated with the pressure gradient and the gravitational acceleration [see (9.68)].

The change in material internal energy associated with shock heating is given by

$$E'_{j+1/2} = E_{j+1/2} - \alpha_{j+1/2} \Delta t\, (Q_{L,j+1/2} + Q_{q,j+1/2} + Q_{D,j+1/2}) \frac{(U_{j+1} - U_j)}{U^t \Delta r_{j+1/2}} .$$

(9.105)

As in Chapter 5, the nominal values for the coefficients in the linear and quadratic artificial viscous stress and in the kinetic energy diffusion corrector are $C_L = 1.0$, $C_q = (\Gamma+1)/2$ and $C_D = 0.25$. Shock heating may be calculated following shock acceleration.

An improvement in the time centering of the artificial viscous stress can be obtained as follows. First we advance the momentum density as in (9.104). The new momentum density is used to find the four velocity

$$\tilde{U}_j = S' / \sigma_j \, .$$

(9.106)

The difference in four velocity appearing in the energy equation (9.105) is then replaced by the quantity

$$\frac{1}{2}(U_{j+1} + \tilde{U}_{j+1} - U_j - \tilde{U}_j) \, .$$

(9.107)

As in the nonrelativistic formulation, the equations may become unstable when shocks are present if the time step is too large. For the relativistic artificial viscous stress above an effective time step limit is given by

$$\Delta t_Q = \frac{W \, \Delta r_{j+1/2}}{2 \, (c_S \, Q_L + 2 \, C_q \, \Delta U / W)} \, .$$

(9.108)

The minimum of (9.108) over the grid is used as the relativistic shock time step limit. The actual time step will, as usual, be the lesser of the Courant, transport and shock time steps.

References

Chapter 9

C. R. Evans, "A Method for Numerical Simulation of Gravitational Collapse and Gravitational Radiation Generation" in Numerical Astrophysics, edited by J. M. Centrella, J. M. LeBlanc, and R. L. Bowers (Boston: Jones and Bartlett; 1985).

C. Misner, K. S. Thorne and J. A. Wheeler, *Gravitation* (San Francisco: W. H. Freeman, 1973).

D. S. Kershaw, *Journal of Computational Physics,* **26** (1978): 43.

J. A. Meijerink and H. A. van der Vorst, *Mathematics of Computation,* **31** (1977): 148.

J. V. Narliker, *Introduction to Cosmology* (Boston: Jones and Bartlett, 1983).

S. Weinberg, *Gravitation and Cosmology* (New York: Wiley, 1972).

J. R. Wilson, in *Sources of Gravitational Radiation*, edited by L. L. Smarr (Cambridge: Cambridge University Press, 1979).

10

THERMONUCLEAR AND NEUTRINO REACTIONS

Among the most important sources of energy in stellar astrophysics are thermonuclear and neutrino reactions. In normal stars, such as those on the main sequence, or for stars during most of their lifetime, energy release accompanying nuclear reactions is a primary source of stellar energy, supporting the star's mass against gravitational collapse, and producing from primordial hydrodgen and helium the lighter chemical elements observed in the universe today. Energy loss associated with neutrino reactions, which also accompany many nuclear processes in normal main sequence stars, may be included trivially since they escape freely. They play a fundamental role in the onset of the final stages of stellar evolution and in core collapse supernovae.

Thermonuclear energy generation in stellar interiors usually involves a large number of elements and an even larger number of reaction paths (channels) through which the elements may react. Thus a treatment of thermonuclear reactions becomes in effect a problem of solving a network of equations describing all of the competing processes which are physically allowed by the composition, density and temperature of the stellar material. For example, the principle reaction channels contributing to the proton-proton chain in the solar interior include:

$$
\begin{aligned}
&H^1 + H^1 \leftrightarrow e^+ + \nu_e + H^2 && \tau = 7.9 \times 10^9 \\
&H^2 + p \leftrightarrow \gamma + He^3 && \tau = 4.4 \times 10^{-8} \\
&He^3 + He^3 \leftrightarrow 2p + He^4 && \tau = 2.4 \times 10^5 \\
&He^3 + He^4 \leftrightarrow \gamma + Be^7 && \tau = 9.7 \times 10^5 \\
&Be^7 + e^- \leftrightarrow \nu_e + Li^7 && \tau = 3.9 \times 10^{-1} \\
&Li^7 + p \leftrightarrow He^4 + He^4 && \tau = 1.8 \times 10^{-5} \\
&Be^7 + p \leftrightarrow \gamma + B^8 && \tau = 6.6 \times 10^1 \\
&B^8 + e^+ \leftrightarrow \nu_e + Be^{8*} \leftrightarrow He^4 + He^4 && \tau = 3 \times 10^{-8}.
\end{aligned}
\tag{10.1}
$$

An estimate of the characteristic time scale, τ, in years for each reaction is given on the right. This simple nuclear network consists of eight reactions among seven different nuclei and involves photons and three leptons (electrons, positrons and electron neutrinos). Furthermore, the reaction rates vary by seventeen orders of magnitude. Nuclear networks for more advanced stages of stellar evolution, particularly if detailed models of nucleosynthesis are desired, can include thousands of isotopes. In this chapter we consider two classes of thermonuclear reactions. In the first model we assume that only a single species reacts at a time, and ignore reverse reactions and competing reactions. While this type of model is not adequate for normal stellar evolution or nucleosynthesis, it is useful for some dynamic processes (such as stellar core collapse) when energy release by nuclear reactions is important but detailed abundances are not. We then consider problems associated with a treatment of nuclear networks wherein the rate of nuclear energy release is as important as the detailed nuclear composition. A discussion of thermonuclear reactions under normal stellar conditions can be found in Cox and Guili (1968) and in Clayton (1968).

Although neutrino processes do occur under normal stellar conditions [such as in the proton-proton chain in the sun (10.1)], they do not represent a significant energy source for these stages of evolution. For example, in the proton-proton chain in the sun (10.1) neutrinos carry off only about two percent of the total nuclear energy produced, while ninty-eight percent of the energy is in the form of photons. Furthermore, the mean free path for the neutrinos that are produced in the solar interior is many orders of magnitude greater than the solar radius, and the neutrinos are not able to couple to the material effectively. Neutrino reactions are, however, a more significant source of energy loss in the cores of highly evolved stars. Neutrino reactions can also play a fundamental role in the energetics of supernovae. In the latter case the energy carried away by neutrinos comprises a large fraction of the energy of the system, and the mean free path of the neutrinos may be comparable to, or smaller than, the radius of the system. The diffusion of neutrino radiation has already been discussed in Chapter 6. Here we consider a representative type of neutrino reaction occurring during the onset of stellar core collapse in which high energy electrons are captured by heavy nuclei to produce electron neutrinos. A discussion of neutrino processes in stellar astrophysics may be found in H. Y. Chiu (1968) and in Bowers and Deeming (1984).

10.1 THERMONUCLEAR REACTIONS

A simple model of thermonuclear energy release can be applied to the region surrounding a collapsed stellar core in some models of supernovae. In some instances the release of nuclear energy represents only a small fraction of the total energy produced by hydrodynamic and neutrino processes, but this small amount can represent an important contribution to the dynamics of material which overlies the stellar core and which may be ejected to produce a supernova explosion. For example, consider the collapse of a stellar core which consists predominantly of iron group nuclei surrounded by concentric shells of lower mass nuclei. Thermonuclear fusion of these surrounding layers can result if the material temperature becomes high enough, as, for example, if a sufficiently strong shock moves outward following core bounce. Although the energy released in the process is not by itself sufficient to eject the stellar mantle, it can contribute to the energy released by the core. Detailed reaction networks leading from, for example, carbon nuclei to iron group nuclei involve many competing processes. In some calculations of core collapse the details of the nuclear network may not be needed in order to approximate the energy released during adiabatic compression, or when the shock traverses a region containing a potential nuclear fuel.

A typical binary nuclear reaction can be represented by

$$a + X \leftrightarrow b + Y \tag{10.2}$$

where X denotes a reactant nucleus and Y is a product nucleus. The product nucleus of one reaction may become the reactant for a subsequent reaction. Finally, a and b denote other reactant and product nuclei or light particles such as electrons or neutrinos, their antiparticles or photons. In addition to the forward process in (10.2) the reverse reaction will also occur. If photons are one of the reactants then the process is often called photodissociation. Finally, the product nuclei may be in excited states, whose decay modes may also need to be taken into consideration. A simple expression for the energy released by a reaction of the form (10.2), which does not involve excited states, is given by

$$Q = (M_X + M_a - M_Y + M_b)\, c^2 . \tag{10.3}$$

Q represents the difference between the mass-energy of the reactants and the mass-energy of the products, and is the energy available to the product nuclei and particles. Some reactions can also occur through the intermediary formation of a compound nucleus. Compound nuclear reactions may also be incorporated in nuclear networks.

Let r represent the total number of reactions (10.2a) that take place in a unit volume per unit time; the change in number density of the reactant n_X is then given by

$$\frac{d\, n_X}{d\, t} = -r , \tag{10.4}$$

and the total energy generated by the single reaction (10.2) per unit mass per unit time is

$$\varepsilon_X = \frac{Qr}{\rho} . \tag{10.5}$$

Equation (10.4) may be used to express the energy generation rate (10.5) in terms of the change in composition of the species X.

Simple Thermonuclear Energy Deposition

A simple model which approximates the energy release without incorporating the details of nucleosynthesis will be considered here. It illustrates some aspects of nuclear processes and their coupling to fluid dynamics, and can be applied to some other stellar processes as well. Suppose that the material surrounding a collapsed stellar core can be approximated by concentric shells dominated respectively by C^{12}, O^{16}, Si^{28} and iron group nuclei (the latter will be denoted collectively by Fe). At any point in the fluid the mass fraction of each of these nuclei will be denoted by X_i. Then the rate of change in material internal energy density can be expressed in the form

$$\frac{d\,(\rho C_V T)}{dt} = \sum_i X_i \,\rho\, \varepsilon_i \, . \tag{10.6}$$

The material heat capacity, which is given by (3.14), relates the temperature to the specific internal energy. The mass fraction of each species (whose atomic weight is A_i) is related to its mass density ρ_i or to its number density n_i by

$$\rho_i = A_i \, m_M \, n_i = X_i \, \rho , \tag{10.7a}$$

where ρ is the usual (total) mass density of the plasma. The mass fractions also satisfy the normalization constraint:

$$X_{Fe} + X_{Si} + X_O + X_C = 1 \, . \tag{10.7b}$$

Finally, $\varepsilon_i = \varepsilon_i(\rho,T,X)$, where $i = C, O$ or Si , is the thermonuclear energy released per gram of reactant per second. In general ε_i will be a local function of density, temperature and composition. For some applications it may be useful to approximate the thermonuclear network leading from carbon to iron group elements by the simple three step process

$$C^{12} \to O^{16}$$

$$O^{16} \to Si^{28}$$

$$Si^{28} \to Fe \; . \tag{10.8}$$

This model assumes that each element reacts once the plasma temperature reaches or exceeds the elements ignition temperature. In addition, it ignores all competing processes (alternate channels through which the reactions might occur) and all reverse processes. Expressions for the reaction rates ε_i, the energy released per gram, and the ignition temperatures, T_i, for these elements can be found in the literature (see, for example, Clayton 1968). Finally, the rate of change of the nuclear compositions satisfy

$$\frac{dX_i}{dt} = -\frac{\varepsilon_i}{E_i} = -\frac{dX_{i+1}}{dt} \tag{10.9}$$

where E_i is the energy released when one gram of the ith element reacts completely to form the next element in the series.

If other processes besides thermonuclear reactions occur (such as mechanical work and radiation transport), then they must be added to the right side of (10.6). Thus (10.6) couples the nuclear energy released to the dynamics of the fluid through the hydrodynamic equations. We shall assume that the equations describing thermonuclear processes can be treated separately from the hydrodynamics, gravitation and radiation transport by the method of operator splitting. In this case (10.6) represents the change in internal energy associated only with nuclear reactions, and may be rewritten in the equivalent Lagrangian form

$$\rho C_V \frac{dT}{dt} = \sum_i X_i \rho \varepsilon_i \tag{10.10a}$$

$$\rho C_V \frac{dT}{dt} = -\rho T \frac{dC_V}{dt} . \tag{10.10b}$$

These two equations have the same form as (3.18) and can be solved as described in Section 2.2. Equations (10.6) or (10.10), and (10.7) and (10.9), along with definitions of the thermonuclear reaction parameters represent a complete set of equations describing the net energy release for the simple series of processes (10.8). Simple models such as this one are not intended to approximate nucleosynthesis, nor are they capable of reproducing features which depend on competing reactions [such as (10.1)]. Nevertheless they can be useful in modelling some aspects of stellar evolution and supernovae explosions in which the net energy release is of importance.

Thermonuclear Reaction Networks

For all but the simplest applications it is necessary to treat at least the principle nuclear species (which we will denote by the set $\{X_i\}$) and their reactions in detail. This more detailed treatment of thermonuclear reactions leads to a series of coupled, nonlinear differential equations relating the rate of change of the various species and is called a thermonuclear reaction network, or simply a nuclear network. Nuclear networks have been developed and discussed in the literature which include a few hundred to a thousand or more nuclear isotopes (Clayton 1968; Arnett and Truran 1969; Woosley, Arnett and Clayton 1973; Weaver, Zimmerman and

Woosley 1978). Many of the problems encounted in solving such networks can be understood in terms of a simple network representing carbon burning through silicon burning and the production of iron group nuclei in the core of highly evolved, massive stars. Before considering this network, we review the equations defining a nuclear network.

Suppose that Y_i represents the number of moles of the ith nuclear species (whose atomic number is A_i). Then Y_i is related to the mass fraction X_i by

$$Y_i = X_i / A_i . \tag{10.11}$$

The concentration n_i of species i will decrease as the reactant nuclei fuse to form higher mass products [the forward reaction in (10.2)] at a rate proportional to the concentration of n_i and n_a and the rate for the forward process. The reverse reaction of (10.2) will also occur with a rate proportional to the concentrations of reactant nuclei n_Y and n_b that can produce the species i and the rate for the reverse reaction. When these two processes are combined for each reaction channel the net rate of change in Y_i can be expressed in the form

$$\frac{dY_i}{dt} = -\sum_{a,b} Y_i Y_a \lambda_{ab}(i) + \sum_{l,c} Y_l Y_c \lambda_{cd}(l) . \tag{10.12}$$

The indices appearing in (10.12) must satisfy the constraint $l + c = i + d$, where a, b, c, and d may represent protons, neutrons, photons, alpha particles, electrons, positrons, neutrinos. An equation of this form holds for each nuclear species which may be present in the star. Finally, $\lambda_{ab}(i)$ represent the reaction rates which were denoted by r in (10.4). A discussion of these quantities can be found in the literature. The change in material thermal energy is expressed in a form analogous to (10.6). Using (10.12), and denoting the nuclear binding energy of the ith species by $\Delta E_{B,\,i}$, the energy generation rate can be expressed as

$$\varepsilon_i = \frac{dY_i}{dt} \Delta E_{B,i} \,. \tag{10.13}$$

These may then be used in (10.10a) to obtain the change in material temperature. The nuclear reaction rates and energy generation rates represent additional constitutive relations which must be used to complete the mathematical description of the model. Nuclear networks based on equations of the form (10.12) constitute a coupled set of equations for the compositions which must be solved along with the dynamics. Once the rates and compositions are known, the material heating rate may be obtained.

Problem 10.1

Two nuclear species whose mole fractions are Y_1 and Y_2 undergo binary reactions described by the simple rate equations

$$\frac{dY_1}{dt} = -\lambda_1 Y_1 + \lambda_2 Y_2 \tag{10.14a}$$

$$\frac{dY_2}{dt} = -\lambda_2 Y_2 + \lambda_1 Y_1 \tag{10.14b}$$

where the rate factors are λ_1 and λ_2. Solve this set of equations for the ratio Y_2 / Y_1 subject to the boundary conditions $Y_2 = 0$ at $t = 0$. Show that the solution approaches the steady state solution $Y_2 / Y_1 = \lambda_1 / \lambda_2$ on a time scale of order $(\lambda_1 + \lambda_2)^{-1}$.

Minimal Network

During their evolution, stars that are considerably more massive than the sun are believed to burn carbon and oxygen, producing successively higher mass isotopes until the elements in the iron group finally predominate. At this stage no further energy producing (exothermic) nuclear reactions are possible, and eventually the iron group nuclei begin to photodissociate, ultimately producing alpha particles and free nucleons. This is a complex process involving thousands of isotopes and many thousands of possible reaction channels. However, a simple, empirical nine element network gives an approximate representation of the energy release by the process, and can be used to illustrate many of the features of numerical solutions to the nuclear network equations. This simple network consists of nine nuclei $\{He^4, C^{12}, O^{16}, Ne^{20}, Mg^{24}, Si^{28}, Ni^{56}, n, p\}$, starts with the triple alpha process and terminates with the photodissociation of nickel to form alpha particles and free nucleons (Woosley 1986). Two types of reactions are included; alpha-gamma processes, denoted by (α,γ) [and their inverses, denoted by (γ,α)], and three compound nucleus reactions (the last three) whose inverses are to be omitted. The reaction rate for the forward and reverse process is given in the second and third column, respectively. The compound nuclei can decay through several competing channels. The single reaction shown for each in Table 10.1 represents the dominant end product of the decay. The inverse processes have been omitted because they do not contribute significantly to the energy release. As the abundance of Si^{28} increases so does the temperature of the plasma, and photodissociation results in a downward chain of (γ,α) reactions which produces free alpha particles. The latter may be captured by other Si^{28} nuclei and their products producing in effect an upward chain of (α,γ) reactions. This is termed silicon burning, and effectively ends with the production of iron group nuclei. The introduction of Ti^{44} production from Ca^{40} acts as an effective rate controlling reaction for the process, whose net effect is $2\,Si^{28} \rightarrow Si^{28} + 7\alpha \rightarrow Ni^{56}$. The minimal network can be thought of as modelling the process

$$Si^{28} + 3\alpha \rightarrow Ca^{40}$$

$$Ca^{40} + \alpha \leftrightarrow \gamma + Ti^{44}$$

$$Ti^{44} + 3\alpha \rightarrow Ni^{56}.$$

Table 10.1

Minimal Thermonuclear Network. The nuclear reactions used to describe a simple, empirical network starting with the triple alpha process, leading through carbon and oxygen burning to silicon burning, the formation of nickel and terminating with the photodissociation of nickel to alpha particles and free nucleons.

Reactions	Forward Rate	Reverse Rate
$H^2 + p \leftrightarrow \gamma + He^3$	$\lambda_{p,\gamma}(He)$	$\lambda_{\gamma,p}(H^2)$
$3\alpha \leftrightarrow \gamma + C^{12}$	$\lambda_{\alpha,\gamma}(C)$	$\lambda_{\gamma,\alpha}(C)$
$C^{12} + \alpha \leftrightarrow \gamma + O^{16}$	$\lambda_{\alpha,\gamma}(O)$	$\lambda_{\gamma,\alpha}(C)$
$O^{16} + \alpha \leftrightarrow \gamma + Ne^{20}$	$\lambda_{\alpha,\gamma}(Ne)$	$\lambda_{\gamma,\alpha}(O)$
$Ne^{20} + \alpha \leftrightarrow \gamma + Mg^{24}$	$\lambda_{\alpha,\gamma}(Mg)$	$\lambda_{\gamma,\alpha}(Ne)$
$Mg^{24} + \alpha \leftrightarrow \gamma + Si^{28}$	$\lambda_{\alpha,\gamma}(Si)$	$\lambda_{\gamma,\alpha}(Mg)$
$Ca^{40} + \alpha \leftrightarrow \gamma + Ti^{44}$	$\lambda_{\alpha,\gamma}(Ti)$	$\lambda_{\gamma,\alpha}(Ca)$
$C^{12} + C^{12} \rightarrow (Mg^{24})^*$	λ_{1212}	*omit*
$\rightarrow Ne^{20} + \alpha$		
$C^{12} + O^{16} \rightarrow (Si^{26})^*$	λ_{1216}	*omit*
$\rightarrow Mg^{24} + \alpha$		
$O^{16} + O^{16} \rightarrow (S^{32})^*$	λ_{1616}	*omit*
$\rightarrow Si^{28} + \alpha$		

Finally, as the temperature in the nickel core increases, photodissociation produces alpha particles which may then photodissociate into free neutrons and protons.

10.2 THERMONUCLEAR ENERGY DEPOSITION MODEL

The simple thermonuclear energy deposition model (10.6)-(10.10) is relatively easy to implement (Bowers and Wilson 1982). The discussion below can be used with either a Lagrangian or an Eulerian treatment of the remaining physics. In addition to the variables required for the hydrodynamics, radiation and gravitation, the mass fractions or mole fractions of the nuclei must be specified. The terms describing thermonuclear reactions do not contain spatial derivatives, so spatial centering of the finite difference equations is straightforward. Thus, the model will be developed for one zone. The finite difference variables denoting the mass fractions of the nuclei appearing in (10.7), which are naturally zone centered variables, will be denoted by $X_{i,k+1/2}$ (where i = C, O, Si, and Fe).

The thermonuclear energy generation rates for these nuclei depend on the local density, plasma temperature, and on the mass fractions of the reactant nuclei. Simple analytic expressions for these rates are available in the literature (see, for example, Clayton 1968). In practice it is usually possible to operator split energy release by thermonuclear reactions from the other physics units, so the temporal superscipt denotes the value of the energy generation rate based on current values of the plasma variables at the beginning of the energy release calculation. The energy generation rates are calculated for each zone in the grid when needed.

The change in plasma temperature given by (10.10a) has the simple finite difference representation for each species

$$C_{V,\,k+1/2} \frac{T^{\,i}_{k+1/2} - T_{k+1/2}}{\Delta t} = X_{i,\,k+1/2}\; \varepsilon_{i,\,k+1/2} \tag{10.15}$$

where the superscript i denotes the plasma temperature after the fusion of the ith nuclear species, and variables without a superscript denote values used to calculate the energy release. Hereafter the spatial subscripts (which are the same for all variables) will be omitted for notational simplicity.

Because of the strong nonlinear temperature dependence of the energy generation rates the change in thermal energy of the plasma associated with nuclear reactions over a dynamic or radiative time step may be enormous. Ideally, changes of this magnitude should be modelled using iterative implicit methods. Because this model attempts only to represent the net energy release to the plasma, a simpler explicit approach will be used, which may be implemented as follows. For each zone in which the plasma temperature is greater than or equal to the ignition temperature (the temperature at which the rates become significant) T_i for the ith process, the energy generation rate is calculated.

The energy generation rate of most nuclear reactions can be approximated over a limited temperature range centered about T_o by $\varepsilon \approx T^n$. Here n is a weak function of the temperature, but is generally much larger than unity. This fact may be used to approximate each term on the right side of (10.10a), which may be expressed as

$$\rho C_V \frac{dT}{dt} = \rho^2 X_i^2 f_i(T), \tag{10.16}$$

in the form

$$\rho C_V \frac{dT}{dt} = \rho^2 X_i^2 f_{i,o} T^n, \tag{10.17a}$$

where

$$n = \left(\frac{d \ln f}{d \ln T}\right)_{T=T_o}. \tag{10.17b}$$

This equation has the solution (for fixed X_i and ρ)

$$T = \frac{T_o}{\left[1-(n-1)\,K\,T_o^{n-1}\Delta t\right]^{\frac{1}{n-1}}} \tag{10.18a}$$

$$K = \rho X_i^2 f_{o,i} / C_V . \tag{10.18b}$$

Some care must be exercised in using this solution, since if Δt is too large nonphysical result may be obtained. A way to avoid this difficulty when it arises is to subcycle this portion of the calculation using a reduced time step. First, a reaction time step Δt_R is constructed as follows. Let α denote the maximum acceptable change in the temperature per computational step:

$$\alpha \equiv T_{acceptable} / T_o .$$

If T/T_o in (10.18a) is replaced by α, the following expression is obtained for Δt_R:

$$(n-1) K T_o^{n-1} \Delta t_R = 1 - \alpha^{1-n}. \tag{10.19}$$

If the current time step is less than Δt_R, then it is used to evaluate (10.18) and no subcycling is needed. The energy generated in the ith process is given by

$$\rho C_V (T - T_o) \equiv \rho C_V \Delta T_i , \tag{10.20}$$

which may be used in (10.9) to find the change in nuclear composition:

$$X'_i = X_i - \frac{\rho C_V \Delta T_i}{E_i} \tag{10.21a}$$

and

$$X'_i = X_i + \frac{\rho C_V \Delta T_i}{E_i} . \tag{10.21b}$$

This stage of the nuclear energy generation calculation is now complete, and we proceed to the next available nuclear fuel whose ignition temperature is less than or equal to the plasma temperature.

If the current time step $\Delta t > \Delta t_R$, then we subcycle through the equations as follows. The time is advanced by Δt_R, and the new temperature $T = \alpha T_o$ is used to construct the energy release (10.20). The changes in nuclear composition are then found from (10.20) and (10.21). We then repeat the process beginning with (10.16) to find a new value of Δt_R. Subcycling continues until the problem time has been advanced from t to $t + \Delta t$ for the given nuclear species. At the end of each subcycle the temperature T must be less than $T_{max} = T_o + E_i X_i / C_V$, so we take

$$T = min \{T_o + E_i X_i / C_V\} .$$

The simple model above can be extended to include additional nuclear species as long as only forward reactions such as those in (10.8) are allowed. If competing branches occur, or if reverse processes are important, then it is best to use the methods of nuclear networks. Formally (10.2) guarantees that if the initial compositions satisfy (10.7b), then so will all subsequent values. In practice this may not occur. The constraint (10.7b) can be enforced by normalizing the new compositions at the end of each process.

Problem 10.2

A simple power law approximation to the energy generation rate is given by

$$\varepsilon_i = \varepsilon_o T^n$$

where ε_o is independent of the plasma temperature. Derive the relation (10.18) using this approximation.

Advection of Isotopic Abundances

If thermonuclear reactions (whether represented by a simple energy deposition model or a thermonuclear network) are to be included within the framework of an Eulerian representation of the hydrodynamics, then the advection stage of the calculation must be modified to account for the transport of nuclear abundances. Eulerian advection was discussed in Chapter 5 where, in particular, schemes for mass transport were developed. Nuclear species can diffuse relative to one another, but a numerical treatment of such processes lies beyond the scope of this text. Thus we assume that the nuclear abundances characterizing a plasma represent an intrinsic property of the plasma and should be transported with mass.

The mass continuity equation for the total plasma density is given by (5.4). Assuming that each nuclear species moves with the same velocity, which is the velocity of the plasma itself (no relative diffusion of species allowed), then (10.7) may be used to obtain

$$\rho = \sum_i \rho_i = \sum_i X_i \rho \tag{10.22}$$

$$\frac{\partial \rho}{\partial t} = \sum_i \left(X_i \frac{\partial \rho}{\partial t} + \rho \frac{\partial X_i}{\partial t} \right) \tag{10.23a}$$

$$\begin{aligned} \nabla \cdot (\rho \, \mathbf{v}) &= \sum_i \nabla \cdot (X_i \, \rho \, \mathbf{v}) \\ &= \sum_i \left(X_i \, \nabla \cdot (\rho \mathbf{v}) + \rho \, \mathbf{v} \cdot \nabla X_i \right) . \end{aligned} \tag{10.23b}$$

For simplicity the grid velocity terms have been omitted. The results below are trivially extended to include them when needed. Adding (10.23a) and (10.23b) and rewriting terms we find

$$\sum_i \{X_i [\frac{\partial \rho}{\partial t} + \nabla\cdot(\rho \mathbf{v})] + \rho [\frac{\partial X_i}{\partial t} + \mathbf{v}\cdot\nabla X_i]\} = 0.$$

The first term inside the summation is the mass continuity equation, and therefore vanishes identically. Since the nuclear abundances are independent, the summation over the second term in square brackets will vanish only if

$$\frac{\partial X_i}{\partial t} + \mathbf{v}\cdot\nabla X_i = 0 \tag{10.24}$$

for each species. It follows that the compositions are remapped when fluid motion occurs. Note that (10.24) is equivalent to requiring that the Lagrangian derivative of the abundances vanishes (which simply states that they move with the fluid).

The second order remap algorithms discussed in earlier chapters may be used to solve the set of equations (10.24) during the hydrodynamic portion of the computation. In practice it is best to remap the quantities $\delta m X_i$, where δm is the mass change associated with the second order advection scheme used in the hydrodynamics. This reduces relative diffusion of mass and compositions.

10.3 THERMONUCLEAR REACTION NETWORKS

Under many stellar conditions the variation in reaction rates between various species can be so large that explicit methods (or simple modifications of them) cannot be used (see Problem 10.3 and 10.4). We shall assume that the binary reactions lead to changes in the mole numbers Y_i given by (10.12), although other types of reactions may also be included. The rate of change in each mole number can be approximated in the usual way:

$$\frac{dY_i}{dt} \rightarrow \frac{Y'_i - Y_i}{\Delta t} = \frac{\Delta Y_i}{\Delta t} \tag{10.25}$$

where the prime denotes the value at the advanced time, Δt is the computational time step ($\Delta t^{n+1/2}$ in the case of the two time level formulation of the hydrodynamics in Chapters 4 and 5), and the subscript denotes the nuclear species. The rate equations then become

$$\frac{\Delta Y_i}{\Delta t} = -Y'_i \sum_{a,b} Y'_a \lambda_{ab}(l) + \sum_{l,a} Y'_l Y'_a \lambda_{a,l}(l) \tag{10.26}$$

where the mole numbers on the right side are evaluated at the new time (fully implicit). In principle the temperature dependence of the reaction rates should also be evaluated at the advanced time, and the plasma heating equation (10.10) solved simultaneously with (10.26). For a sufficiently small set of reactions this may be possible using a straightforward extension of the implicit methods discussed in Section 4.7. Such an approach accounts for the nonlinearity of the reaction rates and of the quadratic nature of the rate equations themselves.

An approach which has been used extensively in studies of stellar evolution and stellar nucleosynthesis, and which is applicable to very large nuclear networks, solves a linearized form of the implicit equations (10.26). Using (10.25) to define the change in the mole numbers, substitute

$$Y'_i = Y_i + \Delta Y_i \tag{10.27}$$

into the right side of (10.26) and expand all terms, retaining only those of order ΔY_i or smaller. Thus, for example

$$\begin{aligned} Y'_i Y'_a &= Y_a Y_i + Y_i \Delta Y_a + Y_a \Delta Y_i + \Delta Y_a \Delta Y_i \\ &\approx Y_a Y_i + Y_i \Delta Y_a + Y_a \Delta Y_i \,. \end{aligned} \tag{10.28}$$

The result is a linear set of coupled, implicit equations connecting the change in mole numbers of all isotopes which enter as reactants or products for the species i. For example, the nuclear reactions which destroy or produce the isotope O^{16} for the miminal network are summarized in Table 10.2.

Table 10.2

Thermonuclear reactions and rates: Forward nuclear reactions which deplete the isotope O^{16} and reverse reactions which produce it (right column) and the reaction rate (left) for a simple minimal network. The reverse reactions for the compound nuclear reactions involving C^{12} on O^{16} and O^{16} on O^{16} are omitted in the miminal model (Woosley 1986).

Forward Reactions	Reaction Rate
$O^{16} + \alpha \rightarrow Ne^{20} + \gamma$	$\lambda_{\alpha,\gamma}(Ne^{20})$
$O^{16} + \gamma \rightarrow C^{12} + \alpha$	$\lambda_{\gamma,\alpha}(C^{12})$
$O^{16} + C^{12} \rightarrow (Si^{28})^* \rightarrow Mg^{24} + \alpha$	λ_{1216}
$O^{16} + O^{16} \rightarrow (S^{32})^* \rightarrow Si^{28} + \alpha$	λ_{1616}
Reverse Reactions	**Reaction Rate**
$Ne^{20} + \gamma \rightarrow O^{16} + \alpha$	$\lambda_{\gamma,\alpha}(O^{16})$
$C^{12} + \alpha \rightarrow O^{16} + \gamma$	$\lambda_{\alpha,\gamma}(O^{16})$

Using these reactions the equation for the rate of change of Y_{16}, the mole number of O^{16} nuclei, becomes

$$dY_{16} / dt = - Y_{16} Y_\alpha \lambda_{\alpha,\gamma}(Ne^{20}) - 2Y_{16}{}^2 \lambda_{1616} - Y_{16} Y_{12} \lambda_{1216} - Y_{16} \lambda_{\gamma,\alpha}(C^{12}) + Y_{20} \lambda_{\gamma,\alpha}(O^{16}) + Y_{12} Y_\alpha \lambda_{\alpha,\gamma}(O^{16}) . \tag{10.29}$$

Substituting (10.25) into the left side of this equation, using (10.27) on the right side, and dropping terms that are quadratic in the change in mole numbers we obtain

$$\Delta Y_{16}\{1/\Delta t + Y_a\lambda_{\alpha,\gamma}(Ne^{20}) + 4Y_{16}\lambda_{1616} + Y_{12}\lambda_{1216} + \lambda_{\alpha,\gamma}(C^{12})\}$$

$$+ \Delta Y_{20}\{-4\lambda_{\gamma,\alpha}(O^{16})\} + \Delta Y_{12}\{Y_{16}\lambda_{1212} - Y_\alpha\lambda_{\alpha,\gamma}(O^{16})\}$$

$$+ \Delta Y_\alpha\{Y_{16}\lambda_{\alpha,\gamma}(Ne^{20}) - Y_{12}\lambda_{\alpha,\gamma}(O^{16})\}$$

$$-\{Y_\alpha Y_{16}\lambda_{\alpha,\gamma}(Ne^{20}) + Y_{16}{}^2\lambda_{1616} + Y_{12}Y_{16}\lambda_{1216} + Y_{16}\lambda_{\gamma,\alpha}(C^{12})$$

$$- Y_{20}\lambda_{\gamma,\alpha}(O^{16}) - Y_{12}Y_\alpha\lambda_{\alpha,\gamma}(O^{16})\} = 0. \tag{10.30}$$

This is a linear equation in the four unknowns ΔY_{16}, ΔY_{20}, ΔY_{12}, and ΔY_α. Using the nuclear reactions coupling the other species in the network leads to eight more equations of this form, and the set can be written in matrix form:

$$\sum_k A_{ik}\Delta Y_k - B_i = 0. \tag{10.31}$$

The matrix elements A_{ik} represent quantities such as those in braces above, and the constant B_i represents all of the terms in the linearized rate equation which do not depend on the ΔY_i.

For most applications the matrix constructed from (10.31) contains nonzero elements along the diagonal, and in several rows on either side of it. If the reaction network involves only (α,γ) processes then the resulting matrix will be tridiagonal and may be inverted using the methods discussed in Chapter 4 for the implicit hydrodynamic equations. When more complicated couplings are included (such as in Table 10.2) then other matrix inversion techniques may need to be employed. A number of standard methods, such as Gaussian Elimination, are available which can be used to invert matricies of this type (see, for example, Press, Flannery, Teukolsky and Vetterling 1986). Once the equations have been formulated in the form (10.31) the matrix elements can be passed to a subroutine which inverts the matrix and returns the change in the nuclear mole numbers ΔY_i.

Time Step Control

The linearized rate equations (10.31) are a reasonable approximation to the differential equations as long as the quadratic terms which were omitted are small. If the nuclear energy generation is modelled principly as an energy source, then it is usually sufficient to restrict changes in composition such that $\Delta Y_i \leq \xi$, where typically $\xi = 10^{-2}$. A reasonable time step constraint is then

$$\Delta t = \xi_n \, dt / dY_i . \tag{10.32a}$$

If the residual compositions (with minimum mole numbers $Y_{i,\, min}$) of the elements are required for nucleosynthesis studies then the time step constraint

$$\Delta t = C_n \, min \{ Y'_i \, dt / dY'_i \} \tag{10.32b}$$

may be used where the constant C_n is less than unity. In studies of nucleosynthesis a value of 0.15 has been found to work well. A practical minimum for the mole numbers is

$$Y_{i,\, min} = 10^{-8}. \tag{10.33}$$

Isotopes having abundances below this value usually can be omitted from the rate equations.

Finally, an additional source of error can arise because the reaction rates $\lambda_{ab}(l)$ appearing in the rate equations vary strongly with temperature. Iterative solutions, using for example Newton-Raphson iteration, would improve the accuracy of the calculations. However, it may be as efficient to simply reduce the time step further so that the rates do not change appreciably over the computational time step.

Initial Conditions

The initial conditions for a nuclear network are the density and temperature of the plasma (from which the reaction rates are to be calculated), and the initial abundances of the nuclear isotopes $Y_i(0)$. The latter must be normalized such that the corresponding mass fractions satisfy the constraint

$$\sum_i X_i(0) = 1 . \tag{10.34}$$

While the exact solution of the differential rate equations (subject to the initial conditions above) will guarantee that the normalization condition is satisfied for all times, the solutions of the finite difference equations may not. It may be desirable to monitor the normalization constraint, or to automatically renormalize the abundances at the conclusion of each computational cycle.

Problem 10.3

Develop an explicit finite difference representation for the two competing reactions (10.14). What constraints must be imposed on the computational time step if the results are to make physical sense? Discuss the limitations of using such an approach for typical nuclear reactions in stars.

Problem 10.4

Develop a fully implicit finite difference representation for the two processes (10.14), and consider its form for time steps which are large compared with the reaction time scales. What is the asymptotic form of the finite difference representation? Do solutions to your model satisfy the constraint $\Sigma X_i = 1$?

Problem 10.5

Show that (α,γ) reactions lead to a reaction matrix (10.31) which is tridiagonal. Do this by applying the linearized implicit approach discussed above to the three stage process (10.8). What happens if the reverse reactions are allowed?

10.4 NEUTRINO REACTIONS

The neutrino reactions with matter which were included in the discussion of radiative transport in Chapter 6 represent only one class of neutrino-matter interactions of importance for the astrophysics of supernovae. In this section another type of interaction, neutrino emission associated with electron capture by heavy nuclei, will be considered. Additional neutrino processes also occur, such as electron capture by free protons, or positron capture by free neutrons, and these may be modeled using the methods developed below, or included in the diffusion process as outlined in Chapter 6. The notation used for the discussion of neutrinos in Chapter 6 will be followed here.

Electron Capture by Heavy Nuclei

Among the weak interaction processes which occur in the initial collapse of evolved stellar cores are electron and positron capture by a heavy nucleus which may be denoted by $N(A,Z)$. Here A is the atomic weight and Z the atomic charge of the nucleus. The reactions which occur are

$$e^- + N(A,Z) \rightarrow \nu_e + N^*(A,Z-1)$$

$$\nu_e + N(A,Z-1) \rightarrow e^- + N^*(A,Z)$$

$$(10.35a)$$

$$e^+ + N(A,Z) \rightarrow \bar{\nu}_e + N^*(A,Z+1)$$

$$\bar{\nu}_e + N(A,Z+1) \rightarrow e^+ + N^*(A,Z) \, .$$

$$(10.35b)$$

The first reaction, which is the most likely in the collapsing stellar core, leads to the production of neutron rich heavy nuclei, and produces electron neutrinos which can carry some of the gravitational binding energy away from the star.

The first reaction in (10.35a) occurs when a degenerate electron in the stellar core is captured by a proton inside a heavy nucleus, converting it into a neutron with the emission of an electron neutrino. Conservation of energy of the particles involved in electron capture requires that

$$\varepsilon_e + \varepsilon_p = \nu + \varepsilon_n + \varepsilon^*, \tag{10.36}$$

where ε_e is the energy of the electron, ν is the energy of the emitted neutrino, ε_p and ε_n are the energies of the proton in the nucleus $N(A,Z)$ and of the neutron in the nucleus $N(A,Z-1)$ respectively. Finally, ε^* is the nuclear excitation energy of the product nucleus [the excited state is denoted by the star in (10.35)]. Note that since the nucleus passes through an excited state, (10.35) does not represent a reversible reaction. Furthermore, the reactions describe a system that is not in β-equilibrium. In the limit as $\varepsilon^* \rightarrow 0$, the energies ε_p and ε_n reduce to the corresponding chemical potentials. The rate of production of neutrinos by electon capture on heavy nuclei is proportional to the product of three factors. The first is the electron flux, $n_e c$, where for all practical purposes the electron's velocity is equal to the speed of light c. The second is the electron capture cross section σ_{Ae}, which is a function of the neutrino energy ν, times the number density n_A of heavy nuclei. Since we are calculating the number of neutrinos emitted with energies in the range ν to $\nu + d\nu$, the cross section is to be evaluated at the neutrino energy ν. Finally, at the high densities reached late in the collapse stage, the neutrino opacity is large enough that the neutrinos become degenerate. Thus the forward process must include the neutrino blocking factor $1 - f_\nu$. Combining the terms above for the forward process, multiplying by the energy ν of the neutrinos produced, and denoting the electon neutrino spectral energy distribution by F_ν (as was done in Chapter 6), the rate of change in the energy spectrum for neutrinos in the energy range ν to $\nu + d\nu$ is

$$\left(\frac{dF_\nu}{dt}\right)_+ = n_e c n_A \sigma_{eA}(\varepsilon_e)(1 - f_\nu)\,\nu. \tag{10.37}$$

The electron number density is given in terms of the electron distribution function $f_e(\varepsilon_e)$ by

$$n_e = \alpha_e \varepsilon_e^2 f_e(\varepsilon_e)$$

$$f_e(x) = [\exp(x - \mu_e)/kT + 1]^{-1}. \tag{10.38}$$

The factor α_e is the usual phase space factor for electrons and μ_e is the electron chemical potential. A simple model for the capture cross section appearing in (10.37) is

$$\sigma_{eA}(\nu) \approx N^*_p \, \sigma_{ep}(\nu)$$

where N^*_p is the number of protons in an average heavy nucleus capable of capturing an electron, and $\sigma_{e\,p}(\nu)$ is the cross section for electron capture by a proton.

The change in the neutrino energy spectrum associated with the capture of an electron neutrino by a heavy nucleus [the second reaction in (10.35a)] can be constructed in a similar fashion, and is

$$\left(\frac{dF_\nu}{dt}\right)_- = n_\nu \, c \, n_A \, \sigma_{\nu A}(\nu) \, (1 - f_e) \, \varepsilon_e , \tag{10.39}$$

where $\sigma_{\nu A}$ is the neutrino capture cross section on a heavy nucleus, and n_ν is the number density of neutrinos having energies in the range ν to $\nu + d\nu$:

$$n_\nu = \alpha \, \nu^2 f_\nu(\nu) . \tag{10.40a}$$

A simple model for the capture cross section appearing in (10.39) is

$$\sigma_{\nu A}(\nu) \approx N^*_n \, \sigma_{\nu n}(\nu)$$

where N^*_n is the number of neutrons in an average heavy nucleus capable of capturing a neutrino, and $\sigma_{\nu n}(\nu)$ is the cross section for neutrino capture by a neutron. It is to be emphasized that the neutrino distribution function does not necessarily correspond to β-equilibrium. When the neutrinos are in β-equilibrium, the neutrino distribution function becomes

$$f_{\nu, eq}(x) = [\exp(x - \mu_\nu)/k_B T + 1]^{-1}. \tag{10.40b}$$

Here μ_ν is the neutrino chemical potential, and α is the usual neutrino phase space factor.

Problem 10.6

Using an analysis similar to the one preceding (10.37), derive an expression for the capture of an electron neutrino by a heavy nucleus, and show that it can be expressed in the form (10.39).

Using the principle of detailed balance, the expressions for the forward and inverse reactions (10.37) and (10.39) may be combined to obtain the net rate of change in the neutrino spectral energy:

$$\left(\frac{dF_\nu}{dt}\right)_{eA} = \left(\frac{dF_\nu}{dt}\right)_+ - \left(\frac{dF_\nu}{dt}\right)_-$$

$$= n_A c \, \sigma_{eA}(\nu) \, \nu \varepsilon_e^2 \, [f_e - f_\nu] . \tag{10.41}$$

A simplifying approximation will be made by noting that $\sigma_{e\,p}(\nu) = \sigma_o(\nu / m_e c^2)^2$, and by assuming that the number of protons in an average heavy nucleus which are effective in capturing electrons can be written in the form

$$n^*_p = n_A \, / Z - Z_{eq} / \tag{10.42}$$

where Z_{eq} is the β-equilibrium value of the nuclear charge, and $X_A \rho = A m_H n_A$. The simple expression (10.42) may be modified if more detailed nuclear model calculations for electron capture are available. Using (10.42) in (10.41), and noting that $F_\nu = \alpha \nu^3 f_\nu$, we find the simple expression

$$\left(\frac{dF_\nu}{dt}\right)_{eA} = Q\left(f_e - \frac{F_\nu}{\alpha \nu^3}\right) \tag{10.43}$$

where

$$Q \equiv \frac{\sigma_o c}{(m_e c^2)^2} \frac{X_A \rho}{A m_H} \nu^3 \varepsilon_e^2 |Z - Z_{eq}| . \tag{10.44}$$

In addition to satisfying (10.43) and (10.36), electron capture must also conserve lepton number. This requirement leads to a constraint on the rate of change of the distribution of electrons and neutrinos

$$\frac{d}{dt}\int \varepsilon_e^2 f_e d\varepsilon_e + \frac{d}{dt}\int \nu^2 f_\nu d\nu = 0. \tag{10.45}$$

Since the neutrino and the electron distribution functions are independent, the constraint reduces to

$$\varepsilon_e^2 \frac{df_e}{dt} = -\nu^2 \frac{df_\nu}{dt} = -\frac{1}{\nu}\frac{dF_\nu}{dt}. \tag{10.46}$$

It may be recalled (Section 6.1) that $F_\nu d\nu/\nu$ is the number density of neutrinos in the energy range ν to $\nu + d\nu$.

In a fully ionized plasma the number density of electrons is related to the number density of heavy nuclei (assuming a single species, and no free protons) by $n_e = Z\, n_A$. It then follows from (10.46) that

$$n_A \frac{dZ}{dt} = - \int \left(\frac{dF_\nu}{dt}\right)_{el\,cap} \frac{d\nu}{\nu}. \tag{10.47}$$

Electron capture removes thermal energy from the plasma, but leaves heavy nuclei in excited states which can decay by the emission of photons. The deexcitation of these nuclei can lead to plasma heating. A simple model for the change in plasma internal energy which does not follow these processes in detail, but which accounts for the net energy change in the neutrino field (and associates it with a corresponding change in the internal energy of the plasma) is given by

$$\rho \frac{d\varepsilon}{dt} = - \int \left(\frac{dF_\nu}{dt}\right)_{el\,cap} d\nu. \tag{10.48}$$

This is simply a statement of energy conservation applied to the electron capture process (10.35a).

A set of equations describing electon capture by heavy nuclei (and its inverse) are given by (10.43)-(10.45), as well as constitutive relations defining the electon and neutrino chemical potentials, the difference $\mu_{Z-1,A} - \mu_{Z,A}$ and the equilibrium charge Z_{eq}. The latter quantities must be supplied, as will be assumed in the discussion of finite difference methods below.

Problem 10.7

Use the principle of detailed balance to derive the expression (10.41) for the change in the neutrino energy spectrum associated with the two processes given by (10.35a). What effect does the presence of the excited state play in the analysis? In particular, discuss the problem associated with the excitation energy ε^*.

Problem 10.8

In β-equilibrium the electron and neutrino distribution functions are equal: $f_e = f_{\nu,eq}$. What effect does the presence of nuclear exicted states, as represented by (10.36), have on the equilibrium state?

Problem 10.9

Discuss modifications to the electron capture model above that would be required to describe the positron capture process (10.35b). Include expressions for the change in atomic charge and the change in plasma internal energy.

10.5 Finite Difference Model for Electron Capture

The physical variables appearing in the electron capture model described in the previous section are for the most part those needed for a description of neutrino transport (Chapter 6 and Chapter 7). They should be centered on the computational grid in the same way as they were for the transport. In particular, the same energy group structure as discussed in Chapter 6 for neutrinos will be used here. There are no spatial centering difficulties associated with (10.43)-(10.45), so only the temporal and energy group indices of the finite difference variables will be denoted in the discussions below.

The electron capture process may often be treated using the method of operator splitting, and may be solved, for example, at the beginning of the hydrodynamic step (see Table 2.2 for an example of an operator split one dimensional Lagrangian approach). It may be noted that (10.43)-(10.45) do not change the material density or velocity directly, but only change the thermal and neutrino energies and the nuclear charge.

Before discussing the order in which the equations can be solved we construct finite difference representations for them. Consider the rate equation (10.43). If either the electron number density or the neutrino number density were small relative to the other it might be possible to use explicit time differencing for the equations. In general this is not the case, particularly if it is of interest to model the way in which the neutrino distribution approaches β-equilibrium. It is then best to use implicit differencing (as noted in the discussions of operator splitting in Section 2.3). An appropriate finite difference representation of (10.43) is

$$F'_{j+1/2} - F_{j+1/2} = Q_{j+1/2}\,\Delta t\,(f'_e - F'_{j+1/2}/\alpha\nu^3_{j+1/2})\,. \tag{10.49}$$

The electron and neutrino distributions appearing on the right side are evaluated at the advanced time, and all subscripts except for those denoting the energy groups have been omitted for notational convenience. The dependence on the advanced value of the electron distribution can be eliminated by using the lepton number constraint (10.46), which has the finite difference representation

$$\varepsilon_e^2\,\Delta f_e = -\Delta F_\nu/\nu \tag{10.50}$$

where $\Delta f_e = f_e' - f_e$ and $\Delta F_\nu = F_\nu' - F_\nu$. The right side of (10.49) may be rewritten so that f_e' and F_ν' are replaced by Δf_e and ΔF_ν respectively. The lepton number constraint (10.50) may be used to eliminate Δf_e, and the result may then be solved for the change in the neutrino spectral energy. The algebra is straightforward, with the result

$$\Delta F_{j+1/2} = \frac{Q_{j+1/2}\,\Delta t\,[f_e - \dfrac{F_{j+1/2}}{\alpha\nu^3_{j+1/2}}]}{1 + Q_{j+1/2}\,\Delta t\,[\dfrac{1}{\nu^3_{j+1/2}} + \dfrac{1}{\varepsilon_e^2\nu_{j+1/2}}]\,\alpha^{-1}}\,. \tag{10.51}$$

The new value of the neutrino spectral distribution is then given by

$$F'_{j+1/2} = F_{j+1/2} + \Delta F_{j+1/2} \tag{10.52}$$

The change in the nuclear charge associated with electron capture by heavy nuclei (10.47) can be represented in the finite difference form as a series of partial changes of the form

$$- \Delta Z_{A,j+1/2} = \frac{\Delta F_{j+1/2}}{n_A} \frac{\Delta v_{j+1/2}}{v_{j+1/2}} . \tag{10.53a}$$

Here we define

$$\Delta Z_A \equiv Z'_A - Z_A = \sum_j \Delta Z_{A,j+1/2} . \tag{10.53b}$$

Similarly, the change in the plasma internal energy (10.48) can be represented by the finite difference form

$$- \Delta \varepsilon_{j+1/2} = \frac{\Delta F_{j+1/2}}{\rho} \Delta v_{j+1/2} . \tag{10.54}$$

Solution of the Capture Equations

The set of equations (10.51)-(10.52) and (10.53)-(10.54) and the constitutive relations for the energies of the proton and neutron in the heavy nucleus and excitation energy represent a set of coupled equations which must be solved for each energy group. Before we consider the solution method in detail, several remarks about the constitutive relations for the model are in order. The electron chemical potential appearing in (10.38) can be expressed as a function of the plasma density and temperature. The neutrino chemical potential can be obtained from the neutrino distribution function itself (Bowers and Wilson 1982). In practice the initial neutrino spectrum and plasma temperature may be used to construct μ_ν. In principle the energies ε_p and ε_n appearing in (10.36) can be obtained from nuclear theory. In the discussion below these energies will be approximated by the equilibrium chemical potentials: $\varepsilon_p \rightarrow \mu_{Z,A}$ and $\varepsilon_n \rightarrow \mu_{Z-1,A}$. A model for the nuclear excitation energy $\varepsilon*$ is also needed to evaluate the energy constraint on the captured particle (10.36). A simple model which has been used in core collapse simulations is

$$\varepsilon_e = \mu_{Z-1,A} - \mu_{Z,A} + \nu + \frac{\varepsilon*}{1 + (k_B T / \varepsilon*)^2}$$

(10.55)

where $\varepsilon*$ is a constant. The temperature dependent term effectively cuts off deexcitation of the heavy nuclei at high temperatures. The threshold for electron capture according to (10.36) occurs for $\nu = 0$. Numerically, capture is turned off if the electron energy is less than

$$\varepsilon_{th} = \mu_{Z-1,A} - \mu_{Z,A} + \frac{(\varepsilon*)^3}{(\varepsilon*)^2 + (k_B T)^2} .$$

(10.56)

The temperature dependent term on the right cuts off the effect of deexcitation when the temperature becomes large.

The solution of the coupled set of equations describing electron capture may be solved in the following way. First, the initial values of the neutrino spectrum $F_{j+1/2}$ are stored, and (10.43) is solved for the distribution $F'_{3/2}$ in the lowest energy group, $\nu_{3/2}$. The old and new values of the spectrum for $j = 1/2$ are then used to obtain the partial change in the plasma internal temperature using (10.54), which may be denoted by $\Delta\varepsilon_{3/2}$, and from (10.53) the partial change in nuclear charge $\Delta Z_{3/2}$. The new plasma temperature and nuclear charge are then used as input for the solution of the neutrino distribution $F'_{5/2}$ in the next energy group. The process is repeated for each higher energy group until new values for the entire spectrum have been constructed. At the end of the process, the net change in plasma energy and nuclear charge has been obtained as the sum of partial changes in each quantity. This approach is similar in philosophy to the partial temperature method used in Chapter 6 to solve the radiation transport equation, and may be subject to similar problems. In practice it is possible to monitor the change in partial quantities as a way of determining if the procedure is failing.

References

W. D. Arnett and J. W. Truran, 1969, *Ap. J.* **157**, 339.

R. L. Bowers and T. J. Deeming, *Astrophysics,* Vol. I (Boston: Jones and Bartlett Publishers, 1984).

R. L. Bowers and J. R. Wilson, 1982, *Ap. J. Suppl.*, **50**, 115.

H. Y. Chiu, *Stellar Physics* (Waltham, Mass: Blaisdell, 1968).

D. D. Clayton, *Principles of Stellar Evolution and Nucleosynthesis* (New York: McGraw-Hill, 1968).

J. Pl Cox and R. T. Guili, *Stellar Structure,* Vol. I (New York: Gordon and Greach, 1968).

W. H. Press, B. P. Flannery, S. A. Teukolsky and W. T. Vetterling, *Numerical Recipies* (London: Cambridge University Press, 1986).

T. A. Weaver, G. B. Zimmerman and S. E. Woosley, 1978, *Ap. J.*, **225**, 1021.

S. E. Woosley, 1986, *Nucleosynthesis and Chemical Evolution*, 16th Advanced Course of the Swiss Academy of Astronomy and Astrophysics, ed. B. Hauck and A. Maeder (Geneva Observatory: Sauverny, Switzerland).

S. E. Woosley, W. D. Arnett and D. D. Clayton, 1973, *Apl J. Suppl.*, **26**, 231.

EPILOGUE

The development of a computational program involves three equally important steps. First, one must select a numerical method with which to solve the physical equations of interest. Second, these methods must be translated into a suitable programming language. Finally, the accuracy of the method must be established.

In previous chapters we presented a class of finite difference methods which have been found to work well in practice. For heuristic reasons we have used simple examples (such as donor cell advection, the von Neumann artificial viscous stress, and face centered components of the fluid velocity) to introduce some topics. Subsequent discussions have extended these simple concepts in an attempt to improve accuracy. The heuristic examples should not be considered as alternatives (except for situations where good accuracy is not required) to the more modern methods which are presented in the latter part of each chapter. The use of face centered velocity components is not recommended except possibly for problems involving pure hydrodynamics in noncurvilinear coordinates, or which do not require tensor stresses, such as viscosity.

The second step, translating numerical methods into a programming language, has not been discussed here.

The third step is extremely important for any computational project. Since it is difficult at best to formally identify the true accuracy of a numerical method, experiments must be performed to assess it. The first class of experiments makes use of any known analytic (often similarity) solutions to the equations represented by the numerical method. Obvious examples are: solutions for shock tubes, free expansion and shock formation for hydrodynamics; the propagation of a thermal wave (Marshak wave) for radiation diffusion; and polytropic equilibrium configurations for gravitation.

Eulerian advection methods should be tested, for example, by translating slab and Gaussian density profiles through the grid. Schemes in two or more spatial dimensions whould be tested by rotating simple geometric shapes on the grid. They should also be tested to verify that they preserve the symmetry of initial conditions. For example, a spherical mass should collapse spherically in the absence of initial perturbations. Finally, methods that are capable of describing the approach to a steady state should do so in a reasonably accurate manner.

Accurate solution to the examples listed above do not guarantee that a numerical method is accurate (or even correct) when used on more complicated problems of interest. One way to assess a method used for complex problems is to compare the results obtained from it with the results obtained from other numerical methods. This is a nontrivial exercise, and requires that great care be taken so that both methods in fact solve the same problem with the same initial and boundary conditions.

Since finite difference representations are chosen which reduce to the correct partial differential equation in the limit as the time step and grid spacing approach zero, calculations should be repeated using successively finer spatial zoning until the results converge (grid refinement). When this is done for explicit methods that require time step constraint, reducing the grid spacing effectively reduces the time step. If grid refinement is used to test the convergence of implicit methods which do not employ accuracy time step constraints, the maximum allowed time step should be reduced as well.

Conserved quantities such as mass, energy and momentum are associated with many physical processes. Whenever possible the numerical methods should be formulated to conserve these quantities, and their values should be monitored during the course of calculation. An equally important variable whose value should be followed is the entropy. For adiabatic processes entropy is conserved locally. Schemes which conserve total energy by construction need to be checked to verify that they properly account for entropy generation. When dissipative processes occur the production of entropy should satisfy the second law of thermodynamics. Errors in entropy generation, which may not show up as energy errors, may nevertheless have a serious impact on the accuracy of results. In principle, any one of the variables temperature, total energy, thermal energy or entropy could be used as a dependent variable in numerical schemes.

Many systems of partial differential equations include constraint equations, and carefully constructed numerical schemes should preserve them during the course of a calculation. Periodic monitoring to see if the constraint equations are satisfied can also be used as a form of error estimation.

Appendix A

COORDINATE TRANSFORMATION OF THE EQUATIONS OF MOTION

Coordinate transformations of the spatial independent variables in first order partial differential equations will be reviewed in a form applicable to the Eulerian and Lagrangian equations of motion. Consider the two independent coordinates x,t and functions of them that satisfy the following type of first order partial differential equations:

$$\left(\frac{\partial f}{\partial t}\right)_x + \left(\frac{\partial g}{\partial x}\right)_t = 0 \, .$$

(A.1)

Many of the equations of motion considered in previous chapters are of this form, or are simple extensions of it. Under the coordinate transformation

$$\xi = \xi(x, t) \, ,$$

(A.2)

the independent variables x,t are to be replaced by the new independent variables ξ, t, and funtions $f(x,t)$ are transformed into functions denoted by $f'(\xi,t)$. The class of transformations (A.2) will be restricted to those for which

$$f'(\xi, t) = f(x, t) \ . \tag{A.3}$$

Examples of such transformations include Galilean transformations between Cartesian coordinate systems.

In the new coordinate system the form of the equations of motion may be found as follows. In the following discussions the inverse transformation

$$x = x(\xi, t) \tag{A.4}$$

is assumed to exist. Variations in the dependent variables x and ξ are related in the obvious way

$$dx = \left(\frac{\partial x}{\partial \xi}\right)_t d\xi + \left(\frac{\partial x}{\partial t}\right)_\xi dt \ . \tag{A.5}$$

Equations (A.3) and (A.4) may be used to relate partial derivatives in ξ, t, to partial derivatives in x,t . Recalling (A.3), which implies that $df = df'$, we find upon expansion

$$\left(\frac{\partial f}{\partial x}\right)_t dx + \left(\frac{\partial f}{\partial t}\right)_x dt = \left(\frac{\partial f'}{\partial \xi}\right)_t d\xi + \left(\frac{\partial f'}{\partial t}\right)_\xi dt \ . \tag{A.6}$$

Eliminating dx between (A.5) and (A.6), and noting that because dt and $d\xi$ are independent their coefficients must each vanish, we find

$$\left(\frac{\partial f}{\partial x}\right)_t = \left(\frac{\partial f'}{\partial \xi}\right)_t \left(\frac{\partial \xi}{\partial x}\right)_t \tag{A.7}$$

and

$$\left(\frac{\partial f}{\partial t}\right)_x = \left(\frac{\partial f'}{\partial t}\right)_\xi - \left(\frac{\partial x}{\partial t}\right)_\xi \left(\frac{\partial \xi}{\partial x}\right)_t \left(\frac{\partial f'}{\partial \xi}\right)_t \ . \tag{A.8}$$

Equation (A.7) has been used to rewrite the last term on the right hand side of (A.8). These are the desired relations between partial derivatives in the two coordinate systems. If (A.7) and (A.8) are substituted into (A.1) we obtain, after multiplying through by the common factor $(\partial x/\partial \xi)_t$,

$$\left(\frac{\partial f'}{\partial t}\right)_{\xi}\left(\frac{\partial x}{\partial \xi}\right)_t - \left(\frac{\partial x}{\partial t}\right)_{\xi}\left(\frac{\partial f'}{\partial \xi}\right)_t + \left(\frac{\partial g'}{\partial \xi}\right)_t = 0\,. \tag{A.9}$$

The first two terms may now be rewritten using the identity

$$\left(\frac{\partial f'}{\partial t}\right)_{\xi}\left(\frac{\partial x}{\partial \xi}\right)_t - \left(\frac{\partial x}{\partial t}\right)_{\xi}\left(\frac{\partial f'}{\partial \xi}\right)_t = \left[\frac{\partial}{\partial t} f'\left(\frac{\partial x}{\partial \xi}\right)_t\right]_{\xi} - \left[\frac{\partial}{\partial \xi} f'\left(\frac{\partial x}{\partial t}\right)_{\xi}\right]_t \tag{A.10}$$

so that (A.9) takes the form

$$\left[\frac{\partial}{\partial t} f'\left(\frac{\partial x}{\partial \xi}\right)_t\right]_{\xi} + \left\{\frac{\partial}{\partial \xi}\left[g' - f'\left(\frac{\partial x}{\partial t}\right)_{\xi}\right]\right\}_t = 0 \tag{A.11}$$

or in equivalent form

$$\left(\frac{\partial f'}{\partial t}\right)_{\xi} + \left(\frac{\partial \xi}{\partial x}\right)_t\left\{\frac{\partial}{\partial \xi}\left[g' - f'\left(\frac{\partial x}{\partial t}\right)_{\xi}\right]\right\}_t = -\left(\frac{\partial \xi}{\partial x}\right)_t f' \frac{\partial^2 x}{\partial \xi\, \partial t}\,. \tag{A.12}$$

Equation (A.11), which should be compared with (A.1), is the equation of motion in the new coordinate frame. The results above may be used to relate equations of motion in various coordinate frames.

Eulerian Equations in a Moving Frame

The Eulerian fluid equations in a moving grid which are discussed in Chapter 5 can be obtained as a special case of (A.12) and the methods above. Let x represents the fixed coordinate system in which the Eulerian equations are most easily derived, and consider a transformation to the new system ξ such that

$$v_g = \left(\frac{\partial x}{\partial t}\right)_\xi .$$

(A.13)

Substituting (A.13) into (A.5) and integrating subject to the initial conditions $x = 0$ for $t = 0$ gives

$$\int \left(\frac{\partial x}{\partial \xi}\right)_t d\xi = x - v_g\, t .$$

(A.14)

For $(\partial x/\partial \xi)_t = 1$, this reduces to a Galilean transformation from the fixed coordinate frame x to the frame ξ moving with relative velocity v_g. Substituting (A.13) into (A.12) and using (A.7) and (A.3) we easily obtain

$$\left(\frac{\partial f}{\partial t}\right)_\xi + \left[\frac{\partial}{\partial x}(g - f v_g)\right]_t = -f\left(\frac{\partial v_g}{\partial x}\right)_t .$$

(A.15)

Notice that in (A.15) the time derivative is evaluated in the moving coordinate frame while the spatial derivatives are taken with respect to the original (fixed) coordinate frame. If $f = \rho$ and $g = \rho v$, then (A.15) reduces to

$$\left(\frac{\partial \rho}{\partial t}\right)_\xi + \left[\frac{\partial}{\partial x}\rho(v - v_g)\right]_t = -\rho\left(\frac{\partial v_g}{\partial x}\right)_t$$

(A.16)

which is identical to the continuity equation (5.4) (the prime on the density has been omitted for simplicity).

Problem A.1

Use the results above to show that the Eulerian momentum equation has the form (5.6) when expressed in a moving frame of reference. Extend the results to the Eulerian internal energy equation

$$\left(\frac{\partial \rho\varepsilon}{\partial t}\right)_x + \left(\frac{\partial \rho\varepsilon v}{\partial x}\right)_t = -P\left(\frac{\partial v}{\partial x}\right)_t , \tag{A.17}$$

and show that it takes the form (5.10) in the moving frame.

Starting with the hydrodynamic equations in standard Eulerian form, we can show that they take the following form in the new frame ξ, t, (Problem A.1):

mass continuity equation:

$$\left[\frac{\partial}{\partial t}\rho'\left(\frac{\partial x}{\partial \xi}\right)_t\right]_\xi + \left\{\frac{\partial}{\partial \xi}\left[\rho' v' - \left(\frac{\partial x}{\partial t}\right)_\xi \rho'\right]\right\}_t = 0 \tag{A.18}$$

momentum equation:

$$\left[\frac{\partial}{\partial t}\rho' v'\left(\frac{\partial x}{\partial \xi}\right)_t\right]_\xi + \left\{\frac{\partial}{\partial \xi}\left[P' + \rho'(v')^2 - \left(\frac{\partial x}{\partial t}\right)_\xi \rho' v'\right]\right\}_t = 0 \tag{A.19}$$

internal energy equation:

$$\left[\frac{\partial}{\partial t}\rho'\varepsilon'\left(\frac{\partial x}{\partial \xi}\right)_t\right]_\xi + \left\{\frac{\partial}{\partial \xi}\left[\rho'\varepsilon' v' - \rho'\varepsilon'\left(\frac{\partial x}{\partial t}\right)_\xi\right]\right\}_t = -P'\left(\frac{\partial v'}{\partial \xi}\right)_t \tag{A.20}$$

Lagrangian Equations of Motion

The Lagrangian equations of motion discussed in Chapter 4 are obtained from (A.18)-(A.20) by choosing the transformation (A.2) such that ξ represents a spatial coordinate system which moves locally with the velocity of the fluid (comoving frame). In such a frame the fluid velocity v is defined by

$$v = \left(\frac{\partial x}{\partial t}\right)_\xi \tag{A.21}$$

where for convenience here and below we have omitted the prime on quantities defined in the new coordinate system. Substituting this relation into the mass continuity equation (A.18) yields

$$\left[\frac{\partial}{\partial t}\rho\left(\frac{\partial x}{\partial \xi}\right)_t\right]_\xi = 0 \tag{A.22}$$

$$\left(\frac{\partial \rho}{\partial t}\right)_\xi\left(\frac{\partial x}{\partial \xi}\right)_t = -\rho\left[\frac{\partial}{\partial \xi}\left(\frac{\partial x}{\partial t}\right)_\xi\right]_t = -\rho\left(\frac{\partial v}{\partial \xi}\right)_t \tag{A.23}$$

where differentiation with respect to t and ξ have been interchanged, and the definition (A.21) used in the last step. Multiplying through by $(\partial \xi/\partial x)_t$ and recalling that partial differentiation in the two coordinate systems is related by (A.7) we obtain the mass continuity equation in Lagrangian form

$$\left(\frac{\partial \rho}{\partial t}\right)_\xi = -\rho\left(\frac{\partial \xi}{\partial x}\right)_t\left(\frac{\partial v}{\partial \xi}\right)_t = -\rho\left(\frac{\partial v}{\partial x}\right)_t \; . \tag{A.24}$$

The (Lagrangian) partial derivative on the left hand side of this equation has been denoted by $d\rho/dt$ in Chapter 4. As a consequence of the choice of coordinates above, the mass which lies between ξ

and $\xi + d\xi$ at time $t = 0$ must lie between the coordinates x and $x + dx$ at any later time t. This requires that

$$\rho(x, t)\, dx = \rho'(\xi, 0)\, d\xi \tag{A.25}$$

where the right hand side is independent of time. Using (A.25) to evaluate $(\partial\xi/\partial x)_t$, in (A.24) yields the mass continuity equation in Lagrangian coordinates

$$\left(\frac{\partial \rho}{\partial t}\right)_\xi = - \frac{\rho(x, t)^2}{\rho(\xi, 0)} \left(\frac{\partial v}{\partial \xi}\right)_t . \tag{A.26}$$

The Lagrangian form of the momentum and internal energy equations may be obtained in similar fashion (Problem A.2).

Problem A.2

Use the results above to derive the Lagrangian form of the momentum equation and the internal energy equation in one dimensional planar coordinates. Express the results both in terms of the Lagrangian coordinate ξ, and in terms of spatial coordinates x. Compare the answer with the results in Chapter 4.

Appendix B

CURVILINEAR COORDINATE GRIDS

The two dimensional, axisymmetric coordinate grid discussed in the first part of Chapter 5, while not Cartesian, does not reflect all of the difficulties associated with the construction of curvilinear coordinate grids in general. Consider a set of coordinates ξ^i $(i = 1,2,3)$ covering a three dimensional space, and the metric $\gamma_{i\,k}$ defined such that the distance between the two nearby points $\xi^k + d\xi^k$ and ξ^k is given by (Sokolnikoff 1964)

$$ds^2 = \sum_{i,k=1}^{3} \gamma_{i\,k}\, d\xi^i d\xi^k. \tag{B.1}$$

The coordinate system is said to be orthonormal if the off diagonal components of the metric vanish:

$$\gamma_{i\,k} = 0 \qquad \text{if } i \neq k\,. \tag{B.2}$$

Nonorthonormal coordinate systems can be constructed and finite difference approximations based on them. However, for the remainder of this section ξ^k shall be assumed to represent an orthonormal coordinate system. The line element (B.1) is then diagonal.

The metric can be represented in analytic form, or it may represent a set of values defined at grid points (numerical metric).

Table B.1 gives the analytic metric for two simple, orthonormal coodinate systems.

Table B.1

Examples of orthonormal coordinate systems and the components of the associated metric tensor.

Coordinate System	ξ^1	ξ^2	ξ^3	Metric Components		
				γ_{11}	γ_{22}	γ_{33}
Cylindrical	z	r	θ	1	1	r^2
Spherical	r	θ	ϕ	1	r^2	$r^2 sin\theta$

We shall restrict attention here to two dimensional coordinate grids based on analytic metrics only, and will denote by the finite difference $\Delta\xi^i$ the coordinate distance between two grid points along the coordinate line ξ^i. The spherical coordinates (r,θ) will be used as a specific example. The grid points (ζ^1,ζ^2) will be denoted by the subscripts (k,j) respectively. This notation is consistent with that used in the first part of Chapter 5 for the axially symmetric cylindrical coordinate grid.

Geometric Distances

The differential geometric distance (physical distance) measured along any one of the coordinate lines in a curvilinear coordinate system is given in terms of the metric by

$$ds_i \equiv (\gamma_{ii})^{1/2} d\xi^i . \tag{B.3}$$

The distance between two points connected by a given coodinate line is found by integrating (B.3) between the two points:

$$\Delta s_i = \int ds_i = \int_{\xi^i}^{\xi^i + \Delta\xi^i} (\gamma_{ii})^{1/2} d\xi^i . \tag{B.4}$$

Here $\Delta\xi^i$ represents the coordinate separation between the two points. When the metric tensor is available in analytic form, then (B.4) can usually be evaluated in closed form.

Finite Difference Area Elements

The differential geometric area dA_i (physical area) normal to a given coordinate line ξ^i is given in terms of the metric by

$$dA_i \equiv \varepsilon_{ijk} ds_j ds_k \tag{B.5}$$

where ε_{ijk} is zero if any of the indices are equal, and is $+1$ (-1) if they are an even (odd) permutation of the indices. Summation over repeated indices is assumed in (B.5). The finite area bounded by a set of coordinate lines is obtained from the integral of (B.5)

$$\Delta A_i = \int dA_i = \varepsilon_{ijk} \int_{\xi^j}^{\xi^j + \Delta\xi^j} (\gamma_{jj})^{1/2} d\xi^j \int_{\xi^k}^{\xi^k + \Delta\xi^k} (\gamma_{kk})^{1/2} d\xi^k . \tag{B.6}$$

Finite Difference Volume Elements

The differential volume element at a point in a curvilinear coordinate system is give in terms of the metric by

$$dV = (\gamma)^{1/2} d\xi^1 d\xi^2 d\xi^3 \tag{B.7}$$

where $\gamma \equiv \gamma_{11}\,\gamma_{22}\,\gamma_{33}$ is the determinant of the metric tensor. The volume of a finite region between a set of coordinate lines is given by the integral of (B.7) along the coordinate lines:

$$\Delta V = \int dV = \int_{\xi^1}^{\xi^1+\Delta\xi^1} (\gamma_{11})^{1/2} d\xi^1 \int_{\xi^2}^{\xi^2+\Delta\xi^2} (\gamma_{22})^{1/2} d\xi^2 \int_{\xi^3}^{\xi^3+\Delta\xi^3} (\gamma_{33})^{1/2} d\xi^3 . \tag{B.8}$$

Two Dimensional Spherical Coordinates

The results above may be applied to the two dimensional spherical coordinate system (r,θ) discussed at the end of Chapter 5 to obtain finite difference expressions for zonal volumes and for the area of the zone faces. For example, the upper area of a zone $\Delta V_{j+1/2,\ k+1/2}$ normal to the radial direction, which will be denoted by $A^r{}_{j+1,\ k+1/2}$, is given by

$$A^r_{j+1,k+1/2} = \int ds_\theta\, ds_\phi = r^2_{j+1} \int \sin\theta\, d\theta\, d\phi$$

$$= 2\pi r^2_{j+1} (\cos\theta_k - \cos\theta_{k+1}) . \tag{B.9}$$

The area normal to the θ coordinate line is

$$A^\theta_{j+1/2,k} = 2\pi (r^2_{j+1} - r^2_j) \sin\theta_k . \tag{B.10}$$

Problem B.1

(a) Use the metric for cylindrical coordinates (Table B.1) and the results above to find the area of the faces of a zone on the (r,z) grid discussed in Chapter 5.

(b) Use the metric for spherical coordinates to find the area of the faces of a zone on an (r,θ) grid.

Problem B.2

Compare (B.9)-(B.10) with the corresponding differential areas for a zone in spherical coordinates. How do the exact areas compare with the simple approximations

$$A^r_{j,k+1/2} = 2\pi r^2 \sin\theta\,\Delta\theta, \qquad A^\theta_{j+1/2,k} = 2\pi r^2 \Delta r\,. \tag{B.11}$$

Can the centering of quantities on the right side of either of the expressions in (B.11) be defined in such a way as to make these expressions exact?

The volume element on the spherical coordinate grid follows from (B.8), and is

$$\Delta V_{j+1/2,k+1/2} = \frac{2\pi}{3}(r^3_{j+1} - r^3_j)(\cos\theta_k - \cos\theta_{k+1})\,. \tag{B.12}$$

For finite Δr and $\Delta\theta$ this is not the same as the simple approximation

$$\Delta V_{j+1/2,k+1/2} \equiv 2\pi r^2_{j+1/2} \sin\theta_{k+1/2}\,\Delta\theta_{k+1/2}\,\Delta r_{j+1/2}\,. \tag{B.13}$$

Problem B.3

Can the grid coordinates appearing in (B.13) be defined so that the expression is exact?

Problem B.4

Derive a consistent representation on a two dimensional finite difference grid for the areas $A^r{}_{j+1/2,k+1/2}$ and $A^\theta{}_{j+1/2,k+1/2}$, and for the volume elements $\Delta V_{j+1/2,k}$ and $\Delta V_{j,k+1/2}$. How must $r^2{}_{j+1/2}$ be defined in this case? Finally, discuss how the factor $\Delta\theta_{k+1/2}$ is to be defined, and find a representation for $\Delta\theta_k$.

Reference

I. S. Sokolnikoff, *Tensor Analysis*, (Wiley and Sons: New York, 1964).

Appendix C

INTERPOLATION FORMULAE

It is often necessary to find interpolated values of quantities which are defined at grid points, on zone faces, or at the center of zones. This occurs most commonly when higher order representations of spatial derivatives are required. Interpolation can be complicated when the grid is nonuniform. The procedure will be illustrated in one spatial dimension for a typical zone centered variable, and for a typical edge centered variable. In each case quadratic interpolation will be used, but the approach can easily be extended to other fitting schemes. In particular, the results below are used to construct second order interpolation for advection and remapping. It should be noted that quadratic or higher order interpolation can result in values for variables which do not lie between neighboring values (the results are nonmonotonic). For this reason, special care should be exercised to guarantee that the interpolated results are monotonic, as discussed in Chapter 5.

Edge Centered Variables

Suppose that F_k represents a typical edge centered variable, such as the fluid velocity discussed in Chapters 4 and 5, defined on a nonuniform grid with zone widths $x_{k+1} - x_k = \Delta x_{k+1/2}$ (Figure C.1). An interpolated value of F_k, located at an arbitrary point x within the zone is given by the quadratic fit:

$$F(x) = a x^2 + bx + c \,. \tag{C.1}$$

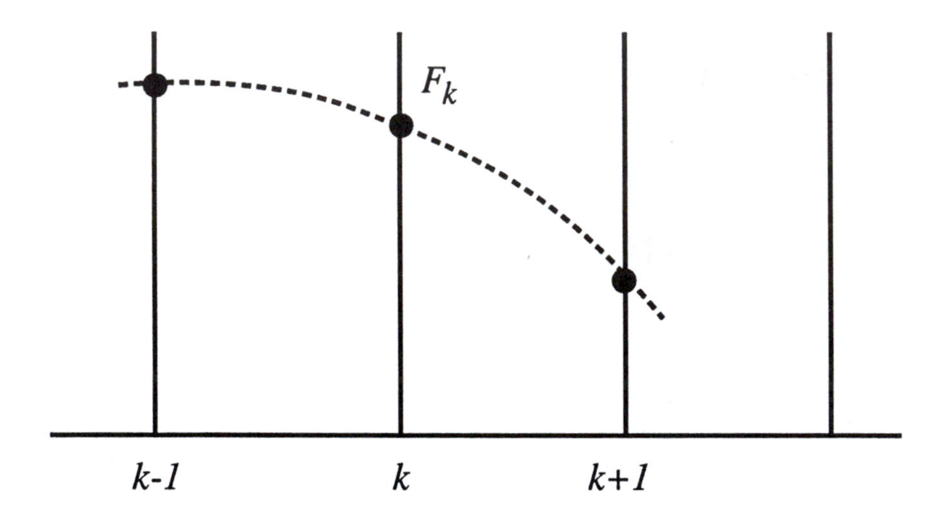

Figure C.1. Quadratic fit (C.1) (dashed curve) through three neighboring, zone edge centered variables on a nonuniform spatial grid. The coefficients are given by (C.4)-(C.6).

The coefficients are fixed so that the quadratic passes through the three neighboring values F_{k-1}, F_k and F_{k+1}. Noting that $F_k \equiv F(x_k)$, (C.1) may be used, for example, to show that

$$F_{k+1} - F_k = a\,(x_{k+1}^2 - x_k^2) + b\,(x_{k+1} - x_k) = \left[a\,(x_{k+1} + x_k) + b\right] \Delta x_{k+1/2} \tag{C.2}$$

which relates the derivative of F at the zone center,

$$D_{k+1/2} = \frac{F_{k+1} - F_k}{\Delta x_{k+1/2}} \tag{C.3}$$

to the coefficients s and b. Using (C.2) with $k \rightarrow k+1$, and subtracting the result from (C.2) itself leads to an expression for the coefficient a appearing in (C.1) of the form

$$a = \frac{D_{k+1/2} - D_{k-1/2}}{2\,\Delta x_k}. \tag{C.4}$$

Substituting into (C.2) gives an expression for the coefficient b

$$b = D_{k+1/2}\left\{1 - \frac{x_k + x_{k+1}}{2\,\Delta x_k}\right\} + D_{k-1/2}\frac{x_k + x_{k+1}}{2\,\Delta x_k}. \tag{C.5}$$

Finally, the expression for c, obtained by substituting (C.4) and (C.5) into (C.1) evaluated at x_k, is

$$c = F_k - D_{k-1/2}\frac{x_k x_{k+1}}{2\,\Delta x_k} - D_{k+1/2}\left\{x_k - \frac{x_k x_{k+1}}{2\,\Delta x_k}\right\}. \tag{C.6}$$

A useful result obtained by differentiating (C.1) is the derivative of the quadratic fit at the grid point k:

$$\left(\frac{\partial F}{\partial x}\right)_k = \frac{D_{k+1/2}\,\Delta x_{k-1/2} + D_{k-1/2}\,\Delta x_{k+1/2}}{2\,\Delta x_k}. \tag{C.7}$$

Notice that for uniform zoning the derivative at the zone edge is just the average of the zone centered derivatives.

Zone Centered Variables

Figure C.2 shows a typical zone centered variable $A_{k+1/2}$ on a nonuniform spatial grid. A quadratic of the form

$$A(x) = \alpha x^2 + \beta x + \gamma \tag{C.8}$$

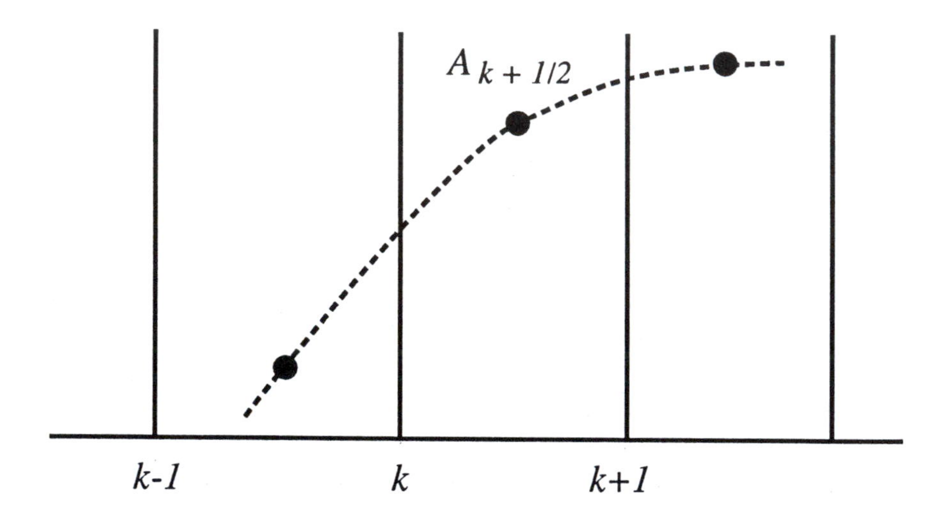

Figure C.2. Quadratic fit (C.8) (dashed curve) through three neighboring, zone centered variables on a nonuniform spatial grid. The coefficients are given by (C.9).

can be fit using any three neighboring points such as $A_{k-1/2}$, $A_{k+1/2}$ and $A_{k+3/2}$, with $A(x_{k+1/2}) = A_{k+1/2}$. The coefficients are found as above, and are listed here for reference:

$$\alpha = \frac{(D_{k+1} - D_k)}{2\Delta}$$

$$\beta = D_k - \frac{(D_{k+1} - D_k)}{2\Delta}(x_{k+1/2} + x_{k-1/2})$$

$$\gamma = A_{k+1/2} - D_k x_{k+1/2} + \frac{(D_{k+1} - D_k)}{2\Delta} x_{k+1/2} x_{k-1/2} \tag{C.9}$$

where

$$D_k = \frac{A_{k+1/2} - A_{k-1/2}}{\Delta x_k}$$

$$2\,\Delta x_k = \Delta x_{k+1/2} + \Delta x_{k-1/2}$$

and

$$4\,\Delta = \Delta x_{k+3/2} + 2\,\Delta x_{k+1/2} + \Delta x_{k-1/2}\,. \tag{C.10}$$

The results above simplify in the case of uniform zoning.

Problem C.1

Use the results above to construct an interpolated mass density at a zone boundary.

Problem C.2

Derive the expression for the derivative of a edge centered variable at a zone boundary (C.7).

Appendix D

FOKKER-PLANCK EQUATION

Consider an isotropic system containing electrons, which are in local thermodynamic equilibrium at material temperature T and an aribitrary but time dependent distribution function $f(\nu,t)$ which is assumed to be positive definite, and which may represent either photons or neutrinos. In the following discussion ν and T will be expressed in units of energy ($k_B=1$). The distribution of electrons in velocity space is Maxwellian. The distribution function $f(\nu,t)$ is related to the total number density of photons or neutrinos by

$$n = \int_0^\infty \alpha\, \nu^2 f(\nu, t)\, d\nu \tag{D.1}$$

where α is a constant. The distribution function is assumed to approach either a Planckian or Fermi-Dirac distribution when the system is in complete thermodynamic equilibrium. When the distribution corresponds to photons, equilibration is achieved through electron-photon scattering,

$$e^- + \gamma <-> e^- + \gamma;$$

for neutrinos it is achieve through electron-neutrino scattering,

$$e^- + \nu_e <-> e^- + \nu_e .$$

In both elastic scattering processes above the total momentum and kinetic energy of the particles is conserved. In order to simplify the discussion we shall consider the specific case of electron-photon scattering, noting when appropriate the modifications required for electron-neutrino scattering. Apart from the difference in cross sections, the primary difference is associated with the particle statistics (photons are bosons and neutrinos are fermions).

Boltzmann Energy Transport Equation

Our principle goal is to construct an expression for the time rate of change of the photon distribution function due to scattering. Since the system is assumed to be isotropic, this expression need only involve the incident and scattered energy of the photon and the transition rate for the process. This may be thought of as an angle averaged version of the Boltzmann transport equation for two body interactions. In this process a photon whose initital energy is ν' scatters into a final state whose energy is ν . The transition rate for this process is defined to be

$$\frac{1}{\tau_{tr}} = \frac{S(\nu', \nu, T)}{\nu^2} d\nu' \tag{D.2}$$

where T is the material temperature. The factor involves an average of the scattering cross section and the electron velocities over a Maxwellian distribution, which explains its dependence on temperature. The increase in the number density of photons in the state ν arises when photons are scattered into it, and is proportional to the product of the transition rate, the number of photons in the initial state ν', and the factor $1 \pm f(\nu)$ which accounts for induced scattering (the plus sign applies to photons, and the minus sign to neutrinos):

$$\left(\frac{df(\nu)}{dt}\right)_{in} \equiv \int_0^\infty [1 \pm f(\nu)] \frac{S(\nu', \nu, T)}{\nu^2} f(\nu')\, d\nu'.$$

(D.3)

The integral is over all possible initial photon states. In addition to (D.3) a decrease in $f(\nu)$ results because photons initially in the state ν may scatter into state ν'; this process is proportional to the transition rate $\nu^{-2} S(\nu, \nu', T)$, the number density of photons in the initial state, and the induced scattering factor $1 \pm f(\nu')$ (the plus sign applies to photons, the minus sign to neutrinos):

$$\left(\frac{df(\nu)}{dt}\right)_{out} \equiv -\int_0^\infty [1 \pm f(\nu')] \frac{S(\nu, \nu', T)}{\nu^2} f(\nu)\, d\nu'.$$

(D.4)

Combining the two terms above leads to the following expression for the net time rate of change of $f(\nu)$ due to scattering

$$\nu^2 \frac{df(\nu)}{dt} = \int_0^\infty d\nu' \{ [1 \pm f(\nu)]\, S(\nu', \nu, T)\, f(\nu')$$

$$- [1 \pm f(\nu')]\, S(\nu, \nu', T)\, f(\nu) \} .$$

(D.5)

The transition rates for forward and reverse elastic scattering are related by the principle of detailed balance, which states that the transition probability for a scattering process is equal to the transition probability for the inverse process. For scattering in an isotropic system this implies that

$$S(\nu', \nu, T) = e^{(\nu' - \nu)/T} S(\nu, \nu', T).$$

(D.6)

Equation (D.5) may now be rewritten in the following form

$$\nu^2 \frac{df(\nu)}{dt} = \int_0^\infty d\nu' \, S(\nu, \nu', T) \left\{ e^{(\nu' - \nu)/T} f(\nu') \, [1 \pm f(\nu)] - [1 \pm f(\nu')] \, f(\nu) \right\}. \tag{D.7}$$

Equation (D.7) is essentially the Boltzmann transport equation for two-body interactions, and the right side is essentially the collision integral. Elastic scattering redistributes photon momenta, and exchanges energy with matter in such a way as to conserve the total number density of photons. This can be verified trivially by integrating (D.5) over the photon energy ν, and using the definition of the number density (D.1). Elastic neutrino scattering also conserves neutrino number density.

Fokker-Planck Equation

The most common example of energy exchange by elastic collisions is photon scattering off of nonrelativistic electrons. In this case the fractional energy change per collision is much less than one, and $S(\nu,\nu',T)$ is strongly peaked in ν' about the value ν. Consequently, the only energies that will make a significant contribution to $df(\nu)/dt$ are those within a narrow range

$$|\Delta\nu| = |\nu' - \nu| << \nu. \tag{D.8}$$

Assuming that the photon distribution function is analytic, and that the energy exchange per collision $\Delta\nu$ is small compared to T, all terms on the right side of (D.7) may be expanded about the energy ν. In particular we have for the distribution function

$$f(v', T) = f(v, t) + \left(\frac{\partial f}{\partial v}\right)_v \Delta v + \frac{1}{2}\left(\frac{\partial^2 f}{\partial v^2}\right)_v \Delta v^2 + \ldots \tag{D.9}$$

and

$$e^{(v' - v)/T} = 1 + \frac{\Delta v}{T} + \frac{\Delta v^2}{2T^2} \ldots . \tag{D.10}$$

Substituting these approximations into (D.7) and retaining terms through order Δv^2 results in

$$v^2 \frac{df(v)}{dt} = \left\{f(v)\,[1 \pm f(v)] + T\frac{\partial f}{\partial v}\right\} \frac{\langle \Delta v \rangle}{T}$$

$$+ \left\{[1 \pm f(v)]\left(T\frac{\partial f(v)}{\partial v} + \frac{f(v)}{2}\right) + \frac{T^2}{2}\frac{\partial^2 f(v)}{\partial v^2}\right\} \frac{\langle (\Delta v)^2 \rangle}{T^2} \tag{D.11}$$

where the nth moment of the collisional energy transfer is denoted by

$$\langle (\Delta v)^n \rangle \equiv \int_0^\infty (v' - v)^n S(v, v', T)\, dv'. \tag{D.12}$$

The first two moments which appear in (D.11) can be related when $\Delta v/v$ and $\Delta v/T$ are small (see Problem D.1):

$$\frac{\partial \langle (\Delta v)^2 \rangle}{\partial v} \approx \frac{\langle (\Delta v)^2 \rangle}{T} + 2\langle \Delta v \rangle. \tag{D.13}$$

If this is used to eliminate $\langle \Delta v \rangle$ in (D.11), the right side may be rewritten in the convenient form:

$$\nu^2 \frac{df(\nu, t)}{dt} = \frac{\partial}{\partial \nu}\{\gamma(\nu, T)\, [T\frac{\partial f}{\partial \nu} + f(1 \pm f)]\}$$

(D.14)

where the plus (minus) sign applies to photons (neutrinos). The coefficient $\gamma(\nu, T)$ is related to the second moment of collisional energy exchange

$$\gamma(\nu, T) \equiv \frac{\langle(\Delta\nu)^2\rangle}{2T},$$

(D.15)

and involves averages over the electron Maxwellian velocity distribution, the collisional cross section and the angles defining the incident and scattered particles. For nonrelativistic electrons and Thomson scattering, it can be shown that

$$\gamma(\nu, T) \equiv \frac{n_e \sigma_T \nu^4}{m_e c}.$$

(D.16)

Here n_e is the number density of free electrons, ν is the photon energy and σ_T is the Thomson cross section.

Problem D.1

Use the principle of detailed balance and the change of variable $x = \nu' - \nu$ to show that the first moment of the collisional energy exchange can be written in the form (note the change in the lower limit of integration):

$$\langle\Delta\nu\rangle \approx \int_{-\infty}^{\infty} dx\; S(x+k, -x, T)\, x\, .$$

Expand the integrand about $x = 0$ and show that (D.13) is valid through first order in small quantities.

Appendix E

MAGNETOHYDRODYNAMIC EQUATIONS FOR A PLASMA

The single fluid mass continuity equation, momentum equation and internal energy equation for a collisional plasma permeated by an electromagnetic field will be derived below in the magnetohydrodynamic approximation. The fluid is assumed to be resistive, completely ionized (consisting of n_e electrons per unit volume, and n_i ions of charge Z and atomic weight A per unit volume), and locally charge neutral:

$$n_e = Z n_i. \tag{E.1}$$

The electronic and ionic components are assumed to be described by separate equations of state (electron-electron and ion-ion collision times assumed to be small compared with all other time scales of interest). In particular, the electron pressure is given in terms of the density and the electron temperature (see Chapter 6) by

$$P_e = P_e(\rho, T_e). \tag{E.2}$$

A similar expression is assumed for the ion pressure in terms of the ion temperature

$$P_i = P_i(\rho, T_i).$$

(E.3)

As we shall see, magnetic energy is converted into thermal energy by Joule heating of the electrons, and $T_e > T_i$. It may be recalled (Chapter 4) that electron and ion temperatures may differ as a result of shock propagation in a fluid, in which case $T_i > T_e$. Electron and ion energies will equilibrate on an electron-ion collisional time scale. If the electon-ion collision time is small compared with all other time scales of interest, then the electron and ion temperatures will be equal and a single equation of state results.

In the presence of electromagnetic fields the mean velocity of the electrons, $\mathbf{v}_e$, and of the ions, $\mathbf{v}_i$, will in general be different. While the difference between the electron and ion velocities is usually small, it gives rise to a net electrical current flowing in the plasma, and thus cannot be neglected. The main purpose of this appendix is to express the dynamical equations in terms of a single fluid velocity in such a way as to properly take into account the flow of electrical currents in the plasma.

Single Fluid Mass Continuity Equation

The mass continuity equation for electrons and ions expresses the fact the change in number of either species in a unit volume is due to the net flux of that species across the surface of the volume element. In differential form, for electrons

$$\frac{\partial n_e}{\partial t} + \nabla \cdot (n_e \mathbf{v}_e) = 0$$

(E.4a)

and for ions

$$\frac{\partial n_i}{\partial t} + \nabla \cdot (n_i \mathbf{v}_i) = 0.$$

(E.4b)

Multiplying (E.4a) by m_e and (E.4b) by m_i, and adding the results gives

$$m_e \frac{\partial n_e}{\partial t} + m_i \frac{\partial n_i}{\partial t} + \nabla\cdot(m_e n_e \mathbf{v}_e + m_i n_i \mathbf{v}_i) = 0. \tag{E.5}$$

But the total mass density is just

$$\rho = n_i m_i + n_e m_e = (m_i + Z m_e)\, n_i \tag{E.6}$$

where charge neutrality has been assumed for the last expression. The mean velocity $\mathbf{v}$ (single fluid velocity) is defined by

$$\rho\, \mathbf{v} = m_i n_i \mathbf{v}_i + m_e n_e \mathbf{v}_e$$

or, for a charge neutral plasma

$$\mathbf{v} = \frac{n_i}{\rho}(m_i \mathbf{v}_i + Z m_e \mathbf{v}_e). \tag{E.7}$$

Substituting (E.6) and (E.7) into (E.5) gives the hydrodynamic result

$$\frac{\partial \rho}{\partial t} + \nabla\cdot(\rho \mathbf{v}) = 0. \tag{E.8}$$

Thus the single fluid mass continuity equation introduced in Chapters 4 and 5 applies to an ionized plasma without modification.

Single Fluid Variables

The two fluid equations, such as the mass continuity equations (E.4) and the equations of motion (see below) are expressed in terms of the four variables $\rho_e = m_e n_e$, $\rho_i = m_i n_i$, $\mathbf{v}_e$ and $\mathbf{v}_i$. The single fluid equation are expressed in terms of an equivalent set of variables which are the total mass density ρ given by (E.6), the mean velocity

$\boldsymbol{v}$ given by (E.7), the electrical current density $\boldsymbol{J}$ and the ionic charge Z. The current density is

$$\boldsymbol{J} = Ze\, n_i \boldsymbol{v}_i - en_e \boldsymbol{v}_e$$

$$= e\,(Zn_i - n_e)\, \boldsymbol{v}_i + e\, n_e (\boldsymbol{v}_i - \boldsymbol{v}_e)$$

In a charge neutral plasma, the first term vanishes because of (E.1), and the current density is

$$\boldsymbol{J} = e\, n_e (\boldsymbol{v}_i - \boldsymbol{v}_e). \tag{E.9}$$

Using (E.9) and (E.6) we find

$$\boldsymbol{v}_i = \boldsymbol{v} + \frac{m_e \boldsymbol{J}}{e\rho}. \tag{E.10a}$$

Eliminating ρ_e from (E.7), and using (E.1) for a neutral plasma yields

$$\boldsymbol{v}_e = \boldsymbol{v} - \frac{m_i \boldsymbol{J}}{Ze\rho}. \tag{E.10b}$$

The ion mass density can be obtained from (E.6) and (E.1), and is

$$\rho_i = \frac{\rho}{1 + Zm_e/m_i}. \tag{E.11a}$$

Finally, combining (E.11a) with (E.6) yields the electron mass density

$$\rho_e = \frac{\rho}{1 + Zm_e/m_i} \frac{Zm_e}{m_i}. \tag{E.11b}$$

Equations (E.10)-(E.11) represent the transformation of variables from the two fluid model to the single fluid model.

Problem E.1

Carry out the derivation of the transformations (E.10)-(E.11) for the single fluid model. Since $m_e/m_i << 1$, it is often possible to ignore the effect of electon inertia. Take the limit $m_e/m_i \rightarrow 0$ in (E.10) and (E.11) above and discuss the results.

Single Fluid Momentum Equation

The Boltzmann equation may be used to derive equations of motion for electrons and ions in an ionized plasma which is permeated by an electric field $\boldsymbol{E}$ and a magnetic field $\boldsymbol{B}$ (Chen 1977; Braginskii 1965). Denote the Maxwellian mean velocity of the electrons by $\mathbf{v}_e$, let f_{ei} represent the force on the electrons due to electron-ion collisions, and let $\boldsymbol{g}$ represent the acceleration due to gravity. Then the equation of motion for the electrons is, in Lagrangian form,

$$n_e m_e \frac{d\mathbf{v}_e}{dt} = -\nabla P_e - e\, n_e(\boldsymbol{E} + \mathbf{v}_e \times \boldsymbol{B}) - f_{ei} + n_e m_e g \ . \tag{E.12a}$$

In particular, the time derivative on the left is with respect to a frame of reference moving with the electrons:

$$\frac{d\mathbf{v}_e}{dt} \equiv \frac{\partial \mathbf{v}_e}{\partial t} + \mathbf{v}_e \cdot \nabla \mathbf{v}_e \ .$$

Similarly, the equation of motion for the ions in Lagrangian form is given by

$$n_i m_i \frac{d\mathbf{v}_i}{dt} = -\nabla P_i + Zen_i(\boldsymbol{E} + \mathbf{v}_i \times \boldsymbol{B}) + f_{ei} + n_i m_i \boldsymbol{g} \tag{E.12b}$$

where $\mathbf{v}_i$ denotes the Maxwellian mean velocity of the ions, and the time derivative is in the comoving frame of the ions:

$$\frac{d\mathbf{v}_i}{dt} \equiv \frac{\partial \mathbf{v}_i}{\partial t} + \mathbf{v}_i \cdot \nabla \mathbf{v}_i \,.$$

The electron-ion collisional force is due to electron-ion Coulomb scattering. In a simple plasma it is often called a frictional force, and can be shown to be proportional to $\mathbf{v}_e - \mathbf{v}_i$ [this property, $f_{ei} = -f_{ie}$, was assumed in writing (E.12) above]. Electron-ion interactions are responsible for the electrical resistivity in a conducting fluid (see, for example, Chen 1977).

A single fluid momentum equation may be obtained by simply adding (E.12a) and (E.12b), and using the assumption of charge neutrality. The result is

$$\rho_e \frac{d\mathbf{v}_e}{dt} + \rho_i \frac{d\mathbf{v}_i}{dt} = -\nabla \cdot (P_e + P_i) + e(Zn_i - n_e)E + \rho g$$

$$+ e\,(Zn_i \mathbf{v}_i - n_e \mathbf{v}_e) \times \mathbf{B}. \tag{E.13}$$

In a charge neutral plasma the term proportional to the electric field vanishes, and the coefficient of the magnetic force is proportional to the current density (E.9). Denoting the total fluid pressure by the sum of the electron and ion pressures,

$$P = P_e + P_i \,, \tag{E.14}$$

the single fluid momentum equation is:

$$\rho_e \frac{d\mathbf{v}_e}{dt} + \rho_i \frac{d\mathbf{v}_i}{dt} = -\nabla \cdot P + \rho g + \mathbf{J} \times \mathbf{B}. \tag{E.15}$$

The electron-ion force f_{ei} enters with opposite signs in (E.12a) and (E.12b), and thus cancels when forming the single fluid momentum equation. Although the right side of this equation is written entirely in terms of single fluid variables, the left side is not [recall the comment below (E.12a)]. Expanding the Lagrangian derivatives on the left side of (E.15)

$$\rho_e \frac{d\mathbf{v}_e}{dt} + \rho_i \frac{d\mathbf{v}_i}{dt} = \rho_e \frac{\partial \mathbf{v}_e}{\partial t} + \rho_e \mathbf{v}_e \cdot \nabla \mathbf{v}_e + \rho_i \frac{\partial \mathbf{v}_i}{\partial t} + \rho_i \mathbf{v}_i \cdot \nabla \mathbf{v}_i.$$

(E.16)

The transformation equations (E.10)-(E.11) may be used to express the right side of (E.16) in terms of the single fluid variables. The transformation is lengthy, but straightforward, and yields

$$\rho_e \frac{d\mathbf{v}_e}{dt} + \rho_i \frac{d\mathbf{v}_i}{dt} = \rho \frac{\partial \mathbf{v}}{\partial t} + \rho \mathbf{v} \cdot \nabla \mathbf{v} + \frac{m_e m_i}{Ze^2} \boldsymbol{J} \cdot \nabla \left(\frac{\boldsymbol{J}}{\rho}\right)$$

$$= \rho \frac{d\mathbf{v}}{dt} + \frac{m_e m_i}{Ze^2} \boldsymbol{J} \cdot \nabla \left(\frac{\boldsymbol{J}}{\rho}\right).$$

(E.17)

The time derivative on the right side of (E.17) is the usual Lagrangian derivative for a single fluid model. Using (E.17) to eliminate the two fluid velocities in (E.15) gives the single fluid momentum equation

$$\rho \frac{d\mathbf{v}}{dt} = -\nabla P + \rho g + \boldsymbol{J} \times \boldsymbol{B} - \frac{m_e m_i}{Ze^2} \boldsymbol{J} \cdot \nabla \left(\frac{\boldsymbol{J}}{\rho}\right).$$

(E.18)

This result should be compared with (4.6). Apart form the last term on the right above, we can identify the Lorentz force, $\boldsymbol{J} \times \boldsymbol{B}$, with a body force in (4.6). Consider the ratio of the last term in (E.18) to the total force per unit volume; since $J \sim e n_e V$ and $\rho \sim n_i m_i$,

$$\frac{m_e m_i}{Ze^2} J \cdot \nabla\left(\frac{J}{\rho}\right) \Big/ \rho \frac{d\mathbf{v}}{dt} \sim \frac{m_e m_i J^2}{Ze^2 \rho\, l_B} \frac{\tau_D}{\rho v} \sim \frac{m_e}{m_p} \frac{Z}{A} \frac{l_D}{l_B}$$

(E.19a)

where l_B represents the magnetic field length scale (which is comparable to the width of the current carrying region), and $l_D \equiv \tau/v$ is the scale length over which hydrodynamic changes occur. The ratio of the last term in (E.18) to the momentum advection terms is of order

$$\frac{m_e m_i}{Ze^2} J \cdot \nabla\left(\frac{J}{\rho}\right) \Big/ \rho\, \mathbf{v} \cdot \nabla \mathbf{v} \sim \frac{n_e m_e}{n_i m_i} \sim \frac{m_e}{m_i} << 1.$$

(E.19b)

In many applications the last term in (E.18) can be neglected, and the momentum equation takes the form

$$\rho \frac{d\mathbf{v}}{dt} = -\nabla P + \rho g + \boldsymbol{J} \times \boldsymbol{B}.$$

(E.20)

Thus, when the ratio (E.19) is small, adding magnetohydrodynamic effects to the momentum equation for a neutral plasma results simply by introducing the Lorentz force term as a body force in (4.6).

Charge Conservation Equation

The net charge density in a plasma is given by

$$Q = Ze\, n_i - e\, n_e\,.$$

(E.21)

The time rate of change of the total charge Q is obtained by multiplying (E.4a) by e, (E.4b) by Ze and subtracting the results.

Using (21) and the definition of the current density, (E.9), gives

$$\frac{\partial Q}{\partial t} + \nabla\cdot\boldsymbol{J} = 0.$$

However, in a charge neutral plasma (E.1), the net charge is zero, and charge conservation reduces to the constraint

$$\nabla\cdot\boldsymbol{J} = 0. \tag{E.22}$$

Electron Internal Energy Equation

In keeping with the hydrodynamics developed in Chapter 4 and Chapter 5, we follow the internal energy of the plasma rather than its total energy. Since an ionized plasma consists of both electrons and ions, the change in total internal energy consists of changes in the internal energy of the electrons and ions.

The first law of thermodynamics applied to the electrons in a plasma may be written in the form (where ρ_e is the electron mass density)

$$\rho_e \frac{dE_e}{dt} = -P_e \nabla\cdot\mathbf{v}_e - \nabla\cdot\boldsymbol{q}_e + Q_{ei} + \boldsymbol{E}\cdot\boldsymbol{J}\,. \tag{E.23}$$

Here E_e is the internal energy associated with electrons, measured per electron, $\boldsymbol{q}_e$ represents the flux of heat energy due to dissipative effects, such as radiation and shocks, $\boldsymbol{E}\cdot\boldsymbol{J}$ represents the work done by currents (Joule heating), and Q_{ei} represents energy exchange between electrons and ions due to scattering between the species. Using the continuity equation (E.4a) the time derivative may be rewritten in the conservative form

$$\frac{\partial(\rho_e E_e)}{\partial t} + \nabla\cdot(\rho_e E_e \mathbf{v}_e) = -P_e \nabla\cdot\mathbf{v}_e - \nabla\cdot\boldsymbol{q}_e + Q_{ei}\,.$$

The quantity $\rho_e E_e$ is just the total internal energy associated with electrons per unit volume so it may be expressed in the equivalent form

$$\rho_e E_e = \rho \, \varepsilon_e$$

where ε_e is the total electron internal energy per gram and the density on the right is the total mass density in the plasma. Thus the electron internal energy equation may be written in the form

$$\frac{\partial(\rho\,\varepsilon_e)}{\partial t} + \nabla\cdot(\rho\,\varepsilon_e \mathbf{v}_e) = -P_e\,\nabla\cdot\mathbf{v}_e - \nabla\cdot\mathbf{q}_e + Q_{ei} \,. \tag{E.24}$$

This represents the two fluid form of the internal energy equation. The adiabatic terms may be transformed into single fluid form using (10b). The result, which contains only the adiabatic terms, is easily shown to be

$$\frac{\partial(\rho\varepsilon_e)}{\partial t} + \nabla\cdot(\rho\varepsilon_e\mathbf{v}) = -P_e\nabla\cdot\mathbf{v} + \frac{m_i}{Ze}\Big[P_e\nabla\cdot\Big(\frac{J}{\rho}\Big) + \nabla\cdot(J\,\varepsilon_e)\Big]. \tag{E.25}$$

The left side is the single fluid Lagrangian derivative, and should be compared with the single fluid form of the momentum equation (E.18).

The ratio of the last term in the right of (E.25) to the mechanical work term is of order

$$\frac{m_i}{Ze}\,\frac{P_e J}{\rho\, l_B}\,\frac{l_D}{P_e v} \sim \frac{l_D}{l_B} \tag{E.26}$$

which does not vanish in the limit $m_e \to 0$. If the scale length for changes in the velocity, l_D, is small relative to the scale length associated with changes in the current distribution, l_B, then the last term in (E.25) may be neglected. We shall assume this to be the case, and take for the single fluid electron internal energy equation

$$\frac{\partial(\rho\varepsilon_e)}{\partial t} + \nabla\cdot(\rho\varepsilon_e \boldsymbol{v}) = -P_e\nabla\cdot\boldsymbol{v} + \eta J^2 + Q_{ei} - \nabla\cdot\boldsymbol{q}_e' . \tag{E.27}$$

The dissipative terms appearing in (E.24) have been included; in particular Joule heating is described by the term $\boldsymbol{E}\cdot\boldsymbol{J} = \eta J^2$, where Ohm's law in the comoving frame (8.1) has been used to eliminate the local electric field (since all terms in (E.27) are defined in the comoving frame, the prime has been omited for notational convenience). The term Q_{ei} on the right side of (E.27) represents all heat flux terms except Joule heating.

Ion Internal Energy Equation

The first law of thermodynamics applied to the ions in a plasma may be written in the form

$$\rho_i \frac{dE_i}{dt} = -P_i\nabla\cdot\boldsymbol{v}_i - \nabla\cdot\boldsymbol{q}_i - Q_{ie} . \tag{E.28}$$

Proceeding as for the electron component, (E.28) may be transformed to single fluid variables. The result, including only the adiabatic terms, is (Problem E.2)

$$\frac{\partial(\rho\varepsilon_i)}{\partial t} + \nabla\cdot(\rho\varepsilon_i \boldsymbol{v}) = -P_i\nabla\cdot\boldsymbol{v} - \frac{m_e}{e}\left[P_i\,\nabla\cdot\left(\frac{J}{\rho}\right) + \nabla\cdot(J\,\varepsilon_i)\right]. \tag{E.29}$$

The ratio of the last term on the right to the energy advection term or the mechanical work term is of order $(m_e/m_i)\,(l_D/l_B)$, and can usually be neglected. Thus the final form of the single fluid ion internal energy equation, including dissipative effects, is

$$\frac{\partial(\rho\varepsilon_i)}{\partial t} + \nabla\cdot(\rho\varepsilon_i \mathbf{v}) = -P_i\nabla\cdot\mathbf{v} - \nabla\cdot\mathbf{q}_i - Q_{ie}\,. \tag{E.30}$$

Notice that there are no magnetohydrodynamic contributions to the ion internal energy in (E.30).

Problem E.2

Derive the single fluid ion internal energy equation (E.30) starting from the two fluid expression (E.28), where E_i is the ion specific internal energy measured per ion, and satisfies $\rho_i E_i = \rho\,\varepsilon_i$. Finally ε_i is the ion internal energy per unit mass.

Electron-Ion Coupling

The electron-ion coupling terms Q_{ie} and Q_{ei} which enter into (E.27) and (E.30) result in energy exchange between electrons and ions associated with Coulomb scattering. The derivation of this effect is straightforward, but tedious (Spitzer 1962; Braginskii 1965). The general form of the coupling for a simple plasma where $Q_{ie} = Q_{ei}$, be understood by considering the following heuristic arguments. Physically the strength of the coupling could be expected to depend on the difference in electron and ion temperatures. Thus the rate of change in the electron internal energy may be expressed as

$$\frac{d(\rho C_{V,e} T_e)}{dt} = -\frac{T_e - T_i}{\tau_{ei}} \tag{E.31}$$

where the electron specific energy and the electron temperature are related by the electron effective heat capacity (Chapter 3)

$$C_{V,e} \equiv \varepsilon_e / T_e. \tag{E.32}$$

The energy exchange time τ_{ei} is to be constructed so that (E.31) is, in fact, correct. To find τ_{ei} consider the scattering of electrons (characterized by a Maxwellian velocity distribution at temperature T_e) off a distribution of ions (characterized by a Maxwellian velocity distribution at temperature T_i). The energy exchange during a typical electron-ion collision is of order (Problem E.3)

$$\left|\frac{\Delta E}{E}\right|_{ei} \sim \frac{m_e m_i}{(m_e + m_i)^2} \sim \frac{m_e}{m_i}. \tag{E.33}$$

The Coulomb cross section for electron-ion scattering is of order $\sigma_{ei} \sim \pi b^2$, where b is the impact parameter. If v_e represents the incident energy of the electrons, the impact parameter satisfies

$$\mu v_e^2 \sim \frac{Ze^2}{b}, \tag{E.34}$$

where $\mu \sim m_e$ is the reduced mass of the electron-ion two body system, and Z is the charge on the ions. Combining results above, the Coulomb cross section is of order:

$$\sigma_{ei} \sim \frac{Z^2}{m_e^2 v_e^4}. \tag{E.35}$$

The electron-ion collision rate, τ_c^{-1}, is given by the product of the incident electron flux, $n_e v_e$, and the cross section (E.35):

$$\tau_c^{-1} \equiv n_e v_e \sigma_{ei} . \tag{E.36}$$

Next the energy exchange rate is defined by

$$\tau_{ei}^{-1} \equiv \frac{1}{E}\left|\frac{dE}{dt}\right| \sim \tau_c^{-1}\left|\frac{\Delta E}{E}\right|_{ei} . \tag{E.37}$$

Finally, it may be noted that the mean electron velocity is related to the electron temperature by

$$m_e v_e^2 \sim T_e . \tag{E.38}$$

The electron-ion energy exchange time scale (E.37) may be expressed as a function of the electron temperature using (E.33), (E.36) and (E.37), and has the form

$$\tau_{ei}^{-1} \sim \frac{m_e}{m_i}\frac{n_e Z^2}{m_e^{1/2} T_e^{3/2}} . \tag{E.39}$$

Problem E.3

Consider the scattering of two particles of mass m and M in one dimension, and show that to order of magnitude the energy exchange is given by:

$$\left|\frac{\Delta E}{E}\right|_{m \to M} \sim \frac{mM}{(m+M)^2}$$

and that the momentum exchange is given by

$$\left|\frac{\Delta p}{p}\right|_{m \to M} \sim \frac{M}{m+M}.$$

Following the analysis above for the electron-ion energy exchange time scale, find order of magnitude expressions for the energy exchange rates due to electron-electron scattering, and ion-ion scattering.

The change in ion internal energy due to electron-ion scattering can be constructed as above. The result is

$$\frac{d(\rho C_{V,i}T_i)}{dt} = -\frac{T_i - T_e}{\tau_{ei}}$$

(E.40)

where the energy exchange time scale is (E.39), and the ion internal energy is related to the ion temperature by

$$C_{V,i} \equiv \varepsilon_i / T_i.$$

(E.40)

Note that there is no net change in energy of a plasma containing electrons and ions, as may be verified by adding (E.40) and (E.31).

Generalized Ohm's Law

A generalization of Ohm's law may be derived from the electron equation of motion. If (E.12a) is divided by the ion mass m_i, and (E.10b) is used to eliminate the electron velocity in the Lorentz force term, we find:

$$n_e \frac{m_e}{m_i}\left(\frac{d\mathbf{v}_e}{dt} - g\right) = -\frac{\nabla P_e}{m_i} - \frac{e n_e}{m_i}(E + \mathbf{v}\times B)$$
$$- \frac{n_e}{Z\rho} J\times B - \frac{f_{ei}}{m_i}.$$

(E.42)

Using (E.9), the electon-ion collisional force may be written in the form

$$f_{ei} = m_e \frac{\mathbf{v}_e - \mathbf{v}_i}{\tau_c} = - \frac{m_e}{m_i \tau_c} \frac{\mathbf{J}}{e n_e},$$

(E.43)

and (E.39) may be used to show that f_{ei} / m_i is proportional to $(m_e / m_i)^{1/2}$. Thus if we drop all terms proportional to (m_e / m_i), (E.42) becomes

$$\eta \mathbf{J} - (\mathbf{E} + \mathbf{v} \times \mathbf{B}) + \frac{\mathbf{J} \times \mathbf{B}}{e n_e} - \frac{\nabla P_e}{e n_e} = 0,$$

(E.44)

where the electrical resistivity of the ionized plasma is

$$\eta = \frac{m_e}{e^2 n_e \tau_{ei}}.$$

(E.45)

Equation (E.44) represents a generalized form of Ohm's law and includes a thermoelectric effect (third term) and the Hall effect (last term). Notice that if these last two terms are omitted we recover the usual form in a moving frame of reference. A more general form of the electron-ion collisional force can be used in place of (E.43) which introduces additional terms which are proportional to the gradient of the electron temperature (see, for example, Braginskii 1965).

References

F. F. Chen, *Plasma Physics,* (New York: Plenum Press, 1977).

S. I. Braginskii, 1965, "Transport Processes in a Plasma," in *Reviews of Plasma Physics*, Vol. 1.

L. Spitzer, 1962, *Physics of Fully Ionized Gases,* (New York: Wiley Interscience; 1962), Second edition.